Biotic and Abiotic Stress Tolerance in Plants

Sharad Vats
Editor

Biotic and Abiotic Stress Tolerance in Plants

Springer

Editor
Sharad Vats
Department of Bioscience & Biotechnology
Banasthali Vidyapith
Rajasthan, India

ISBN 978-981-10-9028-8 ISBN 978-981-10-9029-5 (eBook)
https://doi.org/10.1007/978-981-10-9029-5

Library of Congress Control Number: 2018941813

Printed on acid-free paper

This Springer imprint is published by the registered company Springer Nature Singapore Pte Ltd.
The registered company address is: 152 Beach Road, #21-01/04 Gateway East, Singapore 189721, Singapore

Foreword

Throughout evolution, plants have faced extreme variations in environment. Yet, they have survived and adapted themselves in different ecological niches. However, it is foreseen that in the ensuing period of present day global climatic changes, the impact on the domesticated crops, which feed the humanity, will result in negative growth and productivity. Scientists will have to develop new varieties, either through classical breeding tools or using new genomic approaches like molecular assisted breeding or developing biotech crops using transgenic or genome editing technologies. To achieve success in this direction it is essential to understand, at the biochemical and molecular level, the mechanisms of plant perception to abiotic and biotic stresses, the signaling pathways, and the identification of genes that respond to confer stress tolerance. The present book is an attempt to line up different chapters to illustrate the knowledge that has accumulated in some of the domains in the area of biotic and abiotic stress tolerance in plants.

One of the chapters broadly cover plant responses to drought in particular, to illustrate how the stress affects the physiology and biochemistry of the plants. How plants can be made to survive short drought conditions is an important aspect of future plant biotechnology studies. Using either phenomics or genomics based approaches one should get plants which can produce more per drop of water, which is going to be more scare for agriculture with increasing population and urbanization. More specifically, one of the chapters deals with impact of abiotic stresses on photosynthesis, which is the fundamental process that needs to be protected in order for the plants to survive and grow. It has been seen that senescence and chlorophyll breakdown ensues following stress conditions, which lowers photosynthesis and hence yield. A few chapters deal with the role of signaling molecules like nitric oxide, reactive oxygen species, and salicylic acid in modulating and adapting to stress environment and also in inducing cell death. These signals are produced in addition to changes in abscisic acid and calcium, etc., whose role has been well studied in stress physiology. One of the important molecule that also plays a very significant role is glutathione. Modulation of GSH and GSSG seems to be one of the key parameters that senses and transduces stress signals. In view of this, the role of glutathione transferases and phosphite in adaptation is also discussed in two chapters.

Air, water, and soil pollution influence plant growth and development. Two chapters are devoted to pollution as a stress for plants where effect of insecticides and also biomonitoring have been presented. Among other changes that occur in plants following stress perception, role of bioactive compounds has been presented in a separate chapter. In order to assess the overall molecular changes under stress environment, a chapter deals with changes in miRNA and another on the availability of bioinformatic resources. One chapter on breeding for stress has been included using *Capsicum* as a test case.

Overall, the editor has effectively used his experience and knowledge to incorporate experts from various parts of the globe to write chapters covering important aspects of plant stress biology. The information compiled in this volume will be useful to students and researchers of molecular plant physiology in general and to those working in stress physiology in particular.

Arturo Falaschi Emeritus Scientist,
ICGEB, New Delhi, India

S. K. Sopory

Former Vice Chancellor, JNU,
New Delhi, India

Preface

Plants, being sedentary, are highly exposed to environmental stress (biotic and abiotic). However, they have developed several mechanisms to tolerate adverse conditions, which are rather complex to decipher. Global climatic changes, pollution, ever-increasing population, resistant pests, and other related factors have even worsened the current environmental situation, having a direct negative impact on the world's crop production. Thus, understanding the effects and various tolerance mechanisms of plants under stress is of prime importance to the scientific fraternity. The present work is an attempt to incorporate some of the biochemical, physiological, and molecular aspects of plant stress with latest updates.

The book is organized into 14 chapters written by eminent experts from different parts of the globe. The first chapter focuses on the physiological, biochemical, and molecular response of the plants under drought stress, which is one of the most predominant abiotic stresses. The second chapter highlights the effect of abiotic stress on the photosynthetic apparatus of the plants. The strategies involved to safeguard this apparatus have been discussed, which could help in the development of plants with effective photosynthetic machinery under stress. This is followed by a chapter which emphasizes on the ecotoxicological effects of insecticides on plants with special reference to germination and other phytotoxicity tools. The Chaps. 4 and 5 explore the variations of plant bioactive compounds and the role of salicylic acid in modulating salinity stress. Chapters 6, 7, 8 and 9 bring to light the involvement of beneficial elements, glutathione-S-transferase, phosphite, and nitric oxide, respectively, in the adaptive response of plants under stress and as a stimulator of better plant performance. Stress induced programmed cell death (PCD) in plants as a survival strategy and the role and cross-talk of reactive species of oxygen and nitrogen in activating PCD in plants have been efficiently described in the chapter "Involvement of Reactive Species of Oxygen and Nitrogen in Triggering Programmed Cell Death in Plants." In the Chap. 11, the research progress toward *Capsicum*, a commercially important plant, against stress tolerance has been compiled from classical breeding to the recent use of large-scale transcriptome and genome sequencing technologies. This is followed by a chapter, which underlines the role of small RNAs in the plant development and stress mitigation. Apart from knowing the adaptive mechanisms of the plants it is also very important to identify some biological agents that monitor the level of environmental stress. Viewing the same, Chap. 13 has been included, which specifies the significance of the liliputians of the plant kingdom (Bryophytes) as biomonitors/bioindicators. The last chapter focuses on various general and specialized bioinformatics resources useful for people working in the field of plant stress biology. Overall, the book includes the latest developments in the field of plant stress biology supplemented with related figures and tables, which can be useful for students and research scholars.

I am extremely grateful to the publisher (Springer), contributors, and reviewers for their support and meticulous assessment of the book chapters. I would like to state that the encouragement and unconditional support of my parents, my wife, and my beloved daughter (Vaibhavi) were the guiding factors behind the effective completion of this work. I am also thankful to Prof. S. K. Sopory for providing his guidance and consent to write the foreword of this book.

Rajasthan, India Sharad Vats

Contents

Contributors

Ilyas Ahmad Translational and Evolutionary Genomics Lab, School of Life Sciences, Jawaharlal Nehru University, New Delhi, India

Afroz Alam Department of Bioscience and Biotechnology, Banasthali Vidyapith, Banasthali, Rajasthan, India

Renata Bączek-Kwinta Faculty of Agriculture and Economics, Department of Plant Physiology, University of Agriculture in Cracow, Cracow, Poland

K. V. Bhat Division of Genomic Resources, National Bureau of Plant Genetic Resources (NBPGR), Pusa Campus, New Delhi, India

Irini Bosmali Institute of Agrobiotechnology, CERTH, Thessaloniki, Greece

Idalina Bragança REQUIMTE/LAQV-GRAQ, Instituto Superior de Engenharia do Porto, Politécnico do Porto, Porto, Portugal

Dipjyoti Chakraborty Department of Bioscience and Biotechnology, Banasthali Vidyapith, Banasthali, Rajasthan, India

Shweta Chaudhary Department of Bioscience and Biotechnology, Banasthali Vidyapith, Banasthali, Rajasthan, India

Sushil Satish Chhapekar Translational and Evolutionary Genomics Lab, School of Life Sciences, Jawaharlal Nehru University, New Delhi, India

Parul Chowdhury Dr B Lal Institute of Biotechnology, Jaipur, Rajasthan, India

Evangelia G. Chronopoulou Laboratory of Enzyme Technology, Department of Agricultural Biotechnology, Agricultural University of Athens, Athens, Greece

Luca Lo Cicero Dipartimento di ScienzedelleProduzioniAgrarie e Alimentari (DISPA), Università di Catania, Catania, Italy

Cristina Delerue-Matos REQUIMTE/LAQV-GRAQ, Instituto Superior de Engenharia do Porto, Politécnico do Porto, Porto, Portugal

Valentina F. Domingues REQUIMTE/LAQV-GRAQ, Instituto Superior de Engenharia do Porto, Politécnico do Porto, Porto, Portugal

Rashmi Gaur Translational and Evolutionary Genomics Lab, School of Life Sciences, Jawaharlal Nehru University, New Delhi, India

Fernando Carlos Gómez-Merino Colegio de Postgraduados Campus Córdoba, Amatlán de los Reyes, Veracruz, Mexico

Clara Grosso REQUIMTE/LAQV-GRAQ, Instituto Superior de Engenharia do Porto, Politécnico do Porto, Porto, Portugal

Vandana Jaiswal Translational and Evolutionary Genomics Lab, School of Life Sciences, Jawaharlal Nehru University, New Delhi, India

Dhananjay Kumar Department of Botany, H.N.B. Garhwal University, Srinagar, Uttarakhand, India

Sanjay Kumar Division of Genomic Resources, National Bureau of Plant Genetic Resources (NBPGR), Pusa Campus, New Delhi, India

Sonu Kumar Bioinformatics Programme, Center for Biological Sciences, Central University of South Bihar, Patna, India

Nikolaos E. Labrou Laboratory of Enzyme Technology, Department of Agricultural Biotechnology, Agricultural University of Athens, Athens, Greece

Paulo C. Lemos REQUIMTE/LAQV, Chemistry Department, FCT/Universidade NOVA de Lisboa, Caparica, Portugal

Panagiotis Madesis Institute of Agrobiotechnology, CERTH, Thessaloniki, Greece

Mehak Shaheen Department of Forestry and Range Management, Bahauddin Zakariya University, Multan, Pakistan

Vineet Kumar Maurya Department of Microbiology, H.N.B. Garhwal University, Srinagar, Uttarakhand, India

Irini Nianiou-Obeidat Department of Genetics and Plant Breeding, School of Agriculture, Aristotle University of Thessaloniki, Thessaloniki, Greece

Wasif Nouman Department of Forestry and Range Management, Bahauddin Zakariya University, Multan, Pakistan

Chandramani Pathak Indian Institute of Advanced Research, Gandhinagar, Gujarat, India

Angela Roberta Lo Piero Dipartimento di ScienzedelleProduzioniAgrarie e Alimentari (DISPA), Università di Catania, Catania, Italy

Muhammad Kamran Qureshi Department of Plant Breeding and Genetics, Bahauddin Zakariya University, Multan, Pakistan

Nirala Ramchiary Translational and Evolutionary Genomics Lab, School of Life Sciences, Jawaharlal Nehru University, New Delhi, India

Diana Rede REQUIMTE/LAQV-GRAQ, Instituto Superior de Engenharia do Porto, Politécnico do Porto, Porto, Portugal

Supriya Sachdeva Division of Genetics, Indian Agricultural Research Institute (IARI), New Delhi, India

Sadhana Saingar Department of Bioscience and Biotechnology, Banasthali Vidyapith, Banasthali, Rajasthan, India

Renuka Saraf Department of Bioscience and Biotechnology, Banasthali Vidyapith, Banasthali, Rajasthan, India

Asheesh Shanker Bioinformatics Programme, Center for Biological Sciences, Central University of South Bihar, Patna, India

Jitender Singh National Institute of Plant Genome Research, New Delhi, India

Susana R. Sousa REQUIMTE/LAQV-GRAQ, Instituto Superior de Engenharia do Porto, Politécnico do Porto, Porto, Portugal

i3S – Instituto de Investigação e Inovação em Saúde, Universidade do Porto, Porto, Portugal

INEB – Instituto de Engenharia Biomédica, Universidade do Porto, Porto, Portugal

Evangelia Stavridou Institute of Agrobiotechnology, CERTH, Thessaloniki, Greece

Jitendra K. Thakur National Institute of Plant Genome Research, New Delhi, India

Budhi Sagar Tiwari Indian Institute of Advanced Research, Gandhinagar, Gujarat, India

Libia Iris Trejo-Téllez Colegio de Postgraduados Campus Montecillo, Texcoco, State of Mexico, Mexico

Athanasios Tsaftaris Perrotis College, American Farm School, Thessaloniki, Greece

Sharad Vats Department of Bioscience & Biotechnology, Banasthali Vidyapith, Banasthali, Rajasthan, India

Georgia Voulgari Institute of Agrobiotechnology, CERTH, Thessaloniki, Greece

Muhammad Zubair Department of Forestry and Range Management, Bahauddin Zakariya University, Multan, Pakistan

About the Author

Dr. Sharad Vats is currently working at the Department of Bioscience & Biotechnology, Banasthali Vidyapith, Rajasthan, India. He received his Ph.D. from the University of Rajasthan, India, and has more than a decade of research and teaching experience. He has authored two books and published five book chapters and more than 30 research papers in internationally recognized journals on important aspects of plant secondary metabolites, tissue cultures, antioxidants, and related topics.

Plant Responses to Drought Stress: Physiological, Biochemical and Molecular Basis

Sanjay Kumar, Supriya Sachdeva, K. V. Bhat, and Sharad Vats

Abstract Drought is one of the most serious threats to crop production all over the world and is likely to worsen with anticipated changes in the climate. Drought impairs normal growth, disturbs water relations and reduces water-use efficiency in plants. Plants, however, have a variety of physiological and biochemical responses at cellular and organism levels, making it a more complex phenomenon. Researchers have been trying to understand and dissect the mechanisms of plant tolerance to drought stress using various approaches. The present chapter describes the strategies used by plants to adapt to low water potential at physiological, biochemical and molecular levels. This chapter also describes the strategies involving genetic engineering used by breeders in order to obtain crop varieties with improved drought tolerance, some of which show great promise. Modern genomic and genetic approaches coupled with breeding methodologies are expected to more effectively identify the genes and metabolic pathways that confer drought tolerance in crops.

Keywords Abiotic stress · Photosynthesis · Reactive oxygen species · Regulatory genes · Stress tolerance · Transgenic plants

S. Kumar
Department of Bioscience & Biotechnology, Banasthali Vidyapith, Vanasthali, Rajasthan, India

Division of Genomic Resources, National Bureau of Plant Genetic Resources (NBPGR), Pusa Campus, New Delhi, India

S. Sachdeva
Division of Genetics, Indian Agricultural Research Institute (IARI), New Delhi, India

K. V. Bhat (✉)
Division of Genomic Resources, National Bureau of Plant Genetic Resources (NBPGR), Pusa Campus, New Delhi, India

S. Vats (✉)
Department of Bioscience & Biotechnology, Banasthali Vidyapith, Vanasthali, Rajasthan, India

S. Vats (ed.), *Biotic and Abiotic Stress Tolerance in Plants*,
https://doi.org/10.1007/978-981-10-9029-5_1

1 Introduction

Global climatic change and ever increasing population necessitates the need for developing stress-resistant crops. Drought is one of the major phenomena that limit crop production and yield worldwide. It is estimated that 70% of crop yield loss can be attributed to abiotic stresses, especially drought (Bray et al. 2000). Traditional breeding for drought tolerance has been a basic approach, and success has been achieved in few crops such as maize (Hoisington et al. 1996) and wheat (Zhao et al. 2000). Incorporation of functional, comparative and structural genomics would greatly enhance the success of traditional breeding efforts. Application of modern genomic tools in traditional breeding programmes is becoming common because of its great potential. Modern genetic and genomic technologies and advancement in breeding and phenotyping have helped in identifying candidate genes and metabolic pathways functional in drought-tolerant crops (Ishitani et al. 2004; Cattivelli et al. 2008; Mir et al. 2012). However, a large gap remains between crop yields in ideal and stress conditions.

Drought is a physiological form of water deficit where soil water available to the plant is inadequate, which adversely affects the plant's metabolism. However, plants possess multiple morphological (reduced leaf area, reduced stem length, leaf moulding, wax content, efficient rooting system, stability in yield and number of branches), physiological (transpiration, water-use efficiency, stomatal activity and osmotic adjustment) and biochemical responses (accumulation of proline, polyamine, trehalose, increasing of nitrate reductase activity and storage of carbohydrate at cellular and organism levels) under drought stress, making it a more complex phenomenon to decipher (Haworth et al. 2013; Ammar et al. 2015; Conesa et al. 2016) (Fig. 1). Of various plant responses to water scarcity, enhanced abscisic acid (ABA) accumulation is one of the key mechanisms of adaptation to water stress (Esther et al. 2000; Bano et al. 2012; Brodribb and McAdam 2013). The plant growth regulator, ABA, plays an important role in the response and tolerance against dehydration. It seems that dehydration triggers production of ABA, which induces expression of genes like *rd22* (Abe et al. 1997); *RD29A*, *RD29B*, *KIN2* and *RAB18* (Yao et al. 2012); and *PYL8* (Lim et al. 2013). There are genes that are induced by dehydration and not responsive to exogenous ABA treatments suggesting the existence of ABA independent in addition to ABA-dependent signalling pathways between initial signal of drought stress and expression of specific genes (Shinozaki and Yamaguchi-shinozaki 1997; Yoshida et al. 2010; Ding et al. 2016).

Stomatal response, ROS scavenging, metabolic changes and photosynthesis are majorly affected when plants are exposed to water stress. Thus, in order to acclimatize to abiotic stresses, plants accumulate biomolecules that are harmless and do not interfere with plant processes. They may include protective proteins such as dehydrins; heat shock proteins (HSPs); late embryogenesis abundant (LEA) proteins (Vierling 1991; Lipiec et al. 2013); osmolytes like proline, trehalose and sugars (Zhang et al. 2010; Hayat et al. 2012; Ilhan et al. 2015); glycine; and betaine (Sakamoto and Murata 2002; Wang et al. 2010; Chen and Murata 2011). Some

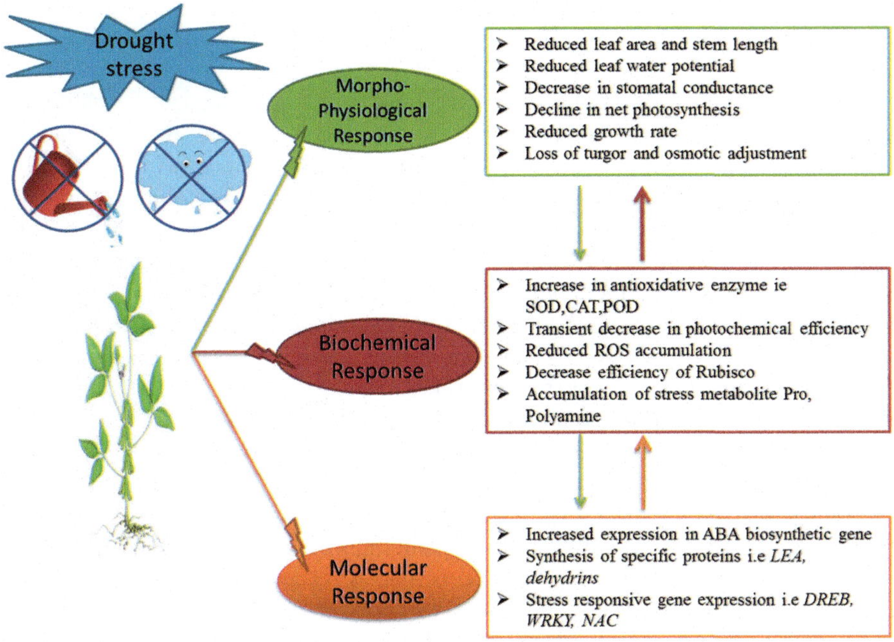

Fig. 1 Plants' responses under drought stress

signalling molecules include polyamines (Roy and Wu 2002; Navakouidis et al. 2003; Capell et al. 2004; Wi et al. 2006; Liu et al. 2007; Wen et al. 2008; Cheng et al. 2009; Gill and Tuteja 2010; Alcazar et al. 2010; Rangan et al. 2014), inositol (Xiong et al. 2001; Sengupta et al. 2008) and hormones like abscisic acid (Davies and Zhang 1991; Saradhi et al. 2000), ethylene (Quan et al. 2010; Xiong et al. 2013) and methyl jasmonate (Bartels and Sunkar 2005; Vincour and Altman 2005; Wu et al. 2008; Jan et al. 2013). Changes in membrane fluidity and protein composition of membranes help to maintain the cellular integrity of plants (Bohnert et al. 1995). Accumulation of LEA proteins is correlated to improved tolerance under drought, salinity and cold (Imai et al. 1996; Close 1996; Xu et al. 1996; Juszczak and Bartels 2017). The accumulating solute appears to act in protein solubilisation (ectoine, glycine, betaine), and uncharged solutes (mannitol, pinitol) may act as scavengers of reactive oxygen species (ROS) (Ashraf and Foolad 2007). Overexpression of Fe-binding ferritin resulted in increased tolerance to removal of free iron which participates in 'Fenton's reaction' and produces hydroxyl radicals (Shen et al. 1996). Nitric oxide has also proved to be protective against oxidative stress conditions. These collective responses are controlled by complex regulatory events intervened by ABA, ion transport and transcription factors (TFs) involved in the regulation of stomatal responses, which are integrated into coordinated molecular networks, enabling plants to adapt and survive.

Osmolytes increase tolerance to environmental stresses in several plants (Wang et al. 2003; Hochberg et al. 2013). Drought-tolerant transgenic rice lines were developed showing tissue or stress-inducible accumulation of trehalose, which accounted for higher levels of soluble carbohydrate, a higher capacity of photosynthesis and concomitant decline in photo-oxidative damages and more favourable mineral balance mutually under stress and non-stress conditions, with no negative effects (Garg et al. 2002). Several studies have identified traits for which presence or expression is linked to plant adaptability to drought conditions (Table 1). Amongst them, traits such as small plant size, reduced leaf area, early maturity and prolonged stomatal closure lead to reduction in the total seasonal evapotranspiration and the yield potential (Fischer and Wood 1979; Karamanos and Papatheohari 1999). Stay-green plants are characterized by a post-flowering drought resistance phenotype that gives plants resistance to premature senescence, stalk rot and lodging when subjected to drought during grain filling. As a result, stay green has been extensively used to improve yield potential and yield stability under water-stressed environments in various breeding programmes (Campos et al. 2004; Tollenar and Wu 1999). Genomics and crop physiology have led to new insights in drought tolerance providing breeders with new tools for plant improvement (Tuberosa and Salvi 2006). The plant drought stress can be managed by adopting strategies such as mass screening and breeding, marker-assisted selection and exogenous application of hormones and osmo-protectants to seeds or plants, as well as engineering for drought resistance. This chapter provides an outline of plant drought stress, its effects on plant's resistance mechanisms and management strategies to cope with this global challenge.

2 Desiccation and Dehydration

Water deficit can affect plants in several ways. A mild water deficit leads to small changes in the water status of plants, and plants cope with such a situation by reducing water loss and/or by increasing water uptake (Bray 1997). The most severe form of water deficit is desiccation, when most of the protoplasmic water is lost and only a very small amount of firmly bound water remains in the cell. Desiccation is drying out of an organism that is exposed to air. Most flowering plants cannot survive exposure to a water deficit equivalent to less than 85–98% (v/v) relative humidity during their vegetative growth phase although desiccation is an essential part of the developmental process of most higher plants with reference to seed formation (Gaff 1971). Desiccation tolerance seemingly depends on the ability of cells to maintain the integrity of cell membranes and to prevent denaturation of proteins. Tolerance in organs such as seeds and pollen is widespread amongst higher plants, and partial desiccation is a precondition for completing lifecycle in most species producing seeds. Desiccation-tolerant plants comprise monocotyledonous and dicotyledonous species within the angiosperms in the so-called resurrection plants (Gaff 1971), and certain ferns, algae, lichens and bryophytes possess

Table 1 Physiological and biochemical responses of plants under drought stress

Sr. No.	Plant traits	Yield-related effects on plant	Variation in stress	References
1.	Net photosynthesis, total leaf area, plant dry weight	To recover the net photosynthesis after well watered	Selected cultivar resistant to drought stress	Fini et al. (2013)
2.	Amino acid, C/N ratio and osmolality	Change in water potential and metabolic changes in plant cause yield decrease under stress	Water scarcity causes the increase in amino acid and osmolality and lowers C/N ratio	Hochberg et al. (2013)
3.	Electrolyte leakage, peroxidase activities	Increase in water stress reduction in budding success	Phenol and peroxidase activities increase, but chlorophyll and relative water content decrease under stress	Bolat et al. (2014)
4.	Membrane stability and chlorophyll content	Reduced membrane stability, relative water content and total carotenoid content in all the cultivars, whereas total chlorophyll content increased	Water deficit stress at pod development stage proved to be more damaging than at pegging stage	Chakraborty et al. (2015)
5.	Root water absorption, leaf relative water content and antioxidative enzyme	Increase tuber yield and activities of antioxidative enzyme higher under water stress condition	Drought resistant increases under stress conditions in selected cultivars	Shia et al. (2015)
6.	Transpiration rate	Variation in leaf area and stomatal conductance	Few landraces show tolerance	Nakhforoosh et al. (2016)
7.	Photosynthesis rate, leaf carbon isotopes	Water-use efficiency increases with stomatal conductance	Increase water deficit tolerance capacity	Bota et al. (2016)
8.	Relative water content, grain yield and leaf area index	Total biomass and yield increase under water deficit in selected genotype	Tolerant under water deficit stress	Panda et al. (2016)
9.	Water-use efficiency	Total yield increase under water deficit	Drought tolerant	Djurovic et al. (2016)
10.	Photosynthetic rate, conductance of stomata	High degree of photosynthetic rate and increased biomass gain under drought	Resistance under stress	Haworth et al. (2017) and Sapeta et al. (2013)
11.	Fruit dry matter, total soluble solids, total ascorbic acids	Increased fruit dry matter and total soluble solid/total ascorbic acid	Improve fruit quality and water deficit capacity	Guida et al. (2017)
12.	Carotenoids and photosynthetic pigments	Decreased amount of chlorophylls, carotenes and neoxanthin, the	Major effect on the concentration of some	Mibei et al. (2017)

(continued)

Table 1 (continued)

Sr. No.	Plant traits	Yield-related effects on plant	Variation in stress	References
		concentration of zea-xanthin increased with water deficit	carotenoids and photo-synthetic pigments	
13.	Leaf area, root length	High leaf area, increased root-to-above ground ratio	Survive under severe drought condition	Silva et al. (2017)
14.	Shoot fresh and dry weights, stomatal conductance and photosynthetic capacity	Less decrease shoot fresh and dry weights, stomatal conductance and photosynthetic capacity	Shows drought stress tolerance in selected species	Aboughadareh et al. (2017)

desiccation-tolerant vegetative tissues. So far, no gymnosperms have been found to be desiccation tolerant (Bartels 2005). Desiccation tolerance is the ability of the plant to survive periods during which the cells are water-stressed and the plant dries up; all its metabolic systems undergo dehydration. The attainment of desiccation tolerance is the result of complex interactions of different cellular processes due to multiple stresses imposed on plant tissues during severe dehydration. The speed of water loss and the events before dehydration appear to be critical for survival, such that if the speed of dehydration is too fast, plants do not acquire tolerance to desiccation. This observation suggests acquisition of desiccation tolerance is an active process and requires explicit biochemical changes and the synthesis of desiccation-related molecules. The intricacy of desiccation tolerance proposes that the gene products induced at the time of dehydration can be correlated with signal transduction pathways and regulation of stress-specific transcription, with carbohydrate metabolism or with cellular protection (Phillips and Bartels 2000). In order to understand the molecular basis of desiccation tolerance, numerous approaches may be used. One strategy is the developing genetic model system to study desiccation tolerance in vegetative tissue. Transposon tagging or insertional mutagenesis via T-DNA could be used in inferring the function of genes in genetic model systems. Secondly, natural allelic variation has been shown to be effective for identifying genes involved in plant development. Quantitative trait locus (QTL) analysis of plant accessions that exhibit extensive variation for desiccation tolerance may be a means of identifying genes in complex regulatory networks.

3 Gene Expression and Dehydration

Molecular responses to unfavourable environment include a series of genes and signal transduction pathways that are highly regulated and enable plants to survive the stress conditions. Although much of this regulation is at transcriptional, post-transcriptional and post-translational levels, the majority of the focus remains at the transcriptional level involving modification and remodelling of chromatin, *cis*-acting elements located upstream and downstream the coding region of the gene and transcription factors (Luo et al. 2012).

Physiological studies on stress responses reveal that the recent progress in plant molecular biology has assisted the detection of many genes governing stress tolerance (Table 2). Functional genes include the cell protection (enzymes for generating protective metabolites and proteins) and regulatory genes which regulate stress response (such as protein kinases and transcription factors). Thus, these genes have been categorized as functional proteins and regulatory proteins (Fig. 2). Functional proteins function in stress tolerance and regulatory proteins function in signal transduction and gene expression to stress response. Variety of drought-inducible genes in plants suggests the complex nature of drought stress. These gene products are involved in drought tolerance and stress response. Mostly the drought-inducible genes respond to cold stress as well except a few. The DNA sequences involved in stress sensing, transduction of the signal and regulation and function of the downstream gene induction and repression mechanism are largely conserved (Serrano 1996; Shinozaki and Yamaguchi-shinozaki 1997; Zhu et al. 1997; Ishitani et al. 1997). A 9 bp conserved sequence, TACCGACAT, named the dehydration responsive element (DRE) is vital for the regulation of induction of *rd29A* under low-temperature, drought and high salt stress conditions, however not as an ABA-responsive element (Kasuga et al. 1999). The *rd29A* promoter which functions in response to ABA also contains ABRE. DRE-related motifs have been found in promoter region of several genes induced under drought and low temperature (Yamaguchi-shinozaki and Shinozaki 1994).

These results show that the DRE-related motifs including C-repeat (CRT) and low-temperature-responsive elements (LTRE) which contain a CCGAC core motif are involved in ABA-independent gene functions in response to drought and cold stress.

Liu et al. (1998) cloned five independent DRE/CRT binding proteins using yeast hybrid assay and classified them into two groups: CBF1/DREB1 and DREB2. The *DREB1A* gene and its two homologs (*DREB1B*=*CBF1*, *DREB1C*) are expressed under low-temperature stress, but the *DREB2A* gene and its homologue (*DREB2B*) are expressed under dehydration (Shinwari et al. 1998). Overproduction of the *DREB1A* and *CBF1/DREB1B* cDNA driven by the 35S CaMV promoter in transgenic plants markedly improved stress tolerance to drought and freezing (Yoshida et al. 2010). However, the *DREB1A* transgenic plants revealed severe growth retardation under normal conditions. The *DREB1A* cDNA driven by the stress-inducible *rd29A* promoter was expressed at low level under unstressed controlled conditions and strongly induced by dehydration, salt and cold stresses (Kasuga et al. 1999). The *rd29A* promoter reduced the negative effects on growth of plants to minimum, whereas the 35S CaMV promoter severely retarded growth under normal growth conditions. Moreover, this stress-inducible promoter enhanced tolerance to drought, salt and freezing as compared to 35S CaMV promoter (Kasuga et al. 1999). Polygenic inheritance of root characters was reported by Ekanayake et al. (1985),

Table 2 Relevant examples of genes conferring drought tolerance

Sr. No	Genes	Function during drought	Mechanism of action	References
1.	*CKX1/ WRKY6*	Enhanced abscisic acid catabolism and regulate stomatal conductance	Modulation of cytokinin with decrease in leaf osmotic potential and proline biosynthetic gene *P5CSA* raised during stress	Mackova et al. (2013)
2.	*GsZFP1*	Relative membrane permeability, malondialdehyde (MDA) content and more free proline and soluble sugars accumulated	Gene overexpression enhanced the salt/drought stress tolerance	Tang et al. (2013)
3.	*SNAC1*	Regulating photosynthesis rate and transpiration rate	Overexpression of *SNAC1* improved tolerance to drought and salt in cotton through enhanced root development and reduced transpiration rates	Liu et al. (2014a)
4.	*DREB2A/ NAC5*	Transcription factors were enhanced by stress	Silicon- and selenium-pretreated plant under water stress showed increase in proline content and glycine betaine in both shoots and roots. Enhanced the expression of drought-specific genes, OsCMO coding rice choline monooxygenase and dehydrin OsRAB16b	Khattab et al. (2014)
5.	*MaPIP1;1*	Reduced expression of ABA-responsive genes and high cytosolic K+/Na+ ratio under stress	Increased primary root elongation, root hair numbers and reduced membrane injury and improved osmotic adjustments due to overexpression of gene in banana	Xu et al. (2014)
6.	*BdWRKY36*	Controlling ROS homeostasis and regulating transcription of stress-related genes	Overexpression enhance lesser ion leakage (IL) and reactive oxygen species (ROS) accumulation, but higher contents of chlorophyll, relative water content (RWC) and activities of antioxidant enzyme under drought condition	Sun et al. (2015)
7.	*popW*	Primed antioxidant responses	Significant increase in peroxidase, superoxide dismutase, catalase activities and ascorbic acid content, and overexpression also enhanced the relative transcript levels of oxidative stress-responsive	Liu et al. (2016)

(continued)

Table 2 (continued)

Sr. No	Genes	Function during drought	Mechanism of action	References
			genes *NtAPX, NtCAT1, NtGST* and *NtCu/Zn-SOD* under drought stress	
8.	*AtWRKY57*	Stress-induced transcription factors	Enhanced drought/salt tolerance by decreased electrolyte leakage, malondialdehyde content; increased proline and reactive oxygen content in transgenic rice	Jiang et al. (2016)
9.	*OsNRRB*	Stress-induced transcription responses	Positively regulate drought stress tolerance through upregulating stress-responsive genes *OsbZIP23*, *OsDREB2A*, *OsP5CS* and *OsLea3* by overexpression of *OsNRRB*, which increase drought tolerance in rice	Zhang and Chen (2017)
10.	*MpCYS4*	ABA hypersensitivity and enhanced stomatal closing	Enhanced stomatal closure and upregulation of the transcriptional levels of ABA and drought-related genes during drought	Tan et al. (2017)
11.	*LEA*	Decrease of photosynthetic activity and activation of antioxidant systems	Increased water deficit stress tolerance	Juszczak and Bartels (2017)
12.	*PgRab7*	Regulating Na + ion homeostasis, altered expression of transporter genes, including OsVHA, maintenance of photosynthetic rate	Overexpression induced the salt/drought stress tolerance	Tripathy et al. (2017)
13.	*ThDREB*	Stress enhancing the antioxidase activity and managing ROS level	Expression of *ThDREB* higher germination rates, fresh weights and root lengths under NaCl and mannitol treatments. The total chlorophyll content, superoxide dismutase (SOD) and peroxidase (POD) activities are also higher	Yang et al. (2017)

where the dominant alleles governed long and more numbers of roots whereas recessive alleles governed the thick root tip (Gaff 1980). Leaf moulding and osmotic adjustment showed monogenic inheritance. Tomar and Prasad (1996) reported a drought resistance gene, *Drt1* in rice, which is linked with pigmentation, hull colour, plant height and pleiotropic effects on the root system.

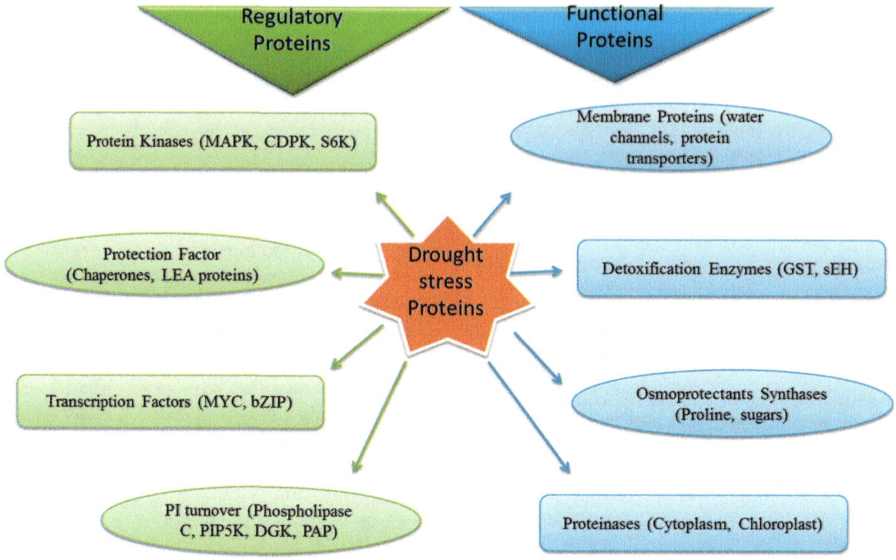

Fig. 2 Drought-inducible proteins in stress tolerance and responses

Numerous stress-related genes have been isolated and characterized in a number of crop species in the last eras (Cattivelli et al. 2002, 2008; Prabha et al. 2011; Joshi et al. 2016). Three coding single nucleotide polymorphisms (SNPs) and one haplotype identified in the *OsDREB1F* gene are likely to be related to drought tolerance in rice (Singh et al. 2015). Six different OsDREB1F protein variants were identified based on translated amino acid residues amongst the orthologs. Deletions in coding region trimmed five protein variants which were found to be susceptible to drought stress. Association study revealed that three coding SNPs of this gene were considerably associated with drought tolerance. One OsDREB1F variant in the activation domain of *OsDREB1F* gene that has an amino acid change from aspartate to glutamate was found to be associated with drought tolerance. The natural allelic variants mined in the *OsDREB1F* gene can be used in translational genomics in the future for improving the water-use efficiency in rice (Singh et al. 2015). Expression of the *SHINE* and *HARDY* genes were found to confer water-use efficiency in rice, although their phenotypic effects have not yet been evaluated under field conditions (Karaba et al. 2007). Transgenic plants with either upregulated stress responses or specific metabolic processes related to drought tolerance have been developed by classical physiological studies (Cattivelli et al. 2008). Few reports on transgenic rice overexpressing NAC1 transcription factor (Hu et al. 2006; Tran et al. 2004) and *OsLEA3* gene (Xiao et al. 2007) showed higher yield under drought conditions due to increased spikelet fertility. Under stress conditions, ectopic expression of *OsCDPK7* gene encoding a calcium-dependent protein kinase improved levels of stress-responsive genes that contribute to improved salt and drought tolerance in rice (Saijo et al. 2000). *CBF3/DREB1A* gene in transgenic rice also increased drought tolerance without affecting growth undesirably (Oh et al. 2005).

Numerous transcription factors initiate stress responses and establish plant stress tolerance by regulating stress-inducible genes. Transcription factors (TFs) are basically proteins that recognize and bind to the cis-acting elements in promoter region and regulate transcription, by activating or inhibiting the expression of particular genes. Overexpression of some transcription factors, including bZIP, ERF/AP2 family, DOF, HDZIP, MYB, NAC, WRKY and Zn-finger (Dubouzet et al. 2003; Yang et al. 2012; Jan et al. 2013), and genes like *CDPKs*, *HAP/CAAT*, *HSPs-LEA* family and *MAPKKK* (Vierling 1991; Kazuko and Shinozaki 2006; Lipiec et al. 2013) have proved to be promising candidates as stress modulators. Members from each transcription factor family show protective phenotypes against multiple stresses such as cold, drought and excess salt (Shukla and Mattoo 2013; Mattoo et al. 2014). For example, in rice, *OsWRKY89* improved tolerance to UV irradiation and fungal infection (Wang et al. 2007), and *OsWRKY45* is found to be highly expressed under cold, heat, salt and dehydration. The overexpression of *OsWRKY11* enhanced heat and drought tolerance (Wu et al. 2008). *AtMYB60* and *AtMYB96* regulate stomata movement in the ABA signalling cascade in response to drought stress (Cominelli et al. 2005). *AtMYB13*, *AtMYB15*, *AtMYB33* and *AtMYB101* are also involved in ABA-mediated responses to environmental stresses (Reyes and Chua 2007). In rice, the *OsISAP1* gene having zinc-finger domain was highly expressed in effect of stress induced by dehydration, cold, salinity and heavy metals (Mukhopadhyay et al. 2004). Several studies reveal transcription factors control various defence mechanisms; therefore, they are being considered of great importance in breeding programmes that aim mechanisms of tolerance to abiotic stresses.

4 Biochemical Aspects of Dehydration Tolerance

Seemingly, most of the plants employ multiple mechanisms to ensure dehydration tolerance. At present, our knowledge on the metabolic changes that lead to dehydration tolerance is partial, but information about the biochemical processes governing dehydration tolerance is essential for successful engineering of dehydration tolerance in crop plants.

4.1 HSPs

HSPs are widely distributed in nature and accumulate during stress. They are commonly known as molecular chaperones involved in protein folding and assembly, removal and disposal of nonfunctional proteins (Wang et al. 2004). HSPs are induced by drought and salinity stress (Alamillo et al. 1995; Campalans et al. 2001), and in vivo evidences propose that HSPs inhibit thermal aggregation of proteins, thus easing the recovery of cell functions after abiotic stress (Lee et al. 1995). They are classified according to their molecular weight: Hsp70 family (family DnaK);

chaperonins, namely, GroEL and Hsp60; the Hsp90 family; the Hsp100 family; and the small Hsp family (Wang et al. 2004). Cyclophilin is a chaperon protein involved in protein folding, highly induced under drought stress; overexpression of *cyclophilin* gene confers manifold abiotic stress tolerance (Gottschalk et al. 2008; Sekhar et al. 2010). An increase in cysteine protease activity has also been observed during drought conditions (Koizumi et al. 1993; Seki et al. 2002). HSF (heat shock factor) family members bind to the promoter region of few chaperones known as heat shock proteins (Pelham 1982). These TFs are located in the cytoplasm when in their inactive state (Baniwal et al. 2004; Hu et al. 2009) and have a C-terminal portion and 3N-terminal portions, besides the amino acid leucine (Schuetz et al. 1991). Various reports suggest the presence of at least 21 HSFs in *A. thaliana* (Baniwal et al. 2004; Nover and Baniwal 2006), 30 in corn, 24 in *Brachypodium*, 25 in rice, 27 in tomato and 52 in soybean (Scharf et al. 2012), supporting the idea that, in plants, there are many duplications, which make HSFs extremely complex. Rice mutants demonstrated the performance of HSFs as the response to abiotic stresses. Overexpression of OsHsfA7 mutant in rice and *A. thaliana* promoted a tolerance of 42 °C, resulting in the survival of more than 50% of the mutants when stressed, twice the value of the results obtained by the control (Liu et al. 2009). Another report established the higher expression of HSPs and HSFs under heat stress in rice, showing that the regulation of abiotic stress induces numerous genes and HSPs that act together in different cascades to combat the problems of abiotic stress (Chandel et al. 2013). These studies highlight the importance of transcription factors and HSFs in the regulation of metabolic pathways responsive to abiotic stress so one can consider them as good candidates in breeding programmes targeting mechanisms of tolerance to abiotic stresses.

4.2 BiP

The bZIP family of TFs is abundant, with its orthologs in several species, which include 17 in yeast, 56 in humans, 75 in *Arabidopsis*, 89 in rice, 92 in sorghum, 125 in maize and 131 in soybean (Jakoby et al. 2002; Wei et al. 2012). Elevated levels of binding protein (BiP) have been associated to a variety of abiotic and biotic stresses such as water stress, fungal manifestations, nutritional stress, cold acclimation, insect attack and elicitors of the plant pathogenesis response (Anderson et al. 1994; Denecke et al. 1995; Kalinski et al. 1995; Fontes et al. 1996, 1999; Figueiredo et al. 1997). The rice gene *OsISAP1*, a bZIP family, when overexpressed in tobacco, conferred tolerance to cold, dehydration and salt stress at the seed germination (Mukhopadhyay et al. 2004). OsbZIP71, a TF in rice, was found to be strongly induced by drought, PEG and ABA treatments and repressed by salinity, signifying its regulatory role in ABA-mediated drought and salt tolerance (Liu et al. 2014b).

4.3 Protein Kinase

Protein kinases belonging to calcium-dependent protein kinase (CDPK), mitogen-activated protein kinase (MAPK) families and calcineurin B-like protein-interacting protein kinases (CIPK) are thought to be majorly involved in drought tolerance. Ca^{2+} cytosolic levels increase rapidly in plant cells in response to environmental stresses, namely, drought and salinity (Sanders et al. 1999). This Ca^{2+} influx is probably mediated by a combination of protein phosphorylation/dephosphorylation cascades involving members of the CDPK family. In rice, overexpression of *OsCDPK7* (under the control of the 35S promoter) resulted in increased seedling recovery rate after a salt treatment (Saijo et al. 2000). Transgenic rice overexpressing three *CIPK* genes (*OsCIPK03*, *OsCIPK12* and *OsCIPK15*) showed enhanced tolerance to cold, drought and salt stress, respectively (Xiang et al. 2007). Overexpression of *OsMAPK5a* gene in rice lead to an increase in kinase activity and enhanced tolerance to drought and salt stresses (Xiong and Yang 2003). Overexpression of *OsMAPK44* gene resulted ERA1 in increased tolerance to salt stress in rice (Jeong et al. 2006). Recently, overexpression in rice of DSM1 (drought-hypersensitive mutant1), a well-accepted MAPK kinase kinase (*MAPKKK*) gene, increased the water stress tolerance at seedling level (Ning et al. 2010). It was suggested that DSM1 might be operating as an early signalling component in controlling mechanisms of ROS scavenging in rice. Expression of a *MAPKKK* gene was proved to trigger an oxidative signal cascade and led to tolerance to environmental stress in transgenic tobacco (Shou et al. 2004). In yeast, the catalytic domain of *Nicotiana* protein kinase 1 (NPK1) activated a bypass of BCK1-mediated signal transduction pathway, which was found to be conserved amongst different organisms (Banno et al. 1993). NPK1 was reported to be upstream of oxidative pathways inducing expression of heat shock proteins and glutathione-S-transferases (GST) (Kovtun et al. 2000). Constitutive overexpression of the tobacco *MAPKKK* in maize enhanced the drought tolerance of the transgenic plants (Shou et al. 2004). The transgenic plants maintained significantly higher photosynthesis rates and kernel weight as compared with wild-type plants under drought conditions. However, the effect of NPK1 on yield components was less apparent.

4.4 Nuclear Factor Y-B Subunit

NF-Y is a conserved hetero-trimeric complex consisting of NF-YA (HAP2), NF-YB (HAP3) and NF-YC (HAP5) subunits (Mantovani 1999). In *Arabidopsis*, AtNF-YB1, a nuclear factor Y (NF-Y complex), was found to regulate transcription through CCAAT DNA elements and confer abiotic stress tolerance when constitutively expressed in *Arabidopsis* (Nelson et al. 2007). In maize, an ortholog of NF-YB gene was found showing similar response to drought (Wei et al. 2012).

4.5 NAC Proteins

Several NAC domain proteins [word derived from the first alphabet of three genes *NAM* (No Apical Meristem), *ATAF* (*Arabidopsis* transcription activation factor) and *CUC* (cup shaped cotyledon)], which are one of the largest plant TF families, have been found to be associated with abiotic stresses (Riechmann et al. 2000). Amongst the 150 members of the NAC family identified in rice that recognizes the cis-acting drought-responsive element NACRS, the expression of about 40 NAC genes increased during drought or salinity stress (Sakuma et al. 2006).

Twofold increase was observed in 20 genes during stress, and a majority of them comprise the SNAC (stress-responsive NAC) group (Fang et al. 2008). Overexpressing *SNAC1* improved biomass accumulation at the vegetative stage in rice plants under both salinity and drought stress due to increased stomatal closure and ABA sensitivity in the transgenic plants (Hu et al. 2006). It was found that the rice genes *ONAC19*, *ONAC55*, *ONAC72* and *ONAC045* were induced by drought and *ONAC045* by high salt, low-temperature and ABA treatment (Zeng et al. 2009). Of late, the overexpression of *OsNAC10* under the control of the constitutive promoter *GOS2* and the root-specific promoter *RCc3* improved tolerance to drought and salinity of the transgenic rice plants at the vegetative stage. However, only the root-specific overexpression of *OsNAC10* enhanced drought tolerance significantly during the reproductive phase, increasing grain yield (25–42%) under drought conditions due to the larger root diameter, which were almost 20% larger than both the wild-type and PGOS2::*OsNAC10* plants (Jeong et al. 2010).

4.6 LEA Proteins

Late embryogenesis abundant (LEA) proteins are low-molecular weight proteins that accumulate at higher levels in embryos (Dure et al. 1981; Galau et al. 1986). LEA proteins accumulate in plants in response to water stress and have various functions in drought tolerance. They act synergistically with trehalose to prevent protein aggregation during water deficit (Goyal et al. 2005). Genes encoding LEA-type proteins are diverse RD (responsive to dehydration), ERD (early response to dehydration), KIN (cold inducible), COR (cold regulated) and RAB (responsive to ABA). Five LEA groups have been identified based on structural domains, group 3 and 5 form dimmers with a coiled-coil conformation that manage the ions during stress (Dure et al. 1989). Dehydrins, also known as group 2 LEA proteins, accumulate in response to dehydration and low temperature (Close 1997). The overexpression of *OsLEA3*-1 under the control of strong constitutive promoters (35S and Actin1) and a stress-inducible promoter (HVA1-like promoter isolated from the upland rice IRAT109) improved drought tolerance in the drought-sensitive Japonica (lowland) rice (Xiao et al. 2007). Increasing *LEA* gene expression under stress, and presumably LEA protein abundance, has also been accomplished indirectly, with the overexpression of NAC genes. The overexpression of the stress-responsive proteins OsNAC5 and OsNAC6 enhanced stress tolerance by upregulating the expression of stress-inducible gene *OsLEA3* in rice.

4.7 Aquaporins

Aquaporins are central membrane proteins that govern the transport of water, small neutral solutes and CO_2 (Tyerman et al. 2002). The regulatory role of aquaporins in cellular water transport had been demonstrated (Knepper 1994). The expression of the aquaporin, RWC3, a member of the plasma membrane intrinsic protein 1 (PIP1) subfamily, induced under stress resulted in improved water status of lowland rice (Lian et al. 2004). Transgenic rice plants constitutively overexpressing a barley plasma membrane aquaporin, HvPIP2, displayed more sensitivity (reduction in growth rate) to salinity stress (Katsuhara et al. 2003).

4.8 Other Transcription Factors

Multiple transcription factors (TFs) have been well characterized in various plant species, but transcriptional reprogramming under drought and stress is not fully understood. Transgenic rice plants overexpressing *AtMYB2* gene conferred salt stress tolerance, with higher biomass and decreased ion leakage under the control of an ABA-inducible promoter (Malik and Wu 2005). Overexpression of WRKY domain containing TF, OsWRKY11 under the control of a *HSP101* promoter, with slower leaf wilting and higher survival rate of green parts of plants conferred heat and drought tolerance at the seedling stage (Wu et al. 2008). It was shown that the constitutive overexpression of bacterial RNA chaperones, CspA and CspB, conferred abiotic stress tolerance to transgenic *Arabidopsis*, rice and maize (Castiglioni et al. 2008). The transgenic maize plants under water-stressed environment showed increase in yield up to 15% (0.75 t/ha) in comparison to the non-transgenic controls that indicates chaperone molecules may be good targets for enhancing abiotic stress tolerance in crop plants (Castiglioni et al. 2008).

5 Oxidative Stress

Free oxygen radicals produced as a consequence of various environmental stresses are very dangerous for cell components and must be regulated precisely. All plants have developed several antioxidant systems to scavenge these deadly compounds which include catalases (CAT), superoxide dismutase (SOD), peroxidases (POD), ascorbate peroxidases (APX), glutathione reductase (GR) and monodehydroascorbate reductase

(MDAR) (Yang et al. 2017). Besides, there are antioxidant molecules such as ascorbic acid (AA), glutathione, tocopherols, flavanones, carotenoids and anthocyanins (Liu et al. 2016). Some osmolytes (e.g. proline), proteins (e.g. peroxiredoxin) and amphiphilic molecules (e.g. tocopherol) have ROS scavenging function and might function as the antioxidant (Mattoo et al. 2014). Non-enzymatic plant antioxidants are either AA-like scavengers or pigments multifunctional in nature acting as the enzyme cofactor and as a donor/acceptor of electron (Chakraborty et al. 2015). The degree of activities of antioxidant systems under drought stress is exceptionally variable owing to variation in plant species, in the cultivars of the same species, development and the metabolism of the plant and the duration and intensity of the stress.

6 Conclusion and Future Prospects

More than 50% yield losses occur in major crop plants owing to different abiotic stresses especially drought (Lipiec et al. 2013). A tremendous effort has to be done to elucidate the stress response pathways, which include interpretation of the function and characterization of various genes and gene families responsible for stress tolerance.

Understanding the molecular basis of plant responses to water stress and their concomitant growth adjustments shall help us to increase plant productivity under water stress conditions. The diversity and specificity of TFs make key components for triggering signalling cascades. Further studies identifying gene variants associated with the significant agronomic traits will assist the molecular engineering of plants with increased tolerance to severe environmental stresses.

In summary, it is vital to integrate crop physiology, genomics and breeding approaches to dissect complex traits, understand the molecular basis of drought tolerance and develop the next-generation crops for our changing climate. Though research is continuing in some major crops, it is predicted that integrated physiology, genomics and breeding approaches will be accelerated in the orphan crops that are essential for food security in many developing countries.

References

Abe H, Yamaguchi-Shinozaki K, Urao T, Iwasaki T, Hosokawa D, Shinozaki K (1997) American society of plant physiologists role of *Arabidopsis* MYC and MYB homologs in drought and abscisic acid-regulated gene expression. Plant Cell 9:1859–1868

Aboughadareh AP, Ahmadi J, Mehrabi AA, Etminan A, Moghaddam M, Siddique KHM (2017) Physiological responses to drought stress in wild relatives of wheat: implications for wheat improvement. Acta Physiol Plant 39:106

Alamillo J, Almogura C, Bartels D, Jordano J (1995) Constitutive expression of small HSP in vegetative tissues of the resurrection plant Craterostigma plantagineum. Plant Mol Biol 29:1093–1099

Alcazar R, Altabella T, Marco F, Bortolotti C, Reymond M, Koncz C, Carrasco P, Tiburcio AF (2010) Polyamines molecules with regulatory functions in plant abiotic stress tolerance. Planta 231(6):1237–1249

Ammar MH, Anwar F, El-Harty EH, Migdadi HM, Abdel-Khalik SM, Al-Faifi AA, Farooq M, Alghamdi S (2015) Physiological and yield responses of Faba bean (*Vicia faba* L.) to drought stress in managed and open field environments. J Agron Crop Sci 201:280–287

Anderson JV, Li QB, Haskell DW, Guy CL (1994) Structural organization of the spinach endoplasmic reticulum luminal 70-kilodalton heat shock genes during cold acclimation. Plant Physiol 104:1395–1370

Ashraf M, Foolad MR (2007) Roles of glycine betaine and proline in improving plant abiotic stress resistance. Environ Exp Bot 59(2):206–216

Baniwal SK, Bharti K, Chan KY et al (2004) Heat stress response in plants: a complex game with chaperones and more than twenty heat stress transcription factors. J Biosci 29(4):471–487

Banno H, Hirano K, Nakamura T, Irie K, Nomoto S, Matsumoto K, Machida Y (1993) NPK1, a tobacco gene that encodes a protein with a domain homologous to yeast BCK1, STE11, and Byr2 protein kinases. Mol Cell Biol 13(8):4745–4752

Bano A, Ullah F, Nosheen A (2012) Role of abscisic acid and drought stress on the activities of antioxidant enzymes in wheat. Plant Soil Environ 58(4):181–185

Bartels D (2005) Desiccation tolerance studied in the resurrection plant *Craterostigma plantagineum*. Integr Comp Biol 45(5):696–701

Bartels D, Sunkar R (2005) Drought and salt tolerance in plants. Crit Rev Plant Sci 24:23–58

Bohnert HJ, Nelson DE, Jensen RG (1995) Adaptations to environmental stresses. Plant Cell 7:1099–1111

Bolat I, Dikilitas M, Ercisli S, Ikinci A, Tonkaz T (2014) The effect of water stress on some morphological, physiological, and biochemical characteristics and bud success on apple and quince rootstocks. Sci World J 769732:8

Bota J, Tomas M, Flexas J, Medrano H, Escalona JM (2016) Differences among grapevine cultivars in their stomatal behaviour and water use efficiency under progressive water stress. Agric Water Manag 164:91–99

Bray EA (1997) Plant responses to water deficit. Trends Plant Sci 2:48–54

Bray EA, Bailey-Serres J, Weretilnyk E (2000) Responses to abiotic stresses. In: Gruissem W, Buchannan B, Jones R (eds) Biochemistry and molecular biology of plants. American Society of Plant Physiologists, Rockville, pp 1158–1249

Brodribb TJ, McAdam SAM (2013) Abscisic acid mediates a divergence in the drought response of two conifers. Plant Physiol 162:1370–1377

Campalans A, Pages M, Messeguer R (2001) Identification of differentially expressed genes by the cDNA AFPL technique during dehydration of almond. Tree Physiol 21:633–643

Campos H, Cooper M, Habben JE, Edmeades GO, Schussler JR (2004) Improving drought tolerance in maize: a view from industry. Field Crop Res 90:19–34

Capell T, Bassie L, Christou P (2004) Modulation of the polyamine biosynthetic pathway in transgenic rice confers tolerance to drought stress. Proc Natl Acad Sci U S A 101:990–991

Castiglioni P, Warner D, Bensen RJ et al (2008) Bacterial RNA chaperones confer abiotic stress tolerance in plants and improved grain yield in maize under water-limited conditions. Plant Physiol 147:446–455

Cattivelli L, Baldi P, Crosatti C et al (2002) Chromosome regions and stress-related sequences involved in resistance to abiotic stress in Triticeae. Plant Mol Biol 48:649–665

Cattivelli L, Rizza F, Badeck FW et al (2008) Drought tolerance improvement in crop plants: an integrated view from breeding to genomics. Field Crop Res 105:1–14

Chakraborty K, Singh AL, Kalariya KA, Goswami N, Zala PV (2015) Physiological responses of peanut (*Arachis hypogaea* L.) cultivars to water deficit stress: status of oxidative stress and antioxidant enzyme activities. Acta Bot Croat 74(1):123–142

Chandel G, Dubey M, Meena R (2013) Differential expression of heat shock proteins and heat stress transcription factor genes in rice exposed to different levels of heat stress. J Plant Biochem Biotechnol 22(3):277–285

Chen TH, Murata N (2011) Glycine betaine protects plants against abiotic stress: mechanisms and biotechnological applications. Plant Cell Environ 34:1–20

Cheng L, Zou YJ, Ding SL, Zhang JJ, Yu XL, Cao JS (2009) Polyamine accumulation in transgenic tomato enhances the tolerance to high temperature stress. J Integr Plant Biol 51:489–499

Close TJ (1996) Dehydrins: emergence of a biochemical role of a family of plant degradation proteins. Physiol Plant 97(4):795–803

Close TJ (1997) Dehydrins: a commonality in response of plants to dehydration and low temperature. Physiol Plant 100:291–296

Cominelli E, Galbiati M, Vavasseur A (2005) A guard-cell specific MYB transcription factor regulates stomatal movements and plant drought tolerance. Curr Biol 15(13):1196–1200

Conesa MR, Rosa JM, Domingo R, Banon S, Perez-Pastor A (2016) Changes induced by water stress on water relations, stomatal behaviour and morphology of table grapes (cv. Crimson seedless) grown in pots. Sci Hortic 202:9–16

Davies WJ, Zhang J (1991) Root signals and the regulation of growth and development of plants in drying soil. Annu Rev Plant Physiol 42:55–76

Denecke J, Carlsson LE, Vidal S, Hoglund AS, Ek B, Zeijl MJV, Sinjorgo KMC, Palva ET (1995) The tobacco homolog of mammalian calreticulin is present in protein complexes in vivo. Plant Cell 7:391–406

Ding W, Fang W, Shi S, Zhao Y, Li X, Xiao K (2016) Wheat WRKY type transcription factor gene TaWRKY1 is essential in mediating drought tolerance associated with an ABA-dependent pathway. Plant Mol Biol Report 34:1111–1126

Djurovic N, Cosic M, Stricevic R, Savic S, Domazet M (2016) Effect of irrigation regime and kaolin application on yield: quality and water use efficiency of tomato. Sci Hortic 201:271–278

Dubouzet JG, Sakuma Y, Ito Y et al (2003) *OsDREB* genes in rice, *Oryza sativa* L., encode transcription activators that function in drought, high salt and cold responsive gene expression. Plant J 33(4):751–763

Dure L, Greenway SC, Galau GA (1981) Developmental biochemistry of cotton seed embryogenesis and germination: changing messenger ribonucleic acid populations as shown in vitro and in vivo protein synthesis. Biochemist 20:4162–4168

Dure L, Crouch M, Harada J, Ho TH, Mundy J, Quatrano R, Thomas T, Sung ZR (1989) Common amino acid sequence domains among the LEA proteins of higher plants. Plant Mol Biol 12 (5):475–486

Ekanayake IJ, Otoole JC, Garrity DP, Masajo TM (1985) Inheritance of root characters and their relations to drought resistance in rice. Crop Sci 25:927–933

Esther M, Gonazalez GL, Arrese-Igor C (2000) Abscisic acid induces a decline in nitrogen fixation that involves leghaemoglobin, but is independent of sucrose synthetase activity. J Exp Bot 25 (355):285–293

Fang Y, You J, Xie K, Xie W, Xiong L (2008) Systematic sequence analysis and identification of tissue-specific or stress-responsive genes of NAC transcription factor family in rice. Mol Gen Genomics 280:547–563

Figueiredo JEF, Cascardo JCM, Carolino SMB, Alvim FC, Fontes EPB (1997) Water stress regulation and molecular analysis of soybean BiP gene family. Braz Plant Physiol 9(2):103–110

Fini A, Bellasio C, Pollastri S, Tattini M, Ferrini F (2013) Water relations, growth, and leaf gas exchange as affected by water stress in Jatropha curcas. J Arid Environ 89:21–29

Fischer RA, Wood JT (1979) Drought resistance in spring wheat cultivars yield association with morpho-physiological traits. Aust J Agric Res 30:1001–1020

Fontes EPB, Silva CJ, Carolino SMB, Figueiredo JEF, Batista DPO (1996) A soybean binding protein (BiP) homolog is temporally regulated in soybean seeds and associated detectably with normal storage proteins in vitro. Braz J Genet 19:306–312

Fontes MA, Otoni WC, Carolino SMB, Brommonschenkel SH, Fontes EPB, Fari M, Louro RP (1999) Hyperhydricity in pepper plants regenerated in vitro involvement of BiP (binding protein) and ultra-structural aspects. Plant Cell Rep 19:81–87

Gaff DF (1971) Desiccation-tolerant flowering plants in Southern Africa. Science 174:1033–1034

Gaff DF (1980) Protoplasmic tolerance of extreme water stress. In: Turner NC, Kramer PJ (eds) Adaptation of plants to water and high temperature stress. Wiley, New York, pp 207–230

Galau GW, Hughes DW, Dure L (1986) Abscisic acid induction of cloned cotton late embryogenesis abundant (LEA) messenger RNAs. Plant Mol Biol 7:155–170

Garg A, Kim J, Owens T, Ranwala A, Choi Y, Kochian L, Wu R (2002) Trehalose accumulation in rice plants confers high tolerance levels to different abiotic stresses. Proc Natl Acad Sci U S A 99:15898–15903

Gill SS, Tuteja N (2010) Polyamines and abiotic stress tolerance in plants. Plant Signal Behav 5:26–33

Gottschalk M, Dolgener E, Xoconostle-Cazares B, Lucas WJ, Komor E, Schobert C (2008) Ricinus communis cyclophilin: functional characterization of a sieve tube protein involved in protein folding. Planta 228:687–700

Goyal K, Walton LJ, Tunnacliffe A (2005) LEA proteins prevent protein aggregation due to water stress. Biochem J 388:151–157

Guida G, Sellami MH, Mistretta C et al (2017) Agronomical, physiological and fruit quality responses of two Italian long-storage tomato landraces under rain-fed and full irrigation conditions. Agric Water Manag 180:126–135

Haworth M, Elliott-Kingston C, McElwain JC (2013) Co-ordination of physiological and morphological responses of stomata to elevated [CO_2] in vascular plants. Oecologia 171:71–82

Haworth M, Cosentino SL, Marino G et al (2017) Physiological responses of Arundo donax ecotypes to drought: a common garden study. Glob Chang Biol Bioenergy 9:132–143

Hayat S, Hayat Q, Alyemeni MN, Wani AS, Pichtel J, Ahmad A (2012) Role of proline under changing environments: a review. Plant Signal Behav 7(11):1456–1466

Hochberg U, Degu A, Toubiana D, Gendler T, Nikoloski Z, Rachmilevitch S, Fait A (2013) Metabolite profiling and network analysis reveal coordinated changes in grapevine water stress response. BMC Plant Biol 13:184

Hoisington D, Jiang C, Khairallah M, Ribaut JM, Bohn M, Melchinger A, Willcox M, Gonzalez-de Leon D (1996) QTL for insect resistance and drought tolerance in tropical maize: prospects for marker assisted selection. Symp Soc Exp Biol 50:39–44

Hu H, Dai M, Yao J, Xiao B, Li X, Zhang Q, Xiong L (2006) Overexpressing a NAM, ATAF, and CUC (NAC) transcription factor enhances drought resistance and salt tolerance in rice. Proc Natl Acad Sci U S A 103(35):12987–12992

Hu W, Hu G, Han B (2009) Genome-wide survey and expression profiling of heat shock proteins and heat shock factors revealed overlapped and stress specific response under abiotic stresses in rice. Plant Sci 176(4):583–590

Ilhan S, Ozdemir F, Bor M (2015) Contribution of trehalose biosynthetic pathway to drought stress tolerance of *Capparis ovata* Desf. Plant Biol 17:402–407

Imai R, Chang L, Ohta A, Bray EA, Takagi M (1996) A lea-class gene of tomato confers salt and freezing tolerance when expressed in Saccharomyces cerevisiae. Gene 170(2):243–248

Ishitani M, Xiong L, Stevenson B, Zhu JK (1997) Genetic analysis of osmotic and cold stress signal transduction in *Arabidopsis*. Interactions and convergence of abscisic acid dependent and independent pathways. Plant Cell 9(11):1935–1949

Ishitani M, Rao I, Wenzl P, Beebe S, Tohme J (2004) Integration of genomics approach with traditional breeding towards improving abiotic stress adaptation: drought and aluminum toxicity as case studies. Field Crop Res 90:35–45

Jakoby M, Weisshaar B, Droge-Laser W, Vicente-Carbajosa J, Tiedemann J, Kroj T, Parcy F (2002) bZIP transcription factors in *Arabidopsis*. Trends Plant Sci 7(3):106–111

Jan A, Maruyama K, Todaka D, Kidokorom S, Abo M, Yoshimura E (2013) OsTZF1, a CCCH-tandem zinc finger protein, confers delayed senescence and stress tolerance in rice by regulating stress-related genes. Plant Physiol 161(3):1202–1216

Jeong MJ, Lee SK, Kim BG et al (2006) A rice (*Oryza sativa* L.) MAP kinase gene, OsMAPK44, is involved in response to abiotic stresses. Plant Cell Tissue Organ Cult 85:151–160

Jeong JS, Kim YS, Baek KH et al (2010) Root-specific expression of *OsNAC10* improves drought tolerance and grain yield in rice under field drought conditions. Plant Physiol 153:185–197

Jiang Y, Qiu Y, Hu Y, Yu D (2016) Heterologous expression of *AtWRKY57* confers drought tolerance in *Oryza sativa*. Front Plant Sci 7:145

Joshi R, Wani SH, Singh B, Bohra B, Dar ZA, Lone AA, Pareek A, Singla-Pareek SL (2016) Transcription factors and plants response to drought stress: current understanding and future directions. Front Plant Sci 7:1029

Juszczak I, Bartels D (2017) LEA gene expression, RNA stability and pigment accumulation in three closely related Linderniaceae species differing in desiccation tolerance. Plant Sci 255:59–71

Kalinski A, Rowley DL, Loer DS, Foley C, Buta G, Herman EM (1995) Binding protein expression is subject to temporal, developmental and stress induced regulation in terminally differentiated soybean organs. Planta 195(4):611–621

Karaba A, Dixit S, Greco R et al (2007) Improvement of water use efficiency in rice by expression of HARDY, an *Arabidopsis* drought and salt tolerance gene. Proc Natl Acad Sci U S A 104 (39):15270–15275

Karamanos AJ, Papatheohari AY (1999) Assessment of drought resistance of crop genotypes by means of the water potential index. Crop Sci 39:1792–1797

Kasuga M, Liu Q, Miura S, Yamaguchi-Shinozaki K, Shinozaki K (1999) Improving plant drought, salt and freezing tolerance by gene transfer of single stress-inducible transcription factor. Nat Biotechnol 17(3):287–291

Katsuhara M, Koshio K, Shibasaka M, Hayashi Y, Hayakawa T, Kasamo K (2003) Over-expression of a barley aquaporin increased the shoot/root ratio and raised salt sensitivity in transgenic rice plants. Plant Cell Physiol 44:1378–1383

Kazuko YS, Shinozaki K (2006) Transcriptional regulatory networks in cellular responses and tolerance to dehydration and cold stresses. Annu Rev Plant Biol 57:781–803

Khattab HI, Emam MA, Emam MM, Helal NM, Mohamed MR (2014) Effect of selenium and silicon on transcription factors *NAC5* and *DREB2A* involved in drought-responsive gene expression in rice. Biol Plant 58(2):265–273

Knepper MA (1994) The aquaporin family of molecular water channels. Proc Natl Acad Sci U S A 91:6255–6258

Koizumi T, Nojima Y, Endo T (1993) Radical ring-opening polymerization of 2-phenyl-3-vinyloxirane derivatives having a methyl group on the vinyl function. J Polym Sci A Polym Chem 31:3489–3492

Kovtun Y, Chiu WL, Ten G, Sheen J (2000) Functional analysis of oxidative stress-activated mitogen-activated protein kinase cascade in plants. Proc Natl Acad Sci U S A 97(6):2940–2945

Lee GJ, Pokala N, Vierling E (1995) Structure and in vitro molecular chaperone activity of cytosolic small HSP from pea. J Biol Chem 270:10432–10438

Lian HL, Yu X, Ye Q, Ding XS, Kitagawa Y, Kwak SS, Su WA, Tang ZC (2004) The role of aquaporin RWC3 in drought avoidance in rice. Plant Cell Physiol 45:481–489

Lim CW, Baek W, Han SW, Lee SC (2013) *Arabidopsis* PYL8 plays an important role for ABA signaling and drought stress responses. Plant Pathol J 29(4):471–476

Lipiec J, Doussan C, Nosalewicz A, Kondracka K (2013) Effect of drought and heat stresses on plant growth and yield: a review. Int Agrophys 27:463–477

Liu Q, Kasuga M, Sakuma Y, Abe H, Miura S, Yamaguchi-shinosaki K, Shinozaki K (1998) Two transcription factors, DREB1 and DREB2 with an ER EBP/AP2 DNA binding domain separate two cellular signal transduction pathways in drought and low temperature responsive gene expression responsible in *Arabidopsis*. Plant Cell 10(8):1391–1406

Liu JH, Kitashiba H, Wang J, Ban Y, Moriguchi T (2007) Polyamines and their ability to provide environmental stress tolerance to plants. Plant Biotechnol 24:117–122

Liu J, Qin Q, Zhang Z (2009) OsHSF7 gene in rice, *Oryza sativa* L., encodes a transcription factor that functions as a high temperature receptive and responsive factor. BMB Rep 42(1):16–21

Liu G, Li X, Jin S, Liu X, Zhu L, Nie Y, Zhang X (2014a) Overexpression of Rice *NAC* gene *SNAC1* improves drought and salt tolerance by enhancing root development and reducing transpiration rate in transgenic cotton. PLoS One 9(1):e86895

Liu C, Mao B, Ou S, Wang W, Liu L, Wu Y, Chu C, Wang X (2014b) OsbZIP71, a bZIP transcription factor, confers salinity and drought tolerance in rice. Plant Mol Biol 84:19–36

Liu H, Wang Y, Zhou X, Wang C, Wang C, Fu J, Wei T (2016) Overexpression of a harpin-encoding gene *popW* from *Ralstonia solanacearum* primed antioxidant defenses with enhanced drought tolerance in tobacco plants. Plant Cell Rep 35:1333–1344

Luo M, Liu X, Singh P, Cui Y, Zimmerli L, Wu K (2012) Chromatin modifications and remodeling in plant abiotic stress responses. Biochim Biophys Acta 1819:129–136

Mackova H, Hronkova M, Dobra J et al (2013) Enhanced drought and heat stress tolerance of tobacco plants with ectopically enhanced cytokinin oxidase/dehydrogenase gene expression. J Exp Bot 64(10):2805–2815

Malik V, Wu R (2005) Transcription factor AtMyb2 increased salt-stress tolerance in rice (*Oryza sativa* L). Rice Genet Newsl 22:63

Mantovani R (1999) The molecular biology of the CCAAT-binding factor NF-Y. Gene 239:15–27

Mattoo AK, Upadhyay RK, Rudrabhatla S (2014) Abiotic stress in crops: candidate genes, osmolytes, polyamines and biotechnological intervention. In: Pandey GK (ed) Elucidation of abiotic stress signaling in plants: a functional genomic perspective. Springer, New York

Mibei EK, Ambuko J, Giovannoni JJ, Onyango AN, Owino WO (2017) Carotenoid profiling of the leaves of selected African eggplant accessions subjected to drought stress. Food Sci Nutr 5 (1):113–122

Mir RR, Zaman-Allah M, Sreenivasulu N, Trethowan R, Varshney RK (2012) Integrated genomics, physiology and breeding approaches for improving drought tolerance in crops. Theor Appl Genet 125:625–645

Mukhopadhyay A, Vij S, Tyagi AK (2004) Overexpression of a zinc-finger protein gene from rice confers tolerance to cold, dehydration, and salt stress in transgenic tobacco. Proc Natl Acad Sci U S A 101(16):6309–6314

Nakhforoosh A, Bodewein T, Fiorani F, Bodner G (2016) Identification of water use strategies at early growth stages in durum wheat from shoot phenotyping and physiological measurements. Front Plant Sci 7:1155

Navakouidis E, Lütz C, Langebartels C, Lütz-Meindl U, Kotzabasis K (2003) Ozone impact on the photosynthetic apparatus and the protective role of polyamines. Biochem Biophys Acta 1621:160–169

Nelson DE, Repetti PP, Adams TR et al (2007) Plant nuclear factor Y (NFY) B subunits confer drought tolerance and lead to improved corn yields on water-limited acres. Proc Natl Acad Sci U S A 104(42):16450–16455

Ning J, Li X, Hicks LM, Xiong L (2010) A Raf-like MAPKKK gene DSM1 mediates drought resistance through reactive oxygen species scavenging in rice. Plant Physiol 152:876–890

Nover L, Baniwal SK (2006) Multiplicity of heat stress transcription factors controlling the complex heat stress response of plants. In: Proceedings of the international symposium on environmental factors. Cellular Stress and Evolution, p 15

Oh SJ, Song SI, Kim YS, Jang HJ, Kim SY, Kim M, Kim YK, Kim NYK, Nahm BH, Kim JK (2005) *Arabidopsis CBF3/DREB1A* and *ABF3* in transgenic rice increased tolerance to abiotic stress without stunting growth. Plant Physiol 138:341–351

Panda RK, Pandit E, Swain A, Mohanty DP, Baig MJ, Kar M, Pradhan SK (2016) Response of physiological and biochemical parameters in deeper rooting rice genotypes under irrigated and water stress conditions. Oryza 53(4):422–427

Pelham HRB (1982) A regulatory upstream promoter element in the *Drosophila* Hsp 70 heat-shock gene. Cell 30(2):517–528

Phillips J, Bartels D (2000) Gene expression during dehydration in the resurrection plant Craterostigma plantagineum. In: Cherry JH, Locy RD, Rychter A (eds) Plant tolerance to abiotic stresses in agriculture: role of genetic engineering, vol 83. NATO Science Series, Brussels, pp 195–199

Prabha R, Ghosh I, Singh DP (2011) Plant stress gene database: a collection of plant genes responding to stress condition. ARPN J Sci Technol 1:1

Quan R, Hu S, Zhang Z, Zhang H, Zhang Z, Huang R (2010) Over-expression of an ERF transcription factor TSRF1 improves rice drought tolerance. Plant Biotechnol J 8:476–488

Rangan P, Subramani R, Kumar R, Singh AK, Singh R (2014) Recent advances in polyamine metabolism and abiotic stress tolerance. Biomed Res Int 239621:9

Reyes JL, Chua N (2007) ABA induction of miR159 controls transcript levels of two MYB factors during *Arabidopsis* seed germination. Plant J 49(4):592–606

Riechmann JL, Heard J, Martin G (2000) *Arabidopsis* transcription factors: genome-wide comparative analysis among eukaryotes. Science 290(5499):2105–2110

Roy M, Wu R (2002) Over-expression of S-adenosylmethionine decarboxylase gene in Rice increases polyamine level and enhances sodium chloride-stress tolerance. Plant Sci 163:987–992

Saijo Y, Hata S, Kyozuka J, Shimamoto K, Izui K (2000) Over-expression of a single Ca2þ--dependent protein kinase confers both cold and salt/drought tolerance on rice plants. Plant 23:319–327

Sakamoto A, Murata N (2002) The role of glycine betaine in the protection of plants from stress: clues from transgenic plants. Plant Cell Environ 25:163–171

Sakuma Y, Maruyama K, Qin F, Osakabe Y, Shinozaki K, Yamaguchi-shinozaki K (2006) Dual function of an *Arabidopsis* transcription factor *DREB2A* in water-stress-responsive and heat-stress-responsive gene expression. Proc Natl Acad Sci U S A 103(49):18822–18827

Sanders D, Brownlee C, Harper JF (1999) Communicating with calcium. Plant Cell 11:691–706

Sapeta H, Costa JM, Lourenco T, Maroco J, Linde PVD, Oliveira MM (2013) Drought stress response in Jatropha curcas: growth and physiology. Environ Exp Bot 85:76–84

Saradhi PP, Suzuki I, Katoh A, Sakamoto A, Sharmilla P, Shi DJ, Murata N (2000) Protection against the photo-induced inactivation of the photosystem II complex by abscisic acid. Plant Cell Environ 23(7):711–718

Scharf KD, Berberich T, Ebersberger I, Nover L (2012) The plant heat stress transcription factor (Hsf) family: structure, function and evolution. Biochim Biophys Acta 1819:104–119

Schuetz TJ, Gallo GJ, Sheldon L, Tempst P, Kingston RE (1991) Isolation of a cDNA for HSF2: evidence for two heat shock factor genes in humans. Proc Natl Acad Sci U S A 88:6911–6915

Sekhar K, Priyanka B, Reddy VD, Rao KV (2010) Isolation and characterization of a pigeon pea cyclophilin (CcCYP) gene, and its overexpression in *Arabidopsis* confers multiple abiotic stress tolerance. Plant Cell Environ 33:1324–1338

Seki M, Narusaka M, Ishida J, Nanjo T, Fujita M, Oono Y, Kamiya A, Nakajima M, Enju A, Sakurai T, Satou M, Akiyama K, Taji T, Yamaguchi-Shinozaki K, Carninci P, Kawai J, Hayashizaki Y, Shinozaki K (2002) Monitoring the expression profiles of 7000 *Arabidopsis* genes under drought, cold and high-salinity stresses using a full-length cDNA microarray. Plant J 31(3):279–292

Sengupta S, Patra B, Ray S, Majumder AL (2008) Inositol methyl transferase from a halophytic wild rice, Porteresia coarctata Roxb. (Tateoka): regulation of pinitol synthesis under abiotic stress. Plant Cell Environ 31:1442–1459

Serrano R (1996) Salt tolerance in plants and microorganisms: toxicity targets and defence responses. Int Rev Cytol 165:1–52

Shen Q, Zhang P, Ho TH (1996) Molecular nature of abscisic acid (ABA) response complexes: composite promoter units are necessary and sufficient for ABA induction of gene expression in barley. Plant Cell 8(7):1107–1119

Shia SH, Fana M, Iwamab K, Lic F, Zhangd Z, Jiaa L (2015) Physiological basis of drought tolerance in potato grown under long-term water deficiency. Int J Plant Prod 9:2

Shinozaki K, Yamaguchi-shinozaki K (1997) Gene expression and salt transduction in water stress response. Plant Physiol 115:327–334

Shinwari ZK, Nakashima K, Miura S, Kasuga M, Seki M, Yamaguchi-shinozaki K, Shinozaki K (1998) An *Arabidopsis* gene family encoding DRE/CRT binding proteins involved in low temperature responsive gene expression. Biochem Biophys Res Commun 250(1):161–170

Shou H, Bordallo P, Wang K (2004) Expression of the Nicotiana protein kinase (NPK1) enhanced drought tolerance in transgenic maize. J Exp Bot 55(399):1013–1019

Shukla V, Mattoo AK (2013) Developing robust crop plants for sustaining growth and yield under adverse climatic changes. In: Tuteja N, Gill S (eds) Climate change and plant abiotic stress tolerance. Wiley-VCH Verlag, Weinheim, pp 27–56

Silva PA, Cosme VS, Rodrigues KCB, Detmann KSC, Leao FM, Cunha RL, Buselli RAF, Damatta FM, Pinheiro HA (2017) Drought tolerance in two oil palm hybrids as related to adjustments in carbon metabolism and vegetative growth. Acta Physiol Plant 39:58

Singh BP, Jayaswal PK, Singh B, Singh PK, Kumar V, Mishra S, Singh N, Panda K, Singh NK (2015) Natural allelic diversity in *OsDREB1F* gene in the Indian wild rice germplasm led to ascertain its association with drought tolerance. Plant Cell Rep 34:993–1004

Sun J, Hu W, Zhou R, Wang L, Wang X, Wang Q, Feng Z, Li Y, Qiu D, He G, Yang G (2015) The *Brachypodium distachyon BdWRKY36* gene confers tolerance to drought stress in transgenic tobacco plants. Plant Cell Rep 34:23–35

Tan Y, Li M, Yang Y, Sun X, Wang N, Liang B, Ma F (2017) Overexpression of *MpCYS4*, a phytocystatin gene from *Malus prunifolia* (Willd.) Borkh., enhances stomatal closure to confer drought tolerance in transgenic *Arabidopsis* and apple. Front Plant Sci 8:33

Tang L, Cai H, Ji W, Luo X, Wang Z, Wu J, Wang X, Cui L, Wang Y, Zhu Y, Bai X (2013) Overexpression of *GsZFP1* enhances salt and drought tolerance in transgenic alfalfa (*Medicago sativa* L.) Plant Physiol Biochem 71:22–30

Tollenar M, Wu J (1999) Yield in temperate maize is attributable to greater stress tolerance. Crop Sci 39:1604–1897

Tomar JB, Prasad SC (1996) Relationship between inheritance and linkage for drought tolerance in upland rice varieties. Indian J Agric Sci 66(8):459–465

Tran LSP, Nakashima K, Sakuma Y, Simpson SD, Fujita Y, Maruyama K, Fujita K, Seki M, Shinozaki K, Yamaguchi-shinozakia K (2004) Isolation and functional analysis of *Arabidopsis* stress-inducible NAC transcription factors that bind to a drought-responsive cis-element in the early responsive to dehydration stress 1 promoter. Plant Cell 16:2481–2498

Tripathy MK, Tiwari BS, Reddy MK, Deswal R, Sopory SK (2017) Ectopic expression of *PgRab7* in rice plants (*Oryza sativa* L.) results in differential tolerance at the vegetative and seed setting stage during salinity and drought stress. Protoplasma 254:109–124

Tuberosa R, Salvi S (2006) Genomics-based approaches to improve drought tolerance of crops. Trends Plant Sci 11:405–412

Tyerman SD, Niemietz CM, Bramley H (2002) Plant aquaporins: multifunctional water and solute channels with expanding roles. Plant Cell Environ 25:173–194

Vierling E (1991) The roles of heat shock proteins in plants. Annu Rev Plant Physiol Plant Mol Biol 42:579–620

Vincour B, Altman A (2005) Recent advances in engineering plant tolerance to abiotic stress: achievements and limitations. Curr Opin Biotechnol 16:123–132

Wang W, Vinocur B, Altman A (2003) Plant response to drought, salinity and extreme temperatures: toward genetic engineering for stress tolerance. Planta 218:1–14

Wang WX, Vincour B, Shoseyov O, Altman A (2004) Role of plant heat-shock proteins and molecular chaperons in the abiotic stress response. Trends Plant Sci 9:244–252

Wang H, Hao J, Chen X (2007) Overexpression of rice WRKY89 enhances ultraviolet B tolerance and disease resistance in rice plants. Plant Mol Biol 65(6):799–815

Wang GP, Zhang XY, Li F, Luo Y, Wang W (2010) Over accumulation of glycine betaine enhances tolerance to drought and heat stress in wheat leaves in the protection of photosynthesis. Photosynthetica 48:117–126

Wei K, Chen J, Wang Y, Chen Y, Chen S, Lin Y, Pan S, Zhong X, Xie D (2012) Genome wide analysis of bZIP encoding genes in maize. DNA Res 19(6):463–476

Wen XP, Pang XM, Matsuda N, Kita M, Inoue M, Hao YJ, Honda C, Moriquchi T (2008) Over-expression of the apple spermidine synthase gene in pear confers multiple abiotic stress tolerance by altering polyamine titers. Transgenic Res 17(2):251–263

Wi SJ, Kim WT, Park KY (2006) Over expression of carnation S-adenosylmethionine decarboxylase gene generates a broad-spectrum tolerance to abiotic stresses in transgenic tobacco plants. Plant Cell Rep 25:1111–1121

Wu X, Shiroto Y, Kishitani S, Ito Y, Toriyama K (2008) Enhanced heat and drought tolerance in transgenic rice seedlings overexpressing OsWRKY11 under the control of HSP101 promoter. Plant Cell Rep 28:21–30

Xiang Y, Huang Y, Xiong L (2007) Characterization of stress-responsive CIPK genes in rice for stress tolerance improvement. Plant Physiol 144:1416–1428

Xiao B, Huang Y, Tang N, Xiong L (2007) Over-expression of a *LEA* gene in rice improves drought resistance under the field conditions. Theor Appl Genet 115:35–46

Xiong L, Yang Y (2003) Disease resistance and abiotic stress tolerance in rice are inversely modulated by an abscisic acid-inducible nitrogen-activated protein kinase. Plant Cell 15:745–759

Xiong L, Lee BL, Ishitani M, Lee H, Zhang C, Zhu JK (2001) FIERY1 encoding an inositol polyphosphate 1-phosphatase is a negative regulator of abscisic acid and stress signaling in *Arabidopsis*. Genes Dev 15:1971–1984

Xiong AS, Jiang HH, Zhuang J, Peng RH, Jin XF, Zhu B, Wang F, Zhang J, Yao QH (2013) Expression and function of a modified AP2/ERF transcription factor from *Brassica napus* enhances cold tolerance in transgenic *Arabidopsis*. Mol Biotechnol 53(2):198–206

Xu D, Duan X, Wang B, Hong B, Ho THD, Wu R (1996) Expression of a late embryogenesis abundant protein gene, HVA1, from barley confers tolerance to water deficit and salt stress in transgenic rice. Plant Physiol 110(1):249–257

Xu Y, Hu W, Liu J, Zhang J, Jia C, Miao H, Xu B, Jin Z (2014) A banana aquaporin gene, *MaPIP1;1*, is involved in tolerance to drought and salt stresses. BMC Plant Biol 14:59

Yamaguchi-shinozaki K, Shinozaki K (1994) A novel cis-acting element in an *Arabidopsis* gene is involved in responsiveness to drought, low-temperature, or high-salt stress. Plant Cell 6:251–264

Yang A, Dai X, Zhang WH (2012) A R2R3-typre MYB gene, OsMYB2, is involved in salt, cold, and dehydration tolerance in rice. J Exp Bot 63(7):2541–2556

Yang G, Yu L, Zhang K, Zhao Y, Guo Y, Gao C (2017) A *ThDREB* gene from *Tamarix hispida* improved the salt and drought tolerance of transgenic tobacco and *T. hispida*. Plant Physiol Biochem 113:187–197

Yao X, Xiong W, Ye T, Wu Y (2012) Overexpression of the aspartic protease ASPG1 gene confers drought avoidance in *Arabidopsis*. J Exp Bot 63:2579–2593

Yoshida T, Fujita Y, Sayama H, Kidokoro S, Maruyama K, Mizoi J, Shinozaki K, Yamaguchi-shinozaki K (2010) AREB1, AREB2, and ABF3 are master transcription factors that cooperatively regulate ABRE-dependent ABA signalling involved in drought stress tolerance and require ABA for full activation. Plant J 61:672–685

Zeng X, Chen B, Lu G, Han B (2009) Overexpression of a NAC transcription factor enhances rice drought and salt tolerance. Biochem Biophys Res Commun 379(4):985–989

Zhang YX, Chen L (2017) Overexpression of the receptor-like kinase gene *OsNRRB* enhances drought- stress tolerance in rice. Euphytica 213:86

Zhang H, Liu W, Wan L, Li F, Dai L, Li D, Zhang Z, Huang R (2010) Functional analyses of ethylene response factor JERF3 with the aim of improving tolerance to drought and osmotic stress in transgenic rice. Transgenic Res 19:809–818

Zhao SH, Wang FZ, Lu L, Zhang HY, Zhang XY (2000) Breeding and selection of drought resistant and salt tolerant wheat variety Cang 6001. Acta Agric Boreal Sin 15:113–117

Zhu JK, Hasegawa PM, Bressan RA, Bohnert HJ (1997) Molecular aspects of osmotic stress in plants. Crit Rev Plant Sci 16(3):253–277

Photosynthesis and Abiotic Stress in Plants

Jitender Singh and Jitendra K. Thakur

Abstract Abiotic stress is a problem of grave concern for the growth and productivity of plants in modern times. Abiotic stresses, such as drought, salinity, and extreme temperatures, are responsible for huge crop losses globally. One of the physiological processes greatly affected by these stresses in plants is photosynthesis. The decline in photosynthetic capacity of plants due to these stresses is directly associated with reduction in yield. Therefore, detailed information on the plant responses and the adaptation methods employed by them to save their photosynthetic apparatus could help in developing new crop plants with more robust photosynthetic machinery capable of higher yields even under stressed environments. In this chapter, effects of four predominant abiotic stresses, i.e., drought, salinity, heat, and high light, on the photosynthetic apparatus of plants have been discussed, and the strategies to overcome the menace of these stresses have been suggested.

Keywords Photosynthesis · Abiotic stress · Chlorophyll · ROS

1 Introduction

Ever-increasing global population, decreasing arable land due to soil degradation and urban encroachment, and the use of agricultural land to grow biofuel crops have dramatically increased the pressure to enhance crop productivity (Godfray et al. 2010; Foley et al. 2011). It is estimated that a leap of 100–110% is required in agricultural production from 2005 to 2050, to meet the per capita caloric demands. The projections are in line with the increase in per capita real income (Tilman et al. 2011). On the contrary, there is stagnation or decrease in the yield of three major crops – maize, rice, and wheat – which constitute 57% of total agriculture output in terms of energy (Ray et al. 2012). This trend is expected to become more alarming due to threatening climate

J. Singh · J. K. Thakur (✉)
National Institute of Plant Genome Research, New Delhi, India
e-mail: jthakur@nipgr.ac.in

S. Vats (ed.), *Biotic and Abiotic Stress Tolerance in Plants*,
https://doi.org/10.1007/978-981-10-9029-5_2

changes. All these observations impose an enormous challenge to our goals of providing food security to all humans on earth without further compromising our ecosystem.

One of the major causes restricting our agricultural produce is abiotic stress. Abiotic stresses, such as drought, high salt, cold, heat, UV radiation, heavy metals, etc., adversely influence plant growth processes. It is estimated that abiotic stresses reduce crop yield by more than 50% (Rodziewicz et al. 2014). Among these, drought, high salt, and extreme temperatures are the most dreadful stresses faced by modern agriculture. Up to 26% of arable land and over 30% of the irrigated land are suffering from drought and salinity problems, respectively (Rehman et al. 2005).

2 Photosynthesis

Plants perform a series of complex reactions that convert light into carbohydrates, a process known as photosynthesis. The carbohydrates, thus produced, serve as the primary source of energy directly or indirectly for heterotrophic organisms, including humans. Photosynthesis is one of the most fundamental components of plant growth and productivity. The photosynthetic apparatus in plants comprises diverse components, such as photosynthetic pigments for light absorption, photosystems and the light reactions for NADPH and ATP generation, and the dark reactions (Calvin–Benson–Bassham or C3 cycle) for CO_2 assimilation. Chlorophyll a, chlorophyll b, pheophytins, and carotenoids are the photosynthetic pigments present in plants. Chlorophyll a is found in all eukaryotes and present at the reaction centers of both photosystems (PS), PSI and PSII (Blankenship 2002). In plants and green algae, chlorophyll b is the principal accessory light-absorbing pigment in light-harvesting complexes. Pheophytins are chlorophyll molecules sans Mg^{2+} at their center and formed in the course of chlorophyll degradation (Blankenship 2002). Pheophytins have high reduction potential and therefore act as the primary electron acceptor of PSII (Fig. 1). Carotenoids are the accessory antenna pigment molecules, which harvest light energy and transfer it to chlorophyll molecules for photosynthesis (Hashimoto et al. 2016). Carotenoids are also involved in photoprotection. Carotenoids are quenchers of the triplet chlorophylls formed due to high light and thus check the formation of the highly toxic singlet oxygen ($^1O_2$). They also quench the $^1O_2$ if it is somehow formed (Blankenship 2002). In this way carotenoids function as powerful antioxidants in plants. Carotenoids have also been implicated in photoprotection of photosynthetic machinery of plants by high light via the xanthophyll cycle (Latowski et al. 2011).

PSI and PSII are the two multi-protein complexes present in the thylakoid membranes and contain the pigments necessary to capture light energy. Both PSII and PSI work in series to initiate the process of electron transport to produce highly reducing compound NADPH, O_2 (Fig. 1), and high-energy substrate ATP (Caffarri et al. 2014). The electron required to reduce the NADP to NADPH is provided by water. Since there is a huge gap in the redox potentials of the ultimate electron donor (water) and the final electron acceptor (NADP), plants use many pigments and

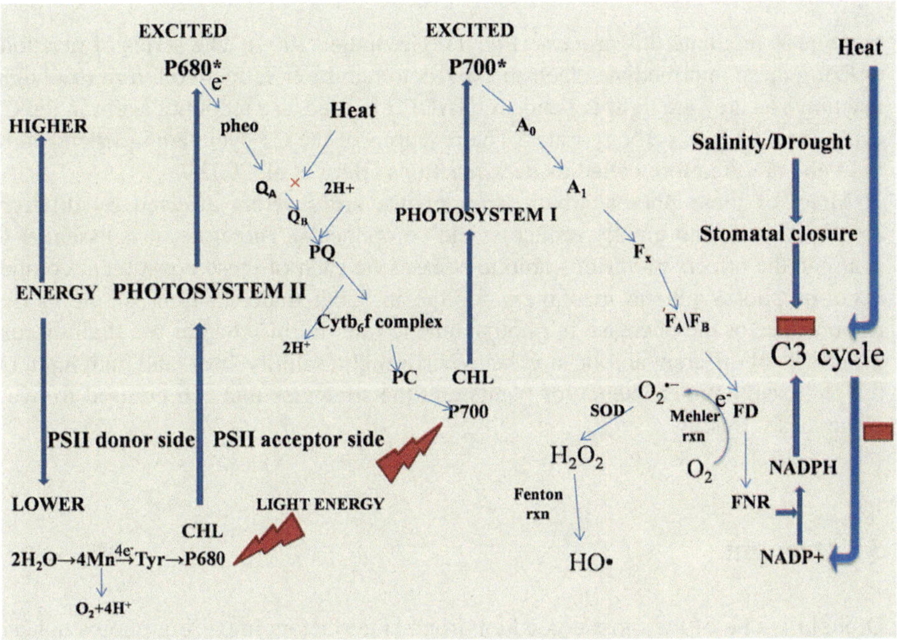

Fig. 1 Representation of the Z scheme of photosynthesis (Govindjee 2004) and the sites of intervention by different abiotic stresses therein. The Z scheme represents the steps in the light reactions involved in the transport of electrons from water to $NADP^+$. A pair of special reaction center chlorophyll a molecules, referred to as P680 and P700 in PSII and PSI, respectively, is simultaneously excited by absorbing light energy. This results in the formation of high energy (excited) P680* and P700*. The P680* and P700* donate an electron to their respective electron acceptors pheo and A_o and become $P680^+$ and $P700^+$. The $P680^+$ returns to its original P600 form by accepting an electron generated by the splitting of water via a tyrosine residue in the PSII complex. The electron accepted by the pheo then traverses a downhill path via Q_A, Q_B, PQ, $Cytb_6f$ complex, and PC to oxidize the reduced $P700^+$ to P700. Q_A and Q_B are two plastoquinone molecules attached at two different sites A and B in PSII. Q_B is loosely bound to PSII and can accept two electrons sequentially from Q_A, whereas Q_A can accept only one electron from pheo. After accepting two electrons from Q_A, Q_B extracts two protons from the stroma and detaches from PSII as PQH_2. PQH_2 is oxidized to PQ by transferring the electrons to $Cytb_6f$ protein complex with concomitant release of two protons into the thylakoid lumen. $Cytb_6f$ complex gives the electrons to PC, which subsequently transfers a single electron to $P700^+$. The electron accepted by A_o from P700* is finally transferred to $NADP^+$ via several intermediate electron carriers as shown in the figure. Heat, drought, and salinity stress inhibit the C3 cycle, which results in over-reduction of ETC. This leads to the leakage of electrons to O_2 and formation of $O_2^{\bullet -}$, H_2O_2, and $HO^{\bullet}$ by the sequential action of Mehler reaction, SOD, and Fenton reaction. Heat stress also blocks the transfer of electron from Q_A to Q_B, which causes over-reduction of PSII. Over-reduction of ETC at any step by any stress such as high light, drought, etc., can lead to the production of $^1O_2$ (not shown in the figure). *Abbreviations*: Chl is chlorophyll; pheo is pheophytin; Q_A, Q_B, and PQ are plastoquinone molecules; $Cytb_6f$ is a multimeric protein complex composed of three main proteins (i) $Cytb_6$, (ii) Rieske Fe–S protein, and (iii) Cytf; PC is plastocyanin, a copper containing mobile protein; A_o is a special chlorophyll a molecule that accepts electron from P700*; A_1 is a phylloquinone (vitamin K) molecule; F_X, F_A, and F_B are three different immobile iron–sulfur protein centers; FD is ferredoxin, a somewhat mobile iron–sulfur protein; FNR is the enzyme ferredoxin–NADP oxidoreductase, which uses FAD (flavin adenine dinucleotide) as its cofactor; $NADP^+$ is the oxidized form of nicotinamide adenine dinucleotide phosphate; SOD is superoxide dismutase; rxn is reaction; – in red indicates inhibition; × in red indicates blockage of electron transfer

proteins to facilitate this process (Fig. 1) (Govindjee 2004). The series of reactions utilizing these intermediate electron carriers to reduce NADP are referred to as light reactions, as they are light dependent. NADPH is used as a reducing agent in the C3 cycle to fix CO_2 as carbohydrates. The reactions of the C3 cycle are independent of light and are therefore called as dark reactions (Berg et al. 2002).

Many of these photosynthetic components are severely affected by different abiotic stresses and greatly reduces yield components. Therefore, it is essential to examine the effects of various abiotic stresses on each of these constituent components of photosynthetic machinery for the in-depth understanding of the causes responsible for the decrease in photosynthetic rate. In this chapter we shall discuss the effects of different abiotic stresses, viz., drought, salinity, heat, and high light, on the photosynthetic efficiency of plants and the strategies that can be used to overcome these stresses.

3 Drought

Drought is one of the most crucial environmental factors impairing photosynthesis and thereby limiting plant growth and yield (Kannan and Kulandaivelu 2011; Rahbarian et al. 2011; Batra et al. 2014; Liu et al. 2016; Meng et al. 2016). The decline in yield is caused due to reduced leaf growth resulting in lower photosynthetic output (Siddique et al. 1999; Kannan and Kulandaivelu 2011). The drought-induced decrease in photosynthesis in plants is mainly ascribed to the decrease in CO_2 conductance through stomata and mesophyll cells. Closure of stomata in response to drought conditions prevents water loss, thereby increasing water use efficiency of plants and decreasing transpiration rate (Ashraf and Harris 2013). Drought stress also suppresses leaf mesophyll CO_2 conductance. The reduction in mesophyll conductance could be an outcome of different factors including restructuring of the intercellular air spaces due to leaf contraction, biochemical changes (bicarbonate to CO_2 conversion), and/or membrane porosity (aquaporins) (Lawlor and Cornic 2002; Chaves et al. 2009). Under prolonged drought stress, decreased leaf CO_2 transport rate leads to lowering of CO_2 concentration in chloroplasts thus weakening photosynthesis. The fall in CO_2 levels inside the cells is found to deactivate Rubisco and reduce the activity of sucrose phosphate synthase and nitrate reductase and capacity for ribulose bisphosphate (RuBP) regeneration (Reddy et al. 2004).

It is widely known that drought induces oxidative stress. Drought stress produces different kinds of reactive oxygen species (ROS) such as superoxide ($O_2^{\bullet -}$), hydroxyl radicals ($HO^{\bullet}$), singlet oxygen ($^1O_2$), hydrogen peroxide (H_2O_2), etc. (Impa et al. 2012). There are multiple pathways that can lead to increased ROS production under drought. The dropdown in chloroplastic CO_2 concentration due to the closure of stomata during water stress results in decreased CO_2 fixation and consequently reduced amounts of $NADP^+$ (Fig. 1). This causes excessive reduction of the electron transport chain (ETC), leading to channeling of electrons to O_2 via Mehler reaction forming $O_2^{\bullet -}$ and subsequently H_2O_2 by the activity of superoxide dismutase (SOD) (Fig. 1). Moreover, severe reduction of the ETC at any step increases the likelihood of $^1O_2$ generation in photosystem II (PSII). Reduced CO_2 concentration also stimulates photorespiration, which is a source of H_2O_2 (Cruz de Carvalho 2008; Noctor et al. 2014). These ROS can decrease the photosynthetic rate significantly under drought stress both (i) by disrupting the photosynthetic machinery, including D1 and D2 proteins of the PSII complex (Fig. 2), thylakoid membranes, and chlorophyll pigments, and (ii) inhibiting the translation of new D1, D2,

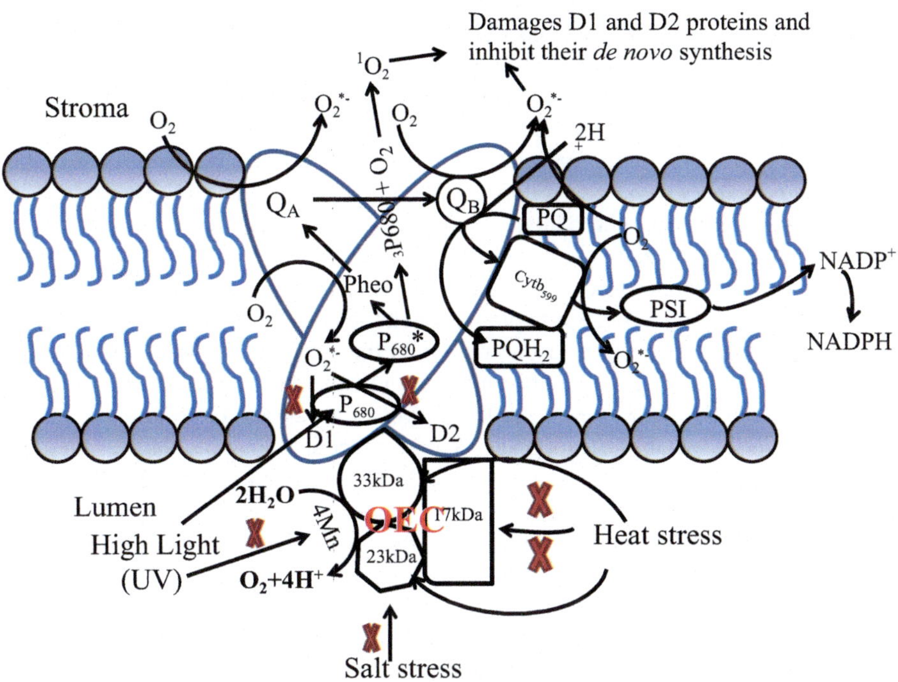

Fig. 2 The sites of $O_2^{\bullet -}$ and $^1O_2$ generation in PSII and the effects of high light, heat, and salt stress on PSII. Reduced forms of pheophytin, Q_A and Q_B, free PQ, and ferrous iron of low-potential form of Cytb_{559} can generate $O_2^{\bullet -}$ by passing on an electron to O_2. The excited P680* and other excited chlorophyll molecules get converted to triplet ^{3}P680 and triplet chlorophyll molecules and react with O_2 to generate $^1O_2$. These ROS can directly damage and/or inhibit the repair process of PSII. Heat stress inactivates the oxygen-evolving complex (OEC) by discharging the 33, 23, and 17 kDa proteins externally associated with PSII. High salt leads to the release of 23 kDa protein subunit. The UV component of the high light discharges two functional Mn of the OEC, thus inactivating OEC. However, visible region of high light leads to ROS production and $^1O_2$ generation by formation of triplet chlorophyll. X sign in red indicates inactivation

and other cell proteins (Reddy et al. 2004; Liu et al. 2006; Zlatev 2009; Anjum et al. 2011).

4 Salinity

Salinity stress evokes two types of distress situations in plants: (i) osmotic stress and (ii) ionic imbalance (Hossain and Dietz 2016). Osmotic stress elicits water deficit like conditions and induces drought-like responses such as closure of stomata, reduced CO_2 fixation, over-reduction of ETC, and stimulation of photorespiration, which can lead to the generation of ROS as described above in the drought section (Abogadallah 2010). In addition to these, salt stress has been found to induce respiratory ETC in the mitochondrion, the activity of plasma membrane-bound respiratory burst oxidase homolog (RBOH) and apoplastic diamine oxidase (Fry et al. 1986; Waie and Rajam 2003; Ben Rejeb et al. 2015; Hossain and Dietz 2016). All of these are a source of ROS. In maize leaves, salt stress increased levels of apoplastic spermidine and spermine. Oxidation of these polyamines by the apoplastic polyamine oxidase produces 1,3-diaminopropane and H_2O_2 (Rodríguez et al. 2009). These ROS can damage the photosynthetic machinery, nucleic acids, and membranes of the cell.

Prolonged salt stress leads to ion toxicity in plants. Plant responses to high salinity are species specific. Some plants check the absorption of Na^+ by selecting K^+ over Na^+; others prevent its accumulation in the cytoplasm by sequestering it in vacuoles, thereby protecting photosynthesis and other basic metabolic processes (Meloni et al. 2003; Munns et al. 2006; Chaves et al. 2009). Excess salt in plants hampers diverse metabolic processes. Toxic concentrations of Na^+ resulted in reduced levels of photosynthetic pigment chlorophyll in salt-sensitive plants such as cotton, tomato, potato, pea, etc., whereas it was increased in salt-tolerant plant species like pearl millet, mustard, and wheat (Ashraf and Harris 2013). The decrease in the amount of chlorophyll could be a result of increased pigment degradation or impaired chlorophyll biosynthesis (Ashraf and Harris 2013). However, in a recent study, salt-tolerant rice cultivar was found to have decreased chlorophyll content. It was suggested that lower amounts of chlorophyll in tolerant cultivars was either due to chlorophyll degradation by ROS or it is an alternative route to produce H_2O_2 by the photocatalytic activity of chlorophyll itself (Kaur et al. 2016). Hydrogen peroxide acts as a secondary messenger in various stress-responsive signaling pathways and is essential for the "salt stress preparedness" in tolerant plant species (Kaur et al. 2016). In *Cucumis sativus*, salt stress also induces degradation of chloroplast structure and abnormalities in thylakoid membranes resulting in decreased net photosynthetic rate. In this study, the salt-induced disruption of photosynthesis was mainly due to the loss of polyamines in the photosynthetic apparatus (Shu et al. 2012). In addition, salt stress inhibits the activity of enzymes involved in photosynthesis. It affects the proteins/enzymes of both the light and dark reactions. Salt stress decreases the ETR by inhibiting the activity of PSII. The PSII activity under salt stress was decreased due to the dissociation of 23 kDa protein extrinsically associated with the PSII complex (Murata et al. 1992) (Fig. 2). It also reduces the activity of Rubisco and other enzymes involved in photosynthesis thereby affecting CO_2 fixation (Sudhir and Murthy 2004; Stepien and Johnson 2008; Chaves et al. 2009; Ashraf and Harris 2013). It has been found that the photosynthetic machinery of salt-tolerant and susceptible rice genotypes responds differently to salinity stress. The 23 kDa protein of the PSII and Rubisco of a salt-tolerant wild rice variety (Pokkali) were less inhibited by salinity stress as compared to the susceptible variety, thereby imparting better photosynthetic performance under salt stress (Lakra et al. 2017).

5 High Temperature

With rising atmospheric CO_2 concentrations, heat stress has become a more prevalent problem checking agricultural yields. Generally, a transitory phase when the temperature exceeds ambient temperature by 10–15 °C is referred to as heat shock or heat stress (Wahid et al. 2007). Transient or persistent high temperature negatively impacts plant growth and development thus limiting productivity (Song et al. 2014). Heat waves particularly during the anthesis and grain-filling stages are detrimental to the photosynthetic system and significantly reduce yields (Feng et al. 2014). PSII complex is the most heat-intolerant part of the photosynthetic apparatus on the light reaction side of photosynthesis. Oxygen-evolving complex (OEC), PSII reaction center, and the light-capturing complexes are the primary components that are damaged by high temperature (Chen et al. 2012; Pastenes and Horton 1996; Mathur et al. 2014). Thermal stress leads to the release of two out of the four Mn present per PSII (Tyystjärvi 2008). It is known that heat stress to chloroplasts liberates 33, 23, and 17 kDa proteins, which are extrinsically attached to the PSII complex (Fig. 2). Discharge of the 33 kDa protein is responsible for the release of Mn and destabilization of the OEC (Enami et al. 1998; Ohnishi et al. 2005; Allakhverdiev et al. 2008). On the electron acceptor side of PSII, heat stress inhibits the transfer of electrons from Q_A to Q_B (Pospíšil 2016) (Fig. 1). This leads to over-reduction of ETC and ROS production. In addition, $^1O_2$ is produced on the PSII electron acceptor side by the reaction of triplet carbonyl with O_2. Triplet carbonyl is formed in the process of lipid peroxidation by heat stress, whereas, on the PSII electron donor side, the two-electron reduction of H_2O by 2 Mn^{2+} present forms H_2O_2, which can be converted to $HO^{\bullet}$ via Fenton reaction (Pospíšil 2016). Heat stress also induces alterations in the ultrastructure of the thylakoid membranes, especially de-stacking of the thylakoid membranes and increase in ion conductivity, which leads to their malfunctioning (Mathur et al. 2014).

Little evidence is present in support of the oxidative damage caused to proteins and membrane lipids by ROS under heat stress. Cleavage of the D1 protein of PSII, however, was observed during mild heat treatment to spinach thylakoids (Yamashita et al. 2008). Furthermore, $^1O_2$ formed at Q_B site during lipid peroxidation caused degradation of the D1 protein as in case of high irradiances (Yamashita et al. 2008). These ROS, however, unequivocally, inhibit the de novo synthesis of the D1 and other proteins during heat stress. Heat stress also affects chlorophyll content in leaves. Low levels of chlorophyll were present in leaves exposed to heat stress (Mathur et al. 2014). This could be due to defective chlorophyll biosynthesis or its enhanced degradation or by both. The impaired chlorophyll biosynthesis under elevated temperatures is a result of the presence of numerous heat-sensitive enzymes in chlorophyll biosynthesis pathway (Mathur et al. 2014).

Another component adversely affected by high temperatures, in photosynthesis, is the carbon assimilation pathway. Enzymes of the C3 cycle are heat sensitive. Thus, the carbon fixation capacity of plants is drastically reduced at elevated temperatures (Berry and Bjorkman 1980; Weis 1981; Feller et al. 1998; Sharkey 2005). Mild increases in leaf temperature lead to deactivation of Rubisco, as its activating enzyme Rubisco activase is a heat-labile enzyme (Salvucci et al. 2001; Sharkey 2005; Allakhverdiev et al. 2008). Moreover, higher temperatures stimulate the oxygenase reaction of Rubisco. This initiates the photorespiratory pathway and results in the generation of H_2O_2, a by-product of the pathway (Sharkey 2005). All these factors significantly reduce the photosynthetic efficiency of plants under moderate heat stress.

6 High Light

Light is fundamental to the process of photosynthesis. However, on the other hand, light intensities above the light saturation point of photosynthesis are harmful to plants and termed as high light stress (Lichtenthaler and Burkart 1999). Light-induced drop in photosynthetic rate is generally referred as photoinhibition. High light mainly impairs the PSII complex, which is the initiation site for linear electron flow and oxygen evolution from water (Fig. 2). The degree of PSII photoinhibition is a measure of the difference between its rate of photodamage and repair (Takahashi and Badger 2011). Multiple mechanisms of PSII photoinhibition occur simultaneously under high light stress. The process, which dominates in photoinhibition, depends on the intensity and quality of light. The primary step in the UV-induced photodamage of PSII is the release of functional Mn^{2+} from the OEC by high light, rendering the OEC inactive (Fig. 2). The dysfunctional OEC is unable to deliver electrons from water to $P680^+$. Since $P680^+$ is a powerful oxidant, it impairs the reaction center by oxidizing proteins in its vicinity, especially the D1 polypeptide (Nishiyama et al. 2006; Murata et al. 2007; Vass 2012). In addition, the direct damaging effects of UV light also extend to Q_A, Q_B, and the D1 polypeptide (Vass 2012).

In the visible region of light, ROS generation is mainly responsible for the PSII photodamage (Vass 2012). Additionally, inactivation of the Mn cluster also occurs due to the weak Mn^{2+} absorption in red part of the visible spectrum (Vass 2012).

When incident light energy absorbed by chlorophyll molecules is not utilized efficiently, conversion of singlet chlorophyll to triplet chlorophyll occurs (Vass 2012). Furthermore, when the rate of electron transport exceeds the electron sink capacity, reduction of Q_A to $Q_A^{-\bullet}$ takes place blocking the forward electron movement from P680. This results in the formation of the triplet excited state of P680 i.e. ^{3}P680 (Pospíšil 2016). The reaction of triplet chlorophyll molecules formed due to excess light energy and ^{3}P680 formed due to restricted electron flow to Q_A from P680, with O_2, produces the highly reactive $^1O_2$, which damages the proteins surrounding it, particularly the D1 protein, thereby impairing PSII complex (Fig. 2) (Pospíšil 2016). Apart from $^1O_2$, other ROS are also produced, especially the $O_2^{\bullet-}$. On the acceptor side of the PSII, $O_2^{\bullet-}$ is formed by the transfer of excess electrons from reduced pheophytin, Q_A, Q_B, free plastoquinone (PQ), and low-potential form of $Cytb_{559}$ to O_2(Pospíšil 2016) (Fig. 2). Further, a sequential reduction of $O_2^{\bullet-}$ by one electron on the acceptor side of PSII gives rise to H_2O_2 and $HO^\bullet$ (Fig. 1). On the donor side of PSII, incomplete oxidation of water produces H_2O_2, which can be oxidized and reduced to produce $O_2^{\bullet-}$ and $HO^\bullet$, respectively (Pospíšil 2016). These ROS prevent the repair of PSII complex by inhibiting the de novo synthesis of D1 and other proteins and induces lipid peroxidation of thylakoid membranes (Nath et al. 2013; Cheng et al. 2016). It has been found that the extent of photoinhibition is inversely related to the Chl a/b ratio of leaves. Leaves with higher Chl a/b ratio are less susceptible to photoinhibition (Aro et al. 1993).

Besides PSII, PSI is also inactivated at high light intensities. PSI photoinhibition is caused due to the ROS-induced specific degradation of one (PSI-B) of the two large subunits of PSI reaction center (Sonoike 1996). This PSI photoinhibition is more pronounced at low temperatures, wherein both the polypeptides, PSI-A and PSI-B, of the reaction center are degraded (Tjus et al. 1999; Cheng et al. 2016).

7 Stress-Induced Effector Molecules and Photosynthesis

7.1 ROS

ROS are normally produced during photosynthesis and other metabolic processes in cells (Foyer and Shigeoka 2011). In limited amounts ROS act as signaling molecules in numerous biological processes including plant growth and development, and plant responses under biotic and abiotic stresses (Tripathy and Oelmüller 2012; Baxter et al. 2014). Stress conditions such as drought, salinity, heat, high light, cold, heavy metal, etc., enhance the production of these ROS (Pandey et al. 2015; Singh et al. 2015). The excess amounts of ROS are scavenged by the enzymatic and nonenzymatic components of plant-antioxidant defense system (Abogadallah 2010). The enzymatic component of the antioxidant defense machinery in plants comprises enzymes SOD, catalase, peroxidase, glutathione reductase, etc., whereas tocopherol, ascorbic acid, carotenoids, glutathione, phenolics, flavonoids etc., are the part of nonenzymatic antioxidant system (for details see review, Gill and Tuteja 2010; Sharma et al. 2012; Kasote et al. 2015).

However, when the rate of ROS generation overwhelms the detoxifying capacity of the plant, it leads to oxidative stress. The oxidative stress caused by higher net ROS formation inflicts damage on DNA, proteins, and lipids, ultimately leading to cell death (Tripathy and Oelmüller 2012).

7.2 Phytohormones

One of the earliest plant responses to many abiotic stresses such as drought, high temperature, chilling, and salinity stress is the change in the levels of abscisic acid (ABA) (Leng et al. 2014). ABA plays a key role in the adaptation of plants to various stresses by initiating different signaling pathways (Bücker-Neto et al. 2017). The higher accumulation of ABA triggers a signaling cascade in guard cells leading to efflux of K^+ ions from guard cells resulting in decreased turgor pressure and subsequently stomata closure (Lim et al. 2015; Salazar et al. 2015). ABA assists plants in adaptation to abiotic stresses by inducing the expression of many gene-encoding enzymes involved in the biosynthesis of osmoprotectants, late embryogenesis abundant proteins, dehydrins, and other defense proteins (Sah et al. 2016; Dar et al. 2017). Increased concentration of ABA in leaves leads to cell wall extensibility, root hydraulic transmission, and tissue turgor. It also leads to feedback inhibition of photosynthesis by accumulating carbohydrates and declining the concentration of photosynthetic enzymes (Vishwakarma et al. 2017).

Another plant hormone acting as a signaling molecule and serving plants to adapt to different stress conditions is ethylene. Whether ethylene is a positive or negative regulator of various environmental stresses is a controversial subject (Kazan 2015; Tao et al. 2015). From different studies, it appears that ethylene production needs to be tightly regulated for plant adaptation in response to different environmental stresses (Kazan 2015; Tao et al. 2015). It is proposed that, in initial stages of plant acclimation to salinity stress, ethylene plays a positive role to help plant endure salinity stress. After the stage of acclimatization, excessive ethylene hinders plant growth and development, which is disadvantageous for the survival of plants (Tao et al. 2015). Ethylene concentration and its interplay with other plant hormones appear to influence plant adaptability and performance under different stress conditions (Iqbal et al. 2017). Salicylic acid and jasmonates are the other important plant hormones playing significant roles in plant defense against abiotic stresses (Kazan 2015).

7.3 Amino Acids

Accumulation of proline in plants in response to stressful environmental conditions such as drought, salinity, heavy metals, etc., is a well-known observation. Interestingly, heat stress does not lead to proline accumulation in plants (tobacco and *Arabidopsis*) and its induction results in increased heat sensitivity in plants

(Krasensky and Jonak 2012). However, proline was accumulated in peach trees when shoots acclimated to cold were exposed to elevated temperatures (Shin et al. 2016). Proline functions as an osmolyte, ROS scavenger, a redox buffer, and a molecular chaperone which helps in stabilizing proteins and membranes, thereby preventing cell damage caused by stress (Krasensky and Jonak 2012; Hossain and Dietz 2016). In addition, proline acts as a nitrogen source during stress recovery phase (Gupta and Huang 2014). It also improves salt tolerance in some plant species such as olive and tobacco by enhancing the activity of enzymes involved in antioxidant defense and photosynthesis (Gupta and Huang 2014). Proline synthesis via the glutamate pathway requires 2 moles of NADPH to form 1 mole of proline and thus draws off electrons from the chloroplast and maintains redox homeostasis of cell (Hossain and Dietz 2016). Proline production in leaves after salt stress facilitates unhindered carbon fixation and reduces photoinhibition and excess ROS generation (Hossain and Dietz 2016).

The level of the non-protein amino acid γ-amino-butyric acid (GABA) rapidly increases in response to different environmental stress conditions including cold, heat, drought, and salt (Kinnersley and Turano 2000). GABA metabolism is associated with the maintenance of cytosolic pH, osmoregulation, carbon–nitrogen balance, and ROS scavenging (Krasensky and Jonak 2012; Li et al. 2016). GABA also enhances the activity of antioxidant enzymes, and very recently, has been suggested to enhance the photosynthetic activity to alleviate heat stress in plants (Li et al. 2016). In some reports, GABA's role as a stress-induced signaling mediator in many signal transduction pathways in plants has been shown (Bouché and Fromm 2004; Yu et al. 2014; Li et al. 2016).

7.4 Glycinebetaine

Glycinebetaine (GB) is an osmolyte overproduced in plants to help them cope with adverse growth conditions such as desiccation, salinity, extreme temperatures, UV radiation, etc. (Ashraf and Foolad 2007). However, GB is not ubiquitously present in all plant species. Some plants, such as spinach and barley, accumulate significantly higher levels of GB in their chloroplasts, while it is completely lacking in *Arabidopsis* and tobacco (Fariduddin et al. 2013). GB protects the cell by osmotic adjustment leading to stabilization of proteins and membranes, protection of PSII apparatus, stabilization of pigment, and mitigation of ROS (Krasensky and Jonak 2012; Gupta and Huang 2014).

7.5 Polyols

Polyols are sugar alcohols containing multiple hydroxyl groups which are formed by the chemical reduction of aldose or ketose carbohydrates (Williamson 2002). Polyols are a class of compatible solutes, which are accumulated in plants after exposure to stress conditions including drought, salinity, and heat (Conde et al. 2015). Polyols constitute a significant part of the photosynthates and are predominant transported sugar form other than sucrose (Williamson 2002). Polyols have been shown to act as osmoprotectants, molecular chaperones, and scavengers of $HO^{\bullet}$. The major polyols in plants are mannitol, sorbitol, and myoinositol (Krasensky and Jonak 2012; Gupta and Huang 2014).

7.6 Polyamines

Polyamines (PAs) are small, aliphatic polycationic species that are ubiquitously present in all living organisms, including bacteria, animals, and plants (Liu et al. 2015). There are many lines of evidences that show a positive correlation between polyamines content and stress resistance in plants. Elevated PA levels are observed in plants exposed to a variety of stress conditions such as desiccation, high salt, cold, heat, and heavy metals (Tiburcio et al. 2014). In addition to their role in plant adaptation to multiple abiotic stresses, PAs have been implicated in various aspects of plant growth and development, such as embryogenic competence, pollen development, apoptosis, fruit ripening, xylem differentiation, as well as biofilm formation and signaling pathways (Tisi et al. 2011; Tiburcio et al. 2014; Liu et al. 2015). PAs exert their positive effects by preserving membrane architecture, controlling the expression of genes involved in the synthesis of compatible solutes acting as osmoprotectants, limiting ROS generation, and balancing Na^+ and Cl^- ions in different cellular compartments (Gupta and Huang 2014). Diamine putrescine, triamine spermidine, and tetra-amine spermine are the most common PAs in higher plants (Krasensky and Jonak 2012).

7.7 Carbohydrates

Carbohydrates are the eventual products of photosynthesis and function as energy providers for different cellular activities. Disaccharides (sucrose, trehalose), raffinose family oligosaccharides (RFOs), and fructans are the three main forms of sugars that are involved in plant stress responses and adaptation (Keunen et al. 2013). These soluble sugars are used for osmotic adjustment and osmoprotection by plants to stabilize membrane structures and maintain cell turgor (Gil et al. 2011). A perfect case is of resurrection plants, which use sugar accumulation as one of the strategies to combat complete dehydration conditions (Djilianov et al. 2011). Fructans accumulation is advantageous to plants during periodic cold and dry conditions as they are easily dissolved in water and are resistant to crystallization at very low temperatures (Krasensky and Jonak 2012). Moreover, fructans can maintain membrane integrity and might help in osmotic adjustment during water stress by serving as a source of hexose sugars (Krasensky and Jonak 2012). It is generally accepted that fructans stabilize tonoplast during stress conditions by integrating themselves in between the head groups of the tonoplast (Keunen et al. 2013). Trehalose is a non-reducing sugar and acts as an osmolyte, and stabilizes proteins and membranes. The exact mechanism of trehalose functioning during abiotic stress is yet to be deciphered (Krasensky and Jonak 2012). RFOs are involved in maintaining membrane integrity and radical scavenging. Disaccharides, galactinol, RFOs, fructans, and sugar alcohols act as antioxidants by scavenging $HO^{\bullet}$. Sucrose at higher concentrations has been shown to act as an antioxidant in sugarbeet and sugarcane plants (Keunen et al. 2013).

8 Strategies to Produce Efficient Stress Tolerant Plants

In the recent past, much success has been achieved in understanding the molecular mechanisms of the plant responses and adaptations to abiotic stresses. Knowing the plant adaptation methods such as changes in the concentration of plant hormones, accumulation of osmolytes, upregulation of antioxidant defense machinery, etc., as described above, against stress conditions, has led to the development of various plants with improved resilience to different stress conditions (Krasensky and Jonak 2012; Keunen et al. 2013; Gupta and Huang 2014; Leng et al. 2014; Bakhsh and Hussain 2015; Kazan 2015; Khan et al. 2015; Tao et al. 2015).

Abiotic stress tolerance being a complex trait is governed by a plethora of genes. Therefore, manipulating a single gene or few genes could result in limited success, particularly when plants are challenged with multiple abiotic stresses. Transcription factors (TFs) are suitable candidates for genetic modifications to develop stress-tolerant crops because of their ability to regulate expression of many stress-responsive genes simultaneously (Verma and Deepti 2016). Many families of TFs (e.g., AP2/EREBP, MYB, WRKY, NAC, and bZIP) have been identified which play vital roles in signaling during various abiotic stresses (Kim 2014). Some of the TF genes belonging to these families have been tweaked to improve stress tolerance in plants (Shao et al. 2015).

Gene expression differs significantly in natural populations. The expression pattern of many genes to stressful conditions is strongly associated with the natural allelic variation (Assmann 2013). Plants could use this versatility in gene expression, either by genotypic variation or an effect of genotype-by-environment interaction, as a mechanism of local adaptation (Assmann 2013). With the emergence of technologies like low-cost and fast high-throughput nucleic acid sequencing, genome-wide association studies and screening of stress-tolerant genotypes in natural populations can reveal novel mechanisms of stress tolerance (Pereira et al. 2013).

Plants perform three types of photosynthesis: C3, C4 (Hatch and Slack pathway), and CAM (Crassulacean acid metabolism) photosynthesis. Plants performing C4 photosynthesis such as maize, sorghum, sugarcane, etc., have better photosynthetic characteristics than plants performing the other two types of photosynthesis. In addition, C4 plants have higher water and nitrogen use efficiency than the C3 plants owing to the presence of a carbon concentrating mechanism (CCM) in them (Singh et al. 2014; Schuler et al. 2016). The increased water and nitrogen use efficiency allow C4 plants to allocate resources in more efficient way under stress conditions such as drought (Sage and Zhu 2011). Carbon fixation process in C4 plants is not inhibited as severely as in C3 plants in dry conditions (Schuler et al. 2016). This prevents the generation of ROS in C4 plants contrary to C3 plants, where over-reduction of ETC occurs under stress conditions, consequently leading to ROS production. C4 plants also exhibit 50% higher radiation use efficiency (RUE) than the C3 plants (Sage and Zhu 2011). The better RUE of C4 plants is useful in preventing ROS generation during high light conditions and results in higher photosynthetic rates and more yields. Therefore, introduction of a C4 or C4-like CCM in C3 crops may be one of the best ways to increase photosynthetic capacity and fitness of C3 crops even under stressful climatic conditions.

Reducing antenna size for light capturing in the photosynthetic apparatus in microalgae mass cultures, has worked as a successful strategy to address the issue of photosynthetic inefficiency associated with the high light intensity (Ort et al. 2011). A decrease in chlorophyll antenna size by 50% and 35% in PSI and PSII, respectively, showed greater radiation conversion efficiencies and higher photosynthetic rate than the wild type in microalgae mass cultures (Polle et al. 2003). However, emulating the same strategy in plants should be meticulously done to ensure that a reduction in the size of light-harvesting system does not constrain photosynthesis in any other way (Ort et al. 2011).

9 Conclusion and Perspectives

Abiotic stress is a serious problem affecting the photosynthetic efficiency of plants and globally reducing agricultural yields by more than 50%. Therefore, there is an urgent need to develop new stress-resilient photosynthetic machinery in order to sustain crop productivity. Recent developments in the field of abiotic stress with the help of transcriptomic, genomic, and proteomic tools have led to the accumulation of vast knowledge about plant adaptation methods to stress conditions vis-a-vis photosynthesis (Gururani et al. 2015; Nouri et al. 2015; Lakra et al. 2017). This huge repertoire of information needs to be applied and translated in the form of new crop varieties to combat adverse environmental conditions.

Although many transgenic crop varieties resistant to different abiotic stresses have been developed, none of them has so far been released for cultivation. Some transgenic varieties of *Brassica*, wheat, rice, and tomato showing tolerance to a range of abiotic stresses, however, got clearance for field testing in some developing countries, including India and China (Verma and Deepti 2016). In this scenario, the development of new genome-editing technologies such as zinc finger nucleases (ZFN), transcription activator-like effector nucleases (TALENs), and clustered regularly interspaced short palindromic repeats/CRISPR-associated (CRISPR/Cas9) system can play a meaningful role. These technologies enable researchers to alter the inherent DNA sequences according to their needs, without incorporating any foreign DNA. Therefore, these products are referred to as genetically edited rather than genetically modified. CRISPR/Cas9 system is the preferred choice over the other two techniques because it is more efficient and user-friendly and can edit multiple genes simultaneously. Many genes implicated in different photosynthetic pathways have been shown to up or downregulate in response to different abiotic stresses (Nouri et al. 2015) and can therefore be tuned and tweaked simultaneously using this technology to suit our needs. Recently, a genetically edited mushroom to resist browning is developed by using CRISPR/Cas9 system (Waltz 2016). The US Department of Agriculture has decided not to regulate the genetically edited mushroom (Waltz 2016). Therefore, genome-editing technologies have enormous future prospects and can be exploited to develop crop plants with better agronomic traits.

Acknowledgment The award of National Post Doctoral Fellowship by SERB-DST, India, to Dr. Jitender Singh is acknowledged. The authors are also thankful to Subhasis Das (PhD student) and Rahul (M.Tech trainee), NIPGR, for extending their help in editing the manuscript. The authors have no conflict of interests to declare.

References

Abogadallah GM (2010) Antioxidative defense under salt stress. Plant Signal Behav 5:369–374. https://doi.org/10.4161/psb.5.4.10873

Allakhverdiev SI, Kreslavski VD, Klimov VV et al (2008) Heat stress: an overview of molecular responses in photosynthesis. Photosynth Res 98:541–550

Anjum S, Xie X, Wang L (2011) Morphological, physiological and biochemical responses of plants to drought stress. Afr J Agric Res 6:2026–2032. https://doi.org/10.5897/AJAR10.027

Aro EM, McCaffery S, Anderson JM (1993) Photoinhibition and D1 protein degradation in peas acclimated to different growth irradiances. Plant Physiol 103:835–843. https://doi.org/10.1104/PP.103.3.835

Ashraf M, Foolad MR (2007) Roles of glycine betaine and proline in improving plant abiotic stress resistance. Environ Exp Bot 59:206–216. https://doi.org/10.1016/j.envexpbot.2005.12.006

Ashraf M, Harris PJC (2013) Photosynthesis under stressful environments: an overview. Photosynthetica 51:163–190

Assmann SM (2013) Natural variation in abiotic stress and climate change responses in *Arabidopsis*: implications for twenty-first-century agriculture. Int J Plant Sci 174:3–26. https://doi.org/10.1086/667798

Bakhsh A, Hussain T (2015) Engineering crop plants against abiotic stress: current achievements and prospects. Emirates J Food Agric 27:24–39

Batra NG, Sharma V, Kumari N (2014) Drought-induced changes in chlorophyll fluorescence, photosynthetic pigments, and thylakoid membrane proteins of *Vigna radiata*. J Plant Interact 9:712–721. https://doi.org/10.1080/17429145.2014.905801

Baxter A, Mittler R, Suzuki N (2014) ROS as key players in plant stress signaling. J Exp Bot 65:1229–1240

Ben Rejeb K, Benzarti M, Debez A et al (2015) NADPH oxidase-dependent H2O2 production is required for salt-induced antioxidant defense in *Arabidopsis thaliana*. J Plant Physiol 174:5–15. https://doi.org/10.1016/j.jplph.2014.08.022

Berg JM, Jeremy M, Tymoczko JL, Stryer L, Stryer L (2002) Biochemistry, 5th edn. WH Freeman, New York

Berry J, Bjorkman O (1980) Photosynthetic response and adaptation to temperature in higher plants. Annu Rev Plant Physiol 31:491–543. https://doi.org/10.1146/annurev.pp.31.060180.002423

Blankenship RE (2002) Photosynthetic pigments: structure and spectroscopy. Mol Mech Photosynth 42–60. https://doi.org/10.1002/9780470758472.ch4

Bouché N, Fromm H (2004) GABA in plants: just a metabolite? Trends Plant Sci 9:110–115

Bücker-Neto L, Paiva ALS, Machado RD et al (2017) Interactions between plant hormones and heavy metals responses. Genet Mol Biol 40:373–386. https://doi.org/10.1590/1678-4685-gmb-2016-0087

Caffarri S, Tibiletti T, Jennings R, Santabarbara S (2014) A comparison between plant photosystem I and photosystem II architecture and functioning. Curr Protein Pept Sci 15:296–331. https://doi.org/10.2174/1389203715666140327102218

Chaves MM, Flexas J, Pinheiro C (2009) Photosynthesis under drought and salt stress: regulation mechanisms from whole plant to cell. Ann Bot 103:551–560

Chen WR, Zheng JS, Li YQ, Guo WD (2012) Effects of high temperature on photosynthesis, chlorophyll fluorescence, chloroplast ultrastructure, and antioxidant activities in fingered citron. Russ J Plant Physiol 59:732–740

Cheng D-D, Zhang Z-S, Sun X-B et al (2016) Photoinhibition and photoinhibition-like damage to the photosynthetic apparatus in tobacco leaves induced by *Pseudomonas syringae* pv. Tabaci under light and dark conditions. BMC Plant Biol 16:29. https://doi.org/10.1186/s12870-016-0723-6

Conde A, Regalado A, Rodrigues D et al (2015) Polyols in grape berry: transport and metabolic adjustments as a physiological strategy for water-deficit stress tolerance in grapevine. J Exp Bot 66:889–906. https://doi.org/10.1093/jxb/eru446

Cruz de Carvalho MH (2008) Drought stress and reactive oxygen species: production, scavenging and signaling. Plant Signal Behav 3:156–165. https://doi.org/10.4161/psb.3.3.5536

Dar NA, Amin I, Wani W et al (2017) Abscisic acid: a key regulator of abiotic stress tolerance in plants. Plant Gene 22:1742

Djilianov D, Ivanov S, Moyankova D et al (2011) Sugar ratios, glutathione redox status and phenols in the resurrection species *Haberlea rhodopensis* and the closely related non-resurrection species *Chirita eberhardtii*. Plant Biol 13:767–776. https://doi.org/10.1111/j.1438-8677.2010.00436.x

Enami I, Kamo M, Ohta H et al (1998) Intramolecular cross-linking of the extrinsic 33-kDa protein leads to loss of oxygen evolution but not its ability of binding to photosystem II and stabilization of the manganese cluster. J Biol Chem 273:4629–4634. https://doi.org/10.1074/JBC.273.8.4629

Fariduddin Q, Varshney P, Yusuf M et al (2013) Dissecting the role of glycine betaine in plants under abiotic stress. Plant Stress 7:8–18

Feller U, Crafts-Brandner SJ, Salvucci ME (1998) Moderately high temperatures inhibit ribulose-1,5-bisphosphate carboxylase/oxygenase (Rubisco) activase-mediated activation of Rubisco. Plant Physiol 116:539–546. https://doi.org/10.1104/pp.116.2.539

Feng B, Liu P, Li G et al (2014) Effect of heat stress on the photosynthetic characteristics in flag leaves at the grain-filling stage of different heat-resistant winter wheat varieties. J Agron Crop Sci 200:143–155. https://doi.org/10.1111/jac.12045

Foley JA, Ramankutty N, Brauman KA et al (2011) Solutions for a cultivated planet. Nature 478:337–342. https://doi.org/10.1038/nature10452

Foyer CH, Shigeoka S (2011) Understanding oxidative stress and antioxidant functions to enhance photosynthesis. Plant Physiol 155:93–100. https://doi.org/10.1104/pp.110.166181

Fry IV, Huflejt M, Erber WWA et al (1986) The role of respiration during adaptation of the freshwater cyanobacterium *Synechococcus* 6311 to salinity. Arch Biochem Biophys 244:686–691. https://doi.org/10.1016/0003-9861(86)90637-5

Gil R, Lull C, Boscaiu M et al (2011) Soluble carbohydrates as osmolytes in several halophytes from a mediterranean salt marsh. Not Bot Horti Agrobot Cluj-Napoca 39:9–17. https://doi.org/10.15835/NBHA3927176

Gill SS, Tuteja N (2010) Reactive oxygen species and antioxidant machinery in abiotic stress tolerance in crop plants. Plant Physiol Biochem 48:909–930

Godfray HCJ, Beddington JR, Crute IR et al (2010) Food security: the challenge of feeding 9 billion people. Science 327:812–818. https://doi.org/10.1126/science.1185383

Govindjee (2004) Chlorophyll a fluorescence: a bit of basics and history. In: Papageorgiou GC, Govindjee (eds) Chlorophyll a fluorescence. Advances in photosynthesis and respiration, vol 19. Springer, Dordrecht

Gupta B, Huang B (2014) Mechanism of salinity tolerance in plants: physiological, biochemical, and molecular characterization. Int J Genomics 2014:Article ID 701596. https://doi.org/10.1155/2014/701596

Gururani MA, Tapan Mohanta K, Bae H (2015) Current Understanding of the Interplay between Phytohormones and Photosynthesis under Environmental Stress. Int J Mol Sci 16:19055–19085. https://doi.org/10.3390/ijms160819055

Hashimoto H, Uragami C, Cogdell RJ (2016) Carotenoids and photosynthesis. In: Stange C (ed) Carotenoids in nature. Subcellular biochemistry, vol 79. Springer, Cham

Hossain MS, Dietz K-J (2016) Tuning of redox regulatory mechanisms, reactive oxygen species and redox homeostasis under salinity stress. Front Plant Sci 7:548. https://doi.org/10.3389/fpls.2016.00548

Impa SM, Nadaradjan S, Jagadish SVK (2012) Drought stress induced reactive oxygen species and anti-oxidants in plants. In: Ahmad P, Prasad M (eds) Abiotic stress responses in plants. Springer, New York

Iqbal N, Khan NA, Ferrante A et al (2017) Ethylene role in plant growth, development and senescence: interaction with other phytohormones. Front Plant Sci 8:475. https://doi.org/10.3389/fpls.2017.00475

Kannan ND, Kulandaivelu G (2011) Drought induced changes in physiological, biochemical and phytochemical properties of *Withania somnifera* Dun. J Med Plant Res 5:3929–3935

Kasote DM, Katyare SS, Hegde MV, Bae H (2015) Significance of antioxidant potential of plants and its relevance to therapeutic applications. Int J Biol Sci 11:982–991

Kaur N, Dhawan M, Sharma I, Pati PK (2016) Interdependency of reactive oxygen species generating and scavenging system in salt sensitive and salt tolerant cultivars of rice. BMC Plant Biol 16:131. https://doi.org/10.1186/s12870-016-0824-2

Kazan K (2015) Diverse roles of jasmonates and ethylene in abiotic stress tolerance. Trends Plant Sci 20:219–229

Keunen E, Peshev D, Vangronsveld J et al (2013) Plant sugars are crucial players in the oxidative challenge during abiotic stress: extending the traditional concept. Plant Cell Environ 36:1242–1255. https://doi.org/10.1111/pce.12061

Khan MS, Ahmad D, Khan MA (2015) Utilization of genes encoding osmoprotectants in transgenic plants for enhanced abiotic stress tolerance. Electron J Biotechnol. https://doi.org/10.1016/j.ejbt.2015.04.002

Kim SY (2014) Transcription factors involved in ABA signaling. In: Zhang DP (ed) Abscisic acid: metabolism, transport and signaling. Springer, Dordrecht

Kinnersley AM, Turano FJ (2000) Gamma aminobutyric acid (GABA) and plant responses to stress. CRC Crit Rev Plant Sci 19:479–509. https://doi.org/10.1080/07352680091139277

Krasensky J, Jonak C (2012) Drought, salt, and temperature stress-induced metabolic rearrangements and regulatory networks. J Exp Bot 63:1593–1608

Lakra N, Kaur C, Anwar K et al (2017) Proteomics of contrasting rice genotypes: identification of potential targets for raising crops for saline environment. Plant Cell Environ. https://doi.org/10.1111/pce.12946

Latowski D, Kuczyńska P, Strzałka K (2011) Xanthophyll cycle – a mechanism protecting plants against oxidative stress. Redox Rep 16:78–90. https://doi.org/10.1179/174329211X13020951739938

Lawlor DW, Cornic G (2002) Photosynthetic carbon assimilation and associated metabolism in relation to water deficits in higher plants. Plant Cell Environ 25:275–294. https://doi.org/10.1046/j.0016-8025.2001.00814.x

Leng P, Yuan B, Guo Y, Chen P (2014) The role of abscisic acid in fruit ripening and responses to abiotic stress. J Exp Bot 65:4577–4588

Li Z, Yu J, Peng Y, Huang B (2016) Metabolic pathways regulated by γ-aminobutyric acid (GABA) contributing to heat tolerance in creeping bentgrass (*Agrostis stolonifera*). Sci Rep 6:30338. https://doi.org/10.1038/srep30338

Lichtenthaler HK, Burkart S (1999) Photosynthesis and high light stress. Bulg J Plant Physiol 25:3–16

Lim CW, Baek W, Jung J et al (2015) Function of ABA in stomatal defense against biotic and drought stresses. Int J Mol Sci 16:15251–15270

Liu NY, Ko SS, Yeh KC, Charng YY (2006) Isolation and characterization of tomato Hsa32 encoding a novel heat-shock protein. Plant Sci 170:976–985. https://doi.org/10.1016/j.plantsci.2006.01.008

Liu J-H, Wang W, Wu H et al (2015) Polyamines function in stress tolerance: from synthesis to regulation. Front Plant Sci 6:827. https://doi.org/10.3389/fpls.2015.00827

Liu EK, Mei XR, Yan CR et al (2016) Effects of water stress on photosynthetic characteristics, dry matter translocation and WUE in two winter wheat genotypes. Agric Water Manag 167:75–85. https://doi.org/10.1016/j.agwat.2015.12.026

Mathur S, Agrawal D, Jajoo A (2014) Photosynthesis: response to high temperature stress. J Photochem Photobiol B Biol 137:116–126. https://doi.org/10.1016/j.jphotobiol.2014.01.010

Meloni DA, Oliva MA, Martinez CA, Cambraia J (2003) Photosynthesis and activity of superoxide dismutase, peroxidase and glutathione reductase in cotton under salt stress. Environ Exp Bot 49:69–76. https://doi.org/10.1016/S0098-8472(02)00058-8

Meng LL, Song JF, Wen J et al (2016) Effects of drought stress on fluorescence characteristics of photosystem II in leaves of *Plectranthus scutellarioides*. Photosynthetica 54:414–421. https://doi.org/10.1007/s11099-016-0191-0

Munns R, James RA, Läuchli A (2006) Approaches to increasing the salt tolerance of wheat and other cereals. J Exp Bot 57:1025–1043

Murata N, Mohanty PS, Hayashi H, Papageorgiou GC (1992) Glycinebetaine stabilizes the association of extrinsic proteins with the photosynthetic oxygen-evolving complex. FEBS Lett 296:187–189. https://doi.org/10.1016/0014-5793(92)80376-R

Murata N, Takahashi S, Nishiyama Y, Allakhverdiev SI (2007) Photoinhibition of photosystem II under environmental stress. Biochim Biophys Acta Bioenerg 1767:414–421

Nath K, Jajoo A, Poudyal RS et al (2013) Towards a critical understanding of the photosystem II repair mechanism and its regulation during stress conditions. FEBS Lett 587:3372–3381

Nishiyama Y, Allakhverdiev SI, Murata N (2006) A new paradigm for the action of reactive oxygen species in the photoinhibition of photosystem II. Biochim Biophys Acta Bioenerg 1757:742–749

Noctor G, Mhamdi A, Foyer CH (2014) The roles of reactive oxygen metabolism in drought: not so cut and dried. Plant Physiol 164:1636–1648. https://doi.org/10.1104/pp.113.233478

Nouri MZ, Moumeni A, Komatsu S (2015) Abiotic stresses: insight into gene regulation and protein expression in photosynthetic pathways of plants. Int J Mol Sci 16:20392–20416

Ohnishi N, Allakhverdiev SI, Takahashi S et al (2005) Two-step mechanism of photodamage to photosystem II: step 1 occurs at the oxygen-evolving complex and step 2 occurs at the photochemical reaction center. Biochemistry 44:8494–8499. https://doi.org/10.1021/bi047518q

Ort DR, Zhu X, Melis A (2011) Optimizing antenna size to maximize photosynthetic efficiency. Plant Physiol 155:79–85

Pandey P, Singh J, Achary VMM, Reddy MK (2015) Redox homeostasis via gene families of ascorbate-glutathione pathway. Front Environ Sci. https://doi.org/10.3389/fenvs.2015.00025

Pastenes C, Horton P (1996) Effect of high temperature on photosynthesis in beans (I. Oxygen evolution and chlorophyll fluorescence). Plant Physiol 112:1245–1251. https://doi.org/10.1104/pp.112.3.1245

Pereira A, Barkla BJ, Vera-Estrella R, Pantoja O (2013) Plant abiotic stress challenges from the changing environment. Proteomics 13:1801–1815. https://doi.org/10.3389/fpls.2016.01123

Polle JEW, Kanakagiri S-D, Melis A (2003) tla1, a DNA insertional transformant of the green alga *Chlamydomonas reinhardtii* with a truncated light-harvesting chlorophyll antenna size. Planta 217:49–59. https://doi.org/10.1007/s00425-002-0968-1

Pospíšil P (2016) Production of reactive oxygen species by photosystem II as a response to light and temperature stress. Front Plant Sci 7:1950. https://doi.org/10.3389/fpls.2016.01950

Rahbarian R, Khavari-Nejad R, Ganjeali A et al (2011) Drought stress effects on photosynthesis, chlorophyll fluorescence and water relations in tolerant and susceptible chickpea (*Cicer arietinum* L.) genotypes. Acta Biol Cracov Ser Bot 53:47–56. https://doi.org/10.2478/v10182-011-0007-2

Ray DK, Ramankutty N, Mueller ND et al (2012) Recent patterns of crop yield growth and stagnation. Nat Commun 3:1293. https://doi.org/10.1038/ncomms2296
Reddy AR, Chaitanya KV, Vivekanandan M (2004) Drought-induced responses of photosynthesis and antioxidant metabolism in higher plants. J Plant Physiol 161:1189–1202
Rehman S, Harris PJC, Ashraf M (2005) Stress environments and their impact on crop production. In: Ashraf M, Harris PJC (eds) Abioticstresses: plant resistance through breeding and molecular approaches. Haworth Press, New York, pp 3–18
Rodríguez AA, Maiale SJ, Menéndez AB, Ruiz OA (2009) Polyamine oxidase activity contributes to sustain maize leaf elongation under saline stress. J Exp Bot 60:4249–4262. https://doi.org/10.1093/jxb/erp256
Rodziewicz P, Swarcewicz B, Chmielewska K et al (2014) Influence of abiotic stresses on plant proteome and metabolome changes. Acta Physiol Plant 36:1–19
Sage RF, Zhu X-G (2011) Exploiting the engine of C4 photosynthesis. J Exp Bot 62:2989–3000. https://doi.org/10.1093/jxb/err179
Sah SK, Reddy KR, Li J (2016) Abscisic acid and abiotic stress tolerance in crop plants. Front Plant Sci 7:571. https://doi.org/10.3389/fpls.2016.00571
Salazar C, Hernandez C, Pino MT (2015) Plant water stress: associations between ethylene and abscisic acid response. Chil J Agric Res 75:71–79. https://doi.org/10.4067/S0718-58392015000300008
Salvucci ME, Osteryoung KW, Crafts-Brandner SJ, Vierling E (2001) Exceptional sensitivity of Rubisco activase to thermal denaturation in vitro and in vivo. Plant Physiol 127:1053–1064. https://doi.org/10.1104/pp.010357
Schuler ML, Mantegazza O, Weber APM (2016) Engineering C4 photosynthesis into C3 chassis in the synthetic biology age. Plant J 87:51–65. https://doi.org/10.1111/tpj.13155
Shao H, Wang H, Tang X (2015) NAC transcription factors in plant multiple abiotic stress responses: progress and prospects. Front Plant Sci 6:902. https://doi.org/10.3389/fpls.2015.00902
Sharkey TD (2005) Effects of moderate heat stress on photosynthesis: importance of thylakoid reactions, rubisco deactivation, reactive oxygen species, and thermotolerance provided by isoprene. Plant Cell Environ 28:269–277
Sharma P, Jha AB, Dubey RS et al (2012) Reactive oxygen species, oxidative damage, and antioxidative defense mechanism in plants under stressful conditions, reactive oxygen species, oxidative damage, and antioxidative defense mechanism in plants under stressful conditions. J Bot J Bot 2012:e217037. https://doi.org/10.1155/2012/217037
Shin H, Oh S, Arora R, Kim D (2016) Proline accumulation in response to high temperature in winter-acclimated shoots of *Prunus persica*: a response associated with growth resumption or heat stress? Can J Plant Sci 96(4):630–638
Shu S, Guo SR, Sun J, Yuan LY (2012) Effects of salt stress on the structure and function of the photosynthetic apparatus in *Cucumis sativus* and its protection by exogenous putrescine. Physiol Plant 146:285–296. https://doi.org/10.1111/j.1399-3054.2012.01623.x
Siddique MRB, Hamid A, Islam (1999) Drought stress effects on photosynthetic rate and leaf gas exchange of wheat. Bot Bull Acad Sin 40:141–145
Singh J, Pandey P, James D et al (2014) Enhancing C3 photosynthesis: an outlook on feasible interventions for crop improvement. Plant Biotechnol J 12:1217–1230
Singh J, Reddy PS, Reddy CS, Reddy MK (2015) Molecular cloning and characterization of salt inducible dehydrin gene from the C4 plant *Pennisetum glaucum*. Plant Gene 4:55–63. https://doi.org/10.1016/j.plgene.2015.08.002
Song Y, Chen Q, Ci D et al (2014) Effects of high temperature on photosynthesis and related gene expression in poplar. BMC Plant Biol 14:111. https://doi.org/10.1186/1471-2229-14-111
Sonoike K (1996) Degradation of psaB gene product, the reaction center subunit of photosystem I, is caused during photoinhibition of photosystem I: possible involvement of active oxygen species. Plant Sci 115:157–164. https://doi.org/10.1016/0168-9452(96)04341-5
Stepien P, Johnson GN (2008) Contrasting responses of photosynthesis to salt stress in the Glycophyte *Arabidopsis* and the halophyte *Thellungiella*: role of the plastid terminal oxidase

as an alternative electron sink. Plant Physiol 149:1154–1165. https://doi.org/10.1104/pp.108.132407

Sudhir P, Murthy SDS (2004) Effects of salt stress on basic processes of photosynthesis. Photosynthetica 42:481–486. https://doi.org/10.1007/S11099-005-0001-6

Takahashi S, Badger MR (2011) Photoprotection in plants: a new light on photosystem II damage. Trends Plant Sci 16:53–60

Tao J-J, Chen H-W, Ma B et al (2015) The role of ethylene in plants under salinity stress. Front Plant Sci 6:1059. https://doi.org/10.3389/fpls.2015.01059

Tiburcio AF, Altabella T, Bitrián M, Alcázar R (2014) The roles of polyamines during the lifespan of plants: from development to stress. Planta 240:1–18

Tilman D, Balzer C, Hill J, Befort BL (2011) Global food demand and the sustainable intensification of agriculture. Proc Natl Acad Sci U S A 108:20260–20264. https://doi.org/10.1073/pnas.1116437108

Tisi A, Federico R, Moreno S et al (2011) Perturbation of polyamine catabolism can strongly affect root development and xylem differentiation. Plant Physiol 157:200–215. https://doi.org/10.1104/pp.111.173153

Tjus SE, Møller BL, Scheller HV (1999) Photoinhibition of photosystem I damages both reaction centre proteins PSI-A and PSI-B and acceptor-side located small photosystem I polypeptides. Photosynth Res 60:75–86. https://doi.org/10.1023/A:1006283618695

Tripathy BC, Oelmüller R (2012) Reactive oxygen species generation and signaling in plants. Plant Signal Behav 7:1621–1633. https://doi.org/10.4161/psb.22455

Tyystjärvi E (2008) Photoinhibition of photosystem II and photodamage of the oxygen evolving manganese cluster. Coord Chem Rev 252:361–376

Vass I (2012) Molecular mechanisms of photodamage in the photosystem II complex. Biochim Biophys Acta Bioenerg 1817:209–217

Verma AK, Deepti S (2016) Abiotic stress and crop improvement: current scenario. Adv Plants Agric Res 4(4):00149. https://doi.org/10.15406/apar.2016.04.00149

Vishwakarma K, Upadhyay N, Kumar N et al (2017) Abscisic acid signaling and abiotic stress tolerance in plants: a review on current knowledge and future prospects. Front Plant Sci 8:161. https://doi.org/10.3389/fpls.2017.00161

Wahid A, Gelani S, Ashraf M, Foolad MR (2007) Heat tolerance in plants: an overview. Environ Exp Bot 61:199–223. https://doi.org/10.1016/j.envexpbot.2007.05.011

Waie B, Rajam MV (2003) Effect of increased polyamine biosynthesis on stress responses in transgenic tobacco by introduction of human S-adenosylmethionine gene. Plant Sci 164:727–734. https://doi.org/10.1016/S0168-9452(03)00030-X

Waltz E (2016) Gene-edited CRISPR mushroom escapes US regulation. Nature 532:293. https://doi.org/10.1038/nature.2016.19754

Weis E (1981) Reversible heat-inactivation of the calvin cycle: a possible mechanism of the temperature regulation of photosynthesis. Planta 151:33–39. https://doi.org/10.1007/BF00384234

Williamson J (2002) Sugar alcohols, salt stress, and fungal resistance: polyols – multifunctional plant protection? J Am Soc Hortic Sci 127:467–473

Yamashita A, Nijo N, Pospíšil P et al (2008) Quality control of photosystem II: reactive oxygen species are responsible for the damage to photosystem II under moderate heat stress. J Biol Chem 283:28380–28391. https://doi.org/10.1074/jbc.M710465200

Yu G-H, Zou J, Feng J et al (2014) Exogenous γ-aminobutyric acid (GABA) affects pollen tube growth via modulating putative Ca^{2+}-permeable membrane channels and is coupled to negative regulation on glutamate decarboxylase. J Exp Bot 65:3235–3248. https://doi.org/10.1093/jxb/eru171

Zlatev Z (2009) Drought-induced changes in chlorophyll fluorescence of young wheat plants. Biotechnol Biotechnol Equip 23:438–441. https://doi.org/10.1080/13102818.2009.10818458

Ecotoxicological Effects of Insecticides in Plants Assessed by Germination and Other Phytotoxicity Tools

Idalina Bragança, Clara Grosso, Diana Rede, Susana R. Sousa, Paulo C. Lemos, Valentina F. Domingues, and Cristina Delerue-Matos

Abstract The management of crop-pests relies largely on conventional insecticides. Farmers around the world use pesticides as an insurance policy against the possibility of a devastating crop loss from pests and diseases. Conversely, the use of insecticides has several drawbacks for agriculture, such as decrease in pollinator population and terrestrial pollution as they are frequently detected in the environment.

Several tests are used to assess phytotoxicity regarding several mechanisms affecting plants, namely, (a) inhibition of biological processes such as photosynthesis, cell division, enzyme function, and root, shoot, and leaf development; (b) interference with the synthesis of pigments, proteins, or DNA; (c) cell membrane instability; and (d) the promotion of uncontrolled growth. Germination tests are extensively used to assess the toxicity induced by pollutants. In these types of tests, the germination indexes and the seedling's growth and development are evaluated in a dose-response manner.

This review evaluates the application of insecticides leading to alteration on germination, in biochemical, physiological, and different enzymatic and nonenzymatic antioxidant levels that may affect the crop yield and insecticide residues in plants. As such, this chapter represents a systematic and integrated picture of insecticide toxicological effects on plants, highlighting germination.

Keywords Insecticides · Germination · Toxicology · Seeds · Bioassay

I. Bragança · C. Grosso (✉) · D. Rede · V. F. Domingues · C. Delerue-Matos
REQUIMTE/LAQV-GRAQ, Instituto Superior de Engenharia do Porto, Politécnico do Porto, Porto, Portugal
e-mail: claragrosso@graq.isep.ipp.pt

S. R. Sousa
REQUIMTE/LAQV-GRAQ, Instituto Superior de Engenharia do Porto, Politécnico do Porto, Porto, Portugal

i3S – Instituto de Investigação e Inovação em Saúde, Universidade do Porto, Porto, Portugal

INEB – Instituto de Engenharia Biomédica, Universidade do Porto, Porto, Portugal

P. C. Lemos
REQUIMTE/LAQV, Chemistry Department, FCT/Universidade NOVA de Lisboa, Caparica, Portugal

S. Vats (ed.), *Biotic and Abiotic Stress Tolerance in Plants*,
https://doi.org/10.1007/978-981-10-9029-5_3

1 Contamination of Soils by Insecticides

There is a myriad of relationships between environment and human health that are often more intricate than can be thought (Fig. 1). Since soil is a crucial component of the environment, studies aiming to evaluate terrestrial toxicity should consider the interactions between chemicals and soil in order to foresee the effect of chemicals on the environment (Kapanen and Itävaara 2001). As it is difficult to evaluate the effect of pollutants on the environment only based on the concentrations of chemical constituents, biotests are also needed (Kapanen and Itävaara 2001).

Agriculture represents an important sector in the worldwide economy. Therefore, the use of pesticides for pest management in agriculture is a common practice to maximize crop production. In the last decades, the use of pesticides has reduced crop losses caused by pests, but their intensive and large-scale application has caused several adverse effects on the environment by remaining in soils and water and by affecting the development of target and nontarget species. Due to adaptation and resistance developed by the target pests to pesticides, every year higher amounts and new chemical compounds are used to protect crops, causing undesired side effects and increasing the costs of food production (Carvalho 2006).

Insecticides are considered a quick, easy, and cheap solution for controlling insect pests; nonetheless, their use comes with a significant environmental cost. Studies included in this section were retrieved from Web of Science core database, between 2007 and 2017, using "soil contamination with insecticides" as a search topic (Table 1). Organochlorine pesticides (OCPs) were those more often found in the literature and also the ones detected in higher concentrations, despite their restricted use and even their production and usage banned in some countries. This type of insecticide is still often detectable in soil and persists in the environment due to their non-biodegradability or very slow degradation. More than half of the reports were

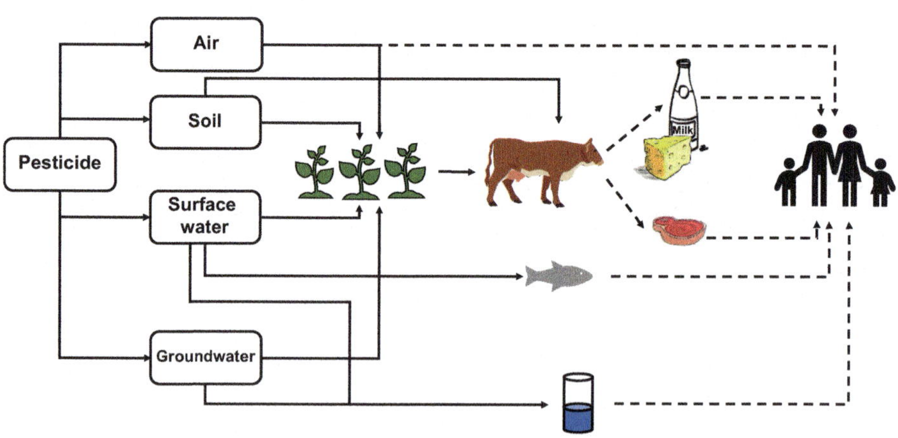

Fig. 1 Exposure routes generally considered in human exposure assessment (Adapted from Kapanen and Itävaara 2001)

Table 1 Residues of insecticides detected in soils/sediments collected from different areas of the planet

Continent	Area	Insecticides analysed	Sampling year	Concentration	References
Africa	Limpopo	DDTs (DDT+DDD+DDE)	February 2008	∑DDTs: village exposed: 5.7–59 μg/kg reference village: 2.1–93 μg/kg	Van Dyk et al. (2010)
	Togo	δ-hexachlorocyclohexane (HCH), heptachlor epoxide, 4,4-DDE, endosulphan (α, β and sulphate), λ-cyalothrin and chlorpyrifos	–	Concentrations of insecticides in different sites and farmers' plots varied from ND–26.93 μg/kg (dw).	Mawussi et al. (2014)
America	Canada	Neonicotinoids: clothianidin and thiamethoxam	April and May of 2014	Clothianidin in particulate matter from (mean value): oat: 9.91 μg/kg canola: 10.20 μg/kg Thiamethoxam: ND	Main et al. (2016)
	Costa Rica	Endosulfan I, endosulfan II, endosulfansulfate	February 2004	Σ Endosulfan (dw): 20–3175 pg/g	Daly et al. (2007)
	Mexico (Southern Sonora)	OCPs: benzene hexachloride (BHC or HCB), lindane, aldrin, endrin, β-endosulfan, methoxychlor, DDTs	2007	BHC: ND–938.5 μg/g Endrin: ND–377.3 μg/g DDTs: ND–679.7 μg/g	Cantu-Soto et al. (2011)
Asia	China	OCPs: HCHs, DDTs chlordane compounds and endosulfans OPs: chlorpyrifos, diazinon, dichlorvos, ethion, ethyl p-nitrophenyl (EPN), malathion, parathion-methyl, profenofos, trichlorfon, and terbufos OPYs: bifenthrin, λ-cyhalothrin, cyfluthrin, cypermethrin(CP), deltamethrin, esfenvalerate, fenpropathrin, permethrin, and tefluthrin Fipronil and its metabolites	1957–2012	Total insecticide concentrations in river sediments (dw): Rural: 67.6–1671 ng/g Suburban: 99.2–231 ng/g.	Sun et al. (2016)

(continued)

Table 1 (continued)

Continent	Area	Insecticides analysed	Sampling year	Concentration	References
	China	OCPs:δ-hexachlorocyclohexanes (HCHs), DDTs, chlorothalonil and dicofol OPs: methamidophos, methylparathion and parathion OPYs: fenvalerate (FEN)	–	Ten OCPs, three OPs and one OPYs detected: OCPs>OPs>OPYs OCPs and OPs concentration were <150 μg/kg. FEN concentration: 1227 μg/kg (maximal value of all pesticides detected).	Guo et al. (2016)
	China (sediments of Nanfei River and of cores from Chaohu Lak estuaries)	Eighteen OCPs: DDTs, HCHs, HCB, aldrin, endrin, heptachlor, heptachlor epoxide A/B, and tetrachloronitrobenzene	July and September of 2012	OCP concentrations: surface sediments: 3.48–121.08 ng/g core sediments: 0.60–39.28 ng/g	Zhang et al. (2016)
	China (Pearl River Delta)	OCPs: DDTs and HCHs OPYs: bifenthrin, fenpropathrin, tefluthrin, λ-cyhalothrin, permethrin, cyfluthrin, CP, esfenvalerate, and deltamethrin OPs: parathion-methyl, malathion, and chlorpyrifos	December 2009 to March 2010	Soil inventories of DDTs and HCHs were 100 and 83 tons, for pyrethroids and organophosphates 39 and 6.2 tons, respectively.	Wei et al. (2015)
	China (Guanting Reservoir area)	DDT, HCH and their metabolites	2003	ΣHCHs: 0.0 to 7.3 ng/g dw ΣDDTs: 0.0 to 76 ng/g dw	Wang et al. (2007)
	India (Unnao district)	OCPs: Aldrin, dieldrin, endrin, HCB, HCHs, DDTs, endosulfan isomers (α and β), endosulfansulfate, heptachlor and its metabolites, α andγ-chlordane and methoxychlor	October 2003	Σ DDT: b.d –74.06 ng/g Σ Endosulfan: b.d –13.07 ng/g ΣHCHs: 0.08 –7.25 ng/g Σ Chlordane: b.d –5.84 ng/g Σ Heptachlor: b.d –1.68 ng/g Σ OCPs: 0.36–104.50 ng/g	Singh et al. (2007)

	India (Kuttanad agroecosystem)	OCPs: BHCs, heptachlor, chlordanes, endosulfans, aldrin, dieldrin and endrins	–	BHC: 0.01–9.55 ng/g DDT: ND–4.55 ng/g DDE: ND–2.18 ng/g DDD: ND–2.07 ng/g α-Endosulfan: 0.74–8.9 ng/g Aldrin: 1.96–2.73 ng/g Dieldrin: 1.29–3.72 ng/g Heptachlor: 1.01–8.52 ng/g	Sruthi et al. (2017)
	India (North East)	OCPs: DDTs and HCHs	2009–2011	District Dibrugarh: 71.2–834 ng/g HCHs 30.1–918 ng/g DDTs District Nagaon: 39.2–743 ng/g HCHs 72.5–932 ng/g DDTs	Mishra et al. (2013)
	India (Ranga Reddy)	Monocrotophos, chlorpyriphos, endosulfan and CP	2008–2009	α-endosulfan: 0.02 μg/g β-endosulfan: 0.02 μg/g	Ratna Kumari et al. (2012)
	India (Nagaon and Dibrugarh)	OCPs: DDTs and HCHs	2009–2010	Nagaon: 98–1945 ng/g HCHs 166–2288 ng/g DDTs Dibrugarh: 178–1701 ng/g HCHs 75–2296 ng/g DDTs	Mishra et al. (2012)
	India (Haryana)	OCPs: HCHs, DDTs, endosulphan, heptachlor, aldrin and chlordane OPYs: CP, fluvalinate, FEN, deltamethrin OPs: chlorpyriphos, monocrotophos, dimethoate, Me-parathion, palathion, quinalphos, triazophos	2002–2003	HCHs: 0.002–0.051 μg/g DDTs: 0.001–0.066 μg/g Endosulfan: 0.002–0.039 μg/g Chlordane: 0.0002–0.019 μg/g CP: 0.001–0.035 μg/g FEN: 0.001–0.022 μg/g Chlorpyriphos: 0.002–0.172 μg/g Malathion: 0.002–0.008 μg/g Quinalphos: 0.001–0.010 μg/g	Kumari et al. (2008)

(continued)

Table 1 (continued)

Continent	Area	Insecticides analysed	Sampling year	Concentration	References
	Japan	Dieldrin	–	0.068–0.125 mg/kg	Saito et al. (2012)
	Tajikistan	DDTs, BHCs, endosulfans, Aldrin, chlordanes, dieldrin, endrins, heptachlors and methoxychlor	2011, 2012, 2013 and 2014	Many soil samples with pesticides concentrations greater than 10 ppm.	Barron et al. (2017)
	Vietnam	DDTs, HCHs, chlordanes, drin compounds, heptachlor, HCB, heptachlor-epoxide and polychlorinated biphenyls (PCBs)	January to October 2002	ΣDDTs: 0.19–140 ng/g Σchlordanes: ND–9.0 ng/g ΣPCBs: 0.11–110 ng/g	Kishida et al. (2007)
Europe	France	Lindane (γ-HCH)	June 2002 to November 2007	γ-HCH: 0.03–4.92 μg/kg	Villanneau et al. (2009)
	Portugal and Spain (Agueda river borders)	Diazinon, chlorpyrifos and dimethoate	January–February 2012	None insecticide detected in soil.	Sánchez-González et al. (2013)
	Serbia (Belgrade)	Deltamethrin, phorate, fenitrothion, chlorpyrifos, carbofuran, γ-HCH, tebupirimfos and terbufos	July–November of 2006	Fenitrothion: b.d–80.5 μg/kg Chlorpyrifos: b.d–47.4 μg/kg	Marković et al. (2010)
	Spain (La Rioja region)	Methoxyfenozide and pirimicarb	March, June and October 2012	Methoxyfenozide: 4.61 μg/kg	Pose-Juan et al. (2015)
	Spain (Valencia)	OPYs: Resmethrin, bifenthrin, fenpropathrin, λ-cyhalothrin, permethrin, cyfluthrin, α-CP, τ-fluvalinate, esfenvalerate and deltamethrin	–	OPYs concentration ≤57.0 ng/g before plow and ≤62.3 ng/g during rice production. Resmethrin and cyfluthrin were the compounds found in higher concentrations.	Aznar et al. (2017)

ND non-detectable, *b.d* below detection limit, and *dw* dry weight found in the Asian continent, mostly from India followed closely by China. These results could be explained by the high pattern of usage of insecticides in India compared with the rest of the world. In India, 80% of the pesticides used are insecticides, in contrast to 29.5% of world use (De et al. 2014). OCPs were massively used in China between 1983 and 2007 for pest and disease control, which explains their high residue level detection in recent years (Guo et al. 2016). Concerning insecticide concentrations, the highest amounts were found in Mexico, more precisely in Southern Sonora, for benzene hexachloride (BHC) and dichloro-diphenyl-trichloroethane (DDT) and its metabolites, dichlorodiphenyldichloroethane (DDD) and dichlorodiphenyldichloroethylene (DDE) in a range from non-detectable (ND) to 938.5 μg/g and ND to 679.7 μg/g, respectively. This study suggests evidence that although DDT was banned, it is still applied in Mexican agricultural soils (Cantu-Soto et al. 2011). Besides OCPs, other insecticide classes were found, namely, neonicotinoids, organophosphates (OPs), and pyrethroids (OPYs).

2 Tests to Assess Phytotoxicity by Insecticides

Pesticides are used globally for the protection of food, and more generally for human health, by killing organisms that cause disease and threaten public health. Insecticides are used to kill, prevent, repel, or harmfully affect insects. Thus, by nature, they present a degree of toxicity for the environment, plant, and animal species. At first thought, plants are not an insecticide target, however, they reach plant tissues when spread.

When an insecticide applied for insect control causes damage to the host-plant tissues, it is said to be phytotoxic. The susceptibility of plants to chemical injury varies greatly among and within species and is influenced by a myriad of factors, such as growth stage, growth rate, temperature, humidity, and other environmental factors (National Research Council 1969). Plants sensitive to harmful substances can be used as bioindicators in toxicity assessment studies, while more resistant plants can be applied in bioremediation processes (Kapanen and Itävaara 2001).

The use of bioindicators to evaluate possible phytotoxic chemical residues affords a direct, inexpensive, and integrated estimation of bioavailability and contaminant toxicity. The advantages of bioindicators consist of:

1. The possibility to detect both toxicity of parent compounds and toxic metabolites;
2. Tests use readily available materials;
3. Tests can be carried out ex situ or in situ;
4. The test period is usually short;
5. Low cost and uncomplicated methodology is used (Maila and Cloete 2005).

However, for environmental evaluation, tests also need to combine some other features, such as (1) to be standardized, (2) to have a defined endpoint, and (3) to be sensitive enough to distinguish differences among sites (Da Silva Júnior et al. 2013).

Phytotoxicity may be chronic, if it induces the immediate death of the affected plant tissue, or acute if it interferes with physiological processes that decrease the performance of the plant (National Research Council 1969). Phytotoxicity can be evaluated in several stages of plant development (seed germination, root elongation, and seedling growth) and can be carried out in pots or in Petri dishes. The inhibition of seed germination and the effects on root elongation or plant growth are the main areas of interest in studies on phytotoxicity, while photosynthesis, respiration, enzyme activities, and tissue cultures are not commonly used as standard tests (Kapanen and Itävaara 2001).

Several bioassay studies demonstrated that the use of insecticides can induce phytotoxic effects in different ways. First, natural pollinators, such as honeybees and butterflies, are very sensitive to pesticides. Pesticides can kill bees and are strongly implicated in pollinator decline (Miller 2004). Besides negatively affecting pollinator populations and their delivery of pollination, the use of insecticide may also have nonlethal impacts that distress the pollination process at pre- or post-pollen deposition stages. For instance, pesticides can make crops unattractive to a major pollinator, or negatively impact post-pollination processes, like pollen germination. Such impacts have received very little attention, and, given the potential for new insecticides to come into use, or for applications to increase in certain crops in response to emergent pests or diseases, a better understanding of these impacts is crucial (Gillespie et al. 2014).

Post-pollination impacts of pesticides could occur through pollen, stigmas, or the interaction of both. Either pollen or the stigmatic tissue may be susceptible to damage by pesticides, which can decrease pollen germination, pollen tube growth, and ovule fertilization, thus resulting in reduced seed set and crop yield (Gillespie et al. 2014).

Seed germination and root elongation tests (Fig. 2) have been used as short-term phytotoxicity test to provide valuable information about inhibition, enzyme activation, hormone production, cell expansion, respiration, and other parameters (Wang et al. 2001). This topic is worthy of a detailed discussion and a full view of the impact of insecticides in seed germination, and seedling development will be provided in Sect. 2.2.

Moreover, insecticides sprayed over the leaves also affect photosynthetic efficiency (Fig. 2) by increasing or decreasing pigment contents and affecting electron transport chain in chloroplasts. The pyrethroid insecticide alphamethrin was shown to induce a decline in chlorophyll a (chl a), chl b, and total chlorophyll in soybeans (*Glycine max* (L.) Merr) at higher concentrations, while carotenoid content increased with increasing concentrations of the insecticide, to protect chlorophylls from photooxidative damage. Regardless of the control group or insecticide-treated groups, the quantities of photosynthetic pigments were maximum at the flowering stage, followed by the pre-flowering and post-flowering stages. The highest concentration of insecticide tested also caused a reduction in the photosynthetic rate and stomatal conductance (Bashir et al. 2014). Chopade et al. (2007) quantified chl a and chl b in ten medicinal plants treated with one of three insecticides (endosulfan, lindane, and dichlorvos) and concluded that, in general, endosulfan was the most

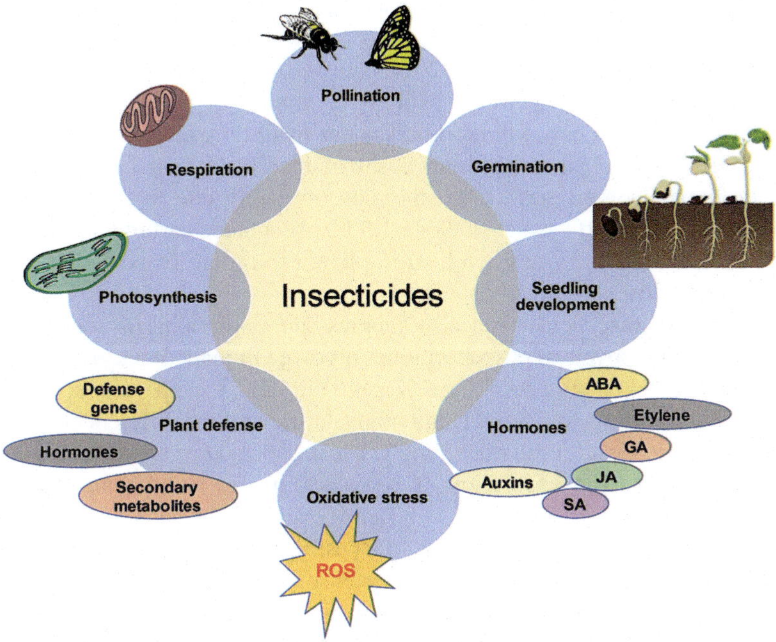

Fig. 2 Impact of insecticides on the development and survival of plant species. *ABA* abscisic acid, *GA* gibberellins, *JA* jasmonic acid, *ROS* reactive oxygen species, *SA* salicylic acid

harmful one. Reduction of photosynthetic pigments may be due to several mechanisms, namely, the inhibition of their biosynthesis or breakdown of pigments or their precursor molecules, changes in chlorophyll fluorescence associated with inhibition of electron transfer chain, the breakdown of the thylakoid and chloroplast envelope, and the decrease in leaf area (Mohamed and Akladious 2017). Chlorophyll breakdown in higher plants occurs in chloroplasts by enzyme-catalyzed processes via pheophorbide a and the red chlorophyll catabolite (RCC) to give primary fluorescent chlorophyll catabolite (*p*FCC). When possessing a propionic acid group, FCCs are translocated to the vacuole, where they spontaneously isomerize to the corresponding nonfluorescent chlorophyll catabolites (NCCs) (Chopade et al. 2007). The opposite effect was described in mung bean (*Vigna radiata* (L.) Wilczek) submitted to foliar spray application of dimecron. Chlorophyll content increased but a deviation in the absorption spectra of chl a and chl b was observed (Siddiqui and Khan 2001). Possible explanations suggested for this phenomenon could be related with one of the following phenomena, the increase of grana and intergrana spaces, NADP/NAD ratios, and NADP and ATP levels, or could be due to increased uptake of K^+, Mg^+, Ca^{2+}, or other ions (Mohamed and Akladious 2017).

Besides interfering with the photosynthesis, insecticides also affect photo and dark respiration in seedlings (Fig. 2) (Mishra et al. 2008). Seedlings of cowpea (*Vigna unguiculata* (L.) Walp.) exposed to higher doses of dimethoate and/or UV-B showed reduced content in chl a and chl b and carotenoids, as well as photosynthetic oxygen yield and photofixation of CO_2. This could be explained based on the direct effect of dimethoate on the activities of photosystems II, I, and whole electron transport chain in chloroplasts which were inhibited by the insecticide or insecticide+UV-B in a concentration-dependent manner (except at the lower dose of dimethoate tested). Similarly, high doses of insecticide and insecticide+UV-B adversely affected photo and dark respiration, probably due to the interaction of dimethoate and UV-B with ribulose-1,5-bisphosphate carboxylase/oxygenase (RubisCO) and other enzymes involved in photorespiration. Opposite effects were verified for low doses of dimethoate.

Application of insecticides can also suppress the expression of important plant defense genes, alter levels of phytohormones involved in plant defense, and decrease plant resistance to unsusceptible herbivores (Fig. 2). Szczepaniec et al. (2013) applied thiamethoxam, clothianidin, and imidacloprid to three different crops, cotton (*Gossypium hirsutum* L.), corn (*Zea mays* L.), and tomato (*Lycopersicon esculentum* Mill.), respectively. The expression of genes coding for phenylalanine ammonia lyase (*PAL*), coenzyme A ligase (*CoA ligase*), and chitinase (*chit*), involved in salicylic acid (SA)-mediated defense, and trypsin proteinase inhibitor (*trypsin PI*), involved in jasmonic acid (JA)-mediated defense, was influenced by the treatment with spider mites and/or insecticide. Spider mites induced the expression of *CoA ligase* and *chitinase* in cotton and elicited the expression of all four genes in corn. *Trypsin PI* was the only gene induced in tomato. When treated with insecticide or insecticide+mites, the expression of *CoA ligase* and *chitinase* increased in cotton exposed to both treatments while none of the genes were induced in corn. Both treatments strongly induced *chitinase* expression in tomato. Besides altering the expression of defense genes, insecticide treatments also changed phytohormone concentrations in these species. All insecticide decreased the levels of 12-oxo-phytodienoic acid (OPDA), a precursor of JA, and imidacloprid enhanced the level of SA and decreased the levels of JA and of a bioactive conjugate of JA (JA-Ile) in tomato. Clothianidin reduced the concentration of abscisic acid (ABA), a hormone also related to the plant defense, and JA in corn (Szczepaniec et al. 2013).

Foliar application of insecticides was shown to induce oxidative stress (Fig. 2). For instance, pre-flowering, flowering, and post-flowering seedlings of mung beans (*Vigna radiata* L.) were treated with chlorpyrifos, and lipid peroxidation rate, proline content, enzyme activity (superoxide dismutase (SOD), ascorbate peroxidase (APX), catalase (CAT), and glutathione reductase (GR)), and ascorbate and glutathione content were determined (Parween et al. 2012). The increase of lipid peroxidation and proline content was age- and dose-dependent. Proline is known to contribute to detoxification of reactive oxygen species (ROS), protection of membrane integrity, stabilization of enzyme or proteins, and tolerance to stresses. In order to detoxify the superoxide radical anion and hydrogen peroxide produced, SOD, APX, CAT, and GR activities increased dose-dependently with maximum activity in the flowering stage, and, after this period, their activity declined. The relatively low activity during post-flowering stage is mainly based on the fact that an older leaf contains lower antioxidants than a younger leaf. APX and GR are components of the ascorbate-glutathione cycle responsible for the recycling of glutathione, and, in this way, they protect chloroplasts by maintaining high reduced/oxidized glutathione (GSH/GSSG) ratio. Similar results were obtained by Bashir et al. (2007) for soybean seedlings treated with deltamethrin.

Concerning the nonenzymatic antioxidants, the decline in the ratio ascorbate/dehydroascorbate is indicative of oxidative stress. Chlorpyrifos induced a dose-dependent reduction in ascorbate as well as in ascorbate+dehydroascorbate contents over time in *V. radiata*, along with a switch of the ratio favoring dehydroascorbate. Concerning GSH, a dose-dependent decrease in GSH and an increase in GSSG and GSH + GSSG contents were observed (Parween et al. 2012). However, in another study, deltamethrin induced an increase of GSH in soybean seedlings, indicating an active GSH participation in the detoxification of ROS, directly (nonenzymatic) as well as through certain enzymes (Bashir et al. 2007).

The synthesis of secondary metabolites in plants, such as phenolic compounds, can be modulated by the application of insecticides. Increase of polyphenol content was observed for mung bean seedlings exposed to profenofos (Mishra et al. 2015) and dimecron (Siddiqui and Khan 2001). These compounds are important for plant defense since they can be toxic for plant pathogens and can also protect plants from oxidative stress.

Therefore, to avoid the negative impact at all levels of the ecosystem, research on insecticides with fewer side effects should be encouraged, and pesticides manufacturers should conduct long-term studies on target and nontarget species to demonstrate that a pesticide has no adverse effects before allowing it to be registered for use in the environment.

2.1 Germination Test

According to the International Rules for Seed Testing (ISTA 1966), germination in a laboratory test is defined as the emergence and development of essential structures from the seed embryo that, for the kind of seed tested, indicate its ability to develop into a normal plant under favorable soil conditions. In other words, it evaluates not only the germination process itself but also the seedling (root and shoot elongation) development. Germination and seedling development of several plant species demonstrated to be affected by insecticide exposure. Studies included in this section were obtained from Web of Science core database, between 2003 and 2017, using "seeds, germination, and insecticides" as a search topic (Table 2).

Some of these bioassays are ruled by specific standards and recommended for soil quality assessment. OECD has developed a plant bioassay (OECD 2006), intended to assess the potential effects of substances on seedling emergence and growth. It does not cover chronic effects or effects on reproduction, such as seed set, flower formation, and fruit maturation. The test assesses the effects on seedling and early growth of higher plants after exposure to the substance in the soil or in another soil

Table 2 Influence of insecticide application on seed germination and seedling development

Plant	Insecticides	Range	Germination conditions	Effect on germination	Other effects	Reference
• **Alliaceae**						
Allium cepa L.	λ –cyhalothrin, Spinetoram Methomyl Acetamiprid Spirotetramat Azadirachtin Urea	–	Field experiment	Insecticides have positive effects on the rate of seed germination. More seeds from treated plants germinated within 5 days.	Decreased flower visitation by honey bees. Can negatively affect multiple stages of the pollination process.	Gillespie et al. (2014)
Allium cepa L.	Me-parathion* *combined with lead, iron, cadmium, chrome, atrazine and 2,4-DB	–	Seeds exposed to sediments elutriatein Petri dishes.	–	Chromosomal aberrations and nuclear abnormalities in roots.	Rambo et al. (2017)
Allium porrum L.	Endosulfans DDTs Diedrin	OCPs mean concentrations (dw) Bulk soil: 36.4 ng/g Rhizosphere 13.7 ng/g	Field experiment	–	Aerial and root tissues bioaccumulate OCPs. High levels of metabolites found in all tissues suggest the ability to metabolize parent compounds.	Gonzalez et al. (2003)
• **Asteraceae**						
Lactuca sativa L.	Chlordanes mixture (*cis*-chlordane, *trans*-chlordane and *trans*-nonachlor and *p, p*'-DDE)	100 ng/mL	–	A statistically non-significant seed germination decrease was noted.	–	Hamdi et al. (2015)

• Brassicaceae						
Brassica rapa var. *chinensis* (L.) Kitamura	CP* * Alone or in combination with Cu^{2+}	Solution (mg/mL): CP: 5.6–17 Cu^{2+}: 1–9; CP + Cu^{2+}: 8 and 16 of CP + 1–9 of Cu^{2+} Soil (mg/kg): CP: 8– 64 Cu^{2+}: 100–300 CP+Cu^{2+}: 50 and 100 of CP+100–300 of Cu^{2+}	Petri dishes with contaminant solutions, vitreous pots with solutions and culture dishes with contaminated soil.	CP reduced germination and root and shoot elongation in solution. Cu^{2+} accelerated germination rate at low doses and inhibited it at high concentrations. CP reduced Cu^{2+} toxicity on seed germination and alleviated Cu^{2+} toxicity on root elongation at low concentrations. Under soil conditions, root and shoot elongation was practically unaffected by CP. CP + Cu^{2+} had less impact on root elongation than that of Cu^{2+} alone.	–	Liu et al. (2009)
Brassica sp.	HCHs	300–12500 mg/kg of total HCHs	Plastic pots containing soil.	Decrease in germination at the highest concentrations of the HCHs. Root length decreased at 10000 mg/kg. SVI: 100< SVI<300 SRL: 50<SRL<100 HCHs induced low productivity and biomass accumulated mainlyin root.	–	Pereira et al. (2010)

(continued)

Table 2 (continued)

Plant	Insecticides	Range	Germination conditions	Effect on germination	Other effects	Reference
Capsella bursa-pastoris L.	Deltamethrin (Decis) Dimethoate (Dimethoate 40)	25 g/L 400 g/L	Pots in controlled greenhouse environment.	No significant effect on germination. Only deltamethrin decreased seedling growth (mean dry weight biomass).	–	Hanley and Whiting (2005)
Raphanus sativus L.	Landfill leachate (LLe) contaminated with Dimethoate and other contaninants* * linuron, bisphenol A, 17α-ethynilestradiol 4-n-nonylphenol.	10 mg/L	Petri dishes with: non-contaminated LLe (NC); contaminated LLe (UN); LLedecontaminated by adsorption and/or ligninolytic fungi.	None of the treatments affected seed germination. UN mainly affected root and shoot elongation while NC inhibited only root elongation. LLe treated with adsorbents or with adsorbents+fungi increased elongation and seedling biomass.	–	Loffredo and Castellana (2015)
• Caryophyllaceae						
Agrostemma githago L.	Deltamethrin (Decis) Dimethoate (Dimethoate 40)	25 g/L 400 g./L	Pots in controlled greenhouse environment.	Insecticide application had no significant effect on germination. Only dimethoate decreased seedling growth (mean dwbiomass).	–	Hanley and Whiting (2005)
• Convolvulaceae						
Ipomea aquatica Forssk	Endosulphansulfate Heptachlor	0.4–40 mg/kg dry soil	Plastic planting containers with contaminated soil.	None of the insecticides significantly decreased seed germination and root length. Both insecticides at 40 mg/kg reduced fresh weight but not dw.	–	Somtrakoon and Pratumma (2012)

• Curcubitaceae						
Cucumis sativus L.	Endosulphansulfate Heptachlor	0.4–40 mg/kg dry soil	Plastic planting containers with contaminated soil.	None of OCPs significantly decrease seed germination or fresh and dw. Endosulphansulfate and heptachor at 40 mg/kg decreased shoot and root length, respectively.	–	Somtrakoon and Pratumma (2012)
Cucumis sativus L.	Permethrin	0.07–0.52 mg/mL	Filter paper hydrated with aqueous suspensions in Petri dish.	Nanopermethrin did not affect germination and root length but permethrin dose-dependently affected these parameters.	–	Kumar et al. (2013)
• Fabaceae						
Cicer arietinum L.	Endosulfan* Chlorpyrifos* *alone or combined with fungicide Captan and *Rhizobium* inoculant.	15 mL/kg 10 mL/kg	Field experiment, conducted at a research farm	No adverse effect on germination were found for Endosulfan and Chlorpyrifos when applied alone or in combination with the recommended fungicide Captan and *Rhizobium* inoculant.	Non-significant differences to controls were observed in nodulation, yield-attributing characters and grain yield.	Cheema et al. (2009)
Glycine max L.	CP (Arrivo 25 EC)	9.6 and 19.2 M	Petri dishes.	Germination percentages of primary roots decreased with increasing CP concentration.	Reduction in the number of different phases of mitosis and in the leaf pigment content.	Aksoy and Deveci (2012)

(continued)

Table 2 (continued)

Plant	Insecticides	Range	Germination conditions	Effect on germination	Other effects	Reference
Glycine max L.	Mepthion (fenitrothion) Myungtaja (etofenprox) Actara (thiamethoxam) Stonate (λ-cyhalothrin + thiamethoxam)	2 mL/L 1 mL/L 0.5 g/L 0.5 g/L	Petri dishes inside an incubator.	Fenitrothion reduced the seed germination, seed and seedling vigour.	Thiamethoxam and λ-cyhalothrin + thiamethoxam showed positive effects on seedling biomass and contents of phenolic compounds. Fenitrothion reduced polyphenol and flavonoid contents.	Dhungana et al. (2016)
Trifolium pratense L. *Trifolium repens* L. *Pisum sativum* L. *Phaseolus vulgaris* L.	HCHs	0–5000 mg/kg soil	Plastic pots containing soil	Germination was not affected as other parameters such as the rate of germination and SVI.	–	Pereira et al. (2010)
Senna obtusifolia (L.) H.S. Irwin & Barneby	Acephate* Carbaryl* Methomyl* Esfenvalerate* Fenpropathrin* λ-Cyhalothrin* Indoxacarb* *Mixture with 2,4-DB herbicide	1.09 kg/ha 1.12 kg/ha 0.67 kg/ha 0.20 kg/ha 0.34 kg/ha 0.36 kg/ha 0.12 kg/ha	Field experiment at a Cherry Farm Unit.	Seed production was not affected by co-application of the 2,4-DB with any of the insecticides to flowering sicklepod.	–	Lancaster et al. (2005)
Vigna sinensis L Savi	Endosulfansulfate Heptachlor	0.4–40 mg/kgdry soil	Plastic planting containers with contaminated soil.	The insecticides did not affect seed germination (%). Heptachlor affected shoot and root length more than endosulfansulfate.	–	Somtrakoon and Pratumma (2012)

• **Poaceae**						
Avena fatua L	Dimethoate Deltamethrin	40, 400 g/L 25 g /L	Pots filled with sterile compost in controlled greenhouse environment.	No significant effect was detected on seed germination.	–	Hanley and Whiting (2005)
Avena sativa L.	HCHs	300–12500 mg/kg of soil	Plastic pots containing soil	For concentration of 300 mg/kg it was observed an increase in germination but for highest concentrations there was a reduction. Reduction of the mean root length was observed. Shoot length, decreased above 1250 mg/kg.	–	Pereira et al. (2010)
Brachiaria brizantha Stapf. cv. Piatã	Thiamethoxam	17.5–40.0 g per 100 kg of seeds	Pots filled with a substrate composed of clay, silt and sand in controlled greenhouse environment.	No effect was detected on seed germination and growth. Increasing doses provoked a slight linear decrease of leaf area and a slight decrease in shoot development. Root growth tends to enhance.	The % of crude protein in the shoots increased with the increased availability of nitrogen in the plant. The activity of nitrate reductase increased.	Macedo et al. (2013)

(continued)

Table 2 (continued)

Plant	Insecticides	Range	Germination conditions	Effect on germination	Other effects	Reference
Cenchrus setigerus Vahl	Chlorpyrifos CP FEN	10–100 mg/kg dry soil	Plastic pots containing soil	Germination rates showed a tendency to decline with pesticide increasing concentration. Higher concentrations of chlorpyrifos caused a significant reduction and delay in seed germination comparatively to CP and FEN.	–	Dubey and Fulekar (2011)
Hordeum vulgare L.	HCHs	300–12500 mg/kg of soil	Plastic pots containing soil.	Reduction in germination (%) for HCHs highest concentrations. Shoot mean length, decreased above 1250 mg/kg.	–	Pereira et al. (2010)
Hordeum vulgare L. cv. karan-16	Chlorpyrifos	0.05 – 0.5%	Seeds were pre-treated with chlorpyrifos and placed on filter paper in Petri dishes.	Seed germination decreased with the increasing of dosage, being the effect more marked for 0.5% of treatment at a duration of 17 h pre-soaking (about 33.8%).	Reduced mitotic index and a gradual increase in % of chromosomal abnormalities for increasing concentrations. Reduction in chlorophyll for the highest concentrations. Carotenoid content slightly increased at 0.05% and 0.1%.	Dubey et al. (2015)

Hordeum vulgare L. cv. karan-16	Profenophos	0.05 - 0.5%	Seeds were pre-treated with profenophos and placed in Petri dishes.	Percentage of seed germination decreased with the increasing of concentration.	Chromosomal abnormalities and induction on chlorophyll mutations were observed.	Srivastava and Singh (2009)
Oryza sativa L.	Diazinon Fipronil λ-cyhalothrin	10 μg/L	Seeds were pre-treated with insecticides and placed on filter paper in Petri dishes.	Fipronil provoked a stimulatory effect onradicle growth.	–	Moore and Kroger (2010)
Oryza sativa L.	Imidacloprid (Gaucho® 600 FS)	250–2000 mg/L	15 cultivars of rice seeds were sown in translucent HDPE trays filled with soil.	Reduction of seed germination provoked by the continuous exposure to 2000 mg/L of imidacloprid. Concentrations of 500 –1000 mg/Lled to an enhancement of seedling growth.	–	Stevens et al. (2008)
Pennisetum pedicellatum Trin.	Chlorpyrifos CP FEN	10–100 mg/kg dry soil	Plastic pots containing soil.	Germination rates showed a tendency to decline with pesticide increasing concentration. Higher concentrations of chlorpyrifos caused a significant reduction and delay in seed germination comparatively to CP and FEN.	–	Dubey and Fulekar (2011)

(continued)

Table 2 (continued)

Plant	Insecticides	Range	Germination conditions	Effect on germination	Other effects	Reference
Poa annua L.	Dimethoate Deltamethrin	40, 400 g/L 25 g/L	Pots filled with sterile compost in controlled greenhouse environment.	No significant effect was detected on seed germination. Deltamethrin caused a reduction on seedling growth.	–	Hanley and Whiting (2005)
Triticum aestivum L.	HCHs	300 – 12500 mg/kg of soil	Plastic pots containing soil.	Decrease in total germination with the increasing concentrations of HCHs. The effect was more marked >5000 mg/kg. Reduction in the roots mean length was observed. Shoot mean length decreased >1250 mg/kg.	–	Pereira et al. (2010)
Zea mays L.	Permethrin	0.07 – 0.52 mg/L	Filter paperhydrated with aqueous suspensions in Petri dishes.	Permethrin significantly decreased germination (%) and root length. Nanopermethrin caused no effect on seed germination and root length.	–	Kumar et al. (2013)

Zea mays var. ceratina	Endosulfansulfate Heptachlor	0.4 – 40 mg/kg dry soil	Plastic planting containers with contaminated soil.	The insecticides did not cause a significant decrease on seed germination. Shoot length was stimulated by the lowest concentrations of insecticides. Fresh and dw increased in the range 0.4-4.0 mg/kg of endosulfansulfate. No significant combined effect on the root and shoot elongation was observed.	–	Somtrakoon and Pratumma (2012)
Zea mays L. *Saccharata* Sturt.	Endosulfansulfate Heptachlor	0.4 – 40 mg/kg dry soil	Plastic planting containers with contaminated soil.	Single toxicants did not cause a significant decrease of seed germination. The highest concentration of heptachlor provoked a reduction in root length. No combined effect on the root and shoot elongation and fresh and dw were observed.	–	Somtrakoon and Pratumma (2012)

(continued)

Table 2 (continued)

Plant	Insecticides	Range	Germination conditions	Effect on germination	Other effects	Reference
Zea mays L. *saccharata* Sturt.	Pyriproxyfen	0.1 – 0.6 ppm	Seeds were pre-treated with pyriproxyfen and placed in Petri dishes with two layers of filter paper.	Germination % decreased with increasing doses of insecticide. Elongation of radicle decreased and was almost inhibited at the highest level of pyriproxyfen. At 0.6 ppm the number of radicles and coleoptile length decreased.	Increasing of pyriproxyfen concentration: decreased the number, width and length of stomata and number of epidermal cells; decreased total chlorophyll and carotenoids; and increased anthocyanin and proline contents. Stomata indexes for both leaf surfaces were affected.	Coskun et al. (2015)
Zea mays L. *saccharata* Sturt.	Deltamethrin	0.01 – 0.5 ppm	Seeds were pre-treated with deltamethrin and placed in Petri dishes with two layers of filter paper.	Germination % decreased with increasing doses of deltamethrin. The length of radicle decreased being the effect more pronounced for the highest concentration.	Increasing of deltamethrin concentration: decreased stoma index and sizes; decreased chlorophylls a and b and carotenoid contents; and increased anthocyanin and proline contents.	Duran et al. (2015)
• Solanaceae						
Lycopersicon esculentum L.	Emamectin benzoate α-CP λ-cyhalothrin Imidacloprid	10 – 160 mg/L 30 – 500 mg/L 15 – 240 mg/L 125 – 2000 mg/L	Seeds were pre-treated with insecticides and placed in Petri dishes with two layers of filter paper.	Germination % and root growth decreased with increasing doses of insecticide. The inhibitory effectdecreased with the exposure time.	Chlorophylls a and b decreased with increasing concentration of pesticides.	Shakir et al. (2016)

				Longest shoots were observed with 60 mg/Lof α-CP while the smallest ones were caused by 500 mg/Lof α-CP. Biomass increased with lower concentrations and decreased with higher concentrations for all the insecticides.	Total carotenoid content was stimulated by lower concentrations of pesticides, and decreased with higher concentrations.	
Lycopersicon esculentum L.	Permethrin	0.07 – 0.52 mg/L	Filter paperhydrated with aqueous suspensions in Petri dishes.	Permethrin caused a decrease in both germination % and root length, being the effect more pronounced for 0.52 mg/L. Nanopermethrin caused no effect on seed germination and root length.	–	Kumar et al. (2013)
• Urticaceae						
Urtica urens L.	Dimethoate (Dimethoate 40 and BASF) Deltamethrin (Decis and Aventis)	40 and 400 g/L 25 g/L	Pots filled with sterile compost in controlled greenhouse environment.	No significant effect was detected on seed germination. The combination of both insecticides caused a reduction on seedling growth.	–	Hanley and Whiting (2005)

SVI seedling vigour index, *SRL* specific root length index, and *dw* dry weight matrix. Seeds in the treatment groups and in the control group are placed in the respective soils, treated or untreated with the test compound, and the evaluation of the effects following around 14–21 days after 50% emergence of the seedlings in the control group is carried out. Endpoints measured in this test include the visual observation of seedling emergence, the measurement of dry shoot weight (alternatively fresh shoot weight) and, in some cases, of the shoot height, as well as an evaluation of visible damaging effects on different parts of the plant. These quantifications and observations are compared to untreated control plants.

Usually, insecticides can influence seed germination by affecting the synthesis and activity of hydrolytic enzymes, such as amylase, ATPase, lipase, and protease (Bashir et al. 2014). Also, the levels of phytohormones involved in germination can be altered. The application of phorate on seeds of mash bean (*Vigna mungo* (L.) Hepper) showed that lower concentrations of insecticide had a stimulatory effect on seed germination, accompanied by amylase activity and increased levels of the phytohormones gibberellin and ethylene, while higher concentrations reduced them (Singh et al. 1982).

Reduction in length of roots and shoots also occurs due to the application of insecticides, and this may be explained due to inhibition of cell division at the meristematic regions (Bashir et al. 2014; Dubey et al. 2015). Soumya et al. (2016) observed that even exposure of onion seeds (*Allium cepa* L.) to relatively low concentrations of the broad-spectrum insecticide Attack had significant effects on mitotic index and structure of chromosomes and disturbed the mitotic spindle formation. Similar results were obtained for the same species with the application of emamectin benzoate and imidacloprid (Al-ahmadi 2013) and for barley (*Hordeum vulgare* L.) treated with chlorpyrifos (Dubey et al. 2015). Mitotic index reduction may occur during the interphase due to (1) inhibition of DNA synthesis at the S phase; (2) blocking of the G_1 phase, thus suppressing the DNA synthesis; or (3) blocking of the G_2 phase, resulting in the prevention of cells from entering the mitosis. The most frequent chromosomal aberrations include chromosome fragmentation, chromosome stickiness, chromosomal bridges, diagonal anaphase, and multipolar anaphase (Dubey et al. 2015). Moreover, the reduced root elongation may also be correlated with the inefficient uptake of the nutrients from the soil (Bashir et al. 2014) or decrease in respiration rate (Lichtenstein et al. 1962).

Seedling development can also be impaired by insecticides that influence the nitrogen metabolism. Mathur et al. (1989) studied the influence of phorate on the activities of the enzymes glutamine synthetase (GS), glutamate dehydrogenase (GDH), and glutamate synthase (GOGAT) in the primary leaves, nitrogenase (N_2-ase) in detached root nodules, and protein concentration in the primary leaves of *V. mungo*. The authors observed that low concentrations of phorate stimulated the activities of these enzymes and protein concentration but behaved in the opposite way at higher concentrations.

To avoid seed destruction, insecticide or fungicide addition is commonly used. However, the addition of these chemicals showed significant lower germination percentages throughout the storage period, with a significant reduction after 8 months. However, it seems that encrusting seeds reduce the negative effect of pesticides on sunflower germination (Szemruch and Ferrari 2013).

From Table 2, it can be seen that most of the studies were performed with species belonging to Poaceae (41.5%) and Fabaceae (22.0%), followed by Brassicaceae (9.8%), Alliaceae (7.3%), Cucurbitaceae (4.9%), Solanaceae (4.9%), Asteraceae (2.4%), Caryophyllaceae (2.4%), Convolvulaceae (2.4%), and Urticaceae (2.4%) (Fig. 3A). Regarding the classes of insecticides tested by plant family, OPYs, OCPs, and OPs were the most used ones (Fig. 3B).

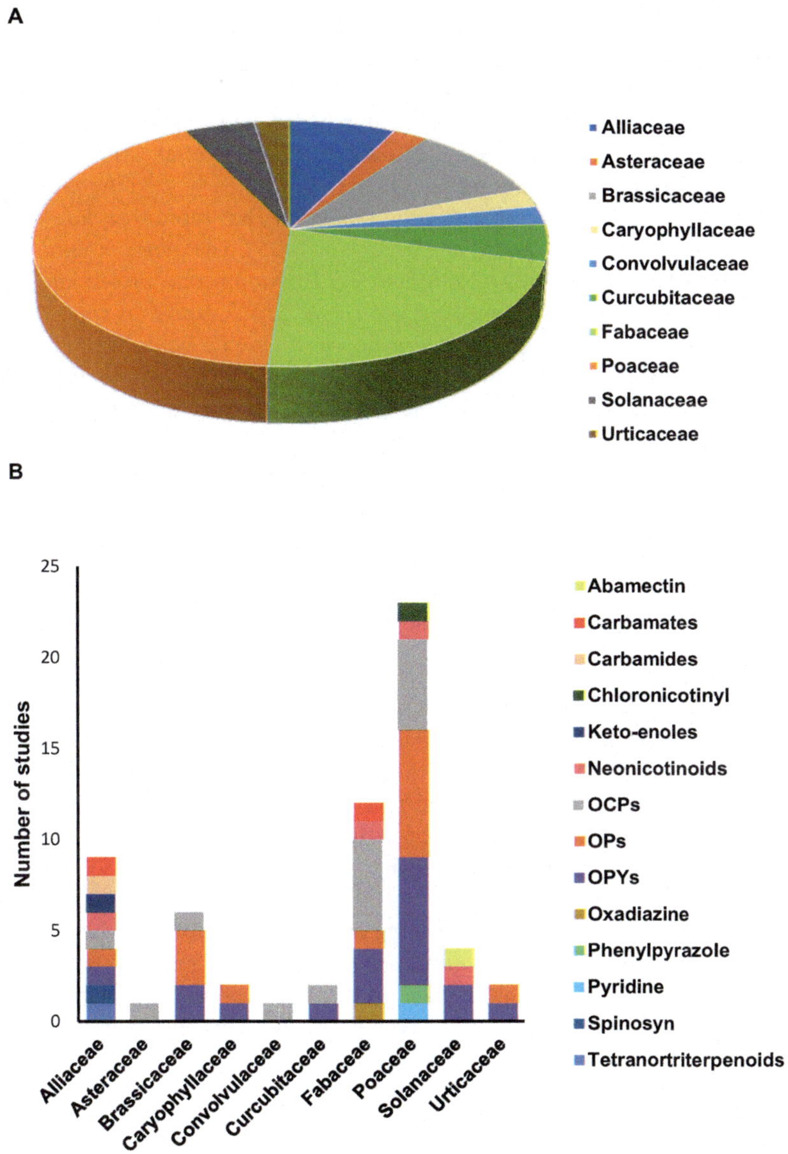

Fig. 3 Prevalence of plant species and insecticides used in germination tests. (**A**) Number of species by plant family, (**B**) classes of pesticides by plant family

2.2 *Germination Tests As a Tool to Assess Phytotoxicity by Insecticides*

The use of higher plants in ecotoxicological studies has increased in the past years since acute germination tests for testing pollutants are fast, sensitive, and cost-effective. Seeds are self-sufficient which means that there is no need to add nutrients to the samples (Wang and Freemark 1995). Thus, the maintenance cost of these tests is minimal and no major equipment is required. Seeds are highly sensitive to pollutants, and the germination test is simple, reproducible, and quick and can be applied in the laboratory (filter paper, Petri dish, and artificial soils) or in field experiments. Moreover, seeds are easily purchased and continue viable for an extended period of time (Wang and Freemark 1995; Priac et al. 2017). Therefore, the application of seed germination, root elongation, and early seedling growth tests to monitoring and assessing environmental conditions has more than a few advantages over animal toxic tests (Wang and Freemark 1995). The most common species recommended by OCDE (OECD 2006) and US Environmental Protection Agency (USEPA 1996) are *Cucumis sativus* L., *Lactuca sativa* L., *Raphanus* spp., *Trifolium pratense* L., *Brassica oleracea* L., and *Triticum aestivum* L., but as it can be seen in Table 2, several other species can be used as test organism. Otherwise, the response to pollutants is closely related to each plant species, and germination period can be insufficient to properly assess phytotoxicity. In field experiments, edaphic factors (e.g., minerals, pH, or salinity) related with soil and sediment can interfere with the toxic effects of the pollutant leading to difficult result interpretation (Wang and Freemark 1995).

3 Final Remarks

From all the classes of insecticides available on the market, OCPs were those more frequently found in the literature. Additionally, OCPs were also detected in higher concentrations in soil, regardless their production and usage banished in some countries. The number of studies demonstrating soil contamination by insecticide application occurs with more incidence in India and China. On the other hand, Mexico showed the highest concentrations of contaminants in soil.

Despite the benefits of insecticides killing or repelling harmful insects, they show a degree of toxicity for both environment and living organisms. Plants are not an insecticide target, but they can target plant tissues when applied, thus impacting their survival and normal development. Moreover, their effects can also negatively impact human beings by spreading along the food chain.

Bioassays allow a global assessment of toxicity, with germination assays being the most useful ones for soil quality evaluation and monitoring. Among their several advantages, minimal maintenance cost and the fact that no major equipment is required can be mentioned. Moreover, seeds are highly sensitive to pollutants;

the methodology is simple, reproducible, and quick and can be applied in situ or in vitro, requiring small amounts of sample.

References

Aksoy O, Deveci A (2012) The investigation of the cytotoxic effects of some pesticides on soybean (*Glycine max* L.) Cytologia 77:475–483

Al-Ahmadi MS (2013) Cytogenetic effects of two synthetic pesticides on mitotic chromosome on root tip cells of *Allium cepa*. Cytologia 78:3–8

Aznar R, Moreno-Ramon H, Albero B et al (2017) Spatio-temporal distribution of pyrethroids in soil in Mediterranean paddy fields. J Soils Sediment 17:1503–1513

Barron MG, Ashurova ZJ, Kukaniev MA et al (2017) Residues of organochlorine pesticides in surface soil and raw foods from rural areas of the Republic of Tajikistan. Environ Pollut 224:494–502

Bashir F, Mahmooduzzafar, Siddiqi TO et al (2007) The antioxidative response system in *Glycine max* (L.) Merr. exposed to Deltamethrin, a synthetic pyrethroid insecticide. Environ Pollut 147:94–100

Bashir F, Zahid F, Iqbal M (2014) Growth performance, photosynthetic efficiency and pigment concentration of *Glycine max* (L.) Merr., as affected by alphamethrin, a synthetic pyrethroid insecticide. Trends Biotechnol Biol Sci 1:29–35

Cantu-Soto EU, Meza-Montenegro MM, Valenzuela-Quintanar AI et al (2011) Residues of organochlorine pesticides in soils from the Southern Sonora, Mexico. Bull Environ Contam Toxicol 87:556

Carvalho FP (2006) Agriculture, pesticides, food security and food safety. Environ Sci Pol 9:685–692

Cheema HK, Sharma P, Singh R et al (2009) Efficacy and compatibility of insecticides, fungicide and Rhizobium inoculant in combination for seed treatment in chickpea (*Cicer arietinum*). Indian J Agric Sci 79:190–194

Chopade AR, Naikwade NS, Nalawade AY et al (2007) Effects of pesticides on chlorophyll content in leaves of medicinal plants. Pollut Res 26:491–494

Coskun Y, Kilic S, Duran RE (2015) The effects of the insecticide pyriproxyfen on germination, development and growth responses of maize seedlings. Fresenius Environ Bull 24:278–284

Da Silva Júnior FMR, Garcia EM, Baisch RM et al (2013) Assessment of a soil with moderate level of contamination using lettuce seed assay and terrestrial isopods assimilation assay. Soil Water Res 8:56–62

Daly GL, Lei YD, Teixeira C et al (2007) Accumulation of current-use pesticides in neotropical montane forests. Environ Sci Technol 41:1118–1123

De A, Bose R, Kumar A et al (2014) Worldwide pesticide use. In: De A, Bose R, Kumar A, Mozumdar S (eds) Targeted delivery of pesticides using biodegradable polymeric nanoparticles. Springer, New Delhi, pp 5–6

Dhungana SK, Kim ID, Kwak HS et al (2016) Unraveling the effect of structurally different classes of insecticide on germination and early plant growth of soybean *Glycine max* (L.) Merr. Pestic Biochem Physiol 130:39–43

Dubey KK, Fulekar MH (2011) Effect of pesticides on the seed germination of *Cenchrus setigerus* and *Pennisetum pedicellatum* as monocropping and co-cropping system: implications for rhizospheric bioremediation. Roum Biotechnol Lett 16:5909–5918

Dubey P, Mishra AK, Shukla P et al (2015) Differential sensitivity of barley (*Hordeum vulgare* L.) to chlorpyrifos and propiconazole: morphology, cytogenetic assay and photosynthetic pigments. Pestic Biochem Physiol 124:29–36

Duran RE, Kilic S, Coskun Y (2015) Response of maize (*Zea mays* L. *saccharata* Sturt) to different concentration treatments of deltamethrin. Pestic Biochem Physiol 124:15–20

Gillespie S, Long R, Seitz N et al (2014) Insecticide use in hybrid onion seed production affects pre- and postpollination processes. J Econ Entomol 107:29–37

Gonzalez M, Miglioranza KSB, Aizpún De Moreno JE et al (2003) Organochlorine pesticide residues in leek *(Allium porrum)* crops grown on untreated soils from an agricultural environment. J Agric Food Chem 51:5024–5029

Guo ZW, Li YC, Yang QP et al (2016) Concentrations, sources and pollution characteristic of organic pesticide in soil from typical Chinese bamboo forest. Environ Prog Sustain Energy 35:729–736

Hamdi H, De La Torre-Roche R, Hawthorne J et al (2015) Impact of non-functionalized and amino-functionalized multiwall carbon nanotubes on pesticide uptake by lettuce (*Lactuca sativa* L.) Nanotoxicology 9:172–180

Hanley ME, Whiting MD (2005) Insecticides and arable weeds: effects on germination and seedling growth. Ecotoxicology 14:483–490

ISTA (1966) International rules for seed testing. Proc Int Seed Test Ass 31:1–152

Kapanen A, Itävaara M (2001) Ecotoxicity tests for compost applications. Ecotoxicol Environ Saf 49:1–16

Kishida M, Imamura K, Maeda Y et al (2007) Distribution of persistent organic pollutants and polycyclic aromatic hydrocarbons in sediment samples from Vietnam. J Health Sci 53:291–301

Kumar RSS, Shiny PJ, Anjali CH et al (2013) Distinctive effects of nano-sized permethrin in the environment. Environ Sci Pollut Res 20:2593–2602

Kumari B, Madan VK, Kathpal TS (2008) Status of insecticide contamination of soil and water in Haryana, India. Environ Monit Assess 136:239–244

Lancaster SH, Jordan DL, Spears JF et al (2005) Sicklepod (*Senna obtusifolia*) control and seed production after 2,4-DB applied alone and with fungicides or insecticides. Weed Technol 19:451–455

Lichtenstein EP, Millington WF, Cowley GT (1962) Insecticide effects on plant growth, effect of various insecticides on growth and respiration of plants. J Agric Food Chem 10:251–256

Liu TF, Wang T, Sun C et al (2009) Single and joint toxicity of cypermethrin and copper on Chinese cabbage (Pakchoi) seeds. J Hazard Mater 163:344–348

Loffredo E, Castellana G (2015) Comparative evaluation of the efficiency of low-cost adsorbents and ligninolytic fungi to remove a combination of xenoestrogens and pesticides from a landfill leachate and abate its phytotoxicity. J Environ Sci Health A 50:958–970

Macedo WR, Fernandes GM, Possenti RA et al (2013) Responses in root growth, nitrogen metabolism and nutritional quality in *Brachiaria* with the use of thiamethoxam. Acta Physiol Plant 35:205–211

Maila MP, Cloete TE (2005) The use of biological activities to monitor the removal of fuel contaminants – perspective for monitoring hydrocarbon contamination: a review. Int Biodeterior Biodegrad 55:1–8

Main AR, Michel NL, Cavallaro MC et al (2016) Snowmelt transport of neonicotinoid insecticides to Canadian prairie wetlands. Agric Ecosyst Environ 215:76–84

Marković M, Cupać S, Đurović R et al (2010) Assessment of heavy metal and pesticide levels in soil and plant products from agricultural area of Belgrade, Serbia. Arch Environ Contam Toxicol 58:341–351

Mathur SN, Singh VK, Mathur M et al (1989) Studies with phorate, an organophosphate insecticide, on some enzymes of nitrogen metabolism in *Vigna mungo* (L.) Hepper. Biol Plant 31:363–369

Mawussi G, Scorza Junior RP, Dossa EL et al (2014) Insecticide residues in soil and water in coastal areas of vegetable production in Togo. Environ Monit Assess 186:7379–7385

Miller GT (2004) Sustaining the earth. Thompson Learning, Pacific Grove

Mishra V, Srivastava G, Prasad SM et al (2008) Growth, photosynthetic pigments and photosynthetic activity during seedling stage of cowpea (*Vigna unguiculata*) in response to UV-B and dimethoate. Pestic Biochem Physiol 92:30–37

Mishra K, Sharma RC, Kumar S (2012) Contamination levels and spatial distribution of organochlorine pesticides in soils from India. Ecotoxicol Environ Saf 76:215–225

Mishra K, Sharma RC, Kumar S (2013) Contamination profile of DDT and HCH in surface sediments and their spatial distribution from North-East India. Ecotoxicol Environ Saf 95:113–122

Mishra IP, Sabat G, Mohanty BK (2015) Phytotoxicity of profenofos 50% EC (Curacron 50 EC) to *Vigna radiata* L. seedlings: III. Studies on secondary metabolites and enzymes. Int J Life Sci 3:351–359

Mohamed HI, Akladious SA (2017) Changes in antioxidants potential, secondary metabolites and plant hormones induced by different fungicides treatment in cotton plants. Pestic Biochem Physiol 142:117–122

Moore MT, Kroger R (2010) Effect of three insecticides and two herbicides on rice (*Oryza sativa*) seedling germination and growth. Arch Environ Contam Toxicol 59:574–581

National Research Council (1969) Insecticides. Insect-pest management and control. National Academies, Washington, DC, pp 64–98

OECD (2006) Terrestrial plant test: seedling emergence and seedling growth test OECD guideline for testing of chemicals. Organization for Economic Cooperation and Development, Paris

Parween T, Jan S, Mahmooduzzafar et al (2012) Evaluation of oxidative stress in *Vigna radiata* L. in response to chlorpyrifos. Int J Environ Sci Technol 9:605–612

Pereira RC, Monterroso C, Macías F (2010) Phytotoxicity of hexachlorocyclohexane: effect on germination and early growth of different plant species. Chemosphere 79:326–333

Pose-Juan E, Sanchez-Martin MJ, Andrades MS et al (2015) Pesticide residues in vineyard soils from Spain: spatial and temporal distributions. Sci Total Environ 514:351–358

Priac A, Badot P-M, Crini G (2017) Treated wastewater phytotoxicity assessment using *Lactuca sativa*: focus on germination and root elongation test parameters. C R Biol 340:188–194

Rambo CL, Zanotelli P, Dalegrave D et al (2017) Hydropower reservoirs: cytotoxic and genotoxic assessment using the *Allium cepa* root model. Environ Sci Pollut Res 24:8759–8768

Ratna Kumari B, Ranga Rao GV, Sahrawat KL et al (2012) Occurrence of insecticide residues in selected crops and natural resources. Bull Environ Contam Toxicol 89:187–192

Saito T, Otani T, Seike N et al (2012) A comparison of dieldrin residues in various vegetable crops cultivated in a contaminated field. J Soil Sci Plant Nutr 58:373–383

Sánchez-González S, Pose-Juan E, Herrero-Hernández E et al (2013) Pesticide residues in groundwaters and soils of agricultural areas in the Águeda River Basin from Spain and Portugal. Int J Environ Anal Chem 93:1585–1601

Shakir SK, Kanwal M, Murad W et al (2016) Effect of some commonly used pesticides on seed germination, biomass production and photosynthetic pigments in tomato (*Lycopersicon esculentum*). Ecotoxicology 25:329–341

Siddiqui ZS, Khan S (2001) Effect of systemic fungicides and insecticides on absorption spectra, chlorophyll and phenolic contents of *Vigna radiata* (L.) Wilczek. Pak J Biol Sci 4:812–814

Singh VK, Mathur M, Mathur SN (1982) Phyto-toxicity of the insecticide phorate on germination of *Vigna mungo*. Agric Biol Chem 46:1681–1682

Singh KP, Malik A, Sinha S (2007) Persistent organochlorine pesticide residues in soil and surface water of northern Indo-Gangetic alluvial plains. Environ Monit Assess 125:147–155

Somtrakoon K, Pratumma S (2012) Phytotoxicity of heptachlor and endosulfan sulfate contaminants in soils to economic crops. J Environ Biol 33:1097–1101

Soumya KR, Teena MT, Sudha S (2016) Evaluation of cytotoxic effects of synthetic pesticide “Attack” on root meristems of *Allium cepa* L. South Indian J Biol Sci 2:35–40

Srivastava AK, Singh AK (2009) Effects of insecticide profenophos on germination, early growth, meiotic behavior and chlorophyll mutation of barley (*Hordeum vulgare* L.) Acta Physiol Plant 31:537–544

Sruthi SN, Shyleshchandran MS, Mathew SP et al (2017) Contamination from organochlorine pesticides (OCPs) in agricultural soils of Kuttanad agroecosystem in India and related potential health risk. Environ Sci Pollut Res 24:969–978

Stevens MM, Reinke RF, Coombes NE et al (2008) Influence of imidacloprid seed treatments on rice germination and early seedling growth. Pest Manag Sci 64:215–222

Sun D, Wei Y, Li H et al (2016) Insecticides in sediment cores from a rural and a suburban area in South China: a reflection of shift in application patterns. Sci Total Environ 568:11–18

Szczepaniec A, Raupp MJ, Parker RD et al (2013) Neonicotinoid insecticides alter induced defenses and increase susceptibility to spider mites in distantly related crop plants. PLoS One 8:e62620

Szemruch CL, Ferrari L (2013) Encrusting offers protection against phytotoxic chemicals and maintains the physiological quality of sunflower (*Helianthus annuus*) seeds. Seed Sci Technol 41:125–132

USEPA (1996) In: Greene JC, Bartels CL, Warren-Hicks WJ, Parkhurst BR, Linder GL, Peterson SA, Miller WEE (eds) Protocols for short term toxicity screening of hazardous waste sites. USEPA, Chicago

Van Dyk JC, Bouwman H, Barnhoorn IEJ et al (2010) DDT contamination from indoor residual spraying for malaria control. Sci Total Environ 408:2745–2752

Villanneau E, Saby NPA, Arrouays D et al (2009) Spatial distribution of lindane in topsoil of Northern France. Chemosphere 77:1249–1255

Wang WC, Freemark K (1995) The use of plants for environmental monitoring and assessment. Ecotoxicol Environ Saf 30:289–301

Wang X, Sun C, Gao S et al (2001) Validation of germination rate and root elongation as indicator to assess phytotoxicity with *Cucumis sativus*. Chemosphere 44:1711–1721

Wang T, Lu Y, Shi Y et al (2007) Organochlorine pesticides in soils around Guanting Reservoir, China. Environ Geochem Health 29:491–501

Wei YL, Bao LJ, Wu CC et al (2015) Assessing the effects of urbanization on the environment with soil legacy and current-use insecticides: a case study in the Pearl River Delta, China. Sci Total Environ 514:409–417

Zhang L, Yin DQ, Wu YT et al (2016) Organochlorine pesticides in sediments around Chaohu Lake: concentration levels and vertical distribution. Soil Sediment Contam 25:195–209

Variation in Plant Bioactive Compounds and Antioxidant Activities Under Salt Stress

Wasif Nouman, Muhammad Kamran Qureshi, Mehak Shaheen, and Muhammad Zubair

Abstract Salinity is one of the major yield-limiting abiotic factors. Under stress conditions, reactive oxygen species (ROS) are produced in plants, which cause reduced productivity and yield. These ROS are scavenged by various bioactive compounds like phenolic acids inducing tolerance in plants to mitigate abiotic stress conditions. In this chapter, the authors have discussed the scientific information related to plants' response under salinity stress conditions, the role of osmoprotectants (polyols, glycine betaine, and proline), polyamines, hormonal modulation, and changes in the concentration of bioactive compounds. Plants undergo several physiological and biochemical changes under salinity stress and show variable expression of the bioactive compounds under stress conditions. Osmoprotectants like glycine betaine and proline are also being applied exogenously to induce tolerance in salt-sensitive plants in order to increase plant productivity.

Keywords Antioxidant defense mechanism · Glycine betaine · Hormonal modulation · Osmoprotectants · Polyamines · Polyols

1 Introduction

Climate change is not only causing respiratory or skin diseases but also causing a serious threat to food security affecting crop productivity. The problems of climate change are not only associated with the increase in global or atmospheric temperature causing glacier melting, desertification, and floods, but these also encompass serious issues of drought and salinity. Developing countries are more vulnerable to these threats, which points out the thirst to improve crop productivity under these changing climatic scenarios. Rise in temperature, soil salinity, drought, and

---

W. Nouman (✉) · M. Shaheen · M. Zubair
Department of Forestry and Range Management, Bahauddin Zakariya University, Multan, Pakistan

M. K. Qureshi
Department of Plant Breeding and Genetics, Bahauddin Zakariya University, Multan, Pakistan

S. Vats (ed.), *Biotic and Abiotic Stress Tolerance in Plants*,
https://doi.org/10.1007/978-981-10-9029-5_4

waterlogging terms are collectively used as abiotic stress to plants. Abiotic stress conditions may induce different changes to the plant's morphology, physiology, and biochemistry resulting in reduced crop yield and quality (Wang et al. 2003).

The problem of soil salinity is widespread not only in arid and semiarid areas, but salt-affected soils also occur extensively in the humid and subhumid climate regions, particularly in the coastal regions where the entrance of seawater through estuaries, rivers, and groundwater movement causes large-scale soil and water salinization. This is also a serious problem in the areas receiving saline water. In such areas, saline water is the only source of irrigation. Unfortunately, such poor-quality saline water and high rate of evaporation ultimately cause the addition of salinity in the soil. High salinity and electrical conductivity of soil cause plant cell dehydration, reduced plant growth, and possibly death. It has been reported by Qadir et al. (2014) that we are facing a loss of $27.3 billion per year in agricultural crop production. High salinity level affects the plants in different ways, such as decrease in osmotic potential of the soil solution, severe ion toxicity, and the interaction of salts with mineral nutrient that may result in nutrients imbalances and deficiencies (Akhkha et al. 2011).

2 Salt Stress Impact on Plants: An Overview

Salinity is a form of physiologically dry habitat under which the plants are unable to uptake the substantial amount of water that results in disturbed metabolic and physiological activities of plants, imbalanced nutrient uptake and impairment of photosystem I and II (PSI and PSII), reduction in leaf expansion, plant growth and development, and yield. The reduction in leaf area index is mainly due to water deficit in the root zone, which causes osmotic stress to plants (Munns and Tester 2008). Mild osmotic stress leads rapidly to growth inhibition of leaves and stems, whereas roots may continue to grow and elongate. The presence of soluble salts in the plant root zone disturbs water uptake by plants and utilization of essential minerals. These adverse effects collectively cause reduced yield and produce quality. In most of the cases, sodium chloride (NaCl), sodium carbonate (Na_2CO_3), and calcium chloride ($CaCl_2$) (neutral soluble salts) cause saline soils. Among these, increased accumulation of sodium (Na^+) and chloride (Cl^-) ions mainly leads toward decreased or stunted plant growth and development as these ions affect seed germination and come up with poor seedling vigor that ultimately results in poor yield. The uptake of Na^+ ions causes nutrient deficiency in plants as it lowers down the availability and absorption of other essential nutrients from soil interfering with the cell metabolism. Consequently, it replaces K^+ ions in key enzymatic reactions in the cytosol and organelles (Anschütz et al. 2014; Benito et al. 2014; Shabala and Pottosin 2014). As K^+ is responsible for activating more than 50 enzymes and is an essential element in protein synthesis binding tRNA to the ribosomes, protein synthesis is damaged by replacing K^+ by Na^+ ions (Blaha et al. 2000; Tester and Davenport 2003). These problems have been observed and reported in various agricultural crops and trees (Farooq et al. 2010; Yasmeen et al.

2013, 2014; Nouman et al. 2012, 2014). Munns and Tester (2008) reported that stunted growth and reduced yield under saline conditions are generally due to salt-induced osmotic stress and specific ion toxicities. Moreover, it has been observed by Debez et al. (2011) that the plants growing under saline conditions show decreased levels of natural osmoprotectants and endogenous plant growth regulators.

3 Plants' Response Under Salt Stress

Under saline conditions, plants undergo a number of morphological and physiological changes even at molecular levels. As a defense against abiotic stress, plants increase osmoprotectants' production to mitigate osmotic and oxidative stress. These osmoprotectants regulate osmotic adjustment, mitigate the impact of reactive oxygen species (ROS), and stabilize proteins and other enzymes (Le and McQueen-Mason 2006; Galvani 2007; Ashraf and Foolad 2007) (Fig. 1). The following is a brief note on how the plants go through physiological and biochemical changes under salinity stress.

3.1 Physiological Changes

Crop yield is mostly dependent on its photosynthetic ability, a main factor responsible for decreased plant growth and crop yield. Tolerant crop varieties are screened

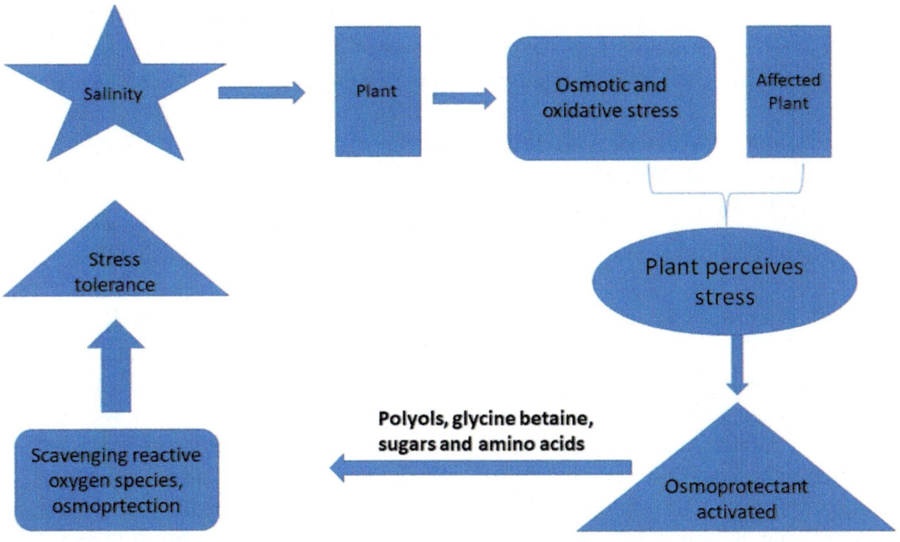

Fig. 1 General scheme of salinity stress tolerance in plants. (Modified from Singh et al. 2015)

based on the capabilities of the strong photosynthetic system under saline conditions and efficacy of plants to exclude or compartmentalize toxic ions. In salt-tolerant plants, it has been observed that leaf surface area, leaf area index, and carbon dioxide assimilation have negative correlation with salinity. The damage to photosynthetic system has been observed in different plants associated with the abundance of chlorophyll contents. Hanaa et al. (2008) reported a salient decrease in chlorophyll *a* and *b* contents in wheat plants growing under saline conditions. The researchers argued that the decrease in chlorophyll contents damages the photosynthetic system which might be attributed to higher Na^+ ions or reduced magnesium ions (Mg^{2+}) in soil or irrigated water as magnesium is an important element in photosynthesis serving as a precursor (Rubio et al. 1995). Moreover, Yasmeen et al. (2013) reported that salinity increases the chances of Na^+ and Cl^- ions accumulation in chloroplasts that is often associated with reduced photosynthetic electron transport activities. In higher plants, salt stress inhibits PSII activity (Munns and Tester 2008). It can be concluded here that salt stress induces damage to photosynthetic system, which results into decreased biomass production and crop yield. Hence, plant's tolerance to saline conditions depends on how the photosynthetic systems may be protected from osmotic and toxic effects of salt stress.

3.2 *Biochemical Changes*

Salinity induces osmotic and ionic imbalances that impose oxidative stress in plants like enhanced generation of ROS and protein denaturation (Mittler 2002; Gill and Tuteja 2010a). So, biomolecules including proteins, DNA, phenolic acids, flavonoids (flavones, flavonols), and hydroxycinnamic acids are damaged by ROS. When the plants face such abiotic stress over a longer period, it may result into peroxidation of membrane lipids of plasmalemma and other cellular organelles causing cell death. Plants have developed various biochemical mechanisms as a tolerance to salinity. These mechanisms include (a) antioxidant defense mechanism, (b) synthesis of osmoprotectants, (c) polyamines synthesis, (d) nitric oxide (NO) generation, and (e) hormonal modulation.

3.2.1 Antioxidant Defense Mechanism

Salinity triggers oxidative burst in plants at a cellular level. The oxidative burst results because under stress condition, oxygen (O_2) molecule acts as an acceptor of electron, which results in the accumulation of ROS in subcellular compartments, especially mitochondria and chloroplast. ROS comprises superoxide radical ($O_2^{\bullet-}$), hydroxyl radical ($^{\bullet}OH$), singlet oxygen ($^1O_2$), and hydrogen peroxide (H_2O_2). These are oxidizing compounds and have the ability to disrupt cell integrity by damaging lipid, protein, and DNA as stated earlier (Quiles and Lopez 2004).

Salt stress triggers the closure of stomata resulting in the reduction of cellular CO_2 concentration. This phenomenon leads to the over-reduction of ferredoxin resulting in the increase of superoxide radicals ($O_2^{\bullet -}$) production by the transfer of electrons from PSI to O_2. Additionally, increase in the photorespiration and other reactions in the cell lead to the overproduction of H_2O_2 and $^{\bullet}OH$ radical (Azevedo et al. 2008). Here, cellular antioxidant metabolism plays a crucial role in ROS detoxification induced by salt stress.

The antioxidant system consists of enzymatic and nonenzymatic components. Some of the antioxidant enzymes, which are directly involved in ROS detoxification induced by salinity, are catalase (CAT), superoxide dismutase (SOD), ascorbate peroxidase (APX), glutathione peroxidase (GPX), and glutathione reductase (GR). The main nonenzymatic antioxidants that are involved in salinity-induced ROS detoxification are ascorbate, glutathione, and flavonoid compounds (van Oosten et al. 2013; Begara-Morales et al. 2014). The expression of these antioxidant compounds varies with germplasms, species, geographical distribution, climatic factors, and seasons. Moreover, Munns and Tester (2008) reported that plants express varying antioxidant activities under saline conditions and such variation might be attributed to the genotypic difference in the expression of these enzymes and the degree of closure of stomata, which modify the rate of CO_2 fixation and avoid photo-inhibition.

It has been previously reported in the literature that plants develop many physiological and biochemical adaptations to adjust themselves under saline conditions. Osmotic adjustments, ionic compartmentalization, variation in K^+/Na^+ ratio, changes in evapotranspiration rate by reducing leaf size, variation in photosynthetic pigments (chlorophyll *a* and *b*, carotenoids, xanthophylls), and stimulation of plant growth regulators, phenolic acids, flavonoids, etc. are salient plant physiological and biochemical adaptations to cope under saline conditions (Sairam and Tyagi 2004). It is important to mention here that increase in antioxidant activities under saline conditions is extremely variable among plants as mentioned earlier. Even different cultivars of same species exhibit variations in the antioxidant activities (Chaitanya et al. 2002). Such response depends on the species, the developmental stage, and the metabolic state of the plant, as well as the duration and intensity of the stress.

3.2.2 Osmoprotectants

Under saline conditions, the plants survive by maintaining their internal water potential lesser than soil water potential. This is meant for ensuring water uptake from the soil (Tester and Davenport 2003). The plants synthesize compatible metabolic solutes to accommodate a cellular ionic balance (Zhifang and Loescher 2003). Osmolytes are a group of organic compounds that are diverse in nature, which are polar, uncharged, and soluble and are not involved in the cellular metabolism. These osmolytes include glycine betaine, proline, and polyols (Bohnert and Jensen 1996; Nounjan et al. 2012; Tahir et al. 2012; Saxena et al. 2013). These osmolytes are chemically diverse in nature and are produced and accumulated in varying

Table 1 Group of osmoprotectants

Group	Compounds		Role in plants	References
Ammonium compound	Polyamines	Putrescine, spermidine, spermine	Detoxify ROS, and improve seed germination, flower initiation, fruit development and maturity	Ashraf and Harris (2004), Ashraf and Foolad (2007), Vinocur and Altman (2005), Groppa and Benavides (2008), Flowers and Colmer (2008), Gill and Tuteja (2010b), and Koyro et al. (2012)
	Betaines	Glycine betaines (GB), β-alanine betaine, proline betaine, choline-*O*-sulfate, dimethyl sulphoniopropionate, hydroxyproline betaine, and pipecolate betaine	Osmotic adjustments	
Sugars and sugar alcohols	Carbohydrate sugars	Fructan, trehalose	Osmotic adjustments, protein stabilization, scavenging ROS, regulation of carbohydrate metabolism	Pilon-Smits et al. (1995), Williamson et al. (2002), Vinocur and Altman (2005), Koyro et al. (2012), Kaya et al. (2013), and Peshev et al. (2013)
	Sugar alcohols	Sorbitol, mannitol, inositol	Osmotic adjustment, limits water loss through transpiration	
Amino acids	Proline, ectoine		Osmotic adjustments, protein stabilization, scavenging ROS, influences cell proliferation and cell death	Bernard et al. (1993) and Kaya et al. (2013)

quantity in various plant species like *Sorghum bicolor*, *Helianthus annuus*, *Eleusine coracana*, *Oryza sativa*, *Medicago sativa*, etc. (Agastian et al. 2000; Ashraf and Harris 2004; Saxena et al. 2013). The basic function of these osmoprotectants is to maintain the cellular structure and osmotic balance through the continuous influx of water (Hasegawa et al. 2000). Osmoprotectants are divided into various groups including ammonium compounds (polyamines and betaines), sugar-related (trehalose, fructans, polyols), sugar alcohols (sorbitol, mannitol, inositol), and amino acids (proline, ectoine) (Table 1 and Fig. 2).

These osmoprotectants play different roles in plants inducing salinity tolerance by scavenging ROS, through osmotic adjustment, protein stabilization, etc. As mentioned earlier and provided in Table 1, polyamines (putrescine, spermidine, spermine) and betaines (glycinebetaines, β-alanine betaine, proline betaine, choline-*O*-sulfate, dimethyl sulphoniopropionate, hydroxyproline betaine, and

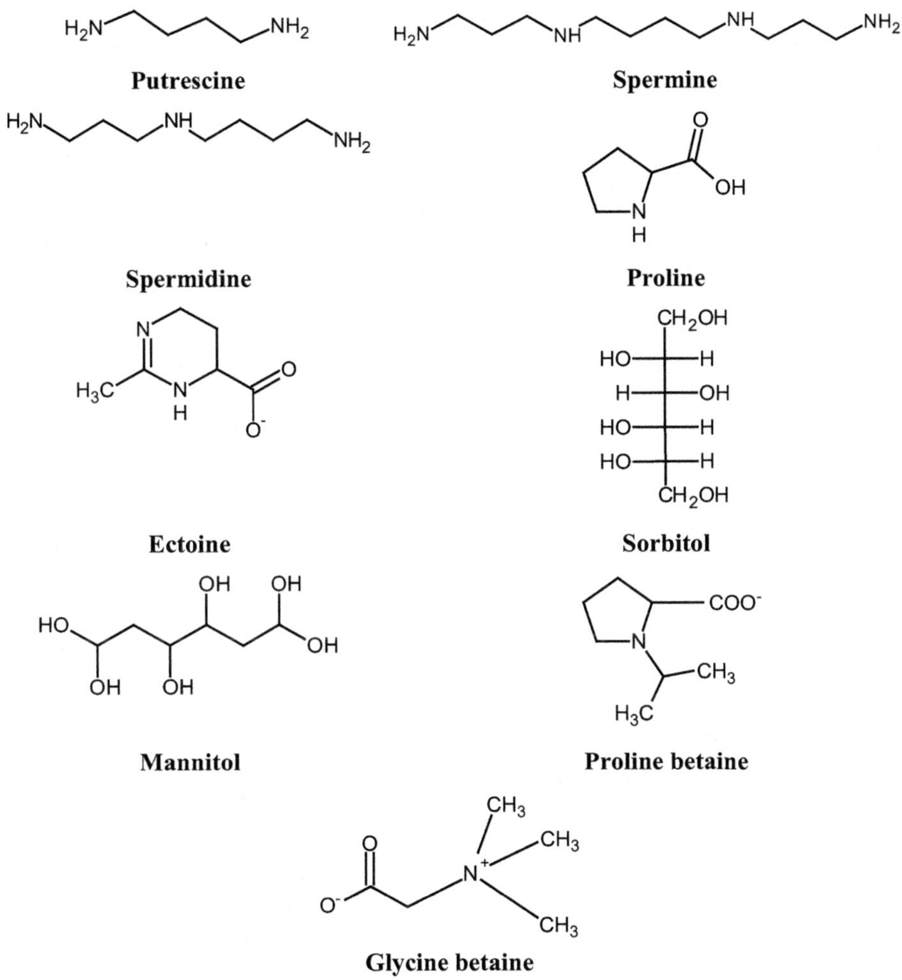

Fig. 2 Chemical structures of a few osmoprotectants which detoxify reactive oxygen species inducing abiotic stress to plants

pipecolate betaine) are grouped into ammonium compounds, which are responsible for detoxifying ROS and osmotic adjustment. Through these processes, the plants are able to mitigate salinity stress and improve seed germination, flower initiation, and fruit development and maturity (Ashraf and Foolad 2007; Groppa and Benavides 2008; Flowers and Colmer 2008; Koyro et al. 2012). Moreover, sugars, sugar alcohols, proline, and ectoine are also responsible osmoprotectants playing important role in protein stabilization, scavenging ROS, etc. (Table 1).

Under abiotic stress, such as salinity, the concentration of proline shows varying expressions in terms of increase or decrease in its concentration. It is, therefore, an accepted parameter for measuring salt stress in plants. Accumulated proline at the intracellular level induces salinity tolerance in plants by initiating antioxidant defense mechanism. Moreover, proline also acts as a source of nitrogen during recovery process, thus improving the growth during stress (Hoque et al. 2008; Ahmed et al. 2010). Proline has been reported as an active compound, which plays a key role in membrane stability under saline conditions, mitigating the impact of NaCl (Mansour 1998). A study was conducted to investigate salinity tolerance (600 mMNaCl stress) in *Arabidopsis thaliana*, and an increase in its proline contents was observed when antisense proline dehydrogenase cDNA was introduced to this plant (Nanjo et al. 2003). Increase in proline content has also been reported in salt-tolerant genotypes of rice and *Sorghum*, while the researchers reported that the increase in proline contents cannot be considered as salt tolerance as it is observed as the result of salt injury (Lutts et al. 1999; De Lacerda et al. 2003). Working on salt-tolerant and salt-sensitive rice genotypes, Lutts et al. (1999) reported more proline contents n salt-sensitive rice genotypes in comparison with salt-tolerant rice genotype. So, more research work is required to investigate the response of proline contents in plants grown under saline conditions (Parvaiz and Satyawati 2008). Increase in proline contents under saline conditions have been observed in *M. sativa*, *Nicotiana tabacum*, *O. sativa*, and *S. bicolor* (Petrusa and Winicov 1997; Lutts et al. 1999; Hong et al. 2000), while a decrease in these contents has been reported in *Solanum lycopersicum* by Aziz et al. (1998) (Table 2). Agastian et al. (2000) conducted a study to evaluate the growth performance and study the variation in different compounds responsible for inducing salinity tolerance in plants. The researchers selected three mulberry (*Morus alba*) accessions, i.e., BC2-59, S-30, and M-5, and cultivated these at five salinity levels (1, 2, 4, 8, and 12 mS cm^{-1}) with one control (no salinity). These salinity levels were prepared by mixing NaCl, Na_2SO_4, and $CaCl_2$ salts and applied to mulberry plants. It was observed that the accumulation of soluble proteins, free amino acids, and soluble sugars increased up to 4 mS cm^{-1} salinity level, while a decrease in these contents was recorded when the plants were subjected to $\geq$8 mS cm^{-1} salinity stress.

Glycine betaine (GB) is a nontoxic ammonium compound that is ubiquitously found in microorganism, plants, and animals. It is a neutral osmolyte that acts over a broad range of pH. It increases osmolarity of the cell during salt stress having an important role in stress mitigation. GB protects cell from stress through stabilizing proteins, osmotic adjustment, ROS scavenging, and protecting photosynthetic apparatus from damage under saline conditions (Ashraf and Foolad 2007; Saxena et al. 2013). The accumulation of GB in various crops like spinach, barley, tomato, potato, rice, carrot, and *Sorghum* has been reported by Yang et al. (2003). The increased accumulation of GB has been noted in salt-tolerant plants in comparison with salt-sensitive ones. Similar studies have been conducted on sorghum, wheat, maize, *Haloxylon recurvum*, and mulberry (Colmer et al. 1995; Saneoka et al. 1995; Agastian et al. 2000; Wang and Nil 2000). In another study, *S. bicolor* plants were grown and maintained under nonsaline conditions up to 4 weeks. After 4 weeks, salinity levels were induced in the growing medium by adding NaCl in a stepwise process, i.e., 50 mM during 4–5 weeks, 100 mM during 5–6 weeks, and 150 mM during 6–7 weeks. After the 7th week, glycine betaine contents were determined in

Table 2 Variation in carbohydrates, proteins, and proline contents in different plants under saline stress

Plant	Experimental conditions	Change in primary metabolites	References
Carthamus tinctorius	Salt-sensitive (199952 and 170274) and salt-tolerant (260622 and 305167) accession of safflower were subjected to 70, 140, and 210 mM NaCl salt concentrations	No significant difference in soluble proteins in salt-tolerant and salt-sensitive genotypes while an increased accumulation of free amino acids was recorded in the experimental plant with the increase in salt stress	Ashraf and Fatima (1995)
Helianthus annuus	Forty-five salt-tolerant and salt-sensitive accessions were studied under saline conditions induced by adding @150 meq l^{-1} of $NaCl_2$+ $CaCl_2$ (1:1 ratio) in half Hoagland's nutrient solution	Increase in soluble sugars, soluble carbohydrates, soluble proteins, total free amino acids and proline in salt-tolerant accessions	Ashraf and Tufail (1995)
Medicago sativa	The increase was observed when the plants were treated with 171 mM NaCl stress	A tenfold increase in proline contents	Petrusa and Winicov (1997)
Solanum lycopersicum	The plants were stressed with 0, 100, 175, 250, 300, and 400 mM NaCl concentrations	Decrease in proline contents in salt-tolerant cultivars	Aziz et al. (1998)
Solanum lycopersicum	The plants were stressed with 100 mM NaCl	Increase in soluble sugars and total saccharides while no significant effect was observed on starch contents	Khavarinejad and Mostofi (1998)
Oryza sativa	Salt-resistant (Nona Bokra) and salt-sensitive genotypes (I Kong Pao/IKP) of *Oryza sativa* were treated with 50 and 100 mM NaCl concentration	Salt-sensitive genotype showed higher levels of proline accumulation than salt-resistant genotype	Lutts et al. (1999)
Morus alba	Three mulberry accessions BC2-59, S-30, and M-5 were grown under saline conditions. The salinity levels (0, 1, 2, 4, 8, and 12 mS cm^{-1}) were prepared using a mixture of NaCl, Na_2SO_4, and $CaCl_2$	Increase in soluble proteins, free amino acids, soluble sugars, sucrose, and starch at low salinity level (1–4 mS cm $^{-1}$) and decrease were observed in these contents at ≥8 ms cm^{-1}	Agastian et al. (2000)
Nicotiana tabacum	Wild and transgenic tobacco seeds were germinated over agar medium containing 0, 150, 200, 250, and 300 mM NaCl concentrations	Transgenic plants showed twofold proline accumulation in comparison to wild-type tobacco plants up to 200 mM NaCl concentration	Hong et al. (2000)
Bruguiera parviflora	The plants were treated with 100, 200, and 400 mm NaCl levels, and the parameters were studied 7, 14, 30, and 45 days after treatment	Decrease in starch and increase in both reducing proteins and nonreducing sugars	Parida et al. (2002)

(continued)

Table 2 (continued)

Plant	Experimental conditions	Change in primary metabolites	References
Sorghum bicolor	*Sorghum* plants were maintained under normal (nonsaline) conditions up to 4 weeks. After 4 weeks, salinity levels were induced in the growing medium by adding NaCl salt in a stepwise manner, i.e., 50 mM during 4–5 week, 100 mM during 5–6 week, and 150 mM during 6–7 week	An increase in glycine betaine was recorded with the increase in NaCl concentration in growing media	Yang et al. (2003)
Sorghum bicolor	Salt-tolerant (CSF20) and salt-sensitive (CSF18) genotypes were grown under 0 and 100 mMNaCl levels	Increase in proline contents and soluble carbohydrates was observed in salt-sensitive genotype with increase in salt stress	De Lacerda et al. (2003)
Prosopis alba	*Prosopis alba* seedlings of 17 days old were subjected to 300 and 600 mmol L^{-1} NaCl treatments which were achieved by adding 50 mmol L $^{-1}$ NaCl salt after every 24 h	Under saline conditions, an increase in glycine betaine content was recorded in both leaves and roots, while increase in total soluble carbohydrates was recorded only in *Prosopis alba* roots, while proline accumulation was not significantly affected by salinity	Meloni et al. (2004)
Olea europaea	One-year-old olive seedlings were grown under 4, 8, and 12 dS m^{-1}) salinity levels which were prepared with NaCl and induced after mixing with half strength Hoagland's solution	Increase in proline contents up to 8 dS m^{-1} was observed after which the contents decrease in olive leaves	Demiral et al. (2011)
Eleusine coracana	*Eleusine coracana* seedlings were subjected to three salts (copper sulfate, cadmium chloride, and zinc sulfate) independently at 100, 150, and 200 μM concentrations. The stress was induced for a period of 3 days	Degradation of stress proteins was observed	Rani (2011)

Sorghum plants, and Yang et al. (2003) reported an increase in the glycine betaine contents with increase in NaCl stress to plants (Table 2).

Polyols are cellular compounds having multiple functional hydroxyl groups for various organic reactions. Sugar alcohols are a group of polyols acting as compatible low molecular weight solutes responsible for ROS scavenging (Ashraf and Foolad 2007). So, the accumulation of these polyols metabolites has a direct correlation with salt stress tolerance.

3.2.3 Polyamines

Polyamines are broadly distributed throughout the kingdom Planta. These are small aliphatic molecules with low molecular weight and are polycationic in nature. Polyamines are involved in somatic embryogenesis, seed germination, plant growth and development, morphogenesis, cell differentiation, bud and flower initiation and development, fruit formation, and leaf senescence (Gupta et al. 2013). In addition to these functions, polyamines also play an important role in plant tolerance against abiotic stress including salt stress. Under salinity stress, the level of endogenous level of polyamine increases, which is associated with the synthesis of solutes for osmotic regulation, reducing ROS production, maintaining membrane integrity, regulating expression of genes, and controlling Na^+ and Cl^- ion accumulation in plant cell and organs (Roychoudhury et al. 2011). The role of polyamine in inducing salinity tolerance to plants has been previously witnessed in various studies. The researchers reported that these compounds play a key role in cell membrane stability, scavenging ROS, modulating ion channels, and stimulation of ATP synthesis (Hartung et al. 2002; Nuttall et al. 2003; Shi and Sheng 2005; Yang et al. 2007). Li et al. (2016) studied the impact of salt stress on two cultivars (salt sensitive, cv. Z081, and salt tolerant, cv. Z057) of zoysia grass (*Zoysia japonica* Steud). The plants of both cultivars were subjected to 200 mM salt stress and were exogenously treated with spermidine treatments for 8 days. The researchers reported an increase in polyamine compounds (spermidine, putrescine, and spermine) was recorded in both cultivars during early stages of plant development.

3.2.4 Nitric Oxide (NO) Generation

NO is gaseous molecule that is involved in plant growth and development like seed germination, root growth, flowering, stomata closure, and respiration. It also acts as signaling molecule under stress and interacts with ROS signaling pathway (Zhao et al. 2009). It triggers the expression of genes involved in redox homeostasis and is involved in the activation of enzymatic antioxidants such as CAT, GPX, APX, SOD, and GR. Moreover, it reacts with lipids avoiding their oxidation (Bajgu 2014). NO is involved in plant response to salinity stress. Peroxisomes, in addition of ROS generation, are the main source of NO generation under saline conditions that also regulates the plasma membrane H^+/ATPase and Na^+/K^+ ratios (Corpas et al. 2009).

3.2.5 Hormonal Modulation

Hormones also play an important role in inducing salt tolerance improving plant growth and development. Among these plant hormones or plant growth regulators, abscisic acid (ABA) is one of the phytohormones, which is upregulated due to water deficit around the root zone. Since salinity triggers drought and osmotic stress, it increased the level of ABA in shoot and root tissues. ABA, in turn, reduces the effect of salt stress on plant growth and photosynthesis and assimilates translocation (Cabot et al. 2009). The interaction between ABA and salinity tolerance might be attributed partially to the accumulation of proline, sugars, and K^+ and Ca^{+2} ions in root vacuoles, which reduces the uptake of Na^+ and Cl^- ions from the soil (Gurmani et al. 2011). Moreover, it also acts as signal molecule and triggers the expression of salinity and drought responsive genes (Fukuda and Tanaka 2011).

Other phytohormones, which also have role in plant response to abiotic stresses, are salicylic acid (SA) and brassinosteroids (BR) (Fragnire et al. 2011). Both SA and BR are also involved in normal growth and development of plant organs. For example, endogenous level of SA increases under salt stress with the increase in the activity of SA biosynthetic enzymes. Moreover, it increases salt tolerance by preventing K^+ efflux induced by salt stress and is involved in the restoration of membrane potential (Jayakannan et al. 2013).

Similarly, BR has a negative impact on salinity stress, as it enhances the activity of enzymatic (GPX, POX, APX, SOD) and nonenzymatic (ascorbate, tocopherol, and glutathione) antioxidants increasing salinity tolerance in plants by reducing ROS generation.

4 ROS Generation and Detoxification

As discussed earlier, ROS are highly reactive forms of molecular oxygen including hydroxyl radical, superoxide, singlet oxygen, and hydrogen peroxide, among which hydrogen peroxide is the most stable reactive form of molecular oxygen (Moller et al. 2007; Shapiguzov et al. 2012). Different studies have identified variation in plant bioactive molecules triggered by salt stress and their possible role in tolerance against salinity. ROS, toxic to cellular integrity by causing oxidative stress, induce antioxidant defense machinery. ROS induced by salt stress triggers the selective regulation of antioxidant enzymes such as SOD and ASC/GSH (glutathione-ascorbate) cycle at the subcellular compartments like mitochondria, peroxisomes, and chloroplasts (Mittova et al. 2003, 2004). An increase in antioxidant enzymes like CAT, GR, SOD, and GSH was also observed in salt-tolerant species (Acosta-Motos et al. 2015). Salt stress triggers the generation of ROS in mitochondria, which initiates the activation of Mn-SOD and Cu/Zn-SOD enzymes in salt-tolerant plants. Additionally, a decrease in peroxisomal ROS especially H_2O_2 was observed when plants were under salt stress. The decrease in ROS level in root peroxisomes might be due to increase in the activities of CAT, SOD, and APX (Mittova et al. 2004). Moreover, increased levels of nonenzymatic antioxidant activities such as glutathione and ascorbate have been observed in salinity-sensitive plants which might be due to the alteration of ASC/DHA (docosahexaenoic acid) ratio in salt-affected plants (Ikbal et al. 2014).

The variation in phytohormones in response to salinity has also been studied and reported in the literature. The response of these hormones depends upon stress condition, growth stage, and plant species. For example, an exogenous application of SA reduces the Na^+ concentration in shoots and roots with the increase in K^+ and Mg^{+2} under salt toxicity. Another phytohormone, BR, is not only involved in plant growth and development; it is also involved in stress tolerance. BR enhances the activity of enzymatic and nonenzymatic antioxidants under salt stress (Ashraf et al. 2010). Similarly, it has been observed that the alleviation of ABA and GA through exogenous application increases plant response to nitrogen fixation, assimilation of ammonium, catabolism of purine, and increase in the activity of antioxidants in saline conditions (Khadri et al. 2006).

Under saline conditions, accumulation of glycine betaine and proline takes place by the modulation of certain salt-responsive genes. These solutes are involved in osmotic adjustments. In addition, glycine betaine interacts with certain enzymatic and protein complexes, which together maintains integrity of membrane structure under salt toxicity. It also improves relative water contents and stomatal conductance (Pruthvi et al. 2014). Likewise, high level of proline causes significant reduction in harmful impacts of salinity on nitrogen fixation in legume crops (Farooq et al. 2017). Polyols, as described earlier, are compounds having hydroxyl groups with multiple functions. Polyols are divided into two major classes as cyclic (pinitol) and acyclic (mannitol). Both pinitol and mannitol are induced during salt toxicity in plant cells. Mannitol functions as a stabilizer of membrane structures, which are sensitive to damage induced by ions or to dehydration that stabilizes certain enzymes to induce tolerance in plants under abiotic stress conditions.

Polyamines are molecules that are involved not only in normal cellular functions of plants but are also involved in stress. Salt stress triggers increase in level of endogenous polyamine in plant cells. This increase in polyamine concentration has a positive role on plant cell in terms of maintaining membrane integrity. Polyamines also regulate gene expression involved in the synthesis of enzymes, which synthesize solutes for osmotic adjustment. These are also involved in reducing ROS production and control Na^+ and Cl^- accumulation in various plant organs (Takahashi and Kakehi 2010). The understanding of the interaction between polyamines and ROS detoxification is complex, although many studies have unfolded these interactions. However, an increased accumulation of polyamine contents have been reported in salt-tolerant plants (Velarde-Buendia et al. 2012; Minocha et al. 2014; Pottosin et al. 2004, 2012). Polyamines detoxify ROS in two different ways, i.e., by scavenging free radicals and activating antioxidant enzymes and by promoting ROS production through polyamine catabolism in the apoplast (Gupta et al. 2013; Campestre et al. 2011). To understand these processes including polyamine biosynthesis and variation in salt stress, impacts on polyamine synthesis and accumulation should be studied as it varies from species to species and even among cultivars of the same species.

5 Variation in the Bioactive Compounds Under Salt Stress

Plants, exposed to salt stress, show many adverse effects as stated earlier. Plants behave differently in response to salinity and have developed different adaptation mechanisms. Such adaptation mechanism depends chiefly on the innate plant tolerance and severity of salt stress (Munns and Tester 2008). One of adaptation processes is the increase or decrease of bioactive compounds associated with salinity levels. Bioactive compounds are the secondary metabolites in plants associated with multiple roles including pharmacological, toxicological effects and antioxidant potential as a defense mechanism (Bernhoft 2010). However, their production is greatly affected when plants are introduced to abiotic stress, such as salinity (Petridis et al. 2012). Production of these bioactive compounds responds differently under varying saline conditions and also varies from plant to plant.

Various studies have been conducted on the response of bioactive compounds of different plants under saline conditions. Bioactive compounds are significantly affected under salt stress as mentioned above. For example, a research conducted on *Portulaca oleracea* L. (purslane) in Malaysia showed irregular response of bioactive compounds at different salinity levels. Bioactive compounds measured were in terms of total phenolic contents (TFC), total carotenoid contents, and total flavonoid contents (TFC) in salt-affected and unaffected plants. Result comparisons showed 124–331% increase of TPC, 164–387% in TCC, and 14–180% in TFC in salt-affected plants as compared to unaffected plants (Alam et al. 2015). This is not only the case of purslane because other studies also showed the similar results. As salinity increases, many plants such as artichoke, rice, strawberry, lettuce, sweet marjoram, etc. activated their defense mechanism to tolerate stress (Keutgen and Pawelzik 2008; Kim et al. 2008; Chutipaijit et al. 2009; Yuan et al. 2010; Baatour et al. 2012; Rezazadeh et al. 2012). Zrig et al. (2011) reported variation in phenolic acids under low and moderate salinity levels in different genotypes of almond. The researchers reported an increase in the phenolic acids at low salinity levels, while no change was observed in leaf phenolics at moderate saline conditions. ROS produced by plants in stress condition changes the metabolism of the plant by oxidation of lipids and proteins (Navarro et al. 2006). So, this increase in concentration of bioactive compounds such as flavonoid, lignin, glutathione, coumarin, proline, and α-tocopherols (Gill and Tuteja 2010a) is a defense mechanism against oxidative stress caused by accumulation of ROS in response to moderate to high salt stress (Wahid and Ghazanfar 2006). Moreover, phenolics play a vital role in reducing the oxidative damage to plants, while the increase of ROS in stress conditions triggers the production of carotenoids, which have scavenging potential for ROS (Rmiki et al. 1999; de Pascale et al. 2001; Netto 2001). Therefore, the increase of phenolic contents, i.e., anthocyanin and flavones in sugarcane varieties, is also considered as a defense mechanism of plants against oxidative stress caused by salinity. It has been observed that soluble phenolics, anthocyanins, and flavones were three times more in sugarcane clones under salt stress as compared to normal condition (Wahid and Ghazanfar 2006).

Similarly, a research conducted on romaine lettuce under NaCl-irrigated water showed an increased amount of bioactive compounds especially carotenoids. During long-term salt stress, among carotenoids, β-carotene content increased to 37%, while lutein contents increased to 80% (Kim et al. 2008). Such variation in the bioactive compounds under saline conditions might be attributed to plants' defense mechanism. Furthermore, release of ABA in response of salt stress also raises the concentration of carotenoids in lettuce (Jia et al. 2002; Zhao et al. 2005; Chen et al. 2006). Similarly, in tomato and pepper, the increase concentration of carotenoids, β-carotene, and lycopene was noted at moderate salinity (De Pascale et al. 2001; Navarro et al. 2006).

By reviewing the literature, it was observed that not all the plants performed well at low and moderate salinity, but high stress conditions are also beneficial to increase the nutritional quality of some plants. Plants such as melon, beetroot, broccoli, and radish (Zapata et al. 2004) require high salt stress for enhancing their bioactive compound activity. For example, in radish and broccoli, a negative behavior of glucosinolates content was observed during germination at low salt stress, while high salt stress condition increased the amount of indole glucosinolates, 4-OH glucobrassicin, glucobrassicin, and 4-methoxyglucobrassicin (Berenguer et al. 2008, 2009; Yuan et al. 2010). This might be attributed to the osmotic adjustment of ions such as Na^+ and Cl^- which become efficient at low water potential or might be due to the genetic makeup of seeds to respond well in high salt stress (Yuan et al. 2010).

Moreover, the increase of bioactive compounds critically depends on the type of species, such as salinity has negative influence on phenolic contents of lettuce and broccoli, while in buckwheat, pepper, and maize, salinity has a positive influence on phenolic contents (Navarro et al. 2006; Lim et al. 2012). The notable increase of phenolic contents (isoorientin, orientin, rutin, and vitexin) and carotenoid in buckwheat was observed from low to high salt stress. Treated sprouts with low salinity showed 57% increase of phenolic content as compared to untreated sprouts, whereas at moderate and high salinity, phenolic contents were increased to 121% and 153%, respectively (Lim et al. 2012). Similar studies have been conducted on quinoa plants growing under saline conditions. Ismail et al. (2016) conducted a study on *Chenopodium quinoa* treated with 400 mM NaCl stress. The researchers studied chlorogenic, caffeic, *p*-coumaric, and rosmarinic acids. Beside these compounds, the stress tolerance also influenced the amount of rutin, feruloyl malate, and *p*-coumaroylmalate (Table 3). They reported an increase in rutin contents and chlorogenic acids when the quinoa plants (cv. Utusaya) were treated 400 mM NaCl, while the other quinoa cultivar, i.e., Titicaca, expressed the increase in all selected phenolic acids except feruloyl malate and *p*-coumaroylmalate. It is worth mentioning here that Titicaca cv. showed 25-fold increase in its rutin contents when

Table 3 Variation in total phenolics and phenolic acids in different plants under saline stress

Plant	Experimental conditions	Change in phenolic compounds	References
Morus alba	Three mulberry accessions BC2-59, S-30, and M-5 were grown under saline conditions. The salinity levels (0, 1, 2, 4, 8, and 12 mS cm^{-1}) were prepared using a mixture of NaCl, Na_2SO_4, and $CaCl_2$	Increase in phenolics	Agastian et al. (2000)
Bruguiera parviflora	The plants were treated with 100, 200, and 400 mm NaCl levels, and the parameters were studied 7, 14, 30, and 45 days after treatment	Increase in polyphenols	Parida et al. (2002)
Capsicum annuum	Pepper fruits grown at 0, 15, and 30 mM NaCl levels were tested for their phenolic compounds	Increase in total phenolics	Navarro et al. (2006)
Allium cepa	The plants were treated in three treatment groups, i.e., irrigated water (300 ppm salts), sea water (2500 and 5000 ppm salts), and irrigation water (300, 2500, 5000 ppm salts) plus alpha-tocopherols combined with KH_2PO_4	Decrease in the total phenolics and total flavonoids was observed with the increase in salts concentration, while a twofold increase was recorded when these plants were sprayed with alpha-tocopherol and alpha-tocopherol $+KH_2PO_4$ under saline conditions	Mohamed and Amina (2008)
Fragaria vesca	Salt-sensitive (Elsanta) and less salt-sensitive (Korona) cultivars were grown at three NaCl/L concentrated media 0 (EC: 0.0013 dS m^{-1}), 40 (EC: 3.9 dS m^{-1}), and 80 mmol (EC: 7.5 dS m^{-1}) concentrations	Increase in total phenolics and anthocyanins	Keutgen and Pawelzik (2008)
Pisum sativum	Nine genotypes of *Pisum sativum* were stressed with 0, 40, 80, and 120 mM NaCl levels	Increase in total phenolics	Noreen and Ashraf (2009)
Lactuca sativa	Two genotypes, i.e., salt sensitive and salt tolerant (Romaine and Verte, respectively), were treated with 0 and 100 mM NaCl levels	Increase in phenolic acids, total phenolics, and flavonoids was recorded in both the genotypes	Mahmoudi et al. (2010)
Raphanus raphanistrum	Radish sprouts were germinated under 0, 10, 50, and 100 mM of NaCl levels, and the sprout samples were collected on 3rd, 5th, and 7th day after sowing for analysis	A decrease in total glucosinolates was recorded when the stress was prolonged from 3 to 7 days, while an increase in these contents was recorded when these sprouts were subjected to 10 and 100 mM NaCl concentration	Yuan et al. (2010)

(continued)

Table 3 (continued)

Plant	Experimental conditions	Change in phenolic compounds	References
Olea europaea	One-year-old olive seedlings were grown under 4, 8, and 12 dS m^{-1}) salinity levels which were prepared with NaCl and induced after mixing with half strength Hoagland's solution	Increase in total phenolic contents up to 8 dS m^{-1}	Demiral et al. (2011)
Moringa oleifera	*Moringa* seeds were sown at 2, 4, 8, and 12 dS m^{-1}. The salinity levels were prepared and maintained with NaCl and induced by mixing with half strength Hoagland's solution	Increase in total phenolics up to 8 dS m^{-1}	Nouman et al. (2012)
Olea europaea	Four olive cultivars (Zard, Iran; Ascolana, Italy; Arbequina, Spain; Koroneiki, Greece) were treated with 0, 75, and 125 mM NaCl contents mixed with half strength Hoagland's solution	Increase in total phenolic contents present in leaves was recorded in all selected olive cultivars, while their roots showed maximum total phenolic contents at 75 mM NaCl stress	Petridis et al. (2012)
Phaseolus vulgaris	Pot and field experiments were conducted to determine the variation in the presence of phenolic compounds. The plants were grown in pot under varying salinity levels from 1.5 to 20.6 dS m^{-1}, while under field conditions the plants were grown from 1.3 to 29 dS m^{-1} salinity levels	An increase in chlorogenic and caffeic acids was recorded at moderate and higher salinity levels	Rezazadeh et al. (2012)
Moringa oleifera	*Moringa* seeds were treated with pre-sowing treatments (moringa leaf extract, MLE and hydropriming, using water). Later the seeds were sown at four salinity levels, i.e., 3, 6, 10, 14 dS m^{-1}	Increase in total phenolics up to 10 dS $m^{-1}$was recorded when the seeds were sown after treating with hydropriming and MLE priming treatments for 12 h	Nouman et al. (2014)
Chenopodium quinoa	Two quinoa cultivars, i.e., Utusaya (salt tolerant) and Titicaca (salt neutral) were grown at 400 mM NaCl level. The salinity stress was induced after 2 weeks of germination and lasted for 5 weeks, till the day of harvest	Increase in rutin contents	Ismail et al. (2016)
Oryza sativa (BC15TB,	Germinated seeds of six rice varieties (OM4900, X7KD, OM8108, BC15TB, BT, and	Increase in vanillin and protocatechuic acid	Minh et al. (2016)

(continued)

Table 3 (continued)

Plant	Experimental conditions	Change in phenolic compounds	References
salt-tolerant cultivar)	Q5) were grown under 0, 5, and 10 NaCl levels		
Oryza sativa (X7KD, salt-sensitive cultivar)		Decrease in vanillin and protocatechuic acid	
Brassica napus var. *oleifera* Del.	Seeds were incubated in plastic trays containing solutions with 0, 25, 50, 100, 200 mM NaCl	Increase in ferulic acid	Falcinelli et al. (2017)

the plants were stressed with 400 mM NaCl stress in comparison to control. Moreover, lack of correlation among rutin concentration and potassium and hydrogen ion fluxes was recorded. These findings suggest that rutin might be responsible for scavenging hydroxyl radical that is formed under saline conditions. Minh et al. (2016) conducted a trial studying the changes in phenolic acids in rice under the application of salt stress. They reported an increase in vanillin and protocatechuic acid in salt-tolerant rice cultivars (BC15TB), while salt-sensitive rice genotypes (X7KD) showed significant decrease in these phenolic acids, while ferulic and *p*-coumaric acids were detected only in salt-tolerant rice cultivars (Table 3). Increased expressions of ferulic acids have also been observed in *Brassica napus* var. *oleifera* Del. under saline conditions (Falcinelli et al. 2017). Beside ferulic acid, the researchers reported that sinapic acid was more abundantly found in *Brassica* sprouts under saline conditions in comparison with ferulic acid. Keeping in view the above studies, it is important to focus on these phenolic acids which might induce salinity tolerance in plants. Many other studies have been carried out on exploring the changes in phenolic acids in various plants like rosemary, black cumin, mint, and basil under saline conditions (Bourgou et al. 2010; Oueslati et al. 2010; Kiarostami et al. 2010; Zahedi et al. 2011; Mehrizi et al. 2012). Salt-sensitive plants showed little variation in expressing phenolic acids. Keutgen and Pawelzik (2008) cultivated strawberry plants under saline conditions and reported only 16% increase in total phenolic contents at 80 mM NaCl salinity level, while another salt-sensitive cultivar of strawberry expressed 14–23% increase in total phenolic contents. These studies suggest that salinity stress stimulates the antioxidant system of plants. Phenolic compounds are responsible for antioxidant system, which are increased under saline conditions (Table 3). As mentioned in Table 3, rosemary plants are good source of phenolics, which serve as antioxidant compounds. An increase in total phenolics was observed in rosemary plants when these were stressed with saline stress. An increase of 134% in total phenolics was observed in rosemary leaves when salinity was induced from 50 to 100 mM NaCl, while a decrease of 17.78% was recorded when NaCl stress was increased from 100 to 150 mM (Kiarostami et al. 2010). In another study, salinity stress in rosemary plants was mitigated by the application of copper as it plays an important role in synthesizing phenolic compounds which are responsible for antioxidant activities (Dicko et al. 2006; Mehrizi et al. 2012). An increase of 12–37% total phenolics was observed in copper-treated NaCl-stressed plant in comparison with those which were not treated with copper. Mitigation of salinity-induced oxidative stress and impacts of metals has previously been reported for other plants by Drzewiecka et al. (2011).

6 Conclusion

Salinity is a major factor that limits plant growth and its productivity. Under stress conditions, the plants undergo several physiological changes which alter the presence and abundance of bioactive compounds. These compounds like phenolic acids and osmoprotectants are responsible for detoxifying reactive oxygen species and provide osmoprotection. The researchers are using a few of these osmoprotectants like proline and glycine betaine to induce salinity tolerance in plants, and the results are encouraging. The studies on the changes in the presence and abundance of osmoprotectants and phenolic acids open new research avenues. However, a lot of research work is needed to explore these bioactive compounds and their mechanism of action during stress tolerance.

References

Acosta-Motos JR, Díaz-Vivancos P, Álvarez S et al (2015) Physiological and biochemical mechanisms of the ornamental *Eugenia myrtifolia* L. plants for coping with NaCl stress and recovery. Planta 242:829–846

Agastian P, Kingsley SJ, Vivekanandan M (2000) Effect of salinity on photosynthesis and biochemical characteristics in mulberry genotypes. Photosynthetica 38:287–290

Ahmed BC, Rouina BB, Sensoy S et al (2010) Exogenous proline effects on photosynthetic performance and antioxidant defense system of young olive tree. J Agric Food Chem 58:4216–4222

Akhkha A, Boutraa T, Alhejely A (2011) The rates of photosynthesis, chlorophyll content, dark respiration, proline and abscicic acid (ABA) in wheat (*Triticum durum*) under water deficit conditions. Int J Agric Biol 13:215–221

Alam AM, Juraimi AS, Rafii MY et al (2015) Effects of salinity and salinity-induced augmented bioactive compounds in purslane (*Portulaca oleracea* L.) for possible economical use. Food Chem 169:439–447

Anschütz U, Becker D, Shabala S (2014) Going beyond nutrition: regulation of potassium homoeostasis as a common denominator of plant adaptive responses to environment. J Plant Physiol 171:670–687

Ashraf M, Fatima H (1995) Responses of some salt tolerant and salt sensitive lines of safflower (*Carthamus tinctorius* L.) Acta Physiol Plant 17:61–71

Ashraf M, Foolad MR (2007) Roles of glycine betaine and proline in improving plant abiotic stress resistance. Environ Exp Bot 59:206–216

Ashraf M, Harris PJC (2004) Potential biochemical indicators of salinity tolerance in plants. Plant Sci 166:3–16
Ashraf M, Tufail M (1995) Variation in salinity tolerance in sunflower (*Helianthus annuus* L.) J Agron Crop Sci 174:351–362
Ashraf M, Akram NA, Arteca RN et al (2010) The physiological, biochemical and molecular roles of brassinosteroids and salicylic acid in plant processes and salt tolerance. Crit Rev Plant Sci 29:162–190
Azevedo NAD, Gomes-Filho E, Prisco JT (2008) Salinity and oxidative stress. In: Khan NA, Singh S (eds) Abiotic stress and plant responses. I K International, New Delhi, pp 57–82
Aziz A, Martin-Tanguy J, Larher F (1998) Stress-induced changes in polyamine and tyramine levels can regulate proline accumulation in tomato leaf discs treated with sodium chloride. Physiol Plant 104:195–202
Baatour O, Tarchoun I, Nasri N et al (2012) Effect of growth stages on phenolics content and antioxidant activities of shoots in sweet marjoram (*Origanum majorana* L.) varieties under salt stress. Afr J Biotechnol 11:16486–16493
Bajgu A (2014) Nitric oxide: role in plants under abiotic stress. In: Physiological mechanisms and adaptation strategies in plants under changing environment. Springer, New York, pp 137–159
Begara-Morales JC, Sanchez-Calvo B, Chaki M et al (2014) Dual regulation of cytosolic ascorbate peroxidase (APX) by tyrosine nitration and S-nitrosylation. J Exp Bot 65:527–538
Benito B, Haro R, Amtmann A et al (2014) The twins K^+ and Na^+ in plants. J Plant Physiol 171:723–731
Berenguer LC, Martinez-Ballesta MC, Garcia-Viguera C (2008) Leaf water balance mediated by aquaporins under salt stress and associated glucosinolate synthesis in broccoli. Plant Sci 174:321–328
Berenguer LC, Martinez-Ballesta MC, Moreno DA et al (2009) Growing hardier crops for better health: salinity tolerance and the nutritional value of broccoli. J Agric Food Chem 57:572–578
Bernard T, Jebbar M, Rassouli Y, Himdi-Kabbab S, Hamelin J, Blanco C (1993) Ectoine accumulation and osmotic regulation in Brevibacterium linens. J Gen Microbiol 139:129–138
Bernhoft A (2010) Bioactive compounds in plants – benefits and risks for man and animals. The Norwegian Academy of Science and Letters, Oslo
Blaha G, Stelzl U, Spahn CMT et al (2000) Preparation of functional ribosomal complexes and effect of buffer conditions on tRNA positions observed by cryoelectron microscopy. Methods Enzymol 317:292–309
Bohnert HJ, Jensen RG (1996) Strategies for engineering water-stress tolerance in plants. Trend Biotechnol 14:89–97
Bourgou S, Pichette A, Marzouk B, Legault J (2010) Bioactivities of black cumin essential oil and its main terpenes from Tunisia. S Afr J Bot 76:210–216
Cabot C, Sibole JV, Barcelo J, Poschenrieder C (2009) Abscisic acid decreases leaf Na^+ exclusion in salt-treated *Phaseolus vulgaris* L. J Plant Growth Regul 28:187–192
Campestre MP, Bordenave CD, Origone AC, Menéndez AB, Ruiz OA, Rodríguez AA et al (2011) Polyamine catabolism is involved in response to salt stress in soybean hypocotyls. J Plant Physiol 168:1234–1240
Chaitanya KV, Sundar D, Masilamani S et al (2002) Variation in heat stress-induced antioxidant enzyme activities among three mulberry cultivars. Plant Growth Regul 36:175–180
Chen H, Jones AD, Howe GA (2006) Constitutive activation of the jasmonate signaling pathway enhances the production of secondary metabolites in tomato. FEBS Lett 580:2540–2546
Chutipaijit S, Cha-um SK, Sompornpailin K (2009) Differential accumulations of proline and flavonoids in indica rice varieties against salinity. Pak J Bot 41:2497–2506
Colmer TD, Epstein E, Dvorak J (1995) Differential solute regulation in leaf blades of various ages in salt sensitive wheat and a salt-tolerant wheat × *Lophopyrum elongatum* (Host.) A. Love amphiploid. Plant Physiol 108:1715–1724

Corpas FJ, Hayashi M, Mano S et al (2009) Peroxisomes are required for in vivo nitric oxide accumulation in the cytosol following salinity stress of *Arabidopsis* plants. Plant Physiol 151:2083–2094

De Lacerda CF, Cambraia J, Oliva MA et al (2003) Solute accumulation and distribution during shoot and leaf development in two sorghum genotypes under salt stress. Environ Exp Bot 49:107–120

de Pascale S, Maggio A, Fogliano V et al (2001) Irrigation with saline water improves carotenoids content and antioxidant activity of tomato. J Hortic Sci Biotechnol 76:447–453

Debez A, Huchzermeyer B, Abdelly C et al (2011) Current challenges and future opportunities for a sustainable utilization of halophytes. In: Öztürk M, Böer B, Barth HJ, Clüsener-Godt M, Khan M, Breckle SW (eds) Sabkha ecosystems, tasks for vegetation science, vol 46. Springer, Dordrecht, pp 59–77

Demiral MA, Deniz AU, Murat U, Erkan K, Arife AK (2011) Biochemical response of *Olea europaea* cv. Gemlik to short-term salt stress. Turk J Biol 35:433–442

Dicko HM, Gruppen H, Traore AS, Voragen AGJ, Berkel WJHV (2006) Phenolic compounds and related enzymes as determinants of sorghum for food use. Biotechnol Mol Biol Rev 1:21–38

Drzewiecka K, Mleczek M, Waśkiewicz A, Goliński P (2011) Oxidative stress and phytoremediation. In: Ahmad P, Prasad MNV (eds) Abiotic stress responses in plants. Springer, New York, pp 425–449

Falcinelli B, Valeria S, Ombretta M et al (2017) Germination under moderate salinity increases phenolic content and antioxidant activity in rapeseed (*Brassica napus* var. *oleifera* Del.) sprouts. Molecules 22:1377–1390

Farooq H, Batool N, Iqbal J et al (2010) Effect of salinity and water types on growth performance and nutrient composition of *Acacia nilotica* L. Int J AgricBiol 12:591–596

Farooq M, Gogoi N, Hussain M et al (2017) Effects, tolerance mechanisms and management of salt stress in grain legumes. Plant Physiol Biochem 118:199–217

Flowers TJ, Colmer TD (2008) Salinity tolerance in halophytes. New Phytol 179:945–963

Fragnire C, Serrano M, Abou-Mansour E et al (2011) Salicylic acid and its location in response to biotic and abiotic stress. FEBS Lett 585:1847–1852

Fukuda A, Tanaka Y (2011) Effects of ABA, auxin, and gibberellin on the expression of genes for vacuolar H^+- inorganic pyrophosphatase, H^+-ATPase subunit A, and Na^+/H^+ antiporter in barley. Plant Physiol Biochem 44:351–358

Galvani A (2007) The challenge of the food sufficiency through salt tolerant crops. Rev Env Sci Biotechnol 6:3–16

Gill SS, Tuteja N (2010a) Reactive oxygen species and antioxidant machinery in abiotic stress tolerance in crop plants. Plant Physiol Biochem 48:909–930

Gill SS, Tuteja N (2010b) Polyamines and abiotic stress tolerance in plants. Plant Signal Behav 5:26–33

Groppa MD, Benavides MP (2008) Polyamines and abiotic stress: recent advances. Amino Acids 34:35–45

Gupta K, Dey A, Gupta B (2013) Plant polyamines in abiotic stress responses. Acta Physiol Plant 35:2015–2036

Gurmani AR, Bano S, Khan U et al (2011) Alleviation of salt stress by seed treatment with abscisic acid (ABA), 6-benzylaminopurine (BA) and chlormequat chloride (CCC) optimizes ion and organic matter accumulation and increases yield of rice (*Oryza sativa* L.) Aus J Crop Sci 5:1278–1285

Hanaa H, El-Baky A, Hussein MM et al (2008) Algal extracts improve antioxidant defense abilities and salt tolerance of wheat plant irrigated with sea water. Elect J Environ Agric Food Chem 7:2812–2832

Hartung W, Leport L, Ratcliffe RG, Sauter A, Duda R, Turner NC (2002) Abscisic acid concentration, root pH and anatomy do not explain growth differences of chickpea (*Cicer arietinum* L.) and lupin (*Lupinus angustifolius* L.) on acid and alkaline soils. Plant Soil 240:191–199

Hasegawa PM, Bressan RA, Zhu JK et al (2000) Plant cellular and molecular responses to high salinity. Ann Rev Plant Biol 51:463–499
Hong Z, Lakkineni K, Zhang Z et al (2000) Removal of feedback inhibition of 1-pyrroline-5-carboxylate synthetase results in increased proline accumulation and protection of plants from osmotic stress. Plant Physiol 122:1129–1136
Hoque MA, Banu MNA, Nakamura Y et al (2008) Proline and glycinebetaine enhance antioxidant defense and methylglyoxal detoxification systems and reduce NaCl-induced damage in cultured tobacco cells. J Plant Physiol 165:813–824
Ikbal F, Hernández JA, Barba-Espín G et al (2014) Enhanced salt-induced antioxidative responses involve a contribution of polyamine biosynthesis in grapevine plants. J Plant Physiol 171:779–788
Ismail H, Maksimovic JD, Maksimovic V et al (2016) Rutin, a flavonoid with antioxidant activity, improves plant salinity tolerance by regulating K^+ retention and Na^+ exclusion from leaf mesophyll in quinoa and broad beans. Funct Plant Biol 43:75–86
Jayakannan M, Bose J, Babourina O et al (2013) Salicylic acid improves salinity tolerance in *Arabidopsis* by restoring membrane potential and preventing salt-induced K^+ loss via a GORK channel. J Exp Bot 64(8):2255–2268
Jia W, Wang Y, Zhang S et al (2002) Salt-stress-induced ABA accumulation is more sensitively triggered in roots than in shoots. J Exp Bot 53:2201–2206
Kaya C, Sonmez O, Aydemir S, Ashraf M, Dikilitas M (2013) Exogenous application of mannitol and thiourea regulates plant growth and oxidative stress responses in salt-stressed maize (*Zea mays* L.) J Plant Interact 3:234–241
Keutgen AJ, Pawelzik E (2008) Quality and nutritional value of strawberry fruit under long term salt stress. Food Chem 107:1413–1420
Khadri M, Tejera NA, Carmen L (2006) Alleviation of salt stress in common bean (*Phaseolus vulgaris* L.) by exogenous abscisic acid supply. J Plant Growth Regul 25:110–119
Khavarinejad RA, Mostofi Y (1998) Effects of NaCl on photosynthetic pigments, saccharides, and chloroplast ultrastructure in leaves of tomato cultivars. Photosynthetica 35:151–154
Kiarostami K, Mohseni R, Saboora A (2010) Biochemical changes of *Rosmarinus officinalis* under salt stress. J Stress Physiol Biochem 6:114–122
Kim HJ, Fonseca JM, Choi JH (2008) Salt in irrigation water affects the nutritional and visual properties of romaine lettuce (*Lactuca sativa* L.) J Agric Food Chem 56:3772–3776
Koyro HW, Ahmad P, Geissler N (2012) Abiotic stress responses in plants: an overview. In: Ahmad P, Prasad MNV (eds) Environmental adaptations and stress tolerance of plants in the era of climate change. Springer, New York, pp 1–28
Le TN, McQueen-Mason SJ (2006) Desiccation-tolerant plants in dry environments. Rev Environ Sci Biotechnol l5:269–279
Li S, Han J, Qiang Z (2016) The effect of exogenous spermidine concentration on polyamine metabolism and salt tolerance in zoysiagrass (*Zoysia japonica* Steud) subjected to short-term salinity stress. Front Plant Sci 7:1–13
Lim JH, Park KJ, Kim BK et al (2012) Effect of salinity stress on phenolic compounds and carotenoids in buckwheat (*Fagopyrum esculentum* M.) sprout. J Food Chem 135:1065–1070
Lutts S, Majerus V, Kinet JM (1999) NaCl effects on proline metabolism in rice (*Oryza sativa*) seedlings. Physiol Plant 105:450–458
Mahmoudi H, Huang J, Gruber MY, Kaddour R, Lachaal M et al (2010) The impact of genotype and salinity on physiological function, secondary metabolite accumulation, and antioxidative response in lettuce. J Agric Food Chem 58:5122–5130
Mansour MMF (1998) Protection of plasma membrane of onion epidermal cells by glycine betaine and proline against NaCl stress. Plant Physiol Biochem 36:767–772
Mehrizi MH, Shariatmadari H, Khoshgoftarmanesh AH, Dehghani F (2012) Copper effects on growth, lipid peroxidation, and total phenolic content of rosemary leaves under salinity stress. J Agric Sci Technol 14:205–212

Meloni DA, Gulotta MR, Martínez CA, Oliva MA (2004) The effects of salt stress on growth, nitrate reduction and proline and glycine betaine accumulation in *Prosopis alba*. Braz J Plant Physiol 116:39–46

Minh LT, Do TK, Pham TTH et al (2016) Effects of salinity stress on growth and phenolics of rice (*Oryza sativa* L.) Int Lett Nat Sci 57:1–10

Minocha R, Majumdar R, Minocha SC (2014) Polyamines and abiotic stress in plants: a complex relationship. Front Plant Sci 5:1–17

Mittler R (2002) Oxidative stress, antioxidants and stress tolerance. Trend Plant Sci 7:405–410

Mittova V, Tal M, Volokita M et al (2003) Up-regulation of the leaf mitochondrial and peroxisomal antioxidative systems in response to salt-induced oxidative stress in the wild salt-tolerant tomato species *Lycopersicon pennellii*. Plant Cell Environ 26:845–856

Mittova V, Guy M, Tal M (2004) Salinity up-regulates the antioxidative system in root mitochondria and peroxisomes of the wild salt-tolerant tomato species *Lycopersicon pennellii*. J Exp Bot 399:1105–1113

Mohamed AA, Amina AA (2008) Alteration of some secondary metabolites and enzymes activity by using exogenous compound in onion plants grown under seawater salt stress. Amer-Euras J Sci Res 3:139–146

Moller IM, Jensen PE, Hansson A (2007) Oxidative modifications to cellular components in plants. Annu Rev Plant Biol 58:459–481

Munns R, Tester M (2008) Mechanisms of salinity tolerance. Ann Rev Plant Biol 59:651–681

Nanjo T, Fujita M, Seki M et al (2003) Toxicity of free proline revealed in an *Arabidopsis* T-DNA-tagged mutant deficient in proline dehydrogenase. Plant Cell Physiol 44:541–548

Navarro J, Flores MP, Garrido C et al (2006) Changes in the contents of antioxidant compounds in pepper fruits at different ripening stages, as affected by salinity. Food Chem 96:66–73

Netto LES (2001) Oxidative stress response in sugarcane. Gen Mol Biol 24:93–102

Noreen Z, Ashraf M (2009) Assessment of variation in antioxidative defense system in salt treated pea (*Pisum sativum*) cultivars and its putative use as salinity tolerance markers. J Plant Physiol 166:1764–1774

Nouman W, Siddiqui MT, Basra SMA et al (2012) Response of *Moringa oleifera* to saline conditions. Int J AgricBiol 14:757–762

Nouman W, Basra SMA, Yasmeen A et al (2014) Seed priming improves the emergence potential, growth and antioxidant system of *Moringa oleifera* under saline conditions. Plant Growth Regul 73:267–278

Nounjan N, Nghia PT, Theerakulpisut P (2012) Exogenous proline and trehalose promote recovery of rice seedlings from salt-stress and differentially modulate antioxidant enzymes and expression of related genes. J Plant Physiol 169:596–604

Nuttall G, Armstrong RD, Connor DJ (2003) Evaluating physicochemical constraints of Calcarosols on wheat yield in the Victorian southern Mallee. Aust J Agric Res 5:487–497

Oueslati S, Karray-Bouraoui N, Attia H et al (2010) Physiological and antioxidant responses of *Mentha pulegium* (Pennyroyal) to salt stress. Acta Physiol Plant 32:289–296

Parida AK, Das AB, Das P (2002) NaCl stress causes changes in photosynthetic pigments, proteins and other metabolic components in the leaves of a true mangrove, *Bruguiera parviflora*, in hydroponic cultures. J Plant Biol 45:28–36

Parvaiz A, Satyawati S (2008) Salt stress and phyto-biochemical responses of plants – a review. Plant Soil Enviorn 54:89–99

Peshev D, Vergauwen R, Moglia A, Hideg E, Ende WVD (2013) Towards understanding vacuolar antioxidant mechanisms: a role for fructans? J Exp Bot 64:1025–1038

Petridis A, Therios I, Samouris G et al (2012) Salinity-induced changes in phenolic compounds in leaves and roots of four olive cultivars (*Olea europaea* L.) and their relationship to antioxidant activity. Environ Exp Bot 79:37–43

Petrusa LM, Winicov I (1997) Proline status in salt tolerant and salt sensitive alfalfa cell lines and plants in response to NaCl. Plant Physiol Biochem 35:303–310

Pilon-Smits E, Ebskamp M, Paul MJ, Jeuken M, Weisbeek PJ, Smeekens S (1995) Improved performance of transgenic fructan-accumulating tobacco under drought stress. Plant Physiol 107:125–130

Pottosin II, Martínez-Estévez M, Dobrovinskaya OR, Muñiz J, Schönknecht G (2004) Mechanism of luminal Ca^{2+} and Mg^{2+} action on the vacuolar slowly activating channels. Planta 219:1057–1070

Pottosin I, Velarde-Buendía AM, Zepeda-Jazo I, Dobrovinskaya O, Shabala S (2012) Synergism between polyamines and ROS in the induction of Ca^{2+} and K^{+} fluxes in roots. Plant Signal Behav 7:1084–1087

Pruthvi V, Narasimhan R, Nataraja KN (2014) Simultaneous expression of abiotic stress responsive transcription factors, AtDREB2A, AtHB7 and AtABF3 improves salinity and drought tolerance in peanut (*Arachis hypogaea* L.) PLoS One 9:1–21

Qadir M, Quillérou E, Nangia V et al (2014) Economics of salt-induced land degradation and restoration. Nat Resour Forum 38:282–295

Quiles MJ, López NI (2004) Photoinhibition of photosystems I and II induced by exposure to high light intensity during oat plant growth effects on the chloroplast NADH dehydrogenase complex. Plant Sci 166:815–823

Rani RJ (2011) Salt stress tolerance and stress proteins in pearl millet (*Pennisetum glaucum* (L.) R. Br.) J Appl Pharm Sci 1:185–188

Rezazadeh A, Ghasemnezhad A, Barani M et al (2012) Effect of salinity on phenolic composition and antioxidant activity of Artichoke (*Cynara scolymus* L.) leaves. Res J Med Plant 6:245–252

Rmiki KE, Lemoine Y, Schoeff B (1999) Carotenoids and stress in higher plants and algae. In: Pessarakli M (ed) Handbook of plant and crop stress. Marcel Dekker Press, New York, pp 465–482

Roychoudhury A, Basu S, Sengupta DN (2011) Amelioration of salinity stress by exogenously applied spermidine or spermine in three varieties of indica rice differing in their level of salt tolerance. J Plant Physiol 168:317–328

Rubio F, Gassmann W, Schroeder JI (1995) Sodium-driven potassium uptake by the plant potassium transporter HKT1 and mutations conferring salt tolerance. Science 270:1660–1663

Sairam RK, Tyagi A (2004) Physiology and molecular biology of salinity stress tolerance in plants. Curr Sci 86:407–421

Saneoka H, Nagasaka C, Hahn DT et al (1995) Salt tolerance of glycine betaine-deficient and containing maize lines. Plant Physiol 107:631–638

Saxena SC, Kaur H, Verma P et al (2013) Osmoprotectants: potential for crop improvement under adverse conditions. In: Plant acclimation to environmental stress. Springer, New York, pp 197–232

Shabala S, Pottosin I (2014) Regulation of potassium transport in plants under hostile conditions: implications for abiotic and biotic stress tolerance. Physiol Plant 151:l257–l279

Shapiguzov A, Vainonen JP, Wrzaczek M, Kangasjarvi J (2012) ROS- talk—how the apoplast, the chloroplast, and the nucleus get the message through. Front Plant Sci 3:292

Shi D, Sheng Y (2005) Effect of various salt–alkaline mixed stress conditions on sunflower seedlings and analysis of their stress factors. Environ Exp Bot 54:8–21

Singh M, Jitendra K, Samiksha S, Vijay PS, Sheo MP (2015) Roles of osmoprotectants in improving salinity and drought tolerance in plants: a review. Rev Environ SciBiotechnol 14:407–426

Tahir MA, Aziz T, Farooq M et al (2012) Silicon induced changes in growth, ionic composition, water relations, chlorophyll contents and membrane permeability in two salt stressed wheat genotypes. Arch Agron Soil Sci 58:247–256

Takahashi T, Kakehi JI (2010) Polyamines: ubiquitous polycations with unique roles in growth and stress responses. Ann Bot 105:1–6

Tester M, Davenport R (2003) Na^{+} tolerance and Na^{+} transport in higher plants. Ann Bot 91:503–527 van Oosten MJ, Sharkhuu A, Batelli G et al (2013) The *Arabidopsis thaliana* mutant air implicates SOS3 in the regulation of anthocyanins under salt stress. Plant Mol Biol 83:405–415

Velarde-Buendia AM, Shabala S, Cvikrova M, Dobrovinskaya O, Pot-tosin I (2012) Salt-sensitive and salt-tolerant barley varieties differ in the extent of potentiation of the ROS-induced K(+) efflux by polyamines. Plant Physiol Biochem 61:18–23

Vinocur B, Altman A (2005) Cellular basis of salinity tolerance in plants. Environ Exp Bot 52:113–122

Wahid A, Ghazanfar A (2006) Possible involvement of some secondary metabolites in salt tolerance of sugarcane. J Plant Physiol 163:723–730

Wang Y, Nil N (2000) Changes in chlorophyll, ribulose bisphosphate carboxylase-oxygenase, glycine betaine content, photosynthesis and transpiration in *Amaranthus tricolor* leaves during salt stress. J Hortic Sci Biotechnol 75:623–627

Wang W, Vinocur B, Altman A (2003) Plant responses to drought, salinity and extreme temperatures: towards genetic engineering for stress tolerance. Planta 218:1–14

Williamson JD, Jennings DB, Guo WW, Pharr DM, Ehrenshaft M (2002) Sugar alcohols, salt stress and fungal resistance: polyols: multifunctional plant protection? J Am Soc Hortic Sci 127:467–473

Yang WJ, Rich PJ, Axtell JD et al (2003) Genotypic variation for glycine betaine in sorghum. Crop Sci 43:162–169

Yang C, Chong J, Li C, Kim C, Shi D, Wang D (2007) Osmotic adjustment and ion balance traits of an alkali resistant halophyte *Kochia sieversiana* during adaptation to salt and alkali conditions. Plant Soil 294:263–276

Yasmeen A, Basra SMA, Farooq M et al (2013) Exogenous application of moringa leaf extract modulates the antioxidant enzyme system to improve wheat performance under saline conditions. Plant Growth Regul 69:225–233

Yasmeen A, Nouman W, Basra SMA, Wahid A, Rehman HU, Hussain N, Afzal I (2014) Morphological and physiological response of tomato (Solanum lycopersicum L.) to natural and synthetic cytokinin sources: a comparative study. Acta Physiol Plant 36:3147–3155

Yuan G, Wang X, Guo R et al (2010) Effect of salt stress on phenolic compounds, glucosinolates, myrosinase and antioxidant activity in radish sprouts. Food Chem 121:1014–1019

Zahedi SM, Nabipour M, Azizi M et al (2011) Effect of kinds of salt and its different levels on seed germination and growth of basil plant. World Appl Sci J 15:1039–1045

Zapata PJ, Serrano M, Pretel AT et al (2004) Polyamines and ethylene changes during germination of different plant species under salinity. Plant Sci 167:781–788

Zhao J, Davis LC, Verpoorte R (2005) Elicitor signal transduction leading to production of plant secondary metabolites. Biotechnol Adv 23:283–333

Zhao MG, Chen L, Zhang LL et al (2009) Nitric reductase-dependent nitric oxide production is involved in cold acclimation and freezing tolerance in Arabidopsis. Plant Physiol 151:755–767

Zhifang G, Loescher WH (2003) Expression of a celery mannose 6-phosphate reductase in *Arabidopsis thaliana* enhances salt tolerance and induces biosynthesis of both mannitol and a glucosyl-mannitol dimmer. Plant Cell Environ 26:275–283

Zrig A, Tounekti T, Vadel AM, Ben Mohamed H, Valero D, Serrano M (2011) Possible involvement of polyphenols and polyamines in salt tolerance of almond rootstocks. Plant Physiol Biochem 49:1313–1322

Response of Plants to Salinity Stress and the Role of Salicylic Acid in Modulating Tolerance Mechanisms: Physiological and Proteomic Approach

Renuka Saraf, Sadhana Saingar, Shweta Chaudhary, and Dipjyoti Chakraborty

Abstract Salinity is one of the most consequential stresses, which limits the productivity of agricultural crops and affects germination, plant strength, and crop yield. High salinity affects plants in several ways, such as water stress, ion toxicity, oxidative stress, alteration of metabolic processes, nutritional disorders, membrane disorganization, and reduction of cell division, expansion, and genotoxicity.

Together all these effects reduce plant growth, development and survival. The mechanisms of genetic control by which plants tolerate the salt stress are very complex and have not yet properly understood. Plants have evolved several mechanisms to acclimatize to salinity. Several biomolecules have been discovered within plants that modulate mechanisms to effectively deal with salinity stress. One such compound is salicylic acid which has been extensively studied for its role in biotic stress and recently is in focus for abiotic stress tolerance research.

A fundamental biological understanding and knowledge of the effects of salt stress on plants is necessary to provide additional information for the dissection of the plant response to salinity. The present chapter reviews plant response to salinity stress and the role of salicylic acid in modulating tolerance mechanisms especially at the molecular level. Recent advances in proteomic studies for elucidation of plant response are discussed.

Keywords Salinity · Salicylic acid · Stress tolerance · Proteomic studies

1 Introduction

Plants are commonly affected by several stresses. Plant productivity depends on its ability to adapt or resist unfavorable environmental stress. Biotic stress involves bacteria, virus, and fungi infections. On the other hand, abiotic stresses

---

R. Saraf · S. Saingar · S. Chaudhary · D. Chakraborty (✉)
Department of Bioscience and Biotechnology, Banasthali Vidyapith, Banasthali, Rajasthan 304 022, India
e-mail: cdipjyoti@banasthali.in

S. Vats (ed.), *Biotic and Abiotic Stress Tolerance in Plants*,
https://doi.org/10.1007/978-981-10-9029-5_5

include unfavorable environmental conditions like drought, salinity, ultraviolet rays, and heavy metal toxicity (Hasanuzzaman et al. 2014).

Salinity limits the production of crops, inhibits seed germination, and affects the crop quality (Munns and Tester 2008). It alters plant growth through ionic imbalance, oxidative alteration, metabolic regulation, nutritional disorders, membrane disorganization, low cell differentiation rate, and genotoxicity (Zhu 2007). According to FAO, almost 6% of the world's land is affected by salinity which covers mainly Mediterranean countries. In India, approximately 8 Mh areas are saline prone which cover major areas of arid and semiarid region and Indo-Gangetic Plain (Yadav et al. 2011). However, the range of salinity-prone land is approximately 900×10^6 ha which is quite competent to pose agriculture threat (Flower and Yeo 1995; Flower 2004). Most of the crops are not able to grow in the highly saline field, except halophytes, which are able to survive in a high salt concentration of about 400 mM. Saline soil and drought affect almost 20–50% of the overall crop production (Shrivastava and Kumar 2015).

Salt stress affects the life cycle of the plant, resulting in a low photosynthetic rate, thereby an alteration in the metabolism deficit water supply to the plant system. This in turn affects the plant productivity as well as soil fertility (Parida and Das 2005). Further, excess salinity affects the osmotic property of plants, resulting in closure of stomata as well as restricted cell expansion and cell division (Flowers and Colmer 2008). Long incubation in salinity causes ionic imbalance, which leads to early aging of adult leaves, reduction in photosynthetic mechanism, and toxicity symptoms in mature leaves by high [Na^+], which cease protein synthesis and interfere with enzyme regulation in plants (Munns 2002a). The presence of high NaCl concentration intervenes nutritional homeostasis of the plants, which in turn accelerates Na^+/K^+, Na^+/Mg^{2+}, Na^+/Ca^{2+}, Cl^-/H_2PO^{-4}, and Cl^-/NO^{-3} (Grattan and Grieve 1999; Fahad and Bano 2012). The urgency for breeding for salt tolerance in crop plants was felt in the early 1980s, and Epstein (1977); Epstein et al. (1980) gave the genetic basis of difference in salt tolerance ability between different species. However, only three salinity-resistant cultivars were developed in the next three decades (Owen et al. 1994; Al-Doss and Smith 1998; Dierig et al. 2001). The mechanisms of genetic control by which plants tolerate the salt stress are very complex and have not yet properly understood. Plants have mechanisms to acclimatize to salinity (Flowers 2004). These tolerance mechanisms can be distinguished in three types: (a) osmotic potential tolerance, (b) [Na^+] exclusion, and (c) tissue resistance (Munns and Tester 2008). Osmotic potential tolerance involves the plant's skill to endure drought aspects of salinity and sustain leaf increment and stomatal conducting process (James et al. 2008). Under salinity in many plants, Na^+ concentration reaches to toxic levels prior [Cl^-]. On the other hand, tissue tolerance leads to improved survival of mature leaves by dividing [Na^+] and [Cl^-] at the cellular or the intracellular level to reduce lethal concentration into the cytoplasm (Munns and Tester 2008). Salt tolerance can be determined by biomass generation under salinity versus control state of time or in terms of survival, which is quite convenient for perennial species (Munns 2002b).

Several biomolecules have been discovered within plants that modulate mechanisms to effectively deal with salinity stress. One such compound is salicylic acid (SA), which has been studied extensively for its role in biotic stress and recently is in focus for abiotic stress tolerance research. "Salicylic" is derived from the Latin word *Salix* which means willow bark. It is a mono-hydroxybenzoic acid, a phenolic acid, or a beta-hydroxyl acid. It has the chemical formula $C_7H_6O_3$. It is a colorless crystalline organic acid, widely used in organic synthesis, and also performs as a plant hormone (Rivas-San Vicente and Plasencia 2011; Derikvand and Azadbakht 2017).

Salicylic acid performs a vital role in plant propagation, photosynthetic activity, transpiration process, ion flux and transportation. It induces changes in leaf appearance and chloroplast representation (Mimouni et al. 2016; Afran et al. 2007). It regulates endogenous signaling pathways and intervenes in the plant defense mechanism against pathogens (Hayat and Ahmad 2007; Kundu et al. 2013). SA has a crucial role in systematic acquired resistance (SAR) (Taiz and Zeiger 2002). SA triggers in response to various devastating biotic and abiotic stresses, which further stimulate systemic acquired resistance (SAR) via hypersensitivity response-involved endogenous pathway (Naylor et al. 1998; Mateo et al. 2006; Cueto-Ginzo et al. 2016a, b). SA mediates immune response as well as gene resistance through positive interaction between small interfering RNA (siRNA) and SA-intervened defense. SA regulates three phases of pathogenicity cycle comprising of long path movement, replication of viral pathogen, and cell-cell movement (Tian et al. 2015). Moreover, SA is also involved in the cross talk with RNA silencing. RNA polymerase I- and RNA-dependent component of the RNA silencing machinery is stimulated by the treatment with SA in a variety of plants (Liu et al. 2009; Tian et al. 2015). Secondly, viral silenced suppressor proteins repressed SA-induced gene expression (Alamillo et al. 2006). Detailed evidence implicates the role of salicylic acid in pathogenesis-related gene expression, SAR, or immune response (Shah 2003). Besides pathogenesis-related resistance, it has role in response to abiotic stresses such as salt and osmotic, ozone, drought, UV exposure, heat, cold, and metal stress (Metwally et al. 2003; Pandey and Chakraborty 2015). SA also takes part in stress-influenced developmental transitions such as flowering, tuberization, and senescence (Morris et al. 2000).

Incidentally, both lower and higher than optimum concentrations of SA increase plants' sensitivity to abiotic stress. For most plants, the optimal range for high stress tolerance is 0.1–0.5 mM (Yuan and Lin 2008). However, at certain level, SA performs differently in moderate and unfavorable abiotic conditions. In plants, the equivalent concentration of SA increases resistance to certain stress but is sensitive to other kinds of unfavorable conditions (Nemeth et al. 2002). High concentration of SA causes growth retardation in *Vigna radiata* (Khan et al. 2010). SA treatment exhibited larger plant growth and maintained membrane integrity in strawberry plants and barley plants (El–Tayeb 2005). In *Arabidopsis*, SA performs both roles, i.e., it is an essential component to induce antioxidant defenses and maintain redox potential of glutathione pool (Sharma et al. 1996). Secondly, this accumulation can stimulate a performed cell death that initiates an immune response against O_3 (Rao and Davis 1999).

The present work reviews the plant response to salinity stress and the role of salicylic acid to generate tolerance in plants under the unfavorable environment.

2 Abiotic Stress in Plants

Abiotic stress is essentially unavoidable in plants. Abiotic stress influences the development and production of crops all over the world (Gao et al. 2007). Abiotic stresses are classified into various forms. As compared to easily identified stresses, less considerable factors of abiotic stress influence the environment continuously (Palta and Farag 2006). The most basic factor includes high winds, extreme temperatures, drought, flood, salinity, etc. On the other hand, less-known stresses generally include poor edaphic conditions and simultaneous dehydration at the period of seed germination. Being a part of ecosystem, abiotic stresses adversely manipulate organisms in many ways, which may be foredeal as well as afterdeal. The most afterdeal concerns of abiotic stress include farming. It is observed that abiotic stress causes reduction in crop productivity by 50% more than from their actual yield (Wang et al. 2007). When the soil is competent and biologically diverse, the plant retains high probability of survival in unfavorable conditions (Brussaard et al. 2007). As compared to optimal condition, the high-stress region induces an enhanced level of facilitation, which concludes that the plants require a wide network among species (such as cross-pollination) to withstand against their undesirable habitat (Maestre et al. 2007).

An adaptation to a particular environment varies among plant species. Thus, the response of various plant species to a variety of different stress signals varies even though the plants have become accustomed to similar environment (Mittler 2006). Rice plants (*Oryza sativa*), which is an essential food all over the world, especially in China and India, suffer slightly distinct abiotic stresses which lead to negative impact on rice production (Gao et al. 2007; Breviario and Genga 2013). In plants, salinity is an undesirable condition, where high salt concentrations cease plant yield or cause apoptosis. Salinity presents an inevitable threat to plant production globally. Improper irrigation along with inadequate drainage is the most severe condition among various soil salinity sources, because it results in infertility of productive agricultural land. This leads to secondary salinization (Kitamura et al. 2014). The excessive salt concentration in the soil has two demerits. Firstly, a high salt ion concentration is toxic to plant cells. Typically, NaCl constitutes the majority of the salts. This concludes that high sodium ion concentration is injurious to most plants, while others are affected by high [Cl^-]. Secondly, an extravagant salt concentration decreases the osmotic strength of the soil, which results in water deficit or osmotic imbalance (Negrão et al. 2016).

2.1 Effects of High Salinity on Plants

Salt stress manifests distinct characteristics such as growth holdup, aging, and apoptosis due to longer exposure. High salinity causes simultaneous hyperionic and hyperosmotic stresses which ultimately cease plant growth (Hasegawa et al. 2000). Plants which are grown under salt stress are affected in three ways: (a) decreased water supply in root zone leading to drought, (b) phytotoxin of ions such as Na^+ and Cl^-, and (c) nutrient retardation which inhibits influx of nutrients. Moreover, there is competition among Na^+ and K^+ for the binding site of carrier proteins (Munns 2002a). High salt concentration also influences the osmotic regulation of soil matrix, which limits the water influx in plants and disturbs stomatal regulation which ultimately causes necrosis (Negrão et al. 2016).

Salinity induces the production of abscisic acid (ABA), which results in stomatal closure during transportation to guard cells, leading to low photosynthesis rate that positively feedbacks photosynthesis and reduces oxidative imbalance (Zeinolabedin 2012).

Salinity increases ROS by modulating plant metabolism. This leads to the activation of cellular redox reaction in the oxidized state, thereby generating oxidative stress that interrupts the cellular and enzymatic activity of plants (Khan and Weber 2008; Negrão et al. 2016). Salinity also infers a state of dormancy at low concentration while ceasing germination of seeds at high level (Khan and Weber 2008).

2.2 Germination

Salinity stress decreases germination of seeds of many crops, including *Glycine max* (Essa 2002), *Brassica* spp. (Ibrar et al. 2003), *Vigna* spp. (Jabeen et al. 2003), *Helianthus annuus* (Mutlu and Bozcuk 2007), *Zea mays* (Carpici et al. 2009), *O. sativa* (Xu et al. 2011), and *Triticum aestivum* (Akbarimoghaddam et al. 2011). Seed germination is inhibited by high level of salt stress as compared to low salt concentration, which induces quiescence (Läuchli and Grattan 2007; Khan and Weber 2008). Under salinity stress, seeds are not able to imbibe water due to low osmotic potential (Khan and Weber 2008). Moreover, salt alters the enzymatic mechanisms of various metabolic pathways, protein functioning, and hormonal homeostasis (Yupsanis et al. 1994; Promila and Kumar 2000; Gomes-Filho et al. 2008). It also deteriorates the internal structure of plant organ, tissue, and cell (Koyro 2002). Salt alters the seed germination by damaging the seed coat, increasing seed aging and dormancy, and decreasing seed vigor index and polymorphism (Panuccio et al. 2014).

2.3 Plant Growth

Reduction in the growth rate of the plant is the primary effect of salt stress, which results in many ways. Firstly, saline soil reduces the water holding level in the plant, which in turn decreases the growth of the plant by desisting the osmotic potential (Munns 2002b). In the second phase, ion toxicity increases, which causes early senescence of the leaves. It also increases the senescence of leaves as compared to emerging of new leaves. The earlier phase is more rapid and unfavorable than the second phase (Munns 2002a, 2005). Both the phases adversely affect the photosynthetic and transpiration rate of the plant. In plants, the increased influx of Na^+ and Cl^- ions causes leaf injury which causes leaf death (Munns and Tester 2008). However, some crops are sensitive to salinity during vegetative and reproductive stage, while some are less sensitive (Lauchli and Grattan 2007).

2.4 Photosynthesis

The reduced photosynthetic rate in plants under high salt concentration is due to imbalance in the osmotic pressure. It further increases the toxicity of ions which leads to accumulation of Na^+ and Cl^- in the chloroplasts. Further, high salt concentration affects the photosynthetic electron transport chain by either hindering the carbon metabolism or photon phosphorylation mechanism (Sudhir and Murthy 2004; Arfan et al. 2007; Farahbakhsh et al. 2017).

In fact, many reports have described that photosynthetic rate depends on the salt concentration as well as on the plant species (Rogers and Noble 1992). However, low salt concentration stimulates the photosynthesis, which in turn increases the chlorophyll content in the plant (Chutipijit et al. 2011). In *O. sativa*, the chlorophyll a and b content of the leaves was measured under 200 mM NaCl treatment after 14 days (Amirijani 2011). It was found that the chlorophyll b content was reduced by 41%, while chlorophyll a content was reduced by 33%. Saha et al. (2010) reported a linear reduction in the intensity of total chlorophyll, chlorophyll a, chlorophyll b, carotenoid, and xanthophylls in *V. radiata* under increasing NaCl concentration. Parida et al. (2004) observed that in *Bruguiera parviflora*, the photosynthesis rate gets stimulated at low salt concentration. Salinity also reduces enzyme activity, alters cytoplasmic structure, and increases senescence. The alteration in stomatal conductance reduces the CO_2 rate for carboxylation reaction (Brugnoli and Bjorkman 1992; Maxwell and Johnson 2000).

2.5 Water Relation

Salinity has a major influence on the root region of the plant, which further decreases the leaf water uptake ability and various other metabolic pathways. Osmotic imbalance is a consequence of the low water uptake potential of the plant which in turn increases salinity concentration in the root zone (Munns 2005). Significant decrease in the relative water content was observed in the sugar beet varieties under salinity stress (Ghoulam et al. 2002). The low relative water content results in decreased turgor pressure and hydrostatic pressure gradient, which limits the water availability for cell extension mechanism (Vysotskaya et al. 2010).

2.6 Nutrient Imbalance

Crop productivity decreases by salinity-induced nutrient imbalance. According to Grattan and Grieve (1999), the relation between salt stress and nutritional content is very complex. The increased salt concentration affects the nutrient level, transport of ions, and additional competitive mineral uptake within the plant. Several reports describe the reduced uptake of nutrients as well as accumulation of nutrient in the plants under high salinity stress (Rogers et al. 2003). The extent of the effect of salinity stress on crop yield depends upon the range of ion toxicity, salt composition, crop variety, and environmental conditions (Grattan and Grieve 1999). Moreover, high osmotic potential reduces the nitrogen content in plants by causing interaction either between NH_4^+ and Na^+ or between NO_3^- and Cl^- (Lea-Cox and Syvertsen 1993; Rozeff 1995; Bar et al. 1997). The phosphorous content is also reduced by high ionic level, low solubility of Ca-P minerals, and tightly regulated sorption processes (Qadir and Schubert 2002). The elevated level of Na^+ in the plant also causes decrease in the level of influx of Ca^+ and K^+ ions in the plant (Suhayda et al. 1990; Hu and Schmidhalter 1997; Asch et al. 2000). The presence of micronutrients in the saline soil depends on its solubility level, pH range, redox potential, and nature of the binding region on the organic and inorganic particle surface (Oertli 1991; Zhu et al. 2004).

2.7 Yield

The saline soil has majorly reduced the yield of the crops. The induction of tolerance ability by producing mutagenic traits is a difficult task under salt stress (Yokoi et al. 2002). However, the relative yield of crop species has been distinguished in terms of their salt tolerance ability. The parameters used for measuring the tolerance level are threshold electrical conductivity and the percent decrease in relative yield per unit of electrical conductivity (dS m^{-1}) (Mass 1986). Nahar and Hasanuzzaman (2009) described the influence of high salt concentration on *V. radiata* by measuring the number of pods per plants, seed weight, and seeds per pod which has negative correlation with high salt concentration.

2.8 Salinity Induced Oxidative Stress

Salinity causes reduction of water potential and initiation of oxidative stress (Munns 2005; Munns and Tester 2008). It also induces the stomatal closure, which reduces the CO_2 level in the leaves (Munns and Tester 2008; Chutipijit et al. 2011). It decreases the carbon fixation ability which in turn leads to the generation of excessive excitation energy and reactive oxygen species (ROS) (Halliwell and Gutteridge 1985; Hasegawa et al. 2000; Parida and Das 2005). The generated reactive oxygen species are very reactive and cause damage at cellular and molecular level (Pastori and Foyer 2002; Miller et al. 2010). ROS-mediated cellular damage under salt stress has been observed in many crops including tomato, pea, mustard, citrus, and rice (Gueta-Dahan et al. 1997). In *Brassica napus* and *T. aestivum*, the increase in lipid peroxidation and H_2O_2 level was observed under the high salt concentration (Hasanuzzaman and Fujita 2011a, b).

3 Plant Mechanisms to Tolerate Salt Stress

The mechanism of the genetic influx of salt resistance in plants has not still properly understood due to its complexity. Genetic variations may be evaluated by limiting the response of various genotypes. Among various responses, both growth and yield are the most common ways to measure moderate salinities (Allen et al. 1994). Resistance to salinity can be restricted by a sudden exposure to salinity, even if the plant species is a halophyte (Albert 1975). The salinity sensitivity among species may change during ontogeny. Salt tolerance depends on plant species as well as environmental conditions. For certain species, salinity tolerance may be higher at germination stage or at the reproduction stage (Munns et al. 2006).

Plants have generated many mechanisms to reduce salinity by:

- Osmotic stress tolerance
- Na^+ exclusion from leaf blades
- Tissue tolerance

3.1 Osmotic Stress Tolerance

Plant growth is reduced under excessive salt stress due to generation of high osmotic potential (James et al. 2011). The altered osmotic stress further causes ion toxicity. This increases the senescence of old leaves as well as reduces the appearance of new leaves. Therefore, the sustainable cropping involves osmotic balance for overall productivity of the plant. Osmotic tolerance involves the plant's potential to tolerate high salt stress and maintain stomatal conductance and leaf expansion (Rajendran et al. 2009).

3.2 Na^+ Exclusion

In plants, salinity induces the generation of superoxides (O_2^-), singlet oxygen ($^1O_2$), hydrogen peroxide (H_2O_2), and hydroxyl radical (OH^-), which in turn induce reduction of K^+ influx via activated ROS channels that trigger programmed cell death (Shabala et al. 2007).

According to the study of plant species under salinity, Na^+ appears to reach toxic level prior to Cl^- (Munns and Tester 2008), which increases ion toxicity in the leaves of the plants. Thus, it becomes necessary to reduce the accumulated Na^+ level in the cytosol of the cells. The tolerance can be achieved by upregulation or downregulation of the expression of specific transporters and ionic channels, which triggers Na^+ transport (Rajendran et al. 2009; Munns and Tester 2008). The tolerance from Na^+ exclusion has been reported in various cereal crops such as durum wheat, barley, and rice (James et al. 2011). The low level of Na^+ has been regulated in the root cortex, which decreases its accumulation in the leaf blades (Davenport et al. 2005). An effective cytosolic Na^+ exclusion has been conducted via vacuolar Na^+/H^+ antiport, which induces influx of toxic ions from cytosol into vacuolar compartment. These ions appear as osmoticum, which allow plants to survive even in high saline environment by maintaining the water flow within the cellular region. In *Arabidopsis*, the AtNHX1 antiporter has been reported to be localized in the tonoplast, which is involved in the balance of osmotic potential (Apse et al. 1999). Durum wheat is a salt-sensitive species as compared to bread wheat due to reduced potential to exclude Na^+ from the leaf blades (Gorham et al. 1987; Flagella et al. 2006).

A novel source of Na^+ exclusion has been identified in the durum wheat genotype named Line 149, *NAX1* and *NAX2* (Munne-Bosch and Penuelas 2003). *NAX1* gene restricts low Na^+ transport from root to shoot and then in the leaf sheath, while *NAX2* also confers low Na^+ transport from root to shoot and higher K^+ level in the leaf (James et al. 2006).

Salt overly sensitive (SOS) stress signaling pathway is involved in salt tolerance and ion homeostasis (Hasegawa et al. 2000). It comprises of three major proteins, SOS1, SOS2, and SOS3. In plasma membrane, SOS1 encodes Na^+/H^+ antiporter. It controls long-distance Na^+ efflux from root to shoot. It confers salt tolerance during excessive expression. *SOS2* gene encodes serine/threonine kinase. It embraces N-terminal catalytic domain and a C-terminal regulatory domain. In high salt concentration, it is activated by eliciting Ca^+ signals. *SOS3* gene consists of myristoylation site at its N-terminus. It is a myristoylated Ca^+ binding protein and confers salt tolerance. FISL motif is a long sequence of about 21 amino acids, which is present in the C-terminal regulatory site of SOS2 protein. Its interaction with Ca^+ binding SOS3 protein leads to the kinase activation, which phosphorylates SOS1 protein and accelerates its efflux activity (Guo et al. 2004). It also accelerates Na^+ efflux and reduction of Na^+ toxicity (Jiang et al. 2007).

3.3 Tissue Tolerance

Tissue tolerance involves the increased rate of survival of old leaves. It induces the compartmentalization of the ions at the cellular level so as to reduce its toxicity in the mesophyll cells of the leaf (Munns and Tester 2008). It can also be achieved by synthesizing the compatible solute in the cytoplasm (Hasegawa et al. 2000). Compatible solutes are low-molecular-weight chaperones important for plant osmotolerance. They comprise amino acids, betaine, amines, organic acids, and sugars (Mansour 2000). Compatible solute stabilizes cellular membrane or adjusts osmotic potential during elevated salinity level (Ashraf and Foolad 2007). Due to hydrophilic nature, compatible solute can maintain the water homeostasis (Zhu 2001; Sakamoto and Murata 2002).

4 Role of Salicylic Acid in Plant Abiotic Stress

The term "salicylic" was first derived in 1826 by the Italians Brugnatelli and Fontana from willow bark (willow tree, *Salix alba*) as salicin, a glucoside of salicylic alcohol. Salicin was transformed into a sugar and an aromatic compound that on oxidation are converted into salicylic acid (SA), a 2-hydroxybenzoic acid, which is widely used in organic synthesis. Aspirin, also known as acetylsalicylic acid, is a popular derivative used as medicine. Endogenously, within plants, SA exists in minute concentration, as glycosylated, methylated, glucose-ester, or amino acid conjugate form and rarely in the free form (Dempsey et al. 2011).

The phenylpropanoid pathway (Metraux 2002) or isochorismate (IC) pathway is used to synthesize salicylic acid in plants. The end product of the shikimate pathway, chorismic acid in the plastid, is the starting point for SA synthesis (Metraux 2002).

In *Arabidopsis thaliana*, *Nicotiana benthamiana*, and tomato, the IC pathway is the major pathway for SA synthesis (Wildermuth et al. 2001; Uppalapati et al. 2007; Catinot et al. 2008; Zahra et al. 2010). Isochorismate is synthesized by the enzyme isochorismate synthase. Homologs of the enzyme have been identified and characterized from a number of plants including grapevine, pepper, poplar, rice, soybean, tobacco, and tomato (Dempsey et al. 2011; Miura and Tada 2014). *ICS1/SID2* is a vital gene identified in *Arabidopsis* responsible for the pathway (Nawrath and Métraux 1999; Dewdney et al. 2000). It is regulated by both biotic stresses and abiotic stresses (Ogawa et al. 2005; Kilian et al. 2007; Wan et al. 2012). Garcion et al. (2008) reported an ICS2 double mutant producing SA, which provided evidence for an alternate biosynthesis pathway.

The enzyme phenyl ammonia lyase (PAL) plays a crucial role between primary and secondary metabolisms (Weng and Chapple 2010; Dempsey et al. 2011). Trans-cinnamic acid is converted to SA through ortho-coumaric acid or benzoic acid intermediates (El-Basyouni et al. 1964; Ellis and Amrhein 1971; Chadha and Brown 1974; Yalpani et al. 1993).

SA is historically reported to be involved in plant defense response, pathogenesis-related (PR) gene expression, and systemic acquired resistance (Shah 2003; Bari and Jones 2009). SA as a plant growth regulator has a definite role in plant photosynthesis, nitrate metabolism flowering response, and senescence (Hatayama and Takeno 2003; Martinez et al. 2004; Lopez-Delgado and Scott 1997; Stacey et al. 2006; Hayat et al. 2010).

SA is also involved in responses to a variety of abiotic stresses, such as ozone and UV-B, drought, salt and osmotic stress, heat and cold stress, heavy metal stress, etc. (Fariduddin et al. 2003; Kadioglu et al. 2011; Pandey and Chakraborty 2015; Hajer et al. 2016; Yan et al. 2016; Multu et al. 2016; Hasanuzzaman et al. 2017; Machado and Serralheiro 2017).

4.1 Role of SA in Salt Stress Tolerance

As discussed earlier, salinity stress leads to cellular sodium toxicity, destroying the ionic flux as well as causing osmotic stress. More than 50% of agricultural land is influenced by high salt concentrations, especially in the arid and semiarid areas of the world (Patel et al. 2011; Shrivastava and Kumar 2015). During salt stress, SA biosynthesis and concentration are significantly increased. Exogenous application of SA is reported to activate the germination of SA-deficient *NahG* transgenic plants under adverse salinity (Borsani et al. 2001). In SA-deficient *Arabidopsis nahG* plants, necrotic lesions are induced under salt stress (Borsani et al. 2001). SA-treated barley plants showed increase in yield and better photosynthetic performance under salinity stress (El-Tayeb 2005). In maize plants, SA treatment is reported to decrease lipid peroxidation and membrane permeability under salinity stress (Gunes et al. 2007).

SA, generally associated with plant biotic stress, is reported to have considerable cross talk with other plant growth regulators like ABA and is implicated in several drought and salinity tolerance mechanisms (Fujita et al. 2006). Both ABA, a plant growth regulator, and proline, a plant metabolite, are invariably linked to stress conditions, and their mitigation is reported to be activated in SA-treated wheat plants under stress (Sakhabutdinova et al. 2003; Shakirova et al. 2003). Retention of ABA and a better adaptation to salinity stress are also reported in tomato plants treated with SA and grown in hydroponic culture (Szepesi et al. 2009). SA treatment maintains optimal Na^+ and high K^+ concentration, thereby increasing salt tolerance (Kovacik et al. 2009). In common, exogenous application of SA in bean plants improves plant development, but increased endogenous SA decreases production (Palma et al. 2009).

Strawberry plants treated with SA and then subjected to salinity stress show higher chlorophyll concentrations and increased yield with respect to untreated plants (Karlidag et al. 2009). In sunflower plants under salinity stress, exogenous SA treatment is reported to stabilize yield (Noreen and Ashraf 2010). Application of SA at 0.1 or 0.5 mM concentrations to mung bean plants under salinity stress regularized photosynthetic activity and the irregular concentrations of leaf Na^+, Cl^-, and H_2O_2 (Khan et al. 2010). However, a negative feedback mechanism is reported to operate at SA concentration 1.0 mM and higher. Lee et al. (2010) reported similar observations, whereby SA treatment at <50 μM prevented the inhibitory effect of excess salinity, while at high concentration (>100 μM), it elevated the deleterious effect of salinity. Low concentration of exogenous SA (around 0.01–0.05 mM) activates stress-tolerant proteins such as alternative oxidase (AOX), catalase, and heat shock proteins (HSP) to inverse the ROS activity and maintains membrane integrity. The SA response in different plant species has varied optimal concentrations (Yang et al. 2004). In maize cultivars D-1184 and TG-8250, minimal concentration of SA was found to be effective to induce tolerance under salt and drought stress (Manzoor et al. 2015).

Root drenching with SA has also been reported to be beneficial to plants like tomato under salinity stress with subsequent increase in photosynthetic pigments, soluble sugars, and K^+ concentration (Wasti et al. 2012). In *Medicago sativa* inoculated with *Sinorhizobium meliloti*, SA treatment increases photosynthetic capacity and reduces plant growth inhibition under salinity stress (Palma et al. 2013). Nodule biomass decrease that is generally associated with salinity stress was prevented by SA treatment in *Medicago* that led to efficient nitrogen fixation. In *Arabidopsis*, pretreatment with 0.01–0.5 mM SA prevents K^+ efflux from the roots induced by the high salt concentration in the rhizosphere and improves plant growth (Jayakannan et al. 2013). In *Lycopersicon esculentum*, pretreatment of SA along with the polyamine spermidine is reported to have an ameliorative effect under salt stress (Fariduddin et al. 2017). A phytohormone profiling study in wild halophyte tomato species indicated the role of SA in osmotic adjustment under salt stress (Gharbi et al. 2017).

Moderate concentration of exogenous SA (0.1–0.5 mM) positively feedbacks ROS level by inversely regulating antioxidant enzymes. ROS behave as secondary signal to elevate the function of antioxidant enzymes such as ascorbate peroxidase (APX), alternate oxidase (AOX), glutathione reductase (GR), guaiacol peroxidase (GPX), catalase, and superoxide dismutase (SOD). Thus, at optimal level, SA is ROS dependent to induce tolerance against abiotic stress, while high concentration of SA (more than 1 mM) tends to lead oxidative burst and apoptosis (Rao and Davis 1999; Tasgin et al. 2003; Mateo et al. 2006). Earlier, it was proposed that SA downregulate H_2O_2 by inhibiting antioxidant enzyme activity (Durner and Klessig 1996). Later, it was shown that endogenous SA is induced by elevated H_2O_2 levels (Leon et al. 1995). At this concentration, plants positively regulate ROS induction and simultaneously diminish its own strength to eliminate H_2O_2 (Mittler 2002). This leads to accumulation of ROS and activation of apoptosis. NO (nitric oxide), ethylene, and JA (jasmonic acid) are the regulators for apoptosis activity (Dat et al. 2003; Van Breusegem and Dat 2006).

4.2 *Molecular Response to Plants Under Salinity Stress: Responses Similar to SA Treatment*

Although the biochemical mechanisms of SA-induced tolerance to salinity stress in plants have been relatively well studied (Table 1), the physiological effects of osmotic potential deficit as a result of drought and salinity on plant cells are similar, and some common metabolic pathways are either induced or repressed.

Various studies revealed that SA activates defense response by coupling with various receptors. The non-expresser of pathogenesis-related gene 1 (*NPR1* protein) was reported as one of the receptors (Wu et al. 2012). As SA prologues *NPR3* and *NPR4*, which trigger the activation of monomeric *NPR1* in the cytoplasm, the activated *NPR1* then influx the nucleus and function as a transcriptional regulator of defense genes (Fu et al. 2012).

Significantly, *DREB* genes that have been linked to drought stress response are also activated under salinity indicating similar pathways at least in part for seeming divergent stress conditions involving considerable molecular cross talk. *DREB 2A*, which encodes DRE/CRT binding proteins, is induced by SA, dehydration, and high salt stress. *DREB 2B* encoding DRE/CRT binding proteins is also induced by salicylic acid (SA), dehydration, and high salt stress (Nakashima et al. 2000).

RD29A, which encodes a protein with a potential protective function during desiccation, is induced by SA, NaCl, and osmotic stress (Yamaguchi-Shinozaki and Shinozaki 1993a, b). *PR1*, a molecular marker for SA accumulation, generally induced on pathogen attack is also overexpressed on salinity stress. *GPX*, a molecular marker for oxidative stress, is induced both by SA and NaCl stress (Rao and Davis 1999).

*LEA*s containing lysine-rich amino acid domain, and encoding a protein responsible for the "exclusion of solute from the surface of membranes," are induced by salicylic acid (SA); water deficit-related stresses, including salinity and cold stress; and ABA (Gilmour et al. 1992). SA, drought, and salinity stress all induce *HVA1* which encodes group 3LEA protein (Xu et al. 1996).

Table 1 Genes responsive to both salicylic acid and salinity in plants

Sr. No.	Gene	Features	Induced (+)/repressed (−)	References
1.	*RD29A*	Encodes a protein with ability to guard during desiccation	(+) SA, NaCl, osmotic stresses	Borsani et al. (2001)
2.	*PR1*	Molecular marker for SA accumulation	(+) SA, pathogen attack, salinity	Borsani et al. (2001)
3.	*GPX*	Molecular marker for oxidative stress	(+) SA, NaCl stress	Borsani et al. (2001)
4.	*LEA*	Contains lysine-rich amino acid domain, the elimination of solute from the surface of membranes	(+) SA, ABA, and water depletion-related stresses, including salinity and cold	Rajjou et al. (2006)
5.	*DREB 2A*	DRE/CRT binding proteins	(+) SA, high salt stress, dehydration	Nakashima et al. (2000)
6.	*DREB 2B*	DRE/CRT binding proteins	(+) SA, high salt, dehydration	Nakashima et al. (2000)
7.	*TOP2*	Encodes topoisomerase II	(+) salt and drought stress, ABA, SA	Hettiarachchi et al. (2005)
8.	*TaLTP 1*	Facilitate transfer of phospholipids between membranes in vitro	(+) wounding, salt, drought stress, SA	Jang et al. (2004)
9.	*HVA1*	Group 3LEA protein gene	(+), SA, drought, and salinity stress	Kang et al. (2014)
10.	*tWRKY4*	Encode proteins with a single WRKY domain that contain the conserved WRKYGQK sequence	(+) SA, salinity, drought	Chen and Chen (2000)
11.	Ubiquitin	Regulatory protein, produced by UBB, UBC, UBA52, and RPS27A	(+) SA, confer salinity tolerance	Wong et al. (2006)
12.	Cytochrome b6	Part of the electron transport chain	(+) SA, confer salinity tolerance	Wong et al. (2006)
13.	*TaCIPK14*	Encoding a calcineurin B-like protein-interacting protein kinase	(+) SA, confers salinity, cold stress	Deng et al. (2013)
14.	*CtPAL*	Small family of PAL(phenylalanine ammonia lyase) genes	(+) SA (salicylic acid), wounding, and salinity stress	Dehghan, et al. (2014)
15.	CtCHS	Chalcone synthase (CHS), a key enzyme in the synthesis of plant flavonoids	(+) SA (salicylic acid), wounding, and salinity stress	
16.	*Deg2*	Encodes a chloroplast DEG2 protein	(+) SA, salt, wounding, high-temperature and high-irradiance stress	Luciński et al. (2011)
17.	*OsbZIP71*	Encodes a rice bZIP TF	(+) SA, drought, polyethylene glycol (PEG), and ABA (−) salinity	Liu et al. (2014)

(continued)

Table 1 (continued)

Sr. No.	Gene	Features	Induced (+)/repressed (−)	References
18.	*AT4G11175*	Encoding chloroplast translation initiation factor	(+) SA, salinity	Omidbakhshfard et al. (2015)
19.	*ATP5A1*	Encodes ATP synthase complex	(+) SA, (−) salinity	Muneer et al. (2014)
20.	*HEK293*	A specific cell line originally derived from human embry onic kidney cells	(+) SA, (−) salinity	Fiol et al. (2009)
21.	*NPR1*	Redox-sensitive protein, regulator of SA-induced defense genes	(+) SA, confer salinity tolerance	Jayakannan et al. (2015)

Hettiarachchi et al. (2005) reported that a topoisomerase II protein, which is encoded by the gene *TOP2*, is induced by drought and salt stress, ABA, and SA. Wounding, salt, and drought stress and SA induce *TaLTP 1* which aid in the transfer of phospholipids between membranes (Kader 1996; Jang et al. 2004).

Proteins with a single WRKY domain that contain the conserved WRKYGQK sequence are encoded by *tWRKY4* gene, which is induced by SA, salinity, and drought (Chen and Chen 2000). Ubiquitin, which confers salinity tolerance (Wong et al. 2006), is also induced by SA (Amaral et al. 2008). Similarly, Cytochrome b6 is also induced by SA (Amaral et al. 2008) and confers salinity tolerance (Wong et al. 2006). Cytochrome b6-f complex has been induced by salt stress (Xu et al. 2010). Whereas, CYP83B1 and CYP71B7 are downregulated in high salinity concentration (Narusaka et al. 2004).

SA, salinity, and cold stress (Deng et al. 2013) induce TaCIPK14 (Amaral et al. 2008). Both CtPAL and CtCHS are induced by SA (Amaral et al. 2008), wounding, and salinity stress (Dehghan et al. 2014). *Deg2*, which *encodes* a chloroplast DEG2 protein, is also induced by salicylic acid (SA) (Amaral et al. 2008), salinity, wounding and high-temperature, and high-irradiance stress (Luciński et al. 2011). Chloroplast translation initiation factor encoded by AT4G11175 is induced by SA (Amaral et al. 2008) and salinity (Omidbakhshfard et al. 2015).

However, there are also some genes that respond differently under salinity and SA treatment. *HEK293* is induced by SA (Amaral et al. 2008) and repressed by salinity (Fiol et al. 2009). A rice bZIP TF encoded by *OsbZIP71* is induced by SA (Amaral et al. 2008), drought, polyethylene glycol, and ABA whereas repressed by salinity (Liu et al. 2013, 2014). Salinity represses *ATP5A1* which encodes ATP synthase complex (Muneer et al. 2014), while it is induced by SA (Amaral et al. 2008). These might indicate specific responses to salinity or tolerance mechanisms induced by SA treatment. Further studies are required to elucidate the molecular mechanisms to salinity to develop plants better adapted under such stress conditions.

The role of NPR1 protein in SA signaling in biotic and abiotic stress is summarized in Fig. 1. Non-expresser of PR (pathogenesis-related) protein 1 (NPR1) is a SA

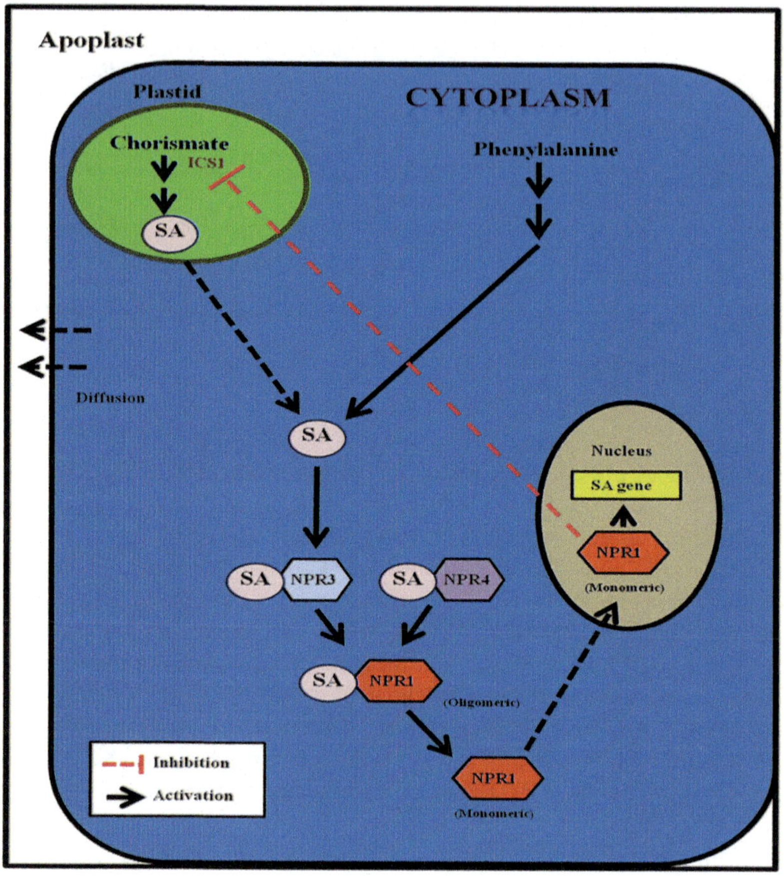

Fig. 1 Role of NPR1 protein in salicylic acid signaling in biotic and abiotic stress

receptor, which acts as a SA-dependent defense regulatory protein of PR gene expression (Vlot et al. 2009; Wu et al. 2012). SA also binds to NPR3 and NPR4 (prologues of NPR1). During low SA concentration, an oligomeric NPR1 is localized as oxidized state in the cytoplasm. When stress level enhances, SA accumulates and alters cellular redox state by activating NPR1 monomers through reduction of oxidized NPR1 oligomer (Dong 2004). The SA-NPR3/NPR4 binding stimulates conversion of oligomeric NPR1 into monomeric form, which transports into nucleus and is complex with specific transcription activators and coactivates SA-responsive PR gene (Fu et al. 2012).

Moreover, the excessive SA concentration can be controlled through negative feedback inhibition of ICS1 (Wildermuth et al. 2001; Zhang et al. 2010). Otherwise, high SA concentration generates hypersensitive response against stresses. Both NPR1-dependent and NPR1-independent mechanisms can control salt tolerance in plants (Jayakannan et al. 2015).

5 Proteomics Study of Plant Salinity Response

Over the last decade, genomics and transcriptomics have found to be important ways to assess the gene expression. A number of genes in response to salinity have been examined in relation to signal transduction, membrane transport, redox reaction, and various other cellular mechanisms. The influx of Na^+ ion and intracellular homeostasis of Na^+/K^+ is maintained by stress-regulating genes including *AtSOS1*, *AtSOS2*, *AVP1*, and *AtNHX1* (Apse et al. 1999; Liu et al. 2000; Shi et al. 2000). In transgenic *Plantago major* plant, *PmSDH1* gene encodes sorbitol dehydrogenase which regulates salinity by accumulating mannitol (Apse et al. 1999). However, there is no relation between mRNA and protein because mRNA is not able to translate into protein and further unable to undergo into posttranscriptional modification (Zorb et al. 2004; Qureshi et al. 2007). Thus, proteomics become an inevitable tool for studying the regulatory mechanism of various proteins which are related to the toxicity and tolerance to various environmental stresses (Table 2).

5.1 Photosynthesis

Ribulose-1,5-bisphosphate carboxylase/oxygenase (Rubisco) constitutes around 50% of protein in water-soluble form, which exists in higher plant leaves (Wostrikoff and Stern 2009; Sudhakar et al. 2016). Rubisco activase catalyzes in vivo activation of Rubisco. It binds to the dormant Rubisco ribulose-1,5-bisphosphate complex and accelerates the disintegration of ribulose-1,5-bisphosphate (RuBP) (Sudhakar et al. 2016). ATP is needed during attachment of Rubisco. Rubisco is the major enzyme involved in the carbon fixation process (Roh et al. 1996). The reduction in Rubisco activase activity under salt stress in *Salvia officinalis* plants deteriorated by salt is corrected by pretreatment of SA (Sahar et al. 2011).

Oxygen-evolving enhancer protein (OEE2) is essential for photosystem (PS) II stability and oxygen evolution processes. OEE1, OEE2, and OEE3 are major proteins, which encode for nuclear genes *PsbO*, *PsbP*, and *PsbQ*, respectively (Miyao and Murata 1989). In higher plants, OEE2 can combine with PSII core complex by OEE1 and OEE3, which expel easily in the presence of salt (Seidler 1996). In the study, it was analyzed that OEE2 could adapt salinity in photoautotrophically cultured green tobacco cells (Murota et al. 1994), rice, and mangrove (Abbasi and Komatsu 2004; Sugihara et al. 2000). Phosphoglycerate kinase catalyzes phosphorylated 3-phosphoglycerate to 1,3-bisphosphoglycerate reaction in Calvin cycle with the utility of ATP. It regulates the initial phase of salinity (Yeo et al. 1991). Porphobilinogen deaminase (PBG deaminase) regulates the synthesis of photosynthetic pigment under stress condition (Cornah et al. 2003).

Table 2 Proteins responsive to both salicylic acid and salinity in plants

Protein	Inducer (+)/repressor (−)	References
Cell wall-related proteins		
Profilin	(+) Salinity	Ramachandran et al. (2000) and Shavrukov et al. (2010)
Germin-like protein	(+) Salinity	Nakata et al. (2002)
V-ATPase	(+) Salinity	Zhou et al. (2010)
SAM	(+) Salinity	Roeder et al. (2009)
Energy		
ATP synthase CF1 beta subunit	(−) Salinity, (+) SA	Parker et al. (2006)
ATP synthase beta subunit	(−) Salinity, (+) SA	Parker et al. (2006)
Metabolism		
Glutamine synthetase isoform GS1c	(−) Salinity, (+) SA	Hoshida et al. (2000)
Ferredoxin-NADP (H) oxidoreductase	(−) Salinity, SA	
Plastid glutamine synthetase 2	(−) Salinity, (−) SA	Hoshida et al. (2000)
Transketolase	(−) Salinity, (+) SA	Bhargava et al. (2008)
Glyceraldehyde-3-phosphate dehydrogenase B	(−) Salinity, (+) SA	Jeong et al. (2001)
Phosphoglycolate phosphatase-like	(+) Salinity, SA	
Glyceraldehyde-3-phosphate dehydrogenase A	(−) Salinity, SA	Jeong et al. (2001) and Zhang et al. (2011)
Isocitrate dehydrogenase	(+) Salinity, SA	Jeong et al. (2001)
Triosephosphate isomerase	(−) Salinity, SA	Gao et al. (2011)
Serine-type peptide	(−) Salinity, SA	
Adenosine diphosphate glucose pyrophosphatase	(+) Salinity, SA (−) salinity + SA	Dong et al. (2011)
Isopentenyl pyrophosphate isomerase	(+) Salinity, SA	Dong et al. (2011)
Photosynthesis		
Oxygen-evolving enhancer protein 2,chloroplastic	(+) Salinity, SA	Sugihara et al. (2000)
Chlorophyll a-b binding protein 8	(+) Salinity, SA	Sugihara et al. (2000)
Ribulose-1,5-bisphosphate carboxylase activase	(+) Salinity, SA	Fatehi et al. (2012)
Thylakoid luminal 19 kDa protein	(+) Salinity, SA	Sugihara et al. (2000)
Photosystem II stability/assembly factor HCF 136	(+) Salinity, SA	Yi et al. (2005)
Ribulose-1,5-bisphosphate carboxylase activase isoform	(+) Salinity, SA	Fatehi et al. (2012)
Putative inner envelope protein	(+) Salinity, SA	Sugihara et al. (2000)
Ribulose-1,5-bisphosphate carboxylase activase isoform 1	(+) Salinity, SA	Fatehi et al. (2012)
Ribulose bisphosphate carboxylase activase B	(+) Salinity, SA	Fatehi et al. (2012)

(continued)

Table 2 (continued)

Protein	Inducer (+)/repressor (−)	References
Oxygen-evolving enhancer protein 1	(+) Salinity, SA	Sugihara et al. (2000)
Ribulose bisphosphate oxygenase activase B	(−) Salinity, (+) SA	Fatehi et al. (2012)
Phosphoglycerate kinase	(+) Salinity	Joshi et al. (2016)
Porphobilinogen deaminase	(−) Salinity	Cornah et al. (2003)
Protein translation and degradation		
RNA binding proteins	(+) Salinity	Gong et al. (2001)
Chloroplast-localized cyclophilins	(+) Salinity	Kumari et al. (2013)
Transcription factor		
NAC	(+/−) Salinity	Chen et al. (2009) and Yan et al. (2005)
Scavenging of ROS		
Glycine decarboxylase	(+) Salinity	Veeranagamallaiah et al. (2008)
Trx	(+) Salinity	Amer and Holmgren (2000)
DHAR	(+) Salinity	Ushimarua et al. (2006)
Signal transduction		
14-3-3 protein	(+) SA, salinity	Guo et al. (2012)
Translationally controlled tumor protein	(+/−) Salinity	Witzel et al. (2010)
Guanine nucleotide binding proteins	(+) Salinity	Assmann (2005) and Neves et al. (2002)
Stress defense		
Ascorbate peroxidase	(+) SA, salinity, SA + salinity	Quiroga et al. (2000)
2-Cys peroxiredoxin BAS1	(+) SA, salinity, SA + salinity	König et al. (2003)
Salt stress root protein RS1-like	(+) SA, salinity	Kang et al. (2014)
Unknown		
Cp31BHv	(+) Salinity, SA	Garcia et al. (1998) and Alikhani et al. (2013)

5.2 *Protein Translation and Degradation*

Moreover, the different expression pattern with rise and drop in chloroplast RNA binding of protein was observed in Afzal and Line 527 genotype of barley, respectively. RNA binding proteins (RBPs) intermediate gene expression with posttranscriptional modifications in RNAs (Curtis et al. 1995; Johnstone and Lasko 2001). Chloroplast-localized cyclophilins are important stress-stimulating proteins surrounding in subcellular compartments, which induce protection in cellular stress condition (Chou and Gasser 1997).

5.3 Signal Transduction

Translationally controlled tumor protein (TCTP) has a high affinity with calcium under salt stress (Sanchez et al. 1997; Gong et al. 2001). In a grain proteomic study, it was observed that TCTP was obtained at different levels in saline-sensitive and saline-tolerant barley genotypes during germination (Witzel et al. 2010). Guanine nucleotide binding proteins (G protein) are involved in trans-membrane signaling and regulate ionic channels, metabolic enzyme activity, motility, secretion, and contractility (Neves et al. 2002; Assmann 2005). $G_{p\beta}L$ induces stamen formation, and maturation of pollen (Peskan-Berghofer et al. 2005) also upregulates the machinery for resisting oxidative stress (Requejo and Tena 2006).

Phosphoserine-binding proteins and 14-3-3 proteins belong to the acidic protein family and are composed of various isoproteins present in plants as well as in mammals (Mhawech 2005; Wang et al. 2008a, b). 14-3-3 protein can either upregulate or downregulate in stress condition. In salt stress, 14-3-3 protein exhibits distinct gene pattern in *Solanum lycopersicum* root (Xu and Shi 2006), downregulates in *Arabidopsis* at 200 mM NaCl (Ndimba et al. 2005), and upregulates in two varieties of wheat (Wang et al. 2008a, b).

5.4 Cell Wall-Related Proteins

The ubiquitous protein, profiling, has high affinity to polymerize or depolymerize actin filaments, which in turn influence the cytoskeleton structure of plant (Ramachandran et al. 2000). In *Suaeda aegyptiaca*, profilin upregulates cellular behavior by adjusting the substantial salt concentration and reduces the subsequent ion toxicity in plant (Shavrukov et al. 2010). During salinity, germin-like protein (GLP) expression was observed in barley root (Hurkman et al. 1994) and *Arabidopsis* root (Jiang et al. 2007). GLP was found to be involved in even biotic stress (Hurkman et al. 1994).

5.5 Nitrogen, Carbon, and Amino Acid Metabolism

In plants, V-ATPase sustains partitioning of ions into different compartments so as to reduce the toxicity of ions (Golldack and Dietz 2001). In transgenic tobacco plants, increased activity of V-ATPase stimulates vacuolar Na^+/H^+ antiporter AtNHX1 so as to improve tolerance to salinity (Zhou et al. 2010). In plants, S-adenosylmethionine (SAM) synthetase behaves as a donor of methyl group in transmethylation of nucleic acids and proteins or a precursor in biosynthesis pathway of biotin, nicotianamine, and polyamines (Roeder et al. 2009). Under drought stress, it stimulates the synthesis of betaine for improving the survival ability of seedlings (Mayne et al. 1996). SAM expression was found to be higher in salt-tolerant cultivar (Apse et al. 1999).

5.6 Transcription Factor

The nascent chain associated complex (NAC) plays a major role in protein sorting and translocation and promotes the targeting of nascent polypeptide chains to the endoplasmic reticulum (Rospert et al. 2002). Upstream regulation of NAC was seen in tomato during salt stress (Chen et al. 2009), while downstream regulation of NAC was observed in roots of rice in salinity (Yan et al. 2005).

5.7 Scavenging of ROS

In mitochondria, glycine decarboxylase (GD) catalyzes glycine to serine conversion in the photorespiratory cycle (Vauclare et al. 1996). GD upregulates in response to salt stress, drought, and chilling (Taylor et al. 2005; Veeranagamallaiah et al. 2008). In *Arabidopsis*, AtTrx h isoform is regulated in response to oxidative stress and pathogens (Laloi et al. 2004). A ubiquitous protein, Trx, is a disulfide reductase and maintains redox reaction through electron donation to enzyme such as peroxiredoxin, ribonucleotide reductase, and MSR (Amer and Holmgren 2000). Rice dehydroascorbate reductase (DHAR) regulates the expression in transgenic *Arabidopsis thaliana* in adverse salt stress (Ushimarua et al. 2006). DHAR also stimulate ozone tolerance and resistance to fungal symbiosis with *Arabidopsis* (Yoshida et al. 2006; Vadasserya et al. 2009).

Plant 2-Cys Prx comprise hem-free peroxidases with a cysteine residue in the active site. It regulates protection and redox potential of thylakoid membrane by activating antioxidant activity (König et al. 2003). Moreover, a tomato peroxidase gene, *TPX1*, was found to be upregulated in 100 mM NaCl by encoding a particular isoenzyme (Botella et al. 1994).

6 Conclusion

Plants are continuously living under a variety of biotic and abiotic stresses. Salinity is one of the most consequential stresses limiting the productivity of agricultural crops and also affects germination, plant strength, and crop yield. Soil salinity alters the physiological and biochemical aspects of around 800 Mha of arable land (Munns 2005), which necessitates to improve the salinity tolerance level in cropping system so as to meet the need of a growing population. Salinity is a major abiotic stress, which causes metabolic disruption either due to imbalance in soil moisture content or excessive ionic accumulation in plants (Yong et al. 2014). Salt stress induces the generation of reactive oxygen species, which in turn affects the photosynthetic activity of the plants. Excessive salt impairs growth and initiates wilting which ultimately leads to plant death.

Plants have the ability to stimulate various genes related to stress and to generate activated protein so as to overcome salt stress. Plants have evolved several mechanisms to acclimatize to salinity. The mechanisms of genetic control by which plants tolerate the salt stress are very complex and have not yet properly understood. Several biomolecules have been discovered within plants that modulate mechanisms to effectively deal with salinity stress. SA is a secondary metabolite belonging to the polyphenols and regulates plant growth even in adverse environmental stresses (Horvath et al. 2007). SA modulates the gene behavior of PR proteins (Klessig and Malamy 1994) and intermediates the ethylene biosynthesis and absorption of K^+ in plants (Leslie and Romani 1986). In addition, it also enhances photosynthesis rate as reported in *B. juncea* where a declined under 50 mM NaCl treatment was improved by 10^{-5} M SA treatment (Yusuf et al. 2008).

Studying the plants' response to salinity stress and with regard to the role of salicylic acid indicates modulating tolerance mechanisms, especially at the molecular level. Although the biochemical mechanisms of SA-induced tolerance to salinity stress in plants have been relatively well studied, not much is known about the molecular mechanisms underlying the process. There are also some genes that respond differently under salinity and SA treatment. These might indicate specific responses to salinity or tolerance mechanisms induced by SA treatment and facilitate to elucidate the molecular responses to salinity to develop plants better adapted under such stress conditions.

References

Abbasi FM, Komatsu S (2004) A proteomic approach to analyze salt-responsive proteins in rice leaf sheath. Proteomics 4:2072–2081

Arfan M, Athar HR, Ashraf M (2007) Does exogenous application of salicylic acid through the rooting medium modulate growth and photosynthetic capacity in two differently adapted spring wheat cultivars under salt stress? J Plant Physiol 164(6):685–694

Akbarimoghaddam H, Galavi M, Ghanbari A, Panjehkeh N (2011) Salinity effects on seed germination and seedling growth of bread wheat cultivars. Trakia J Sci 9:43–50

Alamillo JK, Saénz P, García JA (2006) Salicylic acid-mediated and RNA-silencing defense mechanisms cooperate in the restriction of systemic spread of *Plum pox virus* in tobacco. Plant J 48:217–227

Albert R (1975) Salt regulation in halophytes. Oecologia 21(1):57–71

Al-Doss AA, Smith SE (1998) Registration of AZ-97MEC and AZ-97MEC-ST very non-dormant alfalfa germplasm pools with increased shoot weight and differential response to saline irrigation. Crop Sci 38:568–568

Alikhani M, Khatabi B, Sepehri M, Nekouei MK, Mardi M, Salekdeh GH (2013) A proteomics approach to study the molecular basis of enhanced salt tolerance in barley (*Hordeum vulgare* L.) conferred by the root mutualistic fungus *Piriformospora indica*. Mol Bio Syst 9:1498–1510

Allen JA, Chambers JL, Stine M (1994) Prospects for increasing the salt tolerance of forest trees: a review. Tree Physiol 14(7–8–9):843–853

Amaral DOJ, Lima MMA, Resende LV, Silva M (2008) Differential gene expression, induced by salicylic acid and *Fusarium oxysporum* f. sp. *Lycopersici* infection, in tomato. Pesq Agropec Bras 43(8):1017–1023

Amer ESJ, Holmgren A (2000) Physiological functions of thioredoxin and thioredoxin reductase. Eur J Biochem 267:6102–6109

Amirjani MR (2011) Effect of salinity stress on growth, sugar content, pigments and enzyme activity of rice. Int J Bot 7:73–81

Apse MP, Aharon GS, Snedden WA, Blumwald E (1999) Salt tolerance conferred by overexpression of a vacuolar Na^{+}/H^{+} antiport in *Arabidopsis*. Science 285:1256–1258

Asch F, Dingkuhn M, Miezan K, Doerffling K (2000) Leaf K/Na ratio predicts salinity induced yield loss in irrigated rice. Euphytica 113:109–118

Ashraf M, Foolad MR (2007) Roles of glycine betaine and proline in improving plant abiotic stress resistance. Environ Exp Bot 59(2):206–216

Assmann SM (2005) G proteins Go green: a plant G protein signaling FAQ sheet. Science 310:71–73

Bar Y, Apelbaum A, Kafka fi U, Goren R (1997) Relationship between chloride and nitrate and its effect on growth and mineral composition of avocado and citrus plants. J Plant Nutr 20:715–731

Bari R, Jones JDG (2009) Role of plant hormones in plant defence responses. Plant Mol Biol 69 (4):473–488

Bhargava P, Mishra Y, Srivastava A, Narayan O, Rai L (2008) Excess copper induces anoxygenic photosynthesis in Anabaena doliolum: a homology based proteomic assessment of its survival strategy. Photosynth Res 96:61–74

Borsani O, Valpuesta V, Botella MA (2001) Evidence for a role of salicylic acid in the oxidative damage generated by NaCl and osmotic stress in *Arabidopsis* seedlings. Plant Physiol 126:1024–1030

Botella MA, Quesada MA, Hasegawa PM, Valpuesta V (1994) Nucleotide sequences of two peroxidase genes from tomato (*Lycopersicon esculentum*). Plant Physiol 103:665–666

Breviario D, Genga A (2013) Stress response in rice. J Rice Res 2(1):100e104

Brugnoli E, Björkman O (1992) Growth of cotton under continuous salinity stress: Influence on allocation pattern, stomatal and non-stomatal components of photosynthesis and dissipation of excess light energy. Planta 187(3):335–347

Brussaard L, De Ruiter PC, Brown GG (2007) Soil biodiversity for agricultural sustainability. Agric Ecosys Environ 121:233–244

Carpici EB, Celik N, Bayram G (2009) Effects of salt stress on germination of some maize (*Zea mays* L.) cultivars. Afr J Biotechnol 8:4918–4922

Catinot J, Buchala A, Abou-Mansour E, Metraux JP (2008) Salicylic acid production in response to biotic and abiotic stress depends on isochorismate in *Nicotiana benthamiana*. FEBS Lett 582:473–478

Chadha KC, Brown SA (1974) Biosynthesis of phenolic acids in tomato plants infected with *Agrobacterium tumefaciens*. Can J Bot 52:2041–2047

Chen C, Chen Z (2000) Isolation and characterization of two pathogen- and salicylic acid-induced genes encoding WRKY DNA-binding proteins from tobacco. Plant Mol Biol 42(2):387–396

Chen S, Gollop N, Heuer B (2009) Proteomic analysis of salt-stressed tomato *(Solanum lycopersicum)* seedlings: effect of genotype and exogenous application of glycinebetaine. J Exp Biotechnol 60:2005–2019

Chou IT, Gasser CS (1997) Characterization of the cyclophilins gene family of *Arabidopsis thaliana* and phylogenetics analysis of known cyclophilins proteins. Plant Mol Biol 35:873–892

Chutipaijit S, Chaum S, Sompornpailin K (2011) High contents of proline and anthocyanin increase protective response to salinity in *Oryza sativa* L. spp. indica. Aust J Crop Sci 5:1191–1198
Cornah JE, Terry MJ, Smith AG (2003) Green or red: what stops the traffic in the tetrapyrrole pathway? Trends Plant Sci 85:224–230
Cueto-Ginzo AI, Serrano L, Bostock R, Ferrio JP, Rodríguez R et al (2016a) Salicylic acid mitigates physiological and proteomic changes induced by the SPCP1 strain of *Potato virus* X in tomato plants. Physiol Mol Plant Pathol J 93:1–11
Cueto-Ginzo IA, Serrano L, Sin E, Rodríguez R, Morales JG et al (2016b) Exogenous salicylic acid treatment delays initial infection and counteracts alterations induced by *Maize dwarf mosaic virus* in the maize proteome. Physiol Mol Plant Pathol 96:47–59
Curtis D, Lehmann R, Zamore PD (1995) Translational regulation in development. Cell 81:171–178
Dat JF, Pellinen R, Cotte BVD, Langerbartels C, Kangasjarvi J, Inze D, Van Breusegem F (2003) Changes in hydrogen peroxide homeostasis trigger an active cell death process in tobacco. Plant J 33:621–632
Davenport R, James R, Zakrisson-Plogander A, Tester M, Munns R (2005) Control of sodium transport in durum wheat. Plant Physiol 137:807–818
Dehghan S, Sadeghi M, Pöppel A, Fischer R, Lakes-Harlan R, Kavousi HR, Vilcinskas A, Rahnamaeian M (2014) Differential inductions of phenylalanine ammonia-lyase and chalcone synthase during wounding, salicylic acid treatment, and salinity stress in safflower, *Carthamus tinctorius*. Bio Sci Rep 3:25–34
Dempsey DA, Vlot AC, Wildermuth MC, Klessig DF (2011) Salicylic acid biosynthesis and metabolism. Arabidopsis Book 9:e0156. https://doi.org/10.1199/tab.0156
Deng X, Zhou S, Hu W, Feng J, Zhang F, Chen L, Huang C, Luo Q, He Y, Yang G, He G (2013) Ectopic expression of wheat TaCIPK14, encoding a calcineurin B-like protein-interacting protein kinase, confers salinity and cold tolerance in tobacco. Physiol Plant 149(3):367–377
Derikvand H, Azadbakht A (2017) An Impedimetric sensor comprising magnetic nanoparticles-graphene oxide and carbon nanotube for the electrocatalytic oxidation of salicylic acid. J Inorg Organomet Polym Mater 27(4):901–911
Dewdney J, Reuber TL, Wildermuth MC, Devoto A, Cui J, Stutius LM et al (2000) Three unique mutants of *Arabidopsis* identify eds loci required for limiting growth of a biotrophic fungal pathogen. Plant J 24:205–218
Dierig DA, Shannon MC, Grieve CM (2001) Registration of WCL-SL 1 salt tolerant *Lesquerella fendleri* germplasm. Crop Sci 41:604–605
Dong X (2004) NPR1, all things considered. Curr Opin Plant Biol 7(5):547–552
Dong C, Wang X, Shang Q (2011) Salicylic acid regulates sugar metabolism that confers tolerance to salinity stress in cucumber seedlings. SCI Hortic-Amst 129:629–636
Durner J, Klessig DF (1996) Salicylic acid is a modulator of tobacco and mammalian catalase. J Biol Chem 271:28492–28501
El-Basyouni SZ, Chen D, Ibrahim RK, Neish AC, Towers GHN (1964) The biosynthesis of Hydroxybenzoic acids in higher plants. Phytochemistry 3:485–492
Ellis BE, Amrhein N (1971) The 'NIH-shift' during aromatic ortho-hydroxylation in higher plants. Phytochemistry 10:3069–3072
El-Tayeb MA (2005) Response of barley grains to the interactive effect of salinity and salicylic acid. Plant Growth Regul 45:215–224
Epstein E (1977) Genetic potentials for solving problems of soil mineral stress: adaptation of crops to salinity. In: Wright MJ (ed) Plant adaptation to mineral stress in problem soils. Cornell University Agricultural Experiment Station, Ithaca, pp 73–123
Epstein E, Norlyn JD, Rush DW, Kingsbury R, Kelley DB, Wrana AF (1980) Saline culture of crops: a genetic approach. Science 210:399–404
Essa TA (2002) Effect of salinity stress on growth and nutrient composition of three soybean (*Glycine max* L. Merrill) cultivars. J Agron Crop Sci 88:86–93

Fahad S, Bano A (2012) Effect of salicylic acid on physiological and biochemical characterization of maize grown in saline area. Pak J Bot 44(4):1433–1438

Farahbakhsh H, Pour AP, Reiahi N (2017) Physiological response of henna (*Lawsonia inermis* L.) to salicylic acid and salinity. Plant Prod Sci 20(2):237–247

Fariduddin Q, Hayat S, Ahmad A (2003) Salicylic acid influences net photosynthetic rate, carboxylation efficiency, nitrate reductase activity, and seed yield in *Brassica juncea*. Photosynthetica 41:281–584

Fariduddin Q, Khan TA, Yusuf M Aafaqee ST, Khalil RRAE (2017) Ameliorative role of salicylic acid and spermidine in the presence of excess salt in *Lycopersicon esculentum*. Photosynthetica 1–13

Fatehi F, Hosseinzadeh A, Alizadeh A, Brimavandi T, Sturuik P (2012) The proteome response of salt-resistant and salt-sensitive barley genotypes to long-term salinity stress. Mol Biol Rep 39:6387–6397

Fiol DF, Sanmarti E, Sacchi R, Kültz D (2009) A novel tilapia prolactin receptor is functionally distinct from its paralog. J Exp Biol 212:2007–2015

Flagella Z, Trono D, Pompa M, Di Fonzo N, Pastore D (2006) Seawater stress applied at germination affects mitochondrial function in durum wheat (*Triticum durum*) early seedlings. Funct Plant Biol 33(4):357–366

Flower TJ, Yeo AR (1995) Breeding for salinity resistance in crop plants-where next? Aust J Plant Physiol 22:875–884

Flowers TJ (2004) Improving crop salt tolerance. J Exp Bot 55(396):307–319

Flowers TJ, Colmer TD (2008) Salinity tolerance in halophytes. New Phytol 179(4):945–963

Fu ZQ, Yan S, Saleh A, Wang W, Ruble J, Oka N, Mohan R, Spoel SH, Tada Y, Zheng N (2012) NPR3 and NPR4 are receptors for the immune signal salicylic acid in plants. Nature 486:228–232

Fujita M, Fujita Y, Noutoshi Y, Takahashi F, Narusaka Y et al (2006) Crosstalk between abiotic and biotic stress responses: a current view from the points of convergence in the stress signaling networks. Curr Opin Plant Biol 9:436–442

Gao JP, Chao DY, Lin HX (2007) Understanding abiotic stress tolerance mechanisms: recent studies on stress response in rice. J Integr Plant Biol 49(6):742–750

Gao P, Bai X, Yang L et al (2011) osa-MIR393: a salinity- and alkaline stress-related microRNA gene. Mol Biol Rep 38(1):237–242

Garcia A, Engler A, Claes B, Villarroel R, Montagu M, Gerats T, Caplan A (1998) The expression of the salt-responsive gene salt from rice is regulated by hormonal and developmental cues. Planta 207:172–180

Garcion C, Lohmann A, Lamodiere E, Catinot J, Buchala A, Doermann P, Metraux JP (2008) Characterization and biological function of the ISOCHORISMATE SYNTHASE2 gene of Arabidopsis. Plant Physiol 147(3):1279–1287

Gharbi E, Martinez JP, Benahmed H, Hichri I, Dovrev PI, Motyka V, Quinet M, Lutts S (2017) Phytohormone profiling in relation to osmotic adjustment in NaCl-treated plants of the halophyte tomato wild relative species *Solanum chilense* comparatively to the cultivated glycophyte *Solanum lycopersicum*. Plant Sci 258:77–89

Ghoulam C, Foursy A, Fares K (2002) Effects of salt stress on growth, inorganic ions and proline accumulation in relation to osmotic adjustment in five sugar beet cultivars. Environ Exp Bot 47:39–50

Gilmour SJ, Aetus NN, Thomashow MF (1992) cDNA sequence analysis and expression of two cold-regulated genes of *Arabidopsis thaliana*. Plant Mol Biol 18:13–21

Golldack D, Dietz KJ (2001) Salt-induced expression of the vacuolar H^+-ATPase in the common ice plant is developmentally controlled and tissue specific. Plant Physiol 125:1643–1654

Gomes-Filho E, Machado Lima CRF, Costa JH, Da Silva AC, Da Guia Silva Lima M, De Lacerda CF, Prisco JT (2008) Cowpea ribonuclease: properties and effect of NaCl-salinity on its activation during seed germination and seedling establishment. Plant Cell Rep 27:147–157

Gong Z, Kiowa H, Cushman JC et al (2001) Genes that are uniquely stress-regulated in salt overly sensitive (*SOS*) mutants. Plant Physiol 126:363–375

Gorham J, Hardy C, Jones RGW, Joppa LR, Law CN (1987) Chromosomal location of a K/Na discrimination character in the D-genome of wheat. Theor Appl Genet 74(5):584–588

Grattan SR, Grieve CM (1999) Salinity-mineral nutrient relations in horticultural crops. Sci Hortic 78:127–157

Gueta-Dahan Y, Yaniv Z, Zilinskas BA, Ben-Hayyim G (1997) Salt and oxidative stress: similar and specific responses and their relation to salt tolerance in citrus. Planta 204:460–469

Gunes A, Inal A, Alpaslan M, Eraslan F, Bagci EG, Cicek N (2007) Salicylic acid induced changes on some physiological parameters symptomatic for oxidative stress and mineral nutrition in maize (*Zea mays* L.) grown under salinity. J Plant Physiol 164:728–796

Guo Y, Qin Q, Quintero et al (2004) Transgenic evaluation of activated mutant alleles of SOS2 reveals a critical requirement for its kinase activity and C-terminal regulatory domain for salt tolerance in *Arabidopsis thaliana*. Plant Cell 16(2):435–449

Guo G, Ge P, Ma C, Li X, Lv D, Wang S, Ma W, Yan Y (2012) Comparative proteomic analysis of salt response proteins in seedling roots of two wheat varieties. J Proteome 75:1867–1885

Hajer M, Salma W, Arafet M, Emna G, Abdellah C, Bertrand V, Stanely L, Ben AH (2016) Does salicylic acid (SA) improve tolerance to salt stress in plants? A study of SA effects on tomato plant growth, water dynamics, photosynthesis, and biochemical parameters. OMICS: J Integr Biol 20(3):180–190

Halliwell B, Gutteridge JMC (1985) The importance of free radicals and catalytic metal ions in human diseases. Mol Asp Med 8(2):89–193

Hasanuzzaman M, Fujita M (2011a) Selenium pretreatment upregulates the antioxidant defense and methylglyoxal detoxification system and confers enhanced tolerance to drought stress in rapeseed seedlings. Biol Trace Elem Res 143:1758–1776

Hasanuzzaman M, Fujita M (2011b) Exogenous silicon treatment alleviates salinity-induced damage in *Brassica napus* L. seedlings by up-regulating the antioxidant defense and methylglyoxal detoxification system. In: Proceedings of the annual meeting of the American Society of Plant Biologists, Minneapolis, MN, USA, 6–10 August 2011

Hasanuzzaman M, Nahar K, Gill SS, Gill R, Fujita M (2014) Drought stress responses in plants, oxidative stress, Q1 and antioxidant defense. In: Tuteja N, Gill SS (eds) Climate change and plant abiotic stress tolerance. Wiley-Blackwell, Weinheim, pp 209–237

Hasanuzzaman M, Nahar K, Bhuiyan TF, Anee TI, Inafuku M, Oku H, Fujita M (2017) Salicylic acid: an all-rounder in regulating abiotic stress responses in plants. In: Dr. El-Esawi M (ed) Phytohormones – signaling mechanisms and crosstalk in plant development and stress responses. InTech. https://doi.org/10.5772/intechopen.68213

Hasegawa PM, Bressan RA, Zhu JK, Bohnert HJ (2000) Plant cellular and molecular responses to high salinity. Ann Rev Plant Physiol Plant Mol Biol 51:463–499

Hatayama T, Takeno K (2003) The metabolic pathway of salicylic acid rather than of chlorogenic acid is involved in the stress-induced flowering of *Pharbitis nil*. J Plant Physiol 160:461–467

Hayat S, Ahmad A (2007) Salicylic acid: a plant hormone. In: Janda T, Horvath E, Szalai G, Paldi E (eds) Role of salicylic acid in the induction of abiotic stress tolerance. Springer, Dordrecht, pp 91–150

Hayat Q, Hayat S, Irfan M, Ahmad A (2010) Effect of exogenous salicylic acid under changing environment: a review. Environ Exp Bot 68:14–25

Hettiarachchi GHCM, Reddy MK, Sopory SK, Chattopadhyay S (2005) Regulation of *TOP2* by various abiotic stresses including cold and salinity in pea and transgenic tobacco plants. Plant Cell Physiol 46(7):1154–1160

Horvath E, Szalai G, Janda T (2007) Induction of abiotic stress tolerance by salicylic acid signaling. J Plant Growth Reg 26:290–300

Hoshida H, Tanaka Y, Hibino T, Hayashi Y, Tanaka A, Takabe T (2000) Enhanced tolerance to salt stress in transgenic rice that overexpresses chloroplast glutamine synthetase. Plant Mol Biol 43:103–111

Hu Y, Schmidhalter U (1997) Interactive effects of salinity and macronutrient level on wheat. J Plant Nutr 20:1169–1182

Hurkman WJ, Lane BG, Tanaka CK (1994) Nucleotide sequence of a transcript encoding a germin-like protein that is present in salt-stressed barley (*Hordeum vulgare* L.) roots. Plant Physiol 104:803–804

Ibrar M, Jabeen M, Tabassum J, Hussain F, Ilahi I (2003) Salt tolerance potential of *Brassica juncea* Linn. J Sci Tech Univ Peshawar 27:79–84

Jabeen M, Ibrar M, Azim F, Hussain F, Ilahi I (2003) The effect of sodium chloride salinity on germination and productivity of Mung bean (*Vigna mungo* Linn.) J Sci Tech Univ Peshawar 27:1–5

James R, Davenport R, Munns R (2006) Physiological characterization of two genes for Na^+ exclusion in durum wheat, Nax1 and Nax2. Plant Physiol 142(4):1537–1547

James RA, Caemmerer SV, Condon AG, Zwart AB, Munns R (2008) Genetic variation in tolerance to the osmotic stress component of salinity stress in durum wheat. Funct Plant Biol 35 (2):111–123

James RA, Blake C, Byrt CS, Munns R (2011) Major genes for Na^+ exclusion, Nax1 and Nax2 (wheat HKT1;4 and HKT1;5), decrease Na^+ accumulation in bread wheat leaves under saline and waterlogged conditions. J Exp Bot 62(8):2939–2947

Jang CS, Lee HJ, Chang SJ, Seo YW (2004) Expression and promoter analysis of the *TaLTP1* gene induced by drought and salt stress in wheat (*Triticum aestivum* L.) Plant Sci 167:995–1001

Jayakannan M, Bose J, Babourina O, Rengel Z, Shabala S (2013) Salicylic acid improves salinity tolerance in *Arabidopsis* by restoring membrane potential and preventing salt-induced K^+ loss via a GORK channel. J Exp Bot 64:2255–2268

Jayakannan M, Bose J, Babourina O et al (2015) NPR1-dependent salicylic acid signaling pathway is pivotal for enhanced salt and oxidative stress tolerance in Arabidopsis. J Exp Bot 66 (7):1865–1875

Jeong M, Park S, Byun M (2001) Improvement of salt tolerance in transgenic potato plants by glyceraldehyde-3-phosphate dehydrogenase gene transfer. Mol Cell 12:185–189

Jiang Y, Yang B, Harris NS, Deyholos MK (2007) Comparative proteomic analysis of NaCl stress-responsive proteins in Arabidopsis roots. J Exp Bot 58:3591–3607

Johnstone O, Lasko P (2001) Translational regulation and RNA localization in *Drosophila* oocytes and embryos. Ann Rev Genet 35:365–406

Joshi R, Karan R, Singla-Pareek SL, Pareek A (2016) Ectopic expression of Pokkali phosphoglycerate kinase-2 (OsPGK2-P) improves yield in tobacco plants under salinity stress. Plant Cell Rep 35:27–41

Kader JC (1996) Lipid transfer proteins in plants. Annu Rev Plant Physiol Plant Mol Biol 47:627–654

Kadioglu A, Sağlam A, Terzi R, Acet T, Saruhan N (2011) Exogenous salicylic acid alleviates effects of long term drought stress and delays leaf rolling by inducing antioxidant system. Plant Growth Reg 64(1):27–37

Kang G, Li G, Guo T (2014) Molecular mechanism of salicylic acid-induced abiotic stress tolerance in higher plants. Acta Physiol Plant 36:2287–2297

Karlidag H, Yildirim E, Turan M (2009) Salicylic acid ameliorates the adverse effect of salt stress on strawberry. Sci Agric 66:180–187

Khan MA, Weber DJ (2008) Ecophysiology of high salinity tolerant plants (tasks for vegetation science), 1st edn. Springer, Amsterdam

Khan N, Syeed S, Masood A, Nazar R, Iqbal N (2010) Application of salicylic acid increases contents of nutrients and antioxidative metabolism in mungbean and alleviates adverse effects of salinity stress. Int J Plant Biol 1:1–8

Kilian J, Whitehead D, Horak J, Wanke D, Weinl S, Batistic O et al (2007) The AtGenExpress global stress expression data set: protocols, evaluation and model data analysis of UV-B light, drought and cold stress responses. Plant J 50:347–363

Kitamura Y, Yang SL, Shimizu K (2014) Secondary salinization and its countermeasures. In: Tsunekawa A, Liu G, Yamanaka N, Du S (eds) Restoration and development of the degraded loess plateau, China, Ecological Research Monographs. Springer, Tokyo

Klessig DF, Malamy J (1994) The salicylic acid signal in plants. In: Palme K (ed) Signals and signal transduction pathways in plants. Springer, Dordrecht, pp 203–222

König J, Lotte K, Plessow R, Brockhinke A, Baier M, Dietz K (2003) Reaction mechanism of plant 2-Cys peroxiredoxin: role of the C terminus and the quaternary structure. J Biol Chem 278:24409–24420

Kovacik J, Klejdus B, Hedbavny J, Backor M (2009) Salicylic acid alleviates NaCl-induced changes in the metabolism of *Matricaria chamomilla* plants. Ecotoxicology 18:544–554

Koyro HW (2002) Ultrastructural effects of salinity in higher plants. In: Lauchli A, Luttge U (eds) Salinity: environment – plants – molecules. Kluwer, Amsterdam, pp 139–157

Kumari S, Roy S, Singh P, Singla-Pareek SL, Pareek A (2013) Cyclophilins: proteins in search of function. Plant Signal Behav 8(1):e22734

Kundu A, Patel A, Pal A (2013) Defining reference genes for qPCR normalization to study biotic and abiotic stress responses in *Vigna mungo*. Plant Cell Rep 32(10):1647–1658

Laloi C, Mestres-Ortega D, Marco Y, Meyer Y, Reichheld JP (2004) The Arabidopsis cytosolic thioredoxin h5 gene induction by oxidative stress and its W-box-mediated response to pathogen elicitor. Plant Physiol 134:1006–1016

Läuchli A, Grattan S (2007) Plant growth and development under salinity stress. In: Jenks MA, Hasegawa PM, Jain SM (eds) Advances in molecular breeding toward drought and salt tolerant crops. Springer, Dordrecht, pp 1–32

Lea-Cox JD, Syvertsen JP (1993) Salinity reduces water use and nitrate-N-use efficiency of citrus. Ann Bot 72:47–54

Lee S, Kim SG, Park CM (2010) Salicylic acid promotes seed germination under high salinity by modulating antioxidant activity in *Arabidopsis*. New Phytol 188:626–637

Leon J, Shulaev V, Yalpani N, Lawton MA, Raskin I (1995) Benzoic acid 2-hydroxylase, a soluble oxygenase from tobacco, catalyses salicylic acid biosynthesis. Proc Natl Acad Sci U S A 92:10413–10417

Leslie CA, Romani RJ (1986) Salicylic acid: a new inhibitor of ethylene biosynthesis. Plant Cell Rep 5(2):144–146

Liu J, Ishitani M, Halfter U, Kim CS, Zhu JK (2000) The *Arabidopsis thaliana* SOS2 gene encodes a protein kinase that is required for salt tolerance. Proc Natl Acad Sci U S A 97:3730–3734

Liu Y, Gao Q, Wu B, Ai T, Guo X (2009) NgRDR1, an RNA-dependent RNA polymerase isolated from *Nicotiana glutinosa*, was involved in biotic and abiotic stresses. Plant Physiol Biochem 47:359–368

Liu C, Mao B, Ou S, Wang W, Liu L, Wu Y, Chu C, Wang X (2013) OsbZIP71, a bZIP transcription factor, confers salinity and drought tolerance in rice. Plant Mol Biol 84(1–2):19–36

Liu P, Yin LN, Deng XP, Wang SW, Tanaka K, Zhang SQ (2014) Aquaporin-mediated increase in root hydraulic conductance is involved in silico-induced improved root water uptake under osmotic stress in *Sorghum bicolor* L. J Exp Bot 65:4747–4756

Lopez-Delgado H, Scott IM (1997) Induction of in vitro tuberization of potato microplants by acetylsalicylic acid. J Plant Physiol 151:74–78

Luciński R, Lucyna M, Sławomir S, Grzegorz J (2011) The thylakoid protease *Deg2* is involved in stress-related degradation of the photosystem II light-harvesting protein *Lhcb6* in *Arabidopsis thaliana*. Umultowska 89:61–614

Machado RMA, Serralheiro RP (2017) Soil salinity: effect on vegetable crop growth. Management practices to prevent and mitigate soil salinization. Horticulture 3:30

Maestre FT, Jordi C, Susana B (2007) Mechanisms underlying the interaction between Pinus halepensis and the native late-successional shrub *Pistacia lentiscus* in a semi-arid plantation. Ecography 27:776–786

Mansour MMF (2000) Nitrogen containing compounds and adaptation of plants to salinity stress. Biol Plant 43(4):491–500

Manzoor K, Noshin I, Nazima B, Bashir A, Muhammad A (2015) Effect of salicylic acid on the growth and physiological characteristics of maize under stress conditions. J Chem Soc Pak 37 (3):588–593
Martinez C, Pons E, Prats G, Leon J (2004) Salicylic acid regulates flowering time and links defense responses and reproductive development. Plant J 37:209–217
Mass EV (1986) Salt tolerance of plants. Appl Agric Res 1:12–26
Mateo A, Funck D, Mühlenbock P, Kular B, Mullineaux PM, Karpinski S (2006) Controlled levels of salicylic acid are required for optimal photosynthesis and redox homeostasis. J Exp Bot 57 (8):1795–1807
Maxwell K, Johnson GN (2000) Chlorophyll fluorescence – a practical guide. J Exp Bot 51:659–668
Mayne MB, Coleman JR, Blumwald E (1996) Differential expression during drought conditioning of a root-specific S-adenosylmethionine synthetase from jack pine (*Pinus banksiana* Lamb.) seedlings. Plant Cell Environ 19(8):958–966
Metraux JP (2002) Recent breakthroughs in the study of salicylic acid biosynthesis. Trends Plant Sci 7:332–334
Metwally A, Finkemeier I, Georgi M, Dietz KJ (2003) Salicylic acid alleviates the cadmium toxicity in barley seedlings. Plant Physiol 132:272–281
Mhawech P (2005) 14-3-3 proteins – an update. Cell Res 15:228–236
Miller G, Suzuki N, Ciftci-Yilmaz S, Mittler R (2010) Reactive oxygen species homeostasis and signaling during drought and salinity stresses. Plant Cell Environ 33:453–467
Mimouni H, Wasti S, Manaa A et al (2016) Does salicylic acid (SA) improve tolerance to salt stress in plants? A study of SA effects on tomato plant growth, water dynamics, photosynthesis and biochemical parameters. OMICS 20(3):180–190
Mittler R (2002) Oxidative stress, antioxidants and stress tolerance. Trends Plant Sci 7:405–410
Mittler R (2006) Abiotic stress, the field environment and stress combination. Trends Plant Sci 11 (1):15–19
Miura K, Tada Y (2014) Regulation of water, salinity and cold stress responses by salicylic acid. Front Plant Sci 5(4):1–12
Miyao M, Murata N (1989) The mode of binding of three extrinsic proteins of 33 kDa, 23 kDa and 18 kDa in the photosystem II complex spinach. Biochim Biophys Acta 977:315–321
Morris K, Mackerness SAH, Page T, John CF, Murphy AM, Carr JP, Buchanan-Wollaston V (2000) Salicylic acid has a role in regulating gene expression during leaf senescence. Plant J 23:677–685
Multu S, Atici O, Nalbantoglu B, Mete E (2016) Exogenous salicylic acid alleviates cold damage by regulating antioxidative system in two barley (*Hordeum vulgare* L.) cultivars. Front Life Sci 9 (2):99–109
Muneer S, Park YG, Manivannan A, Soundararajan P, Jeong BR (2014) Physiological and proteomic analysis in chloroplasts of *Solanum lycopersicum* L. under silicon efficiency and salinity stress. Int J Mol Sci 15(12):21803–21824
Munne-Bosch S, Penuelas J (2003) Photo- and antioxidative protection, and a role for salicylic acid during drought and recovery in field-grown *Phillyrea angustifolia* plants. Planta 217:758–766
Munns R (2002a) Comparative physiology of salt and water stress. Plant Cell Environ 25 (2):239–250
Munns R (2002b) Salinity, growth and phytohormones. In: Lauchli A, Luttge U (eds) Salinity: environment – plants – molecules. Kluwer, Dordrecht, pp 271–290
Munns R (2005) Genes and salt tolerance: bringing them together. New Phytol 167:645–663
Munns R, Tester M (2008) Mechanisms of salinity tolerance. Ann Rev Plant Biol 59:651–681
Munns R, James RA, Lauchli A (2006) Approaches to increasing the salt tolerance of wheat and other cereals. J Exp Bot 57(5):1025–1043
Murota KI, Oshshita Y, Watanabe A, Aso S, Sato F, Yamada Y (1994) Changes related to salt tolerance in thylakoid membranes of photoautotrophically cultured green tobacco cells. Plant Cell Physiol 35:107–113

Mutlu F, Bozcuk S (2007) Salinity induced changes of free and bound polyamine levels in Sunflower (*Helianthus annuus* L.) roots differing in salt tolerance. Pak J Bot 39:1097–1102

Nahar K, Hasanuzzaman M (2009) Germination, growth, nodulation and yield performance of three mungbean varieties under different levels of salinity stress. Green Farming 2:825–829

Nakashima K, Shinwari ZK, Sakuma Y, Seki M, Miura S, Shinozaki K, Yamaguchi-Shinozaki K (2000) Organization and expression of two *Arabidopsis DREB2* genes encoding DRE-binding proteins involved in dehydration- and high-salinity-responsive gene expression. Plant Mol Biol 42:657–665

Nakata M, Shiono T, Watanabe Y, Satoh T (2002) Salt stress-induced dissociation from cells of a germin-like protein with Mn-SOD activity and an increase in its mRNA in a moss, *Barbula unguiculata*. Plant Cell Physiol 43(12):1568–1574

Narusaka Y, Narusaka M, Seki M, Ishida TU, Nakajima M (2004) Crosstalk in the responses to abiotic and biotic stresses in *Arabidopsis*: analysis of gene expression in *cytochrome P450* gene superfamily by cDNA microarray. Plant Mol Biol 55:327–342

Nawrath C, Métraux JP (1999) Salicylic acid induction-deficient mutants of *Arabidopsis* express PR-2 and PR-5 and accumulate high levels of camalexin after pathogen inoculation. Plant Cell 11:1393–1404

Naylor M, Murphy AM, Berry JO, Carr JP (1998) Salicylic acid can induce resistance to plant virus movement. Mol Plant Microb Interac 11(9):860–868

Ndimba BK, Chivasa S, Simon WJ, Slabas AR (2005) Identification of Arabidopsis salt and osmotic stress responsive proteins using two-dimensional difference gel electrophoresis and mass spectrometry. Proteomics 5:4185–4196

Negrão S, Schmöckel SM, Tester M (2016) Evaluating physiological responses of plants to salinity stress. Ann Bot 119(1):1–11

Nemeth M, Janda T, Horvath E, Paldi E, Szalai G (2002) Exogenous salicylic acid increases polyamine content but may decrease drought tolerance in maize. Plant Sci 162:569–574

Neves SR, Ram PT, Iyengar R (2002) G protein pathways. Science 296:1636–1639

Noreen S, Ashraf M (2010) Modulation of salt (NaCl)-induced effects on oil composition and fatty acid profile of sunflower (*Helianthus annuus* L.) by exogenous application of salicylic acid. J Sci Food Agric 90:2608–2616

Oertli JJ (1991) Nutrient management under water and salinity stress. In: Proceeding of the symposium on nutrient management for sustained productivity. Dept Soils Punjab Agric Unver Ludhiana, India, pp 138–165

Ogawa D, Nakajima N, Sano T, Tamaoki M, Aono M, Kubo A et al (2005) Salicylic acid accumulation under O_3 exposure is regulated by ethylene in tobacco plants. Plant Cell Physiol 46:1062–1072

Omidbakhshfard MA, Proost S, Fujikura U, Mueller-Roeber B (2015) Growth-regulating factors (GRFs): a small transcription factor family with important functions in plant biology. Mol Plant 8(7):998–1010

Owen PA, Nickell CD, Noel GR, Thomas DJ, Frey K (1994) Registration of 'saline' soyabean. Crop Sci 43:1689

Palma F, Lluch C, Iribarne C, García-Garrido JM, Tejera García NA (2009) Combined effect of salicylic acid and salinity on some antioxidant activities, oxidative stress and metabolite accumulation in *Phaseolus vulgaris*. Plant Growth Regul 58(3):307–316

Palma F, Lluch C, Iribarne C, García-Garrido JM, Tejera García NA (2009) Combined effect of salicylic acid and salinity on some antioxidant activities, oxidative stress and metabolite accumulation in *Phaseolus vulgaris*. Plant Growth Regul 58(3):307-316

Palta JP, Farag K (2006) Methods for enhancing plant health, protecting plants from biotic and abiotic stress related injuries and enhancing the recovery of plants injured as a result of such stresses. United States Patent 7101828

Pandey S, Chakraborty D (2015) Salicylic acid and drought stress response: biochemical to molecular crosstalk. In: Tripathi BN, Muller M (eds) Stress responses in plants. Springer International Publishing, Switzerland, pp 247–265

Panuccio MR, Jacobsen SE, Akhtar SS, Muscolo A (2014) Effect of saline water on seed germination and early seedling growth of the halophyte quinoa. AoB Plant 6(1):plu047
Parida AK, Das AB, Mittra B (2004) Effects of salt on growth, ion accumulation, photosynthesis and leaf anatomy of the mangrove, *Bruguiera parviflora*. Trees 18(2):167–174
Parida AK, Das AB (2005) Salt tolerance and salinity effects on plants: a review. Ecotoxicol Environ Saf 60(3):324–349
Parker R, Flowers T, Moore A, Harpham N (2006) An accurate and reproducible method for proteome profiling of the effects of salt stress in the rice leaf lamina. J Exp Bot 57:1109–1118
Pastori GM, Foyer CH (2002) Common components, networks, and pathways of cross-tolerance to stress. The central role of "Redox" and abscisic acid-mediated controls. Plant Physiol 129:460–468
Patel BB, Patel BB, Dave RS (2011) Studies on infiltration of saline-alkali soils of several parts of Mehsana and Pata districts of North Gujarat. J Appl Technol Environ Sanit 1(1):87–92
Peskan-Berghofer T, Neuwirth J, Kusnetsov V, Oelmuller R (2005) Suppression of heterotrimeric G-protein beta-subunit affects anther shape, pollen development and inflorescence architecture in tobacco. Planta 220:737–746
Promila K, Kumar S (2000) *Vigna radiata* seed germination under salinity. Biol Plant 43:423–426
Qadir M, Schubert S (2002) Degradation processes and nutrient constraints in sodic soils. Land Degrad Dev 13:275–294
Quiroga M, Guerrero C, Botella MA et al (2000) A tomato peroxidase involved in the synthesis of Lignin and Suberin. Plant Physiol 122(4):1119–1128
Qureshi MI, Isra M, Abdin MZ, Iqbal M (2007) Responses of *Artemisia annua* L., to lead salt induced oxidative stress. Environ Exp Bot 53:185–193
Rajendran K, Tester M, Roy SJ (2009) Quantifying the three main components of salinity tolerance in cereals. Plant Cell Environ 32(3):237–249
Rajjou L, Belghazi M, Huguet R et al (2006) Proteomic investigation of the effect of salicylic acid on *Arabidopsis* seed germination and establishment of early defense mechanisms. Plant Physiol 141:910–923
Ramachandran S, Christensen HE, Ishimaru Y et al (2000) Profilin plays a role in cell elongation, cell shape maintenance, and flowering in Arabidopsis. Plant Physiol 124:1637–1647
Rao MV, Davis RD (1999) Ozone-induced cell death occurs via two distinct mechanisms in Arabidopsis: the role of salicylic acid. Plant J 17:603–614
Requejo R, Tena M (2006) Maize response to acute arsenic toxicity as revealed by proteome analysis of plant shoots. Proteomics 6:156–162
Rivas-San Vicente M, Plasencia J (2011) Salicylic acid beyond defence: its role in plant growth and development. J Exp Bot 62(10):3321–3338
Roeder S, Dreschler K, Wirtz M et al (2009) SAM levels, gene expression of SAM synthetase, methionine synthase and ACC oxidase, and ethylene emission from *N. suaveolens* flowers. Plant Mol Biol 70:535–546
Rogers ME, Noble CL (1992) Variation in growth and ion accumulation between two selected populations of *Trifolium repens* L. differing in salt tolerance. Plant Soil 146:131–136
Rogers ME, Grieve CM, Shannon MC (2003) Plant growth and ion relations in lucerne (*Medicago sativa* L.) in response to the combined effects of NaCl and P. Plant Soil 253:187–194
Roh KS, Kim JK, Song SD, Chung HS, Song JS (1996) Decrease of the activation and carbamylation of rubisco by high CO_2 in kidney bean. Korean J Biotech Biosci 11(3):295–302
Rospert S, Dubaquié Y, Gautschi M (2002) Nascent-polypeptide-associated complex. Cell Mol Life Sci 59(10):1632–1639
Rozeff N (1995) Sugarcane and salinity – a review paper. Sugarcane 5:8–19
Saha P, Chatterjee P, Biswas AK (2010) NaCl pretreatment alleviates salt stress by enhancement of antioxidant defense system and osmolyte accumulation in mungbean (*Vigna radiata* L. Wilczek). Indian J Exp Biol 48:593–600
Sahar K, Amin B, Taher NM (2011) The salicylic acid effect on the *Salvia officinalis* L. sugar, protein and proline contents under salinity (NaCl) stress. J Stress Physiol Biochem 7(4):81–87
Sakamoto N, Murata (2002) The role of glycine betaine in the protection of plants from stress: clues from transgenic. Plant Cell Environ 25(2):163–171

Sakhabutdinova AR, Fatkhutdinova DR, Bezrukova MV, Shakirova FM (2003) Salicylic acid prevents the damaging action of stress factors on wheat plants. Bulg J Plant Physiol 29:314–319

Sanchez JC, Schaller D, Ravier E et al (1997) Please, add rest of authors translationally controlled tumor protein: a protein identified in several nontumoral cells including erythrocytes. Electrophoresis 18:150–155

Seidler A (1996) The extrinsic polypeptides of photosystem II. Biochim Biophys Acta 1277:35–60

Shabala S, Cuin TA, Prismall L, Nemchinov LG (2007) Expression of animal CED-9 anti-apoptotic gene in tobacco modifies plasma membrane ion fluxes in response to salinity and oxidative stress. Planta 227:189–197

Shah J (2003) The salicylic acid loop in plant defense. Curr Opin Plant Biol 6:365–371

Shakirova FM, Sakhabutdinova AR, Bezrukova MV, Fatkhutdinova RA, Fatkhutdinova DR (2003) Changes in the hormonal status of wheat seedlings induced by salicylic acid and salinity. Plant Sci 164:317–322

Sharma YK, Leon J, Raskin I, Davis KR (1996) Ozone-induced responses in *Arabidopsis thaliana*: the role of salicylic acid in the accumulation of defense-related transcripts and induced resistance. Proc Natl Acad Sci U S A 93(10):5099–5104

Shavrukov Y, Gupta N, Miyazaki J et al (2010) HvNax3-a locus controlling shoot sodium exclusion derived from wild barley (*Hordeum vulgare* ssp. *Spontaneum*). Funct Integr Genom 10:277–291

Shi H, Ishitani M, Kim C, Zhu J (2000) The Arabidopsis thaliana salt tolerance gene SOS1 encodes a putative Na^+/H^+ antiporter. Proc Natl Acad Sci U S A 97(12):6896–6901

Shrivastava P, Kumar R (2015) Soil salinity: a serious environmental issue and plant growth promoting bacteria as one of the tools for its alleviation. Saudi J Biol Sci 22(2):123–131

Stacey G, McAlvin CB, Kim SY, Olivares J, Soto MJ (2006) Effects of endogenous salicylic acid on nodulation in the model legumes *Lotus japonicas* and *Medicago truncatula*. Plant Physiol 141:1473–1481

Sudhakar P, Latha P, Reddy PV (2016) Phenotyping crop plants for physiological and biochemical traits. Academic, Amsterdam

Sudhir P, Murthy SDS (2004) Effects of salt stress on basic processes of photosynthesis. Photosynthetica 42:481–486

Sugihara K, Hanagata N, Dubinsky Z, Baba S, Karube I (2000) Molecular characterization of cDNA encoding oxygen evolving enhancer protein 1 increased by salt treatment in the mangrove *Bruguiera gymnorrhiza*. Plant Cell Physiol 41:1279–1285

Suhayda CG, Giannini JL, Briskin DP, Shannon MC (1990) Electrostatic changes in *Lycopersicon esculentum* root plasma membrane resulting from salt stress. Plant Physiol 93:471–478

Szepesi Á, Csiszár J, Gémes K, Horváth E, Horváth F, Simon ML et al (2009) Salicylic acid improves acclimation to salt stress by stimulating abscisic aldehyde oxidase activity and abscisic acid accumulation, and increases Na^+ content in leaves without toxicity symptoms in *Solanum lycopersicum* L. J Plant Physiol 166:914–925

Taiz L, Zeiger E (2002) Plant physiology, 3rd edn. Sinauer Associates, Sunderland, p 306

Tasgin E, Atici O, Nalbantoglu B (2003) Effects of salicylic acid and cold on freezing tolerance in winter wheat leaves. Plant Growth Regul 41:231–236

Taylor NL, Heazlewood JL, Day DA, Millar AH (2005) Differential impact of environmental stress on the pea mitochondrial proteome. Mol Cell Proteomics 4:1122–1133

Tian M, Sasvari Z, Gonzalez P, Friso G, Rowland E (2015) Salicylic acid inhibits the replication of tomato bushy stunt virus by directly targeting a host component in the replication complex. Mol Plant-Microbe Interact 28(4):379–386

Uppalapati SR, Ishiga Y, Wangdi T, Kunkel BN, Anand A, Mysore KS et al (2007) The phytotoxin coronatine contributes to pathogen fitness and is required for suppression of salicylic acid accumulation in tomato inoculated with *Pseudomonas syringae* pv. Tomato *DC*3000. Mol Plant-Microbe Interact 20:955–965

Ushimarua T, Nakagawa T, Fujioka Y, Daichoa K, Naitob M et al (2006) Transgenic Arabidopsis plants expressing the rice dehydroascorbate reductase gene are resistant to salt stress. J Plant Physiol 163:1179–1184

Vadasserya J, Tripathib S, Prasadb R, Varmab A, Oelmullera R (2009) Monodehydroascorbate reductase 2 and dehydroascorbate reductase 5 are crucial for a mutualistic interaction between *Piriformospora indica* and *Arabidopsis*. J Plant Physiol 166:1263–1274
Van Breusegem F, Dat JF (2006) Reactive oxygen species in plant cell death. Plant Physiol 141:384–390
Vauclare P, Diallo N, Bourguignon J, Macherel D, Douce R (1996) Regulation of the expression of the glycine decarboxylase complex during pea leaf development. Plant Physiol 112:1523–1530
Veeranagamallaiah G, Jyothsnakumari G, Thippeswamy M et al (2008) Proteomic analysis of salt stress responses in foxtail millet (*Setaria italic* L. cv. Prasad) seedlings. Plant Sci 175:631–641
Vlot A, Dempsey D, Klessig D (2009) Salicylic acid, a multifaceted hormone to combat disease. Annu Rev Phytopathol 47:177–206
Vysotskaya L, Hedley PE, Sharipova G, Veselov D, Kudoyarova G, Morris J, Jones HG (2010) Effect of salinity on water relations of wild barley plants differing in salt tolerance. AoB Plant 2010:1–8
Wan D, Li R, Zou B, Zhang X, Cong J, Wang R et al (2012) Calmodulin-binding protein CBP60g is a positive regulator of both disease resistance and drought tolerance in *Arabidopsis*. Plant Cell Rep 31:1269–1281
Wang W, Vinocur B, Altman A (2007) Plant responses to drought, salinity and extreme temperatures towards genetic engineering for stress tolerance. Planta 218:1–14
Wang C, Ma QH, Lin ZB, He P, Liu JY (2008a) Cloning and characterization of a cDNA encoding 14-3-3 protein with leaf and stem-specific expression from wheat. DNA Seq 19:130–136
Wang MC, Peng ZY, Li CL, Li F, Liu C, Xia GM (2008b) Proteomic analysis on a high salt tolerance introgression strain of *Triticum aestivum/Thinopyrum ponticum*. Proteomics 8:1470–1489
Wasti S, Mimouni H, Smiti S, Zid E, Ben Ahmed H (2012) Enhanced salt tolerance of tomatoes by exogenous salicylic acid applied through rooting medium. OMICS 16:200–207
Weng JK, Chapple C (2010) The origin and evolution of lignin biosynthesis. New Phytol 187:273–285
Wildermuth MC, Dewdney J, Wu G, Ausubel FM (2001) Isochorismate synthase is required to synthesize salicylic acid for plant defense. Nature 414:562–565
Witzel K, Weidneri A, Surabhi G et al (2010) Comparative analysis of the grain proteome fraction in barley genotypes with contrasting salinity tolerance during germination. Plant Cell Environ 33:211–222
Wong CE, Li Y, Labbe A et al (2006) Transcriptional profiling implicates novel interactions between abiotic stress and hormonal responses in *Thellungiella*, a close relative of *Arabidopsis*. Plant Physiol 140:1437–1450
Wostrikoff K, Stern DB (2009) Rubisco. In: Stern DB, Harris E (eds) The chlamydomonas source book, vol 2, 2nd edn. Academic, San Diego, pp 302–332
Wu Y, Zhang D, Chu JY, Boyle P, Wang Y, Brindle ID, De Luca V, Despres C (2012) The *Arabidopsis* NPR1 protein is a receptor for plant defense hormone salicylic acid. Cell Rep 1:639–647
Xu WF, Shi WM (2006) Expression profiling of the 14-3-3 gene family in response to salt stress and potassium and iron deficiencies in young tomato (*Solanum lycopersicum*) roots: analysis by real-time RT-PCR. Ann Bot 98:965–974
Xu D, Duan X, Wang B, Hong B, Ho T, Wu R (1996) Expression of a late embryogenesis abundant protein gene, HVA1, from barley confers tolerance to water deficit and salt stress in transgenic rice. Plant Physiol 110:249–257
Xu C, Sibicky T, Huang B (2010) Protein profile analysis of salt-responsive proteins in leaves and roots in two cultivars of creeping bentgrass differing in salinity tolerance. Palnt Cell Rep 29:595–615
Xu S, Hu B, He Z, Ma F, Feng J, Shen W, Yan J (2011) Enhancement of salinity tolerance during rice seed germination by presoaking with hemoglobin. Int J Mol Sci 12:2488–2501
Yadav S, Irfan M, Ahmad A, Hayat S (2011) Causes of salinity and plant manifestations to salt stress: a review. J Environ Biol 32:667–685

Yalpani N, Leon J, Lawton MA, Raskin I (1993) Pathway of salicylic acid biosynthesis in healthy and virus-inoculated tobacco. Plant Physiol 103:315–321

Yamaguchi-Shinozaki K, Shinozaki K (1993a) *Arabidopsis* DNA encoding two desiccation-responsive *rd29* genes. Plant Physiol 101:1119–1120

Yamaguchi-Shinozaki K, Shinozaki K (1993b) Characterization of the expression of a desiccation-responsive *rd29* gene of *Arabidopsis thaliana* and analysis of its promoter in transgenic plants. Mol Gen Genet 236:331–340

Yan S, Tang Z, Su W, Sun W (2005) Proteomic analysis of salt stress-responsive proteins in rice root. Proteomics 5:235–244

Yan F, Liu Y, Sheng H, Wang Y, Kang H, Zeng J (2016) Salicylic acid and nitric oxide increase photosynthesis and antioxidant defense in wheat under UV-B stress. Biol Plant 60(4):686–694

Yang YN, Qi M, Mei CS (2004) Endogenous salicylic acid protects rice plants from oxidative damage caused by aging as well as biotic and abiotic stress. Plant J 40:909–919

Yeo AR, Lee KS, Izard P, Boursier PJ, Flowers TJ (1991) Short-term and long-term effects of salinity on leaf growth in rice (*Oryza sativa* L.) J Exp Bot 42:881–889

Yi X, McChargue M, Laborde S, Frankel L, Bricker T (2005) The manganese-stabilizing protein is required for photosystem II assembly/stability and photoautotrophy in higher plants. J Biol Chem 280:16170–16174

Yokoi S, Bressan RA, Hasegawa PM (2002) Salt stress tolerance of plants. JIRCAS Working Report, pp 25–33

Yong HY, Zou Z, Kok EP, Kwan BH, Chow K et al (2014) Comparative transcriptome analysis of leaves and rootd response to sudden increase in salinity in *Brassica napus* by RNA-seq. Biomed Res Int 2014:467395

Yoshida S, Tamaoki M, Shikano T, Nakajima N, Ogawa D et al (2006) Cytosolic dehydroascorbate reductase is important for ozone tolerance in *Arabidopsis thaliana*. Plant Cell Physiol 47:304–308

Yuan S, Lin HH (2008) Role of salicylic acid in plant stress. Z Naturforsch C 63(5–6):313–320

Yupsanis T, Moustakas M, Domiandou K (1994) Protein phosphorylation-dephosphorylation in alfalfa seeds germinating under salt stress. J Plant Physiol 143:234–240

Yusuf M, Hasan SA, Ali B, Hayat S, Fariduddin Q, Ahmad A (2008) Effect of salicylic acid on salinity-induced changes in *Brassica juncea*. J Integr Plant Biol 50(9):1096–1102

Zahra S, Amin B, Mehdi Y (2010) The salicylic acid effect on the tomato (*Lycopersicum esculentum* Mill.) germination, growth and photosynthetic pigment under salinty stress (NaCl). J Stress Physiol Biochem 6(2):4–16

Zeinolabedin J (2012) The effects of salt stress on plant growth. Tech J Eng Appl Sci 2(1):7–10

Zhang X, Chen S, Mou Z (2010) Nuclear localization of NPR1 is required for regulation of salicylate tolerance, isochorismate synthase 1 expression and salicylate accumulation in Arabidopsis. J Plant Physiol 167:144–148

Zhang XH, Rao XL, Shi HT, Li RJ, Lu YT (2011) Overexpression of a cytosolic glyceraldehydes-3-phosphate dehydrogenase gene OsGAPC3 confers salt tolerance in rice. Plant Cell Tissue Organ Cult 107:1–11

Zhou S, Zhang Z, Tang Q, Lan H, Li Y, Luo P (2010) Enhanced V-ATPase activity contributes to the improved salt tolerance of transgenic tobacco plants overexpressing vacuolar Na^+/H^+ antiporter AtNHX1. Biotechnol Lett 33:375–380

Zhu JK (2001) Plant salt tolerance. Trends Plant Sci 6(2):66–71

Zhu JK (2007) Plant salt stress. Wiley, Hoboken

Zhu Z, Wei G, Li J, Qian Q, Yu J (2004) Silicon alleviates salt stress and increases antioxidant enzymes activity in leaves of salt-stressed cucumber (*Cucumis sativus* L.) Plant Sci 167:527–533

Zorb C, Schmitt S, Neeb A, Karl S, Linder M, Schubert S (2004) The biochemical reaction of maize (*Zea mays* L.) to salt stress is characterized by a mitigation of symptoms and not by a specific adaptation. Plant Sci 167:91–100

The Role of Beneficial Elements in Triggering Adaptive Responses to Environmental Stressors and Improving Plant Performance

Fernando Carlos Gómez-Merino and Libia Iris Trejo-Téllez

Abstract Aluminum (Al), cerium (Ce), cobalt (Co), iodine (I), lanthanum (La), sodium (Na), selenium (Se), silicon (Si), titanium (Ti), and vanadium (V) are emerging as novel biostimulants that may enhance crop productivity and nutritional quality while improving responses to environmental stimuli and stressors in some plant species. These beneficial elements are not essential for most plants, but when supplied at low dosages, they help improve their growth, development, and yield quality by stimulating different molecular, biochemical, and physiological mechanisms triggering adaptive responses to challenging environments. When plants are exposed to environmental cues such as drought, heavy metal toxicity, low temperatures, saline soils, pest insects, or pathogens, beneficial elements may induce tolerance, resistance, or defense responses that allow plants to achieve acclimation to such stressors. Enhancement of nutrient uptake, synthesis of antioxidants and osmoprotectants, stimulation of secondary metabolism and signaling cascades, and reduction of senescence are among the responses boosted by beneficial elements when applied at low dosages. Nevertheless, beneficial elements may trigger hormesis in plants, a biphasic dose response with at low-dose stimulation or beneficial effect and a high-dose inhibitory or toxic effect. Thus, when properly applied, beneficial elements may have great potential to cope with some of the most daunting challenges facing humanity, such as climate change and food production under restrictive conditions for the growing human population. In this chapter, we mainly focus on the positive effects of beneficial elements on plant performance in restrictive environments and discuss some of the challenges of using these elements as biostimulants.

Keywords Climate change · Environmental stressors · Plant nutrition · Nonessential elements · Biostimulants · Hormesis · Innovation · Beneficial elements

F. C. Gómez-Merino
Colegio de Postgraduados Campus Córdoba, Amatlán de los Reyes, Veracruz, Mexico

L. I. Trejo-Téllez (✉)
Colegio de Postgraduados Campus Montecillo, Texcoco, State of Mexico, Mexico
e-mail: tlibia@colpos.mx

S. Vats (ed.), *Biotic and Abiotic Stress Tolerance in Plants*,
https://doi.org/10.1007/978-981-10-9029-5_6

1 Introduction

Plants need essential elements to ensure successful growth and development during both vegetative and reproductive stages. Essential elements are classified as macronutrients and micronutrients, depending on the amounts contained in plant tissues. Macronutrients are represented by elements which are generally found in plants at concentrations greater than 0.1% of dry matter weight (DMW, >1000 mg kg^{-1}), consisting of nitrogen (N), phosphorus (P), potassium (K), sulfur (S), calcium (Ca), and magnesium (Mg). Micronutrients are represented by chlorine (Cl), boron (B), copper (Cu), iron (Fe), manganese (Mn), molybdenum (Mo), nickel (Ni), and zinc (Zn); these nutrients are typically found at concentrations lower than 0.01% DMW, (<100 mg kg^{-1}DMW) (Pilon-Smits et al. 2009; Alcántar-González et al. 2016). These 14 nutrients, along with the elements carbon (C), hydrogen (H), and oxygen (O), are broadly accepted as essential for all plant species (Kirkby 2012; Alcántar-González et al. 2016).

Beneficial elements are not essential for most plants, but they can promote growth and be essential for some plant species under specific conditions (Pilon-Smits et al. 2009). When supplied at low dosages, they have a favorable impact on some vital processes and can also stimulate the mechanisms of resistance to biotic and abiotic stresses or promote the uptake of other nutrients (Trejo-Téllez et al. 2016). Additionally, beneficial elements can compensate for or remedy the toxic effects of other elements, and they can also, in some cases, provide certain functions of essential nutrients, such as the maintenance of osmotic pressure (Trejo-Téllez and Gómez-Merino 2012), or induce adaptive plant responses to adverse environmental phenomena (Pilon-Smits et al. 2009).

This chapter describes the effects of the beneficial elements identified so far, that is, aluminum (Al), cerium (Ce), cobalt (Co), iodine (I), lanthanum (La), sodium (Na), selenium (Se), silicon (Si), titanium (Ti), and vanadium (V), on the physiology of plants, with special emphasis on the induction of adaptive responses to challenging environments.

2 The Ten Beneficial Elements

The ten beneficial elements recognized until now have been shown to improve plant growth, production, and yield quality, as well as ameliorate the responses of plant to different environmental stress factors. A summary of the main functions of these elements in plants is displayed in Fig. 1.

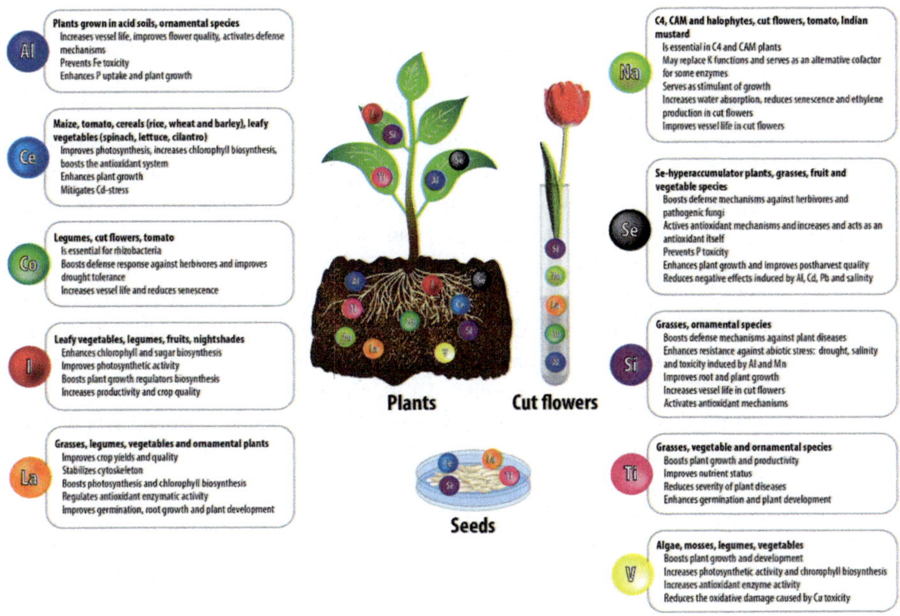

Fig. 1 Overview of the mechanisms responsible for the stimulating effects of the ten beneficial elements Al, Ce, Co, I, La, Na, Se, Si, Ti, and V on plant growth and stress responses. Main groups of plants on which beneficial effects of these elements have been documented are displayed at the heading of each box text

2.1 Aluminum (Al)

Aluminum is the most abundant metal in the Earth's crust (comprising about 7% of its mass), and its solubility increases with decreasing soil pH (Dong et al. 2002). While in acidic soils (pH less than 5) Al may inhibit root growth and display toxic effects to plants, it can be a beneficial element for some plant taxa under certain conditions (Moreno-Alvarado et al. 2017). One of the best-known examples of the beneficial effect of Al is observed in hydrangea (*Hydrangea macrophylla* Thunb. Ser.), since supplying them with different concentrations of Al turns them from pink (50 mg kg^{-1} DBW) to blue (4000 mg kg^{-1} DBW), which is attributed to the formation of a colloidal complex or to the combination of Al with a pigment called delphinidin (Trejo-Téllez et al. 2016), an anthocyanin responsible for the pigments in the plant's epidermal or subepidermal cells.

In rhododendron (*Melastoma malabathricum* L.), supplying Al in the complete nutrient solution enhances root development and plant growth (Watanabe et al. 2005). In rose (*Rosa* spp.) cv. "Cherry Brandy," supplying $Al_2(SO_4)_3$ significantly increased vase life and improved postharvest quality, because it may maintain the flower's fresh matter weight (FMW) and may increase the chlorophyll content in leaves (Jowkar et al. 2012). According to Seyf et al. (2012a), the application of

0, 150, and 300 mg L^{-1} $Al_2(SO_4)_3$ in "Boeing" roses increased vase life from 9 to 12 and 12.3 days, respectively, and increased flower diameter compared to control plants. In lisianthus (*Eustoma grandiflorum* Raf. Shinn.), the application of 150 mg L^{-1}Al sulfate to flowers prolonged vase life from 8 to 15 days, and FMW continued to increase up to 8 days after the start of the experiment (Li-Jen et al. 2001). In tuberose (*Polianthes tuberosa* L.) cv."Single," the application of 50 and 100 mg L^{-1} aluminum sulfate extends the vase life to 11.5 and 12 days, respectively (Mohammadi et al. 2012a). Furthermore, Al increased protein content and reduced FMW losses. The foliar application of 0.5, 1.0, and 1.5 g L^{-1}potassium aluminum sulfate [$KAl(SO_4)_2$] in sampaguita (*Jasminum sambac* L.) twice a day increased vessel life and FMW (Acero et al. 2016). The application of aluminum sulfate and 8% sucrose increases the quality and durability of rose cv. "Maroussia" stems in postharvest (Basaki et al. 2013). Likewise, De la Cruz-Guzmán et al. (2007) found that treatment with 0.6 g L^{-1} $Al_2(SO_4)_3$, in rose cv. "Royalty," reduces FMW loss during the vase period. Therefore, Al has great potential as a senescence retardant in cut flowers and also enhances their quality.

In medicinal plants such as chamomile (*Matricaria chamomilla.*), 60 μM Al increases soluble phenol and flavonoid contents in shoots and free amino acids in roots (Kováčik et al. 2010), which are antioxidant compounds that help plants overcome some stress factors. In silver birch (*Betula pendula* Roth.), the application of 2 and 5 mg L^{-1} Al enhances leaf growth (Kidd and Proctor 2000). In soybean (*Glycine max* L. Merr.) plants exposed to 1.0 μM Cd and 150 μM Al at pH 4.0, the malondialdehyde (MDA, a lipid peroxidation marker) content and superoxide dismutase (SOD) and peroxidase (POD) enzyme activities increased, which shows that Cd and Al are synergistic and stimulate antioxidant mechanisms (Shamsi et al. 2008). In maize (*Zea mays* L.), the application of 48 μM Al increased leaf growth rates, as a consequence of an increase in protein synthesis and a reduction in ubiquitin-mediated proteasomal degradation of growth-repressing proteins, such as DELLA in plants, and consequently promoted growth (Conti et al. 2014; Wang et al. 2015).

2.2 Cerium (Ce)

Cerium levels in soil range from 2 to 150 ppm and average 50 ppm (Trejo-Téllez et al. 2016). As a beneficial element, Morales et al. (2013) reported that adding 125 mg $kg^{-1}$$CeO_2$ nanoparticles ($nCeO_2$) to cilantro (*Coriandrum sativum* L.) plants produces longer roots and increases catalase (CAT) enzyme activity in shoots and that of POD in roots.

Tomato (*Solanum lycopersicum* L.) seeds treated with CeO_2 nanoparticles (<10 mg L^{-1}) develop seedlings with more extensive root hairs than the control. However, substantially higher Ce concentrations were detected in the fruits exposed to 10 mg $n\mathrm{CeO_2L^{-1}}$, compared with controls (Wang et al. 2012), shedding light on the long-term impact of $n\mathrm{CeO_2}$on plant health and its implications for our food safety and security. Importantly, second-generation seedlings grown from seeds collected from treated parent plants with $n\mathrm{CeO_2}$ (treated second-generation seedlings) were generally smaller and weaker, as indicated by their smaller biomass, lower water transpiration, and slightly higher reactive oxygen species content (Wang et al. 2013a).

In spinach (*Spinacia oleracea* L.) plants grown in Mg-deficient medium and treated with $CeCl_3$, Ce stimulated the activity of nitrate reductase, nitrite reductase, glutamate dehydrogenase, glutamate synthase, urease, and glutamic-pyruvic transaminase, which are key for N metabolism, suggesting that Ce may partially replace Mg functions to transform inorganic N to organic N, but the mechanisms underlying such responses need further study (Yin et al. 2009).

In maize and mung bean (*Vigna radiata* L. Wilczek), Ce shows favorable effects on the absorption of other nutrients when applied at concentrations below 0.2 μM (Diatloff et al. 2008). In *Arabidopsis thaliana* L. Heynh., the concentration of Ca^{2+}in protoplasts increased by applying 0.1 mmol Ce^{3+}, showing that this beneficial element can regulate metabolism, growth, and development through changes in the concentration of Ca^{2+} (Liu et al. 2011). Furthermore, Ce stimulates growth of lettuce (*Lactuca sativa* L.) cv. "Regina" seedlings (Barbieri et al. 2013), while in cowpea (*Vigna unguiculata* L. Walp.), the application of 0.713–17.841 μM cerium nitrate [$Ce(NO_3)_3$] increases chlorophyll content, relative yield, and nitrate reductase activity (Shyam and Aery 2012).

In wheat (*Triticum aestivum* L.), the application of 125, 250, and 500 mg $n\mathrm{CeO_2}$ per kg soil improved plant growth, shoot biomass, and grain yield by 9.0%, 12.7%, and 36.6%, respectively, in comparison to the control. As well, Ce nanoparticles increased linolenic acid by up to 6.17% but decreased linoleic acid by up to 1.63%, compared to the other treatments. The findings suggest the potential of nanocerium to modify crop physiology and food quality with unknown consequences for living organisms (Rico et al. 2014). On the other hand, Wu et al. (2014) reported that the alleviation of Cd toxicity by cerium in rice (*Oryza sativa* L.) seedlings is related to improved photosynthesis, elevated antioxidant enzymes, and decreased oxidative stress.

In barley (*Hordeum vulgare* L.), the application of 500 mg kg^{-1} $n\mathrm{CeO_2}$ promoted plant development resulting in a 331% increase in shoot biomass compared with the control, though these plants did not form grains. Moreover, 250 mg kg^{-1} $n\mathrm{CeO_2}$ enhanced grain Ce accumulation by as much as 294%, with a remarkable increases in P, K, Ca, Mg, S, Fe, Zn, Cu, and Al (Rico et al. 2015).

In turfgrass (*Poa pratensis* L.) seedlings, pretreatment with $Ce(NO_3)_3$ decreased the MDA content and electrolyte leakage and increased the FMW and DMW under Cu stress. Furthermore, Ce alleviated oxidative damage by regulating the metabolism of ascorbate and glutathione under Cu stress, and Ce had an important role in the acquisition of Cu tolerance in this species (Liu et al. 2016).

2.3 Cobalt (Co)

The Co concentration in plants normally ranges between 0.1 and 10 ppm considering DMW, although hyperaccumulator plants of the families Lamiaceae, Scrophulariaceae, Asteraceae, and Fabaceae can accumulate more than 1000 ppm of this element in leaves (Pilon-Smits et al. 2009). In higher plants, Co adheres strongly to the roots and is absorbed from the soil solution through passive transport. Since Co shows chemical similarity with nickel (Ni), it is possible that the two elements enter the cell through the same types of membrane transporter proteins (Chen et al. 2009).

Cobalt concentrations may increase under Fe deficiency, since Co can compete with Fe for the active sites of the transporter IRT1 (Baxter et al. 2008). In fact, Gad (2012) proved that a sufficient supply of Co significantly decreases the Fe content in groundnut (*Arachis hypogaea* L.) seeds, and therefore both elements are antagonists and compete for the same transporters.

At low concentrations, Co can have beneficial effects, especially in legumes. In pea (*Pisum sativum* L.), applying 8 ppm Co to the soil increased growth, nodule number and weight, plant nutrient levels, and yield and seed quality, which can be attributed to the importance of Co for *Rhizobium* populations living in the roots of these plants (Gad 2006). Co is a component of cobalamin (vitamin B12), which is required to activate enzymes related to N fixation in symbiotic microorganisms (Palit et al. 1994).

Likewise, the application of 8 ppm Co significantly increased nitrogenase activity and nodule number and weight, especially when N was applied at 75% and 100% in groundnut (Gad 2012). In the plant, Co improved growth and yield indicators and the contents of N, P, K, Mn, and Zn. Cobalt contents in seeds ranged between 2.3 and 3.5 ppm, which is within the range reported for other plants. The application of Co can contribute to a more efficient use of N, with a savings of up to 25% of the N applied (Gad 2012).

In lily (*Lilium* spp.) cv. "Star Fighter," floral stems treated with preservative solutions including 0.1 and 0.2 mM Co increased floral longevity in 61.1% and 44%, respectively, while in the cv. "Star Gazer," the application of 0.1 mM Co enhanced this variable by 19.7% (Mandujano-Piña et al. 2012). In cut marguerite (*Argyranthemum* sp.) flowers, applying 1 and 2 mM Co increased vase life by 5 days compared to the control containing only distilled water (Kazemi 2012). Cobalt and nickel (2.5 mM Co + 2 mM Ni + 2 mM salicylic acid with 2.5% sucrose) increased the vase life of lily cv. "Prato" due to improved membrane stability and reduced oxidative stress damage during flower senescence. In addition, these elements reduce the loss of anthocyanins (Kazemi and Ameri 2012). In carnation (*Dianthus caryophyllus* L.), Co retards senescence since it reduces ethylene production, an effect very similar to that shown with the application of Ni in this species (Jamali and Rahemi 2011).

In tuberose, application of 300 mg L^{-1} cobalt chloride ($CoCl_2$) stimulated vase life (10.66 days) and water uptake (1.53 mL g^{-1} FMW) and reduced FMW losses

(19.99 g). Moreover, the application of 400 mg L^{-1} increased the content of carotenoids (0.40 g) and proteins (31.10%) in petals (Mohammadi et al. 2012b). In tomato, the application of 50 mg kg^{-1} Co to the soil increased the contents of N, P, K, Cu, Fe, Mn, and Zn in leaf tissue (Jayakumar et al. 2013).

In cut roses, cobalt chloride inhibited vascular blockage in the stem and maintained a high water flow rate through stems, leading to significant water uptake by flowers. The best effects were observed with 200 mg L^{-1} $CoCl_2$ in the vessel solution (Aslmoshtaghi et al. 2014).

In gladiola (*Gladiolus grandiflorus* Hort.) cv. Borrega Roja, the application of Co (0, 0.3 and 0.6 mM) significantly increased water absorption in flower stems, while the lowest percentage of weight loss was recorded in fresh rods treated with 0.3 mM Co. Furthermore, the low concentration of Co significantly increased N content in stems and leaf concentration of chlorophyll. The total dry weights of leaves and stems were higher with the treatment of 0.3 mM Co (Trejo-Téllez et al. 2014).

2.4 *Iodine (I)*

In terms of plant nutrition, iodine has been little studied in crop species (Smolen and Sady 2011b), though it can be transported from the underground (roots) to the aboveground (shoots) parts of the plants (Ashworth 2009). As a beneficial element, iodine can promote growth, induce tolerance mechanisms to cope with stress, and trigger antioxidant capacity of the plant (Medrano-Macias et al. 2016). Iodine has the ability to advance the flowering process in fruit tree species, as a result of increased photosynthetic activity, producing a greater accumulation of sugars (Landini et al. 2012).

Blasco et al. (2011) determined that applications of 40 μM of iodate significantly improve nitrogen-use efficiency and nitrogen metabolism, which increase lettuce productivity and quality. The application of 0.05% (w/v) iodine salts in the nutrient solution caused the potato (*Solanum tuberosum* L.) tubers to absorb 272 mg 100 g^{-1} FMW and tomato fruits 527 mg 100 g^{-1} FMW of IO_3^-, respectively (Caffagni et al. 2011).

After the foliar application of 25 g L^{-1} I or soil application of 90 g kg^{-1} I in crops grown in the open field, the maximum iodine content ranged between 9.5 and 14.3 μg 100 g^{-1} for plum and nectarine fruits, to 89.4 and 144.0 μg 100 g^{-1} for potato tubers and tomato fruits, respectively (Caffagni et al. 2012). In hydroponic culture, fresh fruits managed to accumulate up to 2423 μg 100 g^{-1} of iodine. In all cases, iodine was mainly accumulated in the leaves.

In spinach, the application of 1–2 mg dm^{-3} iodine and 1 g dm^{-3} sucrose significantly increased the content of this element, the N and the oxalate content in leaves (Smolen and Sady 2011a). Iodine synergistically improves the uptake of Mg, Na, and Ce, as well as of Fe, and reduces chromium (Cr) uptake. After the application of 2 mg dm^{-3} I in the soil, a greater accumulation of Na, Fe, Zn, and Al and a reduced concentration of P, S, Cu, and barium (Ba) were observed (Smolen and Sady 2011b).

The beneficial effect of foliar-applied iodine was also observed in "Golden Delicious" apple (*Malus domestica* Borkh.) trees grafted on M.9 by applying an organic-mineral liquid fertilizer (Biojodis) at a concentration of 5 L ha^{-1} diluted in water (600 L ha^{-1}), which improved fruit yield, diameter, and uniformity of fruits (Szwonek 2009).

In soybean seeds, the application of 20 μM IO_3^- enhanced the expression of more proteins in comparison to the control. Furthermore, when iodine (20, 40, and 80 μM IO_3^-) was applied to the seeds, the activity of antioxidant enzymes such as SOD, ascorbate peroxidase (APX), and glutathione reductase (GR) was boosted, counteracting the toxic effects of 100 mM Cd (Gupta et al. 2015).

No negative effects of iodine fertilization (5 kg ha^{-1} I, either as KI or as KIO_3) were noted with respect to carrot (*Daucus carota* L.) yield, while higher accumulation and uptake by leaves and storage roots of iodine were obtained after the application of KI than KIO_3, which improved the biofortification of carrot storage roots (Smolen et al. 2016).

A recent review on the use of iodine to biofortify and promote growth and stress tolerance in crops published by Medrano-Macias et al. (2016) stated that this element has strong interactions with Fe, Mn, Cu, and V, either directly in plant metabolism or indirectly through the microbiome of the plant. In general, good results are obtained regarding biofortification when applied to the soil as KIO_3 in concentrations of 7.5 kg ha^{-1}, 10 mg kg^{-1} soil in pots, or 10^{-6}–10^{-5} M in the nutrient solution. Leaf spray with KI at 0.5 kg ha^{-1}gave good results. With higher concentrations, the response is variable: negative, neutral, or positive, depending on the plant species (Medrano-Macias et al. 2016).

2.5 Lanthanum (La)

Lanthanum is considered a beneficial element as it enhances the uptake of essential nutrients such as K, Ca, and Mg (Wahid et al. 2000). In rice, the application of 0.1 mM $La(NO_3)_3$ increased germination rate and biomass accumulation in plantlets (Liu et al. 2012a). Liu et al. (2013) found that applying 0.05 mmol La promotes root growth and that this element tends to accumulate in the cell wall of the root. In addition, La treatments affect the accumulation of K, Mg, Ca, Na, Fe, Mn, Zn, Cu, and Mo in the root and thus plant growth.

The application of 5 up to 50 μM La in the nutrient solution stimulates the growth of maize, mung bean (*Vigna radiata* L. Wilczek), and black gram (*V. mungo* L. Wilczek) plants and the germination percentage, root length, shoot length, as well as FMW and DMW in all three crops (Chaturvedi et al. 2014). In tobacco (*Nicotiana tabacum* L.), the application of 5–20 mg L^{-1} $LaCl_3$ in the Hoagland solution gradually increased dry matter accumulation and chlorophyll content, which decreased when the concentration of $LaCl_3$ exceeded 50 mg L^{-1} (Chen et al. 2001). Best results were observed with 20 mg L^{-1}, since the synthesis and activity of choline, Mg^{2+}-ATPase, and phosphorylation were significantly activated. Diatloff et al. (2008) reported that concentrations of La below 0.2 μM produced positive effects on maize and mung bean, although at higher concentrations it decreases absorption of Ca, Na, Zn, and Mn.

In cucumber (*Cucumis sativus* L.), La^{3+} reduces the levels of Na, Mg, Cl, K, and Ca while increasing Mn and Fe levels. Also, the effects of La^{3+} on ion absorption show similarity to those of Ca^{2+}, indicating that La affects physiological mechanisms in the plant by regulating the levels of Ca; optimal growth occurred at a concentration of 0.02 mM La^{3+} (Zeng et al. 2000).

In cucumber plants, La has been found to be involved in ion transport and also modifies the absorption and distribution of Se, Co, V, and technetium (Tc), affecting growth and absorption of other elements that influence the physiology of cells and biochemical functions in this plant (Huang et al. 2003). In addition, Shi et al. (2005) reported that low concentrations of La (0.002 and 0.02 mM $LaCl_3$) promote growth of cucumber plantlets and increase chlorophyll and carotenoid contents, while La was found to be involved in activating antioxidant enzymes such as POD, CAT, and SOD, as well as in reducing MDA contents.

According to Liu et al. (2012b), the Ca level decreases slightly with 0.2 mM La^{3+}; with 1.0 mMLa^{3+}, oscillations of Ca^{2+} were observed; and at 2.0 mMLa^{3+}, there was an increase in Ca^{2+}, indicating that La^{3+} participates in signal transduction networks mediated by calmodulin (CaM) and that it can enter the root through the cell membrane and intracellular Ca^{2+} channels.

In tulip (*Tulipa gesneriana* L.) cv. "Ile de France," the diameter and length of the floral stem were higher when plants were ferti-irrigated with 10 μM La (Ramírez-Martínez et al. 2009). In the same ornamental species, Ramírez-Martínez et al. (2012) observed that La stimulates the accumulation of Ca, K, and La itself at concentrations of 10 and 20 μM.

Yan et al. (2007) found that adding 20 mg L^{-1} La^{+3} decreased cell membrane permeability and the contents of MDA, H_2O_2, and proline in soybean plants exposed to UV-B radiation (280–320 nm; 0.15–0.45 W cm^{-2}). Moreover, the activity of the enzymes CAT and POD was higher in La-treated plants, demonstrating the activation of antioxidant mechanisms.

In a study aimed at evaluating the bioaccumulation of La and its effects on growth and mitotic index of soybean, plants were exposed to increasing concentrations of La (0, 5, 10, 20, 40, 80, and 160 μM) in the nutrient solution for 28 days. Roots accumulated 60-fold more La than shoots. La deposition occurred mainly in cell walls and in crystals dispersed in the root cortex and in the mesophyll. The application of La resulted in increased contents of essential nutrients such as Ca, P, K, and Mn, whereas Cu and Fe levels decreased. Furthermore, low La concentrations (i.e., 5–10 μM La) stimulated the photosynthetic rate and total chlorophyll content and led to a higher incidence of binucleate cells, resulting in a slight increase in root and shoot biomass. At higher La levels (i.e., 20–160 μM La), soybean growth was reduced, as a result of ultrastructural modifications in the cell wall, thylakoids, and chloroplasts and the appearance of c-metaphases (de Oliveira et al. 2015).

In rangpur lime (*Citrus limonia* Osbeck), 50 mg $LaCl_3$ increased mass and height of plants with a consequent increase of dry matter, suggesting the use of La as fertilizer in citrus, especially when used at low concentrations (Turra et al. 2015).

Recently, García-Jiménez et al. (2017) reported that the application of 10 μM to four sweet pepper (*Capsicum annuum* L.) varieties significantly increased seedling height, shoot diameter, number of flower buds, number of leaves, and leaf area, though it did not affect dry biomass accumulation. Furthermore, La stimulated the biosynthesis of chlorophyll a and b and total chlorophylls, total soluble sugars, and soluble protein concentration.

2.6 Selenium (Se)

In seleniferous soils, most plant species contain from 1 to 10 ppm Se, while hyperaccumulators (such as those belonging to the genera *Stanleya* and *Astragalus*) can accumulate from 1000 to 15,000 ppm Se (0.1–1.5% Se) (Pilon-Smits et al. 2009).

The application of Se at low concentrations can increase tolerance to oxidative stress induced by UV radiation, retard senescence, and promote growth. In addition, Se can regulate water content under drought conditions (Germ et al. 2007). In ryegrass (*Lolium perenne* L.), the application of 0.1 and 1 mg kg^{-1} Se activated antioxidant mechanisms and increased glutathione peroxidase (GPX) activity while reducing senescence processes and promoting plant growth (Hartikainen et al. 2000).

In canola (*Brassica napus* L.), Se can increase seed yield and enhance its nutritional value (Hajiboland and Keivanfar 2012). In carrot, the combination of KI and Na_2SeO_4 stimulated uptake, accumulation, and storage of both I and Se, and the consumption of 100 g FMW carrot produced by plants fertilized with KI + Na_2SeO_3 and KIO_3 + Na_2SeO_3 can provide 100% of the I and Se levels recommended for human nutrition (Smolen et al. 2016).

Because Se increases the absorption of heavy metals like lead (Pb) in common coleus (*Coleus blumei* Benth.), this beneficial element can be useful in stimulating phytoremediation mechanisms in environments contaminated by heavy metals (Yuan et al. 2013).

The foliar application of 10 mg L^{-1}Se as Na_2SeO_4 to soybean cv. "Olna" plants increased respiration potential, especially in young plants (Mechora and Germ 2010). In melon (*Cucumis melo*L.) seedlings under salt stress, supplying 2–8 mM Se improved growth and triggered antioxidant mechanisms by inhibiting lipid peroxidation and increasing the enzymatic activity of SOD and POD (KeLing et al. 2013). Likewise, exogenous Se alleviates salt stress in maize via the improvement of photosynthetic capacity, the activities of antioxidant enzymes, and the regulation of Na^+ homeostasis (Jian et al. 2017).

In maize, applying 5 μM dm^{-3} Se stimulated growth and root elongation (Hawrylak-Nowak 2008). Furthermore, applications of up to 15 μM Se (as selenite or selenate) in lettuce improved plant growth, while no major changes in oxidative state, pigment concentration, and sulfur accumulation were observed (Hawrylak-Nowak 2013). Moreover, the application of Na_2SeO_4 (0, 5 and 10 μM) in cucumber plants treated with Cd (0, 25 and 50 μM) reduced Cd uptake and lipid peroxidation as well, while plasma membrane was more stable, suggesting a beneficial effect of Se in plants exposed to Cd (Hawrylak-Nowak et al. 2014).

In the grass *Stylosanthes humilis* Kunth. grown in acidic soils with high concentrations of toxic aluminum (Al^{3+}), applying up to 1 μM Se activated antioxidant mechanisms, and Se itself contributed to remove reactive oxygen species (Ribeiro et al. 2011).

Selenium has also been associated with fruit maturation. In peach cv. "Suncrest" grafted onto GF 677 (*Prunus persica* L. Batsch. × *Prunus amygdalus* Stokes) and pear (*Pyrus communis* L.) cv. "Conference" grafted onto BA 29 (*Cydonia oblonga* Mill.) receiving 0.1 and 1 mg L^{-1} Na_2SeO_4 in leaves or 1 mg L^{-1} Na_2SeO_4 in fruits, Se kept firmness and delayed fruit maturation, which prolonged fruit storage (Pezzarossa et al. 2012). Similar effects have been reported in tomato, since the application of 1 mg L^{-1} Na_2SeO_4 reduced lipid peroxidation, increased the activity of antioxidant enzymes such as SOD and GPX, and improved fruit quality during storage (Zhu et al. 2016).

2.7 *Silicon (Si)*

Silicon constitutes between 0.1% and 10% of the DMW of higher plants and its accumulation can vary significantly among species. Importantly, Si-deficient plants are brittle and susceptible to fungal infections (Ma and Yamaji 2006).

Silicon can counteract the toxic effects of elements such as Al and Mn, confer resistance against pests and diseases, and even allow the formation of nanostructures using organic compounds, enzymes, or organisms as catalysts (Raya and Aguirre 2009). Silicon is absorbed in a pH range of 2–9, being taken up by the roots in the solution as monosilicic acid ($Si(OH)_4$) to be accumulated in the epidermal cells of leaves (Borda et al. 2007).

The beneficial effects of Si are associated with its high deposition in plant tissues, improving their strength and rigidity (Ma and Yamaji 2006). It is also possible that Si plays an active role in resistance to plant diseases by stimulating defense mechanisms. Also, Si can play an important role in resistance to abiotic stress factors such as heavy metal toxicity, salinity, and drought and can reduce the generation of reactive oxygen species, due to the increased activity of antioxidant enzymes (Balakhnina and Borkowska 2013).

In forage oat (*Avena sativa* L.), applying 100 mg kg^{-1} (116 g per pot) of monosilicic acid in the pre-sowing period increased height and dry matter production as a result of better nutritional absorption promoted by Si (Borda et al. 2007). Moreover, Si stimulated cell elongation and turgor and improved the conversion of assimilates. In bitter gourd (*Momordica charantia* L.) plants under saline stress (50 mM NaCl), the application of increasing concentrations of Si (1–5 mM) stimulated the germination rate and index, seedling vitality, and antioxidant enzyme activities of SOD, POD, and CAT (Wang et al. 2010).

By adding 100, 150, and 200 mg L^{-1} potassium silicate (K_2SiO_3), carnation flower cv. "Harlem" improved vase life as a result of a significant reduction in ethylene production (Jamali and Rahemi 2011). The application of 2.5 mM Si together with 3 mM acetylsalicylic acid reduces wilting in carnation flowers, retards chlorophyll and carbohydrate degradation, and reduces the activity of oxidase enzymes (Kazemi et al. 2012).

The application of Si increases the amount of O_2 in leaves, stems, and roots, which causes the rhizosphere to oxidize. Thus, the elements Fe and Mn are oxidized, which prevents excessive uptake of these elements by the plant (Furcal-Beriguete and Herrera-Barrantes 2013). In common borage(*Borago officinalis* L.), Si plays a detoxifying role when the plant is under aluminum stress because it stimulates the synthesis of phenolic compounds and proline (Shahnaz et al. 2011).

In maize, application of Si by seed priming improved growth of stressed plants (exposed to alkaline stress induced by 0, 25, 50, and 75 mM Na_2CO_3) while enhancing the leaf relative water content and levels of photosynthetic pigments, soluble sugars, soluble proteins, total free amino acids, and K^+, as well as activities of SOD, CAT, and POD enzymes. Moreover, Si supplement resulted in a decrease in the contents of proline, MDA, and Na^+, which together with an enhanced K^+ level led to a favorable adjustment of K^+/Na^+ ratio, in stressed plants relative to plants treated with alkaline stress alone. These findings confirm that Si plays a pivotal role in alleviating the negative effects of alkaline stress on maize (Abdel Latef and Tran 2016). Similarly, Marxen et al. (2016) showed that application of 0.4 and 17.3 t ha^{-1} Si (as silica gel) to rice plants cv. "Khang Dan 18" increased Si contents in plant tissues, as well as biomass production and grain yield.

2.8 *Sodium (Na)*

Sodium is one of the most studied ions in plant biology due to its toxic effects, although at low concentrations its beneficial effect has also been proved. In fact, in some plants with C4 photosynthetic metabolism, Na is considered an essential element (Kronzucker et al. 2013), and its benefits are more evident in conditions of potassium deficiency (Schulze et al. 2012). Sodium also increases the biosynthesis of amino acids, especially proline (Jouyban 2012).

Salt stress can promote growth in some crops such as wheat, while in rice the low yield caused by salinity is mainly associated with the reduction in tillers and an increase in sterile spikelets in some cultivars (Läuchli and Grattan 2007).

According to Lee and van Iersel (2008), salinity induced by NaCl has the potential to act as a growth regulator. In fact, the application of 60–120 mM NaCl increases the height of faba bean (*Vicia faba* L.) plants (Abdul Qados 2011). Importantly, salinity may lead to toughening of tomato fruit skin. Accordingly, Silva et al. (2015) reported a linear correlation between thickness of the subepidermis and salinity of the irrigation water (up to 12.61 dS m^{-1}). Interestingly, the tougher tomato skin obtained under conditions of salinity is attributed to increased number of hypodermal cell layers rather than to changes in cell wall composition.

In a proteomic study using two genotypes of Indian mustard (*Brassica juncea* L. Czern.) displaying contrasting sensitivity to salt stress, Yousuf et al. (2016) reported differential expression of 21 salt stress-responsive proteins associated with various functional processes, including osmoregulation, photosynthesis, carbohydrate metabolism, ion homeostasis, protein synthesis and stabilization, energy metabolism, and antioxidant defense system. Salt-tolerant genotype (CS-52) showed a relatively higher expression of proteins involved in turgor regulation, stabilization of photosystems and proteins, and salt compartmentalization, as compared to salt-sensitive genotype (Pusa Varuna). These results suggest that modulating the expression of salt-responsive proteins can pave the way for developing salt tolerance in the Indian mustard plants.

According to Lee and van Iersel (2008), the quality of *Chrysanthemum* x *morifolium* Ramat. cut flower is improved by applying 1 g L^{-1}NaCl in irrigation water. In cut roses cv. "Avalanche," the application of 250 mg L^{-1} sodium benzoate improved vase life by reducing ethylene production (Imani et al. 2012). Similarly, applying 20 μM of sodium nitroprusside (SNP) in rose and 60 μM in sunflower (*Helianthus annuus* L.) increased vase life (Nazirimoghaddam et al. 2014). In cut rose cv. "Utopia" treated with 50 μM SNP, a higher soluble protein content, an increased solution uptake rate by the flower stems, and an improved FMW ratio were observed, while vase life increased from 11 to 13.3 days compared to the control (Seyf et al. 2012b). Moreover, in giant chincherinchee (*Ornithogalum saundersiae* Bak.) plants grown in pots receiving either 100 or 200 mM NaCl weekly, Na increased chlorophylls and carotenoids contents, as well as N, K, Na, and Cl concentrations in leaves (Salachna et al. 2016). In purpletop vervain (*Verbena bonariensis* L.), the application of 200 mM NaCl enhanced Mn contents in leaves, whereas neither P, K, Mg, Cu, Zn, and Fe contents nor the initiation of flowering was affected (Salachna and Piechocki 2016).

2.9 Titanium (Ti)

Titanium began to gain importance in plant biology studies in the 1930s and is now considered a beneficial element (Carvajal and Alcaraz 1998). This element is not toxic to animals or to humans, and at low concentrations it is beneficial for plants since it triggers physiological mechanisms leading to better growth and development under certain environmental conditions (Jaberzadeh et al. 2013). So far, limited information is available regarding critical levels of Ti in plant toxicity.

Apart from the application of Ti as conventional reagent, the use of Ti nanoparticles (*n*Ti) is currently gaining more importance. A recent review of Ti as a beneficial element described that seeds treated with *n*Ti suspensions exhibited increased germination rates, enhanced root lengths, or improved seedling growth in different plant species. Furthermore, the application of *n*Ti increased plant tolerance to abiotic and biotic stresses, including cold, drought, Cd toxicity, and bacterial spot disease caused by *Xanthomonas perforans* (Lyu et al. 2017).

In oat plants (*Avena sativa* L.)cv. "Zlat'ák," the effect of Ti is considerably weaker if it is applied on leaves than if being added to the nutrient solution (Kuzel et al. 2003). Importantly, the action of Ti on plant physiology can be explained by a hormetic effect. The application of 960 g ha^{-1} Ti to tomato increases the content of N, P, Ca y Mg, and that of K with 80 g ha^{-1} Ti,which demonstrated the importance of Ti in plant nutrition and the quality of this vegetable (Kleiber and Markiewicz 2013).

Under drought stress, application of 0.02% of titanium dioxide nanoparticles in wheat increases gluten and starch content (Jaberzadeh et al. 2013). Furthermore, the application of 1200 and 1500 mg L^{-1}Ti in canola seedlings produced greater root development and bud growth (Mahmoodzadeh et al. 2013). In addition, TiO_2 has the potential to be used as an alternative for the control of bacterial blight caused by *Xanthomonas* in zonal geranium (*Pelargonium x hortorum* L. H. Bailey) and leaf spot in poinsettia (*Euphorbia pulcherrima* Willd. ex Klotzsch), since treatments using this compound at concentrations of 25 and 75 mM showed a reduction in lesions of 85% and 93%, respectively (Norman and Chen 2011).

Two applications of TiO_2 at 125 cm^3 ha^{-1} in cowpea in the 1st and 2nd year of production improved development and yield and reduced the severity of foliar and pod diseases compared to a single application at lower concentration. Application of TiO_2 increased cowpea yield from 8.74% to 36.11% and from 10.33% to 51.31%, respectively, in the 2 years of evaluation (Owolade and Ogunleti 2008).

The response of potato, barley, and wheat plants to titanium application is almost negligible under N deficiency. However, when there is sufficient N, responses are more evident (Tlustos et al. 2005). In mung bean, foliar application of 10 mg L^{-1} TiO_2 increased shoot (17.0%) and root (49.6%) lengths, chlorophylls, as well as total protein contents in leaves. Furthermore, Ti application increased microbial populations in the rhizosphere and the activity of enzymes involved in the P nutrient cycle, such as acid phosphatase, alkaline phosphatase, phytase, and dehydrogenase (Raliya et al. 2015). Foliar application of Ti-ascorbate at concentrations of 25, 50,

75, and 100 mg L^{-1} improved height of geranium cv. "Elite Cherry," petunia (*Petunia x hybrida* hort. ex E.Vilm.) cv. "Celebrity White," pansy (*Viola x wittrockiana* Gams.) cv. "Delta Premium Marina," and snapdragon (*Antirrhinum majus* L.) cv. "Montego Purple" (Whitted-Haag et al. 2014).

Just recently, Andersen et al. (2016) evaluated the application of different concentrations of Ti dioxide nanoparticles (nTiO$_2$) on the germination of ten crop species. They reported that the application of 500 μgmL^{-1} nTiO$_2$ increased the percentage of germination in cabbage, and root length in onion, while applying 1000 μg mL^{-1} enhanced root length both in cucumber and onion. In barley the application of 500 or 1000 mg kg^{-1} nTiO$_2$ to the soil stimulated plant growth and reduced the toxic effects induced by nCeO$_2$ (Marchiol et al. 2016).

It is important to note that Ti and nTi may have neutral or negative effects on plant physiology, which may be attributed to several factors including differences in plant species, physiological status of plants at the time being evaluated, seed quality, nanoparticle sizes and their uniformity, and experimental objectives and methods. Furthermore, attention does need to be given to the fate and consequence of applied nTi within the environment and food chain (Lyu et al. 2017).

2.10 Vanadium (V)

The biological importance of V can be divided into three levels, depending on the daily intakes and tissue contents: nutritional (intakes of μg a day), pharmacological (mg a day), and toxicological (mg kg^{-1} food DMW) (Antal et al. 2009). Because of its toxic, mutagenic, genotoxic, and even carcinogenic potential, V has been the subject of numerous public health studies worldwide (Rodríguez-Mercado and Altamirano-Lozano 2006), while its beneficial effects on plants have been sparsely addressed. In the field of plant physiology, one of those first reports on this element showed that soybean plants grown in oxisols develop normally even at concentrations of 75 mg kg^{-1} V in the soil (Wang and Liu 1999).

Vanadium has been considered as either beneficial or as a secondary metabolism elicitor in plants, but the mechanisms involved are not yet fully understood (Saco et al. 2013). This element is a metal widely distributed both in nature and in biological systems and is also one of the trace elements present in fossil fuels (Rodríguez-Mercado and Altamirano-Lozano 2006).

Antal et al. (2009) studied 56 medicinal plant species to determine V contents, finding the highest V contents in flowering aerial parts, with an average of 763 μg kg $^{-1}$DMW, followed by the leaves (682 μg kg^{-1} DMW), roots (600 μg kg^{-1} DMW), flowers (352 μg kg^{-1} DMW), and fruits (112 μg kg^{-1} DMW). Of the plants analyzed, lemon thyme (*Thymus pulegioides* L.) displays a particular capacity to accumulate this element, while other species like *Geum urbanum* L., *Urtica dioica* L., *Hypericum perforatum* L., and *Valeriana officinalis*. also tend to accumulate this element (Antal et al. 2009).

In soybean, the application of two different sources of V increased chlorophyll contents, as well as fresh and dry biomass (Sozudogru et al. 2001). The application of 240 μM V in common bean (*Phaseolus vulgaris* L.) cv. "Contender" caused thicker roots, where it accumulated more than in leaves (Saco et al. 2013). In mustard (*Brassica campestris* L. ssp. *chinensis* cv. "Parachinensis") and tomato (*Solanum lycopersicum* L.), Vachirapatama et al. (2011) showed that the application of 20 mgL $^{-1}NH_4VO_3$ improves growth in both species; V accumulates mostly in root in comparison to leaf, stem, or fruit. In wheat, the application of 40 μM V effectively improved the antioxidant defense system to alleviate the oxidative damage induced by Cu (Wang et al. 2013b). In pennyroyal (*Mentha pulegium*L.), the application of 10, 20, and 40 mg $L^{-1}NH_4VO_3$ increased root DMW, while V concentration was higher in roots than in shoots. Furthermore, V did not affect K, Ca, Mn, and Zn concentrations in roots, while a reduction of Ca, K, Mg, and Mn concentration was observed in leaves at high V application rates (Akoumianaki-Ioannidou et al. 2015).

Root uptake and translocation efficiency of V do not significantly vary with the species, whereas its translocation to plant aerial parts depends on individual plant response. Future studies are necessary to determine the effect of V-status on different taxa, while the accumulation and transfer of vanadium within the food chain remain a daunting task (Qian et al. 2014).

A summary of the levels at which beneficial elements have been applied in various model and cultivated species is presented in Table 1.

3 Effects of Beneficial Elements on Plant Nutrition as a Mean to Overcome Abiotic Stress

In several modern plant nutrition approaches, beneficial elements are considered part of nutrient management in an increasing number of crop plants. Some of these elements, including Ce, Co, I, Na, Se, Si, and Ti, have been proved to increase crop yield. Some others such as Al and La have been less studied, and their influence on plant yield and nutrient status are little known. Cerium raises P, K, Ca, Mg, S, Fe, Zn, Cu, and Al concentrations in barley leaves and grains (Rico et al. 2015), which boosts plant growth, yield, and grain quality. Cobalt enhances N, P, K, Cu, Fe, Mn, and Zn concentrations in plant tissues not only in tomato (Jayakumar et al. 2013) but also in groundnut (Gad 2012) and pea (Gad 2006), which results in increased yields and harvest quality. Iodine has been found to improve N, Mg, Na, Ce, and Fe uptake in fruits (Szwonek 2009), lettuce (Blasco et al. 2011), spinach (Smolen and Sady 2011a, b), and tomato (Caffagni et al. 2012) while enhancing productivity and crop quality. Lanthane increases Mn and Fe contents in cucumber (Zeng et al. 2000), displays a synergic effect with Ca and K uptake in tulip (Ramírez-Martínez et al. 2012), and increases Ca, P, K, and Mn concentrations in soybean (de Oliveira et al. 2015), which renders better plant growth and productivity. Selenium enhances P and Ca uptake, though it also reduces K absorption in maize (Hawrylak-Nowak 2008).

Table 1 Levels of beneficial elements tested in some model or cultivated plants

Beneficial element	Species studied	Level evaluated	Beneficial level	Application system	Reference
Al	Silver birch (*Betula pendula* Roth.)	0, 2, 5, 10, 15, 25, and 35 mg $Al(NO_3)_3L^{-1}$	2 and 5 mg L^{-1}	Nutrient solution	Kidd and Proctor (2000)
	Lisianthus (*Eustoma grandiflorum* Raf. Shinn.) cv. "Hei Hou"	0, 50, 100, and 150 mg L^{-1} $Al_2(SO_4)_3$	150 mg L^{-1}	Floral preservative solution	Li-Jen et al. (2001)
	Rhododendron (*Melastoma malabathricum* L.)	0 and 0.5 mM $AlCl_3$	0.5 mM	Nutrient solution	Watanabe et al. (2005)
	Rose (*Rosa* spp.) cv. "Royalty"	0 and 600 mg $L^{-1}Al_2(SO_4)_3$+ 4.5% sucrose	600 mg L^{-1}	Preservative solution	De la Cruz-Guzmán et al. (2007)
	Soybean (*Glycine max* L. Merr.)	0 and 1.0 μM $CdCl_2$ + 150 μM $AlCl_3$	150 μM	Culture solution	Shamsi et al. (2008)
	Chamomile (*Matricaria chamomilla* L.)	0, 60, and 120 μM $AlCl_3$	60 μM	Nutrient solution	Kováčik et al. (2010)
	Rose (*Rosa* spp.) cv. "Cherry Brandy"	0, 100, 200, and 300 mg L^{-1} $Al_2(SO_4)_3$	300 mg L^{-1}	Preservative solution	Jowkar et al. (2012)
	Rose (*Rosa hybrida*) cv. "Boeing"	0, 150, and 300 mg L^{-1} $Al_2(SO_4)_3$	150 and 300 mg L^{-1}	Floral preservative solution	Seyf et al. (2012a)
	Tuberose (*Polianthes tuberosa* L.) cv. "Single"	0, 50, 100, and 150 mg L^{-1} $Al_2(SO_4)_3$	100 mg L^{-1}	Floral preservative solution	Mohammadi et al. (2012a)
	Rosa hybrid (*Rosa* spp.) cv. "Maroussia"	0 and 600 mg $L^{-1}Al_2(SO_4)_3$	600 mg L^{-1}+ 8% sucrose	Floral preservative solution	Basaki et al. (2013)
	Maize (*Zea mays* L.)	0 and 48 μM $AlCl_3$	48 μM	Nutrient solution	Wang et al. (2015)
	Sampaguita (*Jasminum sambac* L.)	0, 500, 1000, and 1500 mgL^{-1} $KAl(SO_4)_2$	0.5, 1.0, and 1.5 g L^{-1}sprayed twice a day	Foliar application with spraying solution	Acero et al. (2016)

(continued)

Table 1 (continued)

Beneficial element	Species studied	Level evaluated	Beneficial level	Application system	Reference
Ce	Maize (*Zea mays* L.) cv. "Hycorn 82" and mung bean (*Vigna radiata* L. Wilczek) cv. "Berken"	0, 0.2, 1.0, and 5.0 μM $Ce(NO_3)_3$	< 0.2 μM	Continuously flowing solution culture units	Diatloff et al. (2008)
	Spinach (*Spinacia oleracea* L.)	0 and 15 μM $CeCl_3$	15 μM $CeCl_3$	Culture solution	Yin et al. (2009)
	Arabidopsis thaliana (L.) Heynh.	0, 0.1, 0.5 and 1 mM $Ce(NO_3)_3$	0.1 mM	Protoplasts solution	Liu et al. (2011)
	Cowpea (*Vigna unguiculata* L. Walp.)	0, 0.713, 3.568, 17.841, 89.206, and 446.030 μM $Ce(NO_3)_3$	0.713, 3.568, 17.841 μM	Pots with soil (silty sand)	Shyam and Aery (2012)
	Lettuce (*Lactuca sativa* L.) cv. "Regina"	0, 5, 10, 15, 20, and 25 mg L^{-1} $Ce(NH_4)_2(NO_3)_6$	15 mg L^{-1}	Seed immersed in aqueous solutions	Barbieri et al. (2013)
	Cilantro (*Coriandrum sativum* L.)	0, 62.5, 125, 250, and 500 mg kg^{-1} $nCeO_2$	125 mg kg^{-1}	Organic potting soil	Morales et al. (2013)
	Tomato (*Solanum lycopersicum* L.)	0 and 10 mg L^{-1} $nCeO_2$	< 10 mg L^{-1}	Germination solution and hydroponic solution	Wang et al. (2013a)
	Wheat (*Triticum aestivum* L.)	0, 125, 250, and 500 mg kg^{-1} $nCeO_2$	500 mg kg^{-1}	Potting soil	Rico et al. (2014)
	Rice (*Oryza sativa* L.)	0 and 100 μM $CdCl_2$ + 10 μM $CeCl_3$	10 μM	Nutrient solution and foliar application	Wu et al. (2014)
	Barley (*Hordeum vulgare* L.)	0, 125, 250, and 500 mg kg^{-1} $nCeO_2$	250 mg kg^{-1}	Potting soil	Rico et al. (2015)

Co	Pea (*Pisum sativum* L.)	0 and 8 mg kg^{-1} $CoSO_4$	8 mg kg^{-1}	Potting soil in greenhouse and field experiment	Gad (2006)
	Carnation (*Dianthus caryophyllus* L.) cv. "Harlem"	0, 50, 75, and 100 mg L^{-1} $CoCl_2$	100 mg L^{-1}	Treatment solution	Jamali and Rahemi (2011)
	Groundnut (*Arachis hypogaea* L.)	0 and 8 mg kg^{-1} $CoSO_4$	8 mg kg^{-1}	Field experiments	Gad (2012)
	Lily (*Lilium* spp.) cv. "Star Fighter" and cv. "Star Gazer"	0, 0.1, 0.2, 0.4, and 0.8 mM $CoCl_2$	0.1 and 0.2 mM	Preservative solution	Mandujano-Piña et al. (2012)
	Tuberose (*Polianthes tuberosa* L.)	0, 200, 300, and 400 mg L^{-1} $CoCl_2$	300 and 400 mg L^{-1}	Preservative solution	Mohammadi et al. (2012b)
	Marguerite (*Argyranthemum* sp.)	0, 1, and 2 mM $CoCl_2$	2 mM	Preservative solution	Kazemi (2012)
	Lily (*Lilium* spp.) cv. "Prato"	0, 1, and 2 mM $CoCl_2$	2.5 mM Ni + 2 mM Co + 2 mM salicylic acid with 2.5% sucrose	Preservative solution	Kazemi and Ameri (2012)
	Tomato (*Solanum lycopersicum* L.)	0, 50, 100, 150, 200, and 250 mg kg^{-1} $CoCl_2$	50 mg kg^{-1}	Pot contained soil	Jayakumar et al. (2013)
	Rose (*Rosa* spp.) cv. "Red one"	0, 100, 200, and 300 mg L^{-1} $CoCl_2$	200 mg L^{-1}	Preservative solution	Aslmoshtaghi et al. (2014)
	Gladiola (*Gladiolus grandiflorus* Hort.) cv. Borrega Roja	0, 0.3, and 0.6 mM $CoCl_2$ $6H_2O$	0.3 mM	Preservative solution	Trejo-Téllez et al. (2014)

(continued)

Table 1 (continued)

Beneficial element	Species studied	Level evaluated	Beneficial level	Application system	Reference
I	Apple (*Malus domestica* Borkh.) cv. "Golden Delicious/ M.9"	0, 1.5–2.40 and 3.0–4.8 mg L^{-1}organic-mineral liquidfertilizer	3.0–4.8 mg L^{-1}	Sprayed with foliar treatments	Szwonek (2009)
	Lettuce (*Lactuca sativa* L.) cv. "Longifolia"	0, 20, 40, and 80 μM KI and KIO_3	≤40 μM	Pots with vermiculite as asubstrate	Blasco et al. (2011)
	Potato tubers (*Solanum tuberosum* L.) and tomato (*S. lycopersicum* L.)	0, 0.05, and 0.1% KI (w/v) or 0, 0.05, 0.1, 0.2, and 0.5% KIO_3 (w/v)	0.05% KI, 0.1 and 0.2% KIO_3	Pot contained soil irrigated with treatments	Caffagni et al. (2011)
	Spinach (*Spinacia oleracea* L.) cv. "Olbrzym Zimowy"	0, 1, and 2 mg dm^{-3}KI	2 mg I + 1 g dm^{-3} sucrose	Containers filled with silt loam in plastic tunnel	Smolen and Sady (2011a)
	Spinach (*Spinacia oleracea* L.) cv. "Olbrzym Zimowy"	0, 1, and 2 mg dm^{-3}KI	2 mg dm^{-3}	Containers filled with silt loam in plastic tunnel	Smolen and Sady (2011b)
	Tomato (*Solanum lycopersicum* L.)	Foliar application:2500 mg L^{-1} I; and soil application: 90 g kg^{-1}I crystalline fertilizer	Hydroponic culture: 1, 2, and 5 mM KI	Foliar spray and soil fertilizer in field experiments and hydroponic culture in greenhouse	Caffagni et al. (2012)
	Soybean (*Glycine max* L. Merr.)	0, 20, 40, and 80 μM KIO_3	20 μM	Pots carrying soil and cow dung manure (ratio 3:1)	Gupta et al. (2015)
	Carrot (*Daucus carota* L.) cv. "Kazan F_1"	0 and 2.5 kg ha^{-1} KI and KIO_3, applied twice	2.5 kg ha^{-1} KI + Na_2SeO_3	Field study	Smolen et al. (2016)

La	Cucumber (*Cucumis sativus* L.)	0, 0.02, and 2 mM LaCl3	0.02 mM	Quartz sand irrigated with treatments	Zeng et al. (2000)
	Tobacco (*Nicotiana tabacum* L.)	0, 5, 10, 20, 50, and 100 mg L^{-1} LaCl3	20 mg L^{-1}	Hydroponic with nutrient solution	Chen et al. (2001)
	Cucumber (*C. sativus* L.)	0, 0.002, 0.02, 0.2, and 2 mM	0.2 and 2 mM	Nutrient solution	Huang et al. (2003)
	Cucumber (*C. sativus* L.)	0, 0.002, 0.02, 0.2, and 2 mM LaCl3	0.002 and 0.02 mM	Spraying twice daily	Shi et al. (2005)
	Soybean (*Glycine max* L. Merr.) cv. "Kennong"	0, 10, 20, 30, 40, and 50 mg L^{-1} LaCl3	20 mg L^{-1}	Sprayed solution on leaves	Yan et al. (2007)
	Maize (*Zea mays* L.) cv. "Hycorn 82" and mung bean (*Vigna radiata* L. Wilczek) cv. "Berken"	0, 0.2, 1.0, and 5.0 μM $La(NO_3)_3$	< 0.2 μM	Continuously flowing solution culture units	Diatloff et al. (2008)
	Tulip (*Tulipa gesneriana*) cv. "Ile de France"	0, 5, 10, 20, 30, and 40 μM $La(NO_3)_3$	10 μM	Pots irrigated with treatments	Ramírez-Martínez et al. (2009)
	Tulip (*Tulipa gesneriana*)	0, 5, 10, 20, 30, and 40 μM LaCl3 and $La(NO_3)_3$	10 and 20 μM	Pots irrigated with treatments	Ramírez-Martínez et al. (2012)
	Rice (*Oryza sativa* L.) cv. "Shengdao 16"	0, 0.05, 0.1, 0.5, 1.0, and 1.5 mM $La(NO_3)_3$	0.1 mM	Application in basal medium	Liu et al. (2012a)
	Rice (*Oryza sativa* L.) cv. "Shengdao 16"	0, 0.05, 0.1, 0.5, 1, and 1.5 mM $La(NO_3)_3$	0.05 mM	Application in basal medium	Liu et al. (2013)
	Maize (*Zea mays* L.), mung bean (*Vigna radiata*	0, 5, 10, 15, 20, 25, 30, 35, 40, 45, and 50 μM La_2O_3	50 μM	Pots with nutrient solution	Chaturvedi et al. (2014)

(continued)

Table 1 (continued)

Beneficial element	Species studied	Level evaluated	Beneficial level	Application system	Reference
	L. Wilczek)and black gram (*V. mungo*)				
	Soybean (*Glycine max* L. Merr.) cv. "BRSMG760SRR"	0, 5, 10, 20, 40, 80, and 60 μM $La(NO_3)_3$	≤ 10 μM	Pots containing nutrient solution	de Oliveira et al. (2015)
	Rangpur lime (*Citrus limonia* Osbeck)	0, 50, 100, 200, and 400 mg LaCl3in 100 mL of water	50 mg	Polypropylene tubes with substrates	Turra et al. (2015)
	Horseradish (*Armoracia rusticana* G. Gaertn., B. Mey. & Scherb.)	0, 20, 100, and 300 mg L^{-1} $LaCl_3$	20 mg L^{-1}	Nutrient solution	Zhang et al. (2016)
	Bell pepper (*Capsicum annuum* L.)	0 and 10 μM $LaCl_3$	10 μM	Nutrient solution	García-Jiménez et al. (2017)
Na	Chrysanthemum (*Chrysanthemum* x *morifolium* Ramat.) cv. "Yellow blush"	0, 1000, 3000, 6000, and 9000 mg L^{-1}NaCl	1000 mg L^{-1}	Pots with soilless substrate	Lee and van Iersel (2008)
	Faba bean (*Vicia faba* L.)	0, 60, 120, and 240 mM NaCl	60 mM	Pots with vermiculite	Abdul Qados (2011)
	Rose (*Rosa hybrida*) cv. "Avalanche"	0, 150, 200, and 250 mg L^{-1} sodium benzoate	250 mg L^{-1}	Preservative solution	Imani et al. (2012)
	Rose (*Rosa hybrida*) cv. "Utopia"	0, 50, and 100 μM sodium nitroprusside	50 μM	Pretreatment for 24 h in preservative solution	Seyf et al. (2012b)
	Rose (*Rosa hybrida*)	0, 20, 40, and 60 μM sodium nitroprusside	20 μM	Preservative solution	Nazirimoghaddam et al. (2014)
	Sunflower (*Helianthus annuus* L.)	0, 20, 40, and 60 μM sodium nitroprusside	60 μM	Preservative solution	Nazirimoghaddam et al. (2014)

	Lisianthus (*Eustoma grandiflorum* Raf. Shinn.)	0, 20, 40, and 60 μM sodium nitroprusside	40 μM	Preservative solution	Nazirimoghaddam et al. (2014)
	Cherry tomato (*Solanum lycopersicum* L.)	0, 0.87, and 28.6 mM Na^+, saline water	28.6 mM Na^+	Sandy soil	Silva et al. (2015)
	Purpletop vervain (*Verbena bonariensis* L.)	0 and 200 mM NaCl	200 mM	Pots with deacidified peat	Salachna and Piechocki (2016)
	Giant chincherinchee (*Ornithogalum saundersiae* baker.)	0, 100, and 200 mM NaCl	100 or 200 mM	Pots with a mixture of peat and fertilizer	Salachna et al. (2016)
	Indian mustard (*Brassica juncea* L. Czern.)	0 and 150 mM NaCl	150 mM	Nutrient solution	Yousuf et al. (2016)
Se	Ryegrass (*Lolium perenne* L.)	0, 0.1, 1, 10, and 30 mg kg^{-1} H_2SeO_4	1 mg kg^{-1}	Pot with coarse-textured soil	Hartikainen et al. (2000)
	Maize (*Zea mays* L.) cv. "Złota Karłowa"	0, 5, 25, 50, and100 μMdm^{-3} Na_2SeO_3	5 μM dm^{-3}	Nutrient solution	Hawrylak-Nowak (2008)
	Soybean (*Glycine max* L. Merr.) cv. "Olna"	0 and 10 mg $L^{-1}$$Na_2SeO_4$	10 mg L^{-1}	Foliar sprayed with aqueous solution	Mechora and Germ (2010)
	Townsville stylo (*Stylosanthes humilis* Kunth Hester)	1 μM and 0.1 mM Na_2SeO_4	1 μM	Seedling in Petri dishes with test solution	(2011)
	Peach grafted onto GF 677 (*Prunus persica* L.Batsch x *Prunus amygdalus* Stokes) and pear (*Pyrus communis* L.)	0, 0.1, and 1 mg $L^{-1}$$Na_2SeO_4$ via foliar or 1 mg $L^{-1}$$Na_2SeO_4$ via fruit	1 mg L^{-1}	Via foliar and fruit application	Pezzarossa et al. (2012)
	Canola (*Brassica napus*L.) cv. "RGS"	0, 10, and 20 μg plant^{-1} Na_2SeO_4	10 and 20 μg	Foliar application	Hajiboland and Keivanfar (2012)
	Lettuce (*Lactuca sativa* L.) cv. "Justyna"	0, 2, 4, 6, 15, 20, 30, 40, and 60 μM Na_2SeO_4 and 0, 2, 4, 6, 10, 15, 20, 25, and 30 μM Na_2SeO_3	<15 μM Na_2SeO_4 or Na_2SeO_3	Nutrient solution	Hawrylak-Nowak (2013)

(continued)

Table 1 (continued)

Beneficial element	Species studied	Level evaluated	Beneficial level	Application system	Reference
	Melon (*Cucumis melo* L.)	0, 2, 4, 8, and 16 μM Na_2SeO_3	2, 4, and 8 μM	Nutrient solution	KeLing et al. (2013)
	Coleus (*Coleus blumei* Benth.)	0, 0.1, 0.5, 1, 2.5, and 5 mM Na_2SeO_3	1 mM	Hydroponic system in nutrient solution	Yuan et al. (2013)
	Cucumber (*Cucumis sativus* L.) cv. "Polan F1"	0, 5, and 10 μM Na_2SeO_4	10 μM	Nutrient solution	Hawrylak-Nowak et al. (2014)
	Tomato (*Solanum lycopersicum* L.) cv. "Provence"	0, 0.1, 1, and 10 mg L^{-1} Na_2SeO_4	1 mg L^{-1}	Foliar spray	Zhu et al. (2016)
	Carrot (*Daucus carota* L.) cv. "Kazan F1"	0 and 0.5 kg $ha^{-1}$$Na_2SeO_4$ or Na_2SeO_3	0.5 kg ha^{-1} Na_2SeO_3	Soil fertilization	Smolen et al. (2016)
Si	Forage oat (*Avena sativa L.*)	0, 50, 100, 150, and 200 mg $kg^{-1}$$H_4O_4Si$	100 mg kg^{-1}	Soil fertilization in pots	Borda et al. (2007)
	Bitter gourd (*Momordica charantia*)	0, 1, 2, 3, and 5 mM K_2SiO_3	3 mM	Aerated solution	Wang et al. (2010)
	Carnation (*Dianthus caryophyllus* L.) cv. "Harlem"	0, 100, 150, and 300 mg L^{-1} K_2SiO_3	300 mg L^{-1}	Floral preservative solution	Jamali and Rahemi (2011)
	Starflower (*Borago officinalis* L.)	0, 0.5, 1, 1.5, and 2 mM $Na_2(SiO_2)_3$	0.5 and 1.5 mM	Pots with vermiculite	Shahnaz et al. (2011)
	Carnation (*Dianthus caryophyllus* L.)	0, 1.5, 2.5, and 3.5 mM Si	2.5 mM	Floral preservative solution	Kazemi et al. (2012)
	Rice (*Oryza sativa* L.) cv. "Dongjin"	0, 0.5, 1, and 2 mM Na_2SiO_3	0.5 mM	Nutrient solution	Kim et al. (2014)
	Snapdragon (*Antirrhinum majus* L.) cv. "Montego Purple"	0, 50, 100, 150, and 200 mg $L^{-1}$$NaSiO_3$	150 mg L^{-1}	Foliar application	Whitted-Haag et al. (2014)

	Maize (*Zea mays* L.)	0 and 1.5 mM Na_2SiO_3	1.5 mM	Seed priming with Si solution	Abdel Latef and Tran (2016)
	Rice (*Oryza sativa* L.) cv. "Khang Dan 18"	0, 0.4, and 17.3 Mg ha^{-1}silica gel	0.4 and 17.3 Mg ha^{-1}	Soil fertilization and pots with soil	Marxen et al. (2016)
Ti	Cowpea (*Vigna unguiculata* L. Walp.) cv. "Ife Brown"	0, 62, and 125 cm^3 $ha^{-1}$$TiO_2$	125 cm^3 ha^{-1}	Foliar application	Owolade and Ogunleti (2008)
	Geranium (*Pelargonium x hortorum*L. H. Bailey) cv. "Patriot Bright Violet"	0, 25, and 75 mM TiO_2	75 mM	Foliar application	Norman and Chen (2011)
	Poinsettia (*Euphorbia pulcherrima* Willd. ex Klotzsch) cv. "Snowcap"	0, 25, and 75 mM TiO_2	25 and 75 mM	Spraying onto plants	Norman and Chen (2011)
	Tomato (*Solanum lycopersicum* L.) cv. "ISI 68249"	0, 80, 240, 480, and 960 g ha^{-1} Tytanit® fertilizer	960 g ha^{-1}	Rockwool and fertigation system	Kleiber and Markiewicz (2013)
	Canola (*Brassica napus*L.) cv. "RGS003"	10, 100, 1000, 1200, 1500, 1700, and 2000 mg $L^{-1}$$nTiO_2$	2000 mg L^{-1}	Seeds in suspensions with treatments	Mahmoodzadeh et al. (2013)
	Wheat (*Triticum aestivum* L.) cv. "Pishtaz"	0, 0.01, 0.02, and 0.03% $nTiO_2$ and bulk Ti	0.02%	Foliar application	Jaberzadeh et al. (2013)
	Geranium(*Pelargonium x hortorum* L. H. Bailey)cv. "Elite Cherry"	0, 25, 50, 75, and 100 mg L^{-1} Ti-ascorbate (Tytanit®)	50 and 75 mg L^{-1}	Foliar application	Whitted-Haag et al. (2014)
	Snapdragon (*Antirrhinum majus* L.) cv. "Montego Purple"	0, 25, 50, 75, and 100 mg L^{-1} Ti-ascorbate (Tytanit®)	75 mg L^{-1}	Foliar application	Whitted-Haag et al. (2014)
	Mung bean (*Vigna radiata*L. Wilczek)	0 and 10 mg L^{-1}ordinary TiO_2 or $nTiO_2$	10 mg L^{-1} 10 mg L^{-1}	Foliar application	Raliya et al. (2015)

(continued)

Table 1 (continued)

Beneficial element	Species studied	Level evaluated	Beneficial level	Application system	Reference
	Timothy (*Phleum pratense*L.)	0, 0.2, 0.4, and 0.8 $dm^3\ ha^{-1}$ Tytanit®	0.4 and 0.8 $dm^3\ ha^{-1}$	Foliar fertilization in field	Radkowski et al. (2015)
	Cabbage (*Brassicaoleracea* L.) cv. "Dutch premium"	0, 250, 500, and 1000 μg mL^{-1} *n*TiO_2	500 μg mL^{-1}	Seeds in treatment suspension	Andersen et al. (2016)
	Cucumber (*Cucumis sativus* L.) cv. "Straight eight"	0, 250, 500, and 1000 μg mL^{-1} *n*TiO_2	500 and 1000 μg mL^{-1}	Seeds in treatment suspension	Andersen et al. (2016)
	Oat (*Avena sativa* L.)	0, 250, 500, and 1000 μg mL^{-1} *n*TiO_2	500 and 1000 μg mL^{-1}	Seeds in treatment suspension	Andersen et al. (2016)
	Barley (*Hordeum vulgare* L.)	0, 500, and 1000 mg kg^{-1} *n*TiO_2	1000 mg kg^{-1}	Mixture of soil and nTiO_2	Marchiol et al. (2016)
V	Soybean (*Glycine max* L. Merr.)	0, 5, 10, 15, 30, 50, or 75 mg $kg^{-1}NH_4VO_3$	15 mg kg^{-1} in fluvo-aquic soil and 50 mg kg^{-1} in red earth soil	Mixture of soil and V	Wang and Liu (1999)
	Soybean (*Glycine max* L. Merr.) cv. "Corsoy"	0, 0.5, 1.0, and 2.0 mg kg^{-1} Na_3VO_4	1 mg kg^{-1} with farmyard manure	Mixture of soil with V	Sozudogru et al. (2001)
	Chinese green mustard (*Brassica campestris* L. subsp. *Chinensis*) cv. "Parachinensis"	0, 1, 5, 10, 20, 40, and 80 mg $L^{-1}NH_4VO_3$	1–20 mg L^{-1}	Nutrient solution	Vachirapatama et al. (2011)
	Tomato (*Solanum lycopersicum* L.)	0, 1, 10, 20, 40, and 80 mg L^{-1} NH_4VO_3	1 and 10 mg L^{-1}	Hydroponics solution	Vachirapatama et al. (2011)
	Common bean (*Phaseolus vulgaris* L.) cv. "Contender"	0, 160, 240, 320, and 400 μM $VOSO_4$	240 μM	Vermiculite and watered with treatments	Saco et al. (2013)
	Wheat (*Triticum aestivum* L.) cv. "Liaochum 9"	0 and 40 μM Na_3VO_4	40 μM	Nutrient solution	Wang et al. (2013b)
	Pennyroyal (*Mentha pulegium* L.)	0, 5, 10, 20, and 40 mg L^{-1} NH_4VO_3	10–40 mg L^{-1}	Pots with peat and perlite	Akoumianaki-Ioannidou et al. (2015)

Application of Se increases its concentrations in fruit trees (Pezzarossa et al. 2012), lettuce (Hawrylak-Nowak 2013), and carrot (Smolen et al. 2016), which in turn improves plant growth and development. Under saline stress, Si reduces Na uptake and increases that of K, in order to maintain a better K/Na ratio in maize (Abdel Latef and Tran 2016), while in rice, Si increases P, N, and Mg uptake, though reduces that of K, which dramatically improves plant growth, yield, and grain quality (Marxen et al. 2016). Sodium increases N and K contents in *Ornithogalum saundersiae* Baker. (Salachna et al. 2016), which improves vessel life and flower quality. Titanium may improve crop performance through stimulating the activity of certain enzymes, enhancing chlorophyll content and photosynthesis, promoting nutrient uptake, strengthening stress tolerance, and improving crop yield and quality. These benefits lie in its interaction with other nutrient elements, especially Fe. Fe and Ti have synergistic and antagonistic relationships, depending on the Fe status of the plant. When plants experience Fe deficiency, Ti may enhance Fe uptake and utilization and subsequently improving plant growth. When Ti concentration is high in plants, Ti competes with Fe for ligands or proteins. The competition could be severe, resulting in Ti phytotoxicity (Lyu et al. 2017). Moreover, Ti has been shown to increase N, P, K, Ca, and Mg contents in tomato (Kleiber and Markiewicz 2013). Vanadium does not affect Ca, K, Mn, and Zn concentrations in roots, though it does reduce Ca, K, Mg, and Mn concentrations in leaves of *Mentha pulegium* L. (Akoumianaki-Ioannidou et al. 2015). Importantly, its role in productivity, nutrient uptake, and mobility within the plant deserves further investigation. All these effects of beneficial elements ameliorate the responses of plants to environmental stressors. A summary of the main effects of the beneficial elements on plant nutrition and their functions on plant performance is presented in Fig. 2.

4 Conclusions and Perspectives

Aluminum, cerium, cobalt, iodine, lanthanum, sodium, selenium, silicon, titanium, and vanadium have been shown to have beneficial effects in some species of model and cultivated plants. The positive effects of these elements on plants include improved yield and postharvest quality, absorption of other nutrients, and activation of mechanisms of defense against pests and diseases and resistance or tolerance to abiotic stress factors such as heavy metals, drought, and salinity. In some cases, these elements can replace some biological functions of other essential nutrients in plant metabolism.

In every case, the beneficial effects of these elements are always observed when they are used at low concentrations. Since they trigger hormetic effects, at high doses, these elements can disrupt the homeostasis of the plant and cause deleterious effects.

The great challenges facing humanity such as population growth, climate change, pollution, and the depletion and degradation of natural resources make it necessary

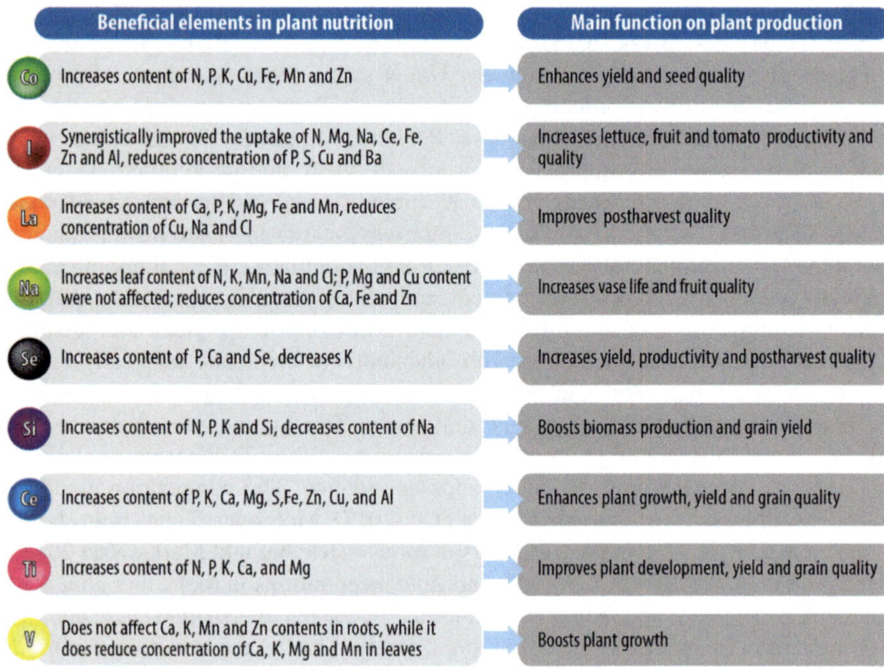

Fig. 2 Beneficial effects of Al, Ce, Co, I, La, Na, Se, Si, Ti, and V on plant nutrition and their impacts on plant production

to find new methods of sustainable crop production. In this context, beneficial elements offer an alternative little explored until now.

To date, well-supported experimental evidence demonstrates the positive effects of beneficial elements in different plant species. Nevertheless, more in-depth research is still needed in order to know their action based on the plant genotypes used, the production systems, the chemical forms in which they should be applied, and, above all, the optimal doses and phenological stages in which their beneficial effects are more evident and their application cost-effective. Importantly, nanotechnology has the potential to positively impact the agrifood sector, minimizing adverse problems of agricultural practices on environment and human health and improving food security and productivity while promoting social and economic equity. However, acquisition of knowledge and developments of methods for risk and life-cycle assessment of nanomaterials, nanopesticides, and nanofertilizers, as well as assessment of the impacts on nontarget organisms (i.e., other plants, soil microbiota, and bees), and the regulations about the use of nanomaterials require further attention (Amenta et al. 2015; Fraceto et al. 2016).

In addition to the ten beneficial elements described herein, recent reports indicate that other elements such as silver (Ag), chromium (Cr), fluorine (F), and tungsten (W) may also have potential benefits in crops, but research on them is still in its infancy.

Because the responses that trigger beneficial elements differ among families, genera, and species of plants, it is crucial to explore the underlying genetic and molecular foundations that explain the positive effect of these elements, which constitute an area of great interest for future studies. Food security, sustainability, and efficient use of current inputs eminently justify such an approach.

References

Abdel Latef AA, Tran LSP (2016) Impacts of priming with silicon on the growth and tolerance of maize plants to alkaline stress. Front Plant Sci 7:243. https://doi.org/10.3389/fpls.2016.00243

Abdul Qados AMS (2011) Effect of salt stress on plant growth and metabolism of bean plant *Vicia faba* (L.) J Saudi Soc Agric Sci 10:7–15. https://doi.org/10.1016/j.jssas.2010.06.002

Acero LH, Tuy FS, Virgino JS (2016) Potassium aluminum sulfate solution on the vase life of sampaguita (*Jasminum sambac*) flowers. J Med Bioeng 5:33–36. https://doi.org/10.12720/jomb.5.1.33-36

Akoumianaki-Ioannidou A, Barouchas PE, Kyramariou A et al (2015) Effect of vanadium on dry matter and nutrient concentration in pennyroyal (*Mentha pulegium* L.) Bull UASVM Hortic 72:295–298. https://doi.org/10.15835/buasvmcn-hort:11348

Alcántar-González G, Trejo-Téllez LI, Fernández-Pavía L et al (2016) Elementos esenciales. In: Alcántar-González G, Trejo-Téllez LI, Gómez-Merino FC (eds) Nutrición de cultivos, 2nd edn. Colegio de Postgraduados, Mexico City, pp 23–55

Amenta V, Aschberger K, Arena M et al (2015) Regulatory aspects of nanotechnology in the agri/feed/food sector in EU and non-EU countries. Regul Toxicol Pharmacol 73:463–476. https://doi.org/10.1016/j.yrtph.2015.06.016

Andersen CP, King G, Plocher M et al (2016) Germination and early plant development of ten plant species exposed to titanium dioxide and cerium oxide nanoparticles. Environ Toxicol Chem 35:1–7. https://doi.org/10.1002/etc.3374

Antal DS, Dehelean CA, Canciu CM et al (2009) Vanadium in medicinal plants: new data on the occurrence of an element both essential and toxic to plants and man. An Univ Oradea Fasc Biol 2:5–10

Ashworth DJ (2009) Transfers of iodine in the soil–plant–air system: solid-liquid partitioning, migration, plant uptake and volatilization. In: Preedy VR, Burrow GN, Watson R (eds) Comprehensive handbook of iodine. Oxford Academic Press, San Diego, pp 107–118

Aslmoshtaghi E, Jafari M, Rahemi M (2014) Effects of daffodil flowers and cobalt chloride on vase life of cut rose. J Chem Health Risks 4:1–6

Balakhnina T, Borkowska A (2013) Effects of silicon on plant resistance to environmental stresses: review. Int Agrophys 27:225–232. https://doi.org/10.2478/v10247-012-0089-4

Barbieri APP, Espíndola GMC, de Menezes NL et al (2013) Lettuce seeds treated with cerium and lanthanum aqueous solutions. Pesq Agropec Trop 43:104–109. https://doi.org/10.1590/S1983-40632013000100013

Basaki T, Faraji S, Nejat MA, Azimi MH (2013) The effect of chemical treatments on cut flower longevity of *Rosa* hybrid cultivar Maroussia. Int J Agron Plant Prod 4:450–453

Baxter IR, Vitek O, Lahner B et al (2008) The leaf ionome as a multivariable system to detect a plant's physiological status. Proc Natl Acad Sci U S A 105:12081–12086. https://doi.org/10.1073/pnas.0804175105

Blasco B, Rios JJ, Cervilla LM et al (2011) Iodine application affects nitrogen-use efficiency of lettuce plants (*Lactuca sativa* L.) Soil Plant Sci 61:378–383. https://doi.org/10.1080/09064710.2010.492782

Borda OA, Barón FH, Gómez MI (2007) Silicon as a beneficial element in forage oat (*Avena sativa* L.): physiological responses of growth and management. Agron Colomb 25:273–279

Caffagni A, Arru L, Meriggi P et al (2011) Iodine fortification plant screening process and accumulation in tomato fruits and potato tubers. Commun Soil Sci Plant Anal 42:706–718. https://doi.org/10.1080/00103624.2011.550372

Caffagni A, Pecchioni N, Meriggi P et al (2012) Iodine uptake and distribution in horticultural and fruit tree species. Ital J Agron 7:229–236. https://doi.org/10.4081/ija.2012.e32

Carvajal M, Alcaraz CF (1998) Why titanium is a beneficial element for plants. J Plant Nutr 21:655–664. https://doi.org/10.1080/01904169809365433

Chaturvedi N, Gannavarapu R, Dhal NK (2014) Effect of lanthanum on the growth and physiological activities of *Zea mays, Vigna radiata* and *Vigna mungo*. Int J Environ Sci 4:653–659. https://doi.org/10.6088/ijes.2014040404505

Chen WJ, Tao Y, Gu YH et al (2001) Effect of lanthanide chloride on photosynthesis and dry matter accumulation in tobacco seedlings. Biol Trace Elem Res 79:169–176. https://doi.org/10.1385/BTER:79:2:169

Chen Z, Watanabe T, Shinano T et al (2009) Rapid characterization of plant mutants with altered ion-profile: a case study using *Lotus japonicus*. New Phytol 181:795–801. https://doi.org/10.1111/j.1469-8137.2008.02730.x

Conti L, Nelis S, Zhang C et al (2014) Small ubiquitin-like modifier protein SUMO enables plants to control growth independently of the phytohormone gibberellins. Dev Cell 28:102–110. https://doi.org/10.1016/j.devcel.2013.12.004

de la Cruz-Guzmán GH, Arriaga-Frías A, Mandujano-Piña M et al (2007) Effect of three longevity preservatives on post-harvest life of *Rosa* cv. Royalty. Rev Chapingo Ser Hortic 13:109–113

de Oliveira C, Ramos SJ, Siqueira JO et al (2015) Bioaccumulation and effects of lanthanum on growth and mitotic index in soybean plants. Ecotoxicol Environ Saf 122:136–144. https://doi.org/10.1016/j.ecoenv.2015.07.020

Diatloff E, Smith FW, Asher CJ (2008) Effects of lanthanum and cerium on the growth and mineral nutrition of corn and mungbean. Ann Bot 101:971–982. https://doi.org/10.1093/aob/mcn021

Dong B, Sang WL, Jiang X et al (2002) Effects of aluminum on physiological metabolism and antioxidant system of wheat (*Triticum aestivum* L.) Chemosphere 47:87–92. https://doi.org/10.1016/S0045-6535(01)00210-7

Fraceto LF, Grillo R, de Medeiros GA et al (2016) Nanotechnology in agriculture: which innovation potential does it have? Front Environ Sci 4:20. https://doi.org/10.3389/fenvs.2016.00020

Furcal-Beriguete P, Herrera-Barrantes A (2013) Efecto del silicio y plaguicidas en la fertilidad del suelo y rendimiento del arroz. Agron Mesoam 24:365–378

Gad N (2006) Increasing the efficiency of nitrogen fertilization through cobalt application to pea plants. Res J Agric Biol Sci 2:433–442

Gad N (2012) Role and importance of cobalt nutrition on groundnut (*Arachis hypogaea*) production. World Appl Sci J 20:359–367. https://doi.org/10.5829/idosi.wasj.2012.20.03.2819

García-Jiménez A, Gómez-Merino FC, Tejeda-Sartorius O, Trejo-Téllez LI (2017) Lanthanum affects bell pepper seedling quality depending on the genotype and time of exposure by differentially modifying plant height, stem diameter and concentrations of chlorophylls, sugars, amino acids, and proteins. Front Plant Sci 8:308. https://doi.org/10.3389/fpls.2017.00308

Germ M, Stibilj V, Kreft I (2007) Metabolic importance of selenium for plants. Eur J Plant Sci Biotech 1:91–97

Gupta N, Bajpai MS, Majumdar RS et al (2015) Response of iodine on antioxidant levels of *Glycine max* L. grown under Cd^{2+} stress. Adv Biol Res 9:40–48. https://doi.org/10.5829/idosi.abr.2015.9.1.9183

Hajiboland R, Keivanfar N (2012) Selenium supplementation stimulates vegetative and reproductive growth in canola (*Brassica napus* L.) plants. Acta Agric Slov 99:13–19. https://doi.org/10.2478/v10014-012-0002-7

Hartikainen H, Xue T, Piironen V (2000) Selenium as an anti-oxidant and pro-oxidant in ryegrass. Plant Soil 225:193–200. https://doi.org/10.1023/A:1026512921026

Hawrylak-Nowak B (2008) Effect of selenium on selected macronutrients in maize plants. J Elem 13:513–519

Hawrylak-Nowak B (2013) Comparative effects of selenite and selenate on growth and selenium accumulation in lettuce plants under hydroponic conditions. Plant Growth Regul 70:149–157. https://doi.org/10.1007/s10725-013-9788-5

Hawrylak-Nowak B, Dresler S, Wójcik M (2014) Selenium affects physiological parameters and phytochelatins accumulation in cucumber (*Cucumis sativus* L.) plants grown under cadmium exposure. Sci Hortic 172:10–18. https://doi.org/10.1016/j.scienta.2015.12.002

Huang Z, Chen G, Du J (2003) Influence of lanthanum on the uptake of trace elements in cucumber plant. Biol Trace Elem Res 95:185–195. https://doi.org/10.1385/BTER:95:2:185

Imani MH, Hashemabadi D, Kaviani B et al (2012) Effect of sodium benzoate on longevity and ethylene production in cut rose (*Rosa hybrida* L. cv. Avalanche). Eur J Exp Biol 2:2485–2488

Jaberzadeh A, Moaveni P, Tohidimoghadam HR et al (2013) Influence of bulk and nanoparticles titanium foliar application on some agronomic traits, seed gluten and starch contents of wheat subjected to water deficit stress. Not Bot Hortic Agrobo 41:201–207

Jamali B, Rahemi M (2011) Carnation flowers senescence as influenced by nickel, cobalt and silicon. J Biol Environ Sci 5:147–152

Jayakumar K, Rajesh M, Baskaran L, Vijayarengan P (2013) Changes in nutritional metabolism of tomato (*Lycopersicon esculantum* Mill.) plants exposed to increasing concentration of cobalt chloride. Int J Food Nutr Saf 4:62–69

Jian C, Zu C, Lu D et al (2017) Effect of exogenous selenium supply on photosynthesis, Na^{+} accumulation and antioxidative capacity of maize (*Zea mays* L.) under salinity stress. Sci Rep 7:42039. https://doi.org/10.1038/srep42039

Jouyban Z (2012) The effects of salt stress on plant growth. Tech J Eng App Sci 2:7–10

Jowkar MM, Khalighi A, Kafi M, Hasanzadeh N (2012) Evaluation of aluminum sulfate as vase solution biocide on postharvest microbial and physiological properties of 'Cherry Brandy' rose. Acta Hortic 1012:1132–1144. https://doi.org/10.17660/ActaHortic.2013.1012.83

Kazemi M (2012) Effect of cobalt, silicon, acetylsalicylic acid and sucrose as novel agents to improve vase-life of *Argyranthemum* flowers. Trends Appl Sci Res 7:579–583. https://doi.org/10.3923/tasr.2012

Kazemi M, Ameri A (2012) Effect of Ni, Co, SA and sucrose on extending the vase-life of lilycut flower. Iranica J Energy Environ 3:162–166. https://doi.org/10.5829/idosi.ijee.2012.03.02.0258

Kazemi M, GHolami M, Bahmanipour F (2012) Effect of silicon and acetylsalicylic acid on antioxidant activity, membrane stability and ACC-oxidase activity in relation to vase life of carnation cut flowers. Biotechnology 11:87–90. https://doi.org/10.3923/biotech.2012.87.90

KeLing H, Ling Z, JiTao W et al (2013) Influence of selenium on growth, lipid peroxidation and antioxidative enzyme activity in melon (*Cucumis melo* L.) seedlings under salt stress. Acta Soc Bot Pol 82:193–197. https://doi.org/10.5586/asbp.2013.023

Kidd PS, Proctor J (2000) Effects of aluminium on the growth and mineral composition of *Betula pendula* Roth. J Exp Bot 51:1057–1066. https://doi.org/10.1093/jexbot/51.347.1057

Kim YH, Khan AL, Waqas M et al (2014) Silicon application to rice root zone influenced the phytohormonal and antioxidant responses under salinity stress. Plant Growth Regul 33:137–149. https://doi.org/10.1007/s00344-013-9356-2

Kirkby E (2012) Introduction, definition and classification of nutrients. In: Marschner P (ed) Marschner's mineral nutrition of higher plants. Elsevier, Amsterdam, pp 3–5
Kleiber T, Markiewicz B (2013) Application of "Tytanit" in greenhouse tomato growing. Acta Sci Pol 12:117–126
Kováčik J, Klejdus B, Hedbavny J (2010) Effect of aluminium uptake on physiology, phenols and amino acids in *Matricaria chamomilla* plants. J Hazard Mater 178:949–955. https://doi.org/10.1016/j.jhazmat.2010.02.029
Kronzucker HJ, Coskun D, Schulze LM et al (2013) Sodium as nutrient and toxicant. Plant Soil 369:1–23. https://doi.org/10.1007/s11104-013-1801-2
Kuzel S, Hruby M, Cígler P et al (2003) Mechanism of physiological effects of titanium leaf sprays on plants grown on soil. Biol Trace Elem Res 91:179–190
Landini M, Gonzali S, Kiferle C (2012) Metabolic engineering of the iodine content in *Arabidopsis*. Sci Rep 2:338. https://doi.org/10.1038/srep00338
Läuchli A, Grattan SR (2007) Plant growth and development under salinity stress. In: Jenks MA, Hasegawa PM, Jain SM (eds) Advances in molecular breeding toward drought and salt tolerant crops. Springer, Dordrecht, pp 1–32. https://doi.org/10.1007/978-1-4020-5578-2_1
Lee MK, van Iersel MW (2008) Sodium chloride effects on growth, morphology, and physiology of chrysanthemum (*Chrysanthemum x morifolium*). Horticult Sci 43:1888–1891
Li-Jen L, Yu-Han L, Kuang-Liang H et al (2001) Vase life of *Eustoma grandiflorum* as affected by aluminumsulfate. Bot Bull Acad Sin 42:35–38
Liu D, Wang X, Lin Y et al (2011) Analysis of the effects of cerium on calcium ion in the protoplasts of *Arabidopsis thaliana* with confocal microscopy. Afr J Biotechnol 10:10781–10785. https://doi.org/10.5897/AJB11.946
Liu D, Lin Y, Wang X (2012a) Effects of lanthanum on growth, element uptake, and oxidative stress in rice seedlings. J Plant Nutr Soil Sci 175:907–911. https://doi.org/10.1002/jpln.201200016
Liu D, Wang X, Chen X et al (2012b) Effects of lanthanum on the change of calcium level in the root cells of rice. Commun Soil Sci Plant Anal 43:1994–2003. https://doi.org/10.1080/00103624.2012.693231
Liu D, Wang X, Zhang X et al (2013) Effect of lanthanum on growth and accumulation of rice seedlings. Plant Soil Environ 59:19–200
Liu R, Shan C, Gao Y et al (2016) Cerium improves the copper tolerance of turf grass *Poa pratensis* by affecting the regeneration and biosynthesis of ascorbate. Braz J Bot 39:779–785. https://doi.org/10.1007/s40415-015-0246-7
Lyu S, Wei X, Chen J et al (2017) Titanium as a beneficial element for crop production. Front Plant Sci 8:597. https://doi.org/10.3389/fpls.2017.00597
Ma JF, Yamaji N (2006) Silicon uptake and accumulation in higher plants. Trends Plant Sci 11:392–397. https://doi.org/10.1016/j.tplants.2006.06.007
Mahmoodzadeh H, Nabavi M, Kashefi H (2013) Effect of nanoscale titanium dioxide particles on the germination and growth of canola (*Brassica napus*). J Orn Hort Plants 3:25–32
Mandujano-Piña M, Colinas-León MT, Castillo-González AM et al (2012) Cobalt as senescence retardant in postharvest of oriental hybrid *Lilium*. Rev Chapingo Ser Hortic 18:239–252. https://doi.org/10.5154/r.rchsh.2010.09.034
Marchiol L, Mattiello A, Poscic F et al (2016) Changes in physiological and agronomical parameters of barley (*Hordeum vulgare*) exposed to cerium and titanium dioxide nanoparticles. Int J Environ Res Public Health 13:332. https://doi.org/10.3390/ijerph13030332
Marxen A, Klotzbücher T, Jahn R et al (2016) Interaction between silicon cycling and straw decomposition in a silicon deficient rice production system. Plant Soil 398:153–163. https://doi.org/10.1007/s11104-015-2645-8
Mechora S, Germ M (2010) Selenium induced lower respiratory potential in *Glycine max* (L.) Merr. Acta Agric Slov 95:29–34. https://doi.org/10.2478/v10014-010-0004-2

Medrano-Macias J, Leija-Martínez P, González-Morales S et al (2016) Use of iodine to biofortify and promote growth and stress tolerance in crops. Front Plant Sci 7:1146. https://doi.org/10.3389/fpls.2016.01146

Mohammadi M, Hashemabadi D, Kaviani B (2012a) Improvement of vase life of cut tuberose (*Polianthes tuberosa* cv. 'Single') with aluminum sulfate. Ann Biol Res 3:5457–5461

Mohammadi M, Hashemabadi D, Kaviani B (2012b) Effect of cobalt chloride on vase life and postharvest quality of cut tuberose (*Polianthes tuberosa* L.) Eur J Exp Biol 2:2130–2133

Morales MI, Rico CM, Hernández-Viezcas JA et al (2013) Toxicity assessment of cerium oxide nanoparticles in cilantro (*Coriandrum sativum* L.) plants grown in organic soil. J Agric Food Chem 61:6224–6230. https://doi.org/10.1021/jf401628v

Moreno-Alvarado M, García-Morales S, Trejo-Téllez LI et al (2017) Aluminum enhances growth and sugar concentration, alters macronutrient status and regulates the expression of *NAC* transcription factors in rice. Front Plant Sci 8:73. https://doi.org/10.3389/fpls.2017.00073

Nazirimoghaddam N, Hashemabadi H, Kaviani B (2014) Improvement of vase life of cut rose, sunflower and lisianthus with sodium nitroprusside. Eur J Exp Biol 4:162–165

Norman DJ, Chen J (2011) Effect of foliar application of titanium dioxide on bacterial blight of geranium and *Xanthomonas* leaf spot of poinsettia. Horticult Sci 46:426–428

Owolade OF, Ogunleti DO (2008) Effects of titanium dioxide on the diseases, development and yield of edible cowpea. J Plant Prot Res 48:329–335. https://doi.org/10.2478/v10045-008-0042-5

Palit S, Sharma A, Talukder G (1994) Effect of cobalt on plants. Bot Rev 60:149–181. https://doi.org/10.1007/BF02856575

Pezzarossa B, Remorini D, Gentile ML et al (2012) Effects of foliar and fruit addition of sodium selenate on selenium accumulation and fruit quality. J Sci Food Agric 92:781–786. https://doi.org/10.1002/jsfa.4644

Pilon-Smits EA, Quinn CF, Tapken W et al (2009) Physiological functions of beneficial elements. Curr Opin Plant Biol 12:267–274. https://doi.org/10.1016/j.pbi.2009.04.009

Qian Y, Gallagher FJ, Feng H et al (2014) Vanadium uptake and translocation in dominant plant species on an urban coastal brownfield site. Sci Total Environ 476–477:696–704. https://doi.org/10.1016/j.scitotenv.2014.01.049

Radkowski A, Radkowska I, Lemek T (2015) Effects of foliar application of titanium on seed yield in timothy (*Phleum pratense* L.) Ecol Chem Eng S 22:691–701. https://doi.org/10.1515/eces-2015-0042

Raliya R, Biswas P, Tarafdar JC (2015) TiO_2 nanoparticle biosynthesis and its physiological effect on mung bean (*Vigna radiata* L.) Biotechnol Rep 5:22–26. https://doi.org/10.1016/j.btre.2014.10.009

Ramírez-Martínez M, Gómez-Merino FC, Trejo-Téllez LI et al (2009) Effect of lanthane on quality of tulip flower 'Ile de France'. Acta Hortic 847:295–300. https://doi.org/10.17660/ActaHortic.2009.847.39

Ramírez-Martínez M, Trejo-Téllez LI, Gómez-Merino FC et al (2012) Bioaccumulation of potassium, calcium and lanthanum in tulip treated with lanthanum. Terra Latin 30:229–238

Raya PJC, Aguirre MCL (2009) Elemental composition of several wild Mexican plants. Rev Chapingo Ser Cien For Ambient 15:95–99

Ribeiro MD, Mapeli AM, Antunes WC, Barros RS (2011) A dual role of selenium in the growth control of seedlings of *Stylosanthes humilis*. Agric Sci 2:78–85. https://doi.org/10.4236/as.2011.22012

Rico CM, Lee SC, Rubenecia R et al (2014) Cerium oxide nanoparticles impact yield and modify nutritional parameters in wheat (*Triticum aestivum* L.) J Agric Food Chem 62:9669–9675. https://doi.org/10.1021/jf503526r

Rico CM, Barrios AC, Tan W et al (2015) Physiological and biochemical response of soil-grown barley (*Hordeum vulgare* L.) to cerium oxide nanoparticles. Environ Sci Pollut Res 22:10551–10558. https://doi.org/10.1007/s11356-015-4243-y

Rodríguez-Mercado JJ, Altamirano-Lozano MA (2006) Vanadio: contaminación, metabolismo y genotoxicidad (Vanadium: pollution, metabolism and genotoxicity). Rev Int Contam Ambient 22:173–189

Saco D, Martín S, San José P (2013) Vanadium distribution in roots and leaves of *Phaseolus vulgaris*: morphological and ultrastructural effects. Biol Plant 57:128–132. https://doi.org/10.1007/s10535-012-0133-z

Salachna P, Piechocki R (2016) Effects of sodium chloride on growth and mineral nutrition of purpletop vervain. J. Ecol Eng 17:148–152. https://doi.org/10.12911/22998993/62311

Salachna P, Zawadzinska A, Podsiadlo C (2016) Response of *Ornithogalum saundersiae* Bak. to salinity stress. Acta Sci Pol-Hortoru 15:123–134

Schulze LM, Britto DT, Li M et al (2012) A pharmacological analysis of high-affinity sodium transport in barley (*Hordeum vulgare* L.): a $^{24}Na^{+}/^{42}K^{+}$ study. J Exp Bot 63:2479–2489. https://doi.org/10.1093/jxb/err419

Seyf M, Khalighi A, Mostofi Y et al (2012a) Study on the effect of aluminum sulfate treatment on postharvest life of the cut rose 'Boeing' (*Rosa hybrida* cv. Boeing). J Hortic Biotech 16:128–132

Seyf M, Khalighi A, Mostofi Y et al (2012b) Effect of sodium nitroprusside on vase life and postharvest quality of a cut rose cultivar (*Rosa hybrida* 'Utopia'). J Agric Sci 4:174–181. https://doi.org/10.5539/jas.v4n12p174

Shahnaz G, Shekoofeh E, Kourosh D et al (2011) Interactive effects of silicon and aluminum on the malondialdehyde (MDA), proline, protein and phenolic compounds in *Borago officinalis* L. J Med Plant Res 5:5818–5827

Shamsi IH, Wei K, Zhang GP et al (2008) Interactive effects of cadmium and aluminum on growth and antioxidative enzymes in soybean. Biol Plant 52:165–169. https://doi.org/10.1007/s10535-008-0036-1

Shi P, Chen GC, Huang ZW (2005) Effects of La^{3+} on the active oxygen-scavenging enzyme activities in cucumber seedling leaves. Russ J Plant Physiol 52:294–297

Shyam R, Aery NC (2012) Effect of cerium on growth, dry matter production, biochemical constituents and enzymatic activities of cowpea plants [*Vigna unguiculata* (L.) Walp.] J Soil Sci Plant Nutr 12:1–14. https://doi.org/10.4067/S0718-95162012000100001

Silva RM, Yasuor H, Ben-Gal A et al (2015) Salinity induced fruit hypodermis thickening alters the texture of tomato (*Solanum lycopersicum* Mill.) fruits. Sci Hortic 192:244–249. https://doi.org/10.1016/j.scienta.2015.06.002

Smolen S, Sady W (2011a) Influence of iodine fertilization and soil application of sucrose on the effectiveness of iodine biofortification, yield, nitrogen metabolism and biological quality of spinach. Acta Sci Pol Hortoru 10:51–63

Smolen S, Sady W (2011b) Influence of soil application of iodine and sucrose on mineral composition of spinach plants. Acta Sci Pol Hortoru 10:3–13

Smolen S, Skoczylas Ł, Ledwozyw-Smolen I et al (2016) Biofortification of carrot (*Daucus carota* L.) with iodine and selenium in a field experiment. Front. Plant Sci 7:730. https://doi.org/10.3389/fpls.2016.00730

Sozudogru S, Kutuk AC, Halilova H (2001) Effects of vanadium on the growth, chlorophyll, and mineral content of soybean (*Glycine max* (L.) Merr). Ciencia 9:88–95

Szwonek E (2009) Impact of foliar fertilizer containing iodine on "Golden Delicious" apple trees. Proceedings of the International Plant Nutrition Colloquium XVI. University of California at Davis. pp 1–5. Available online at: https://escholarship.org/uc/item/8bp5w7z7

Tlustos P, Cígler P, Hruby M, Kuzel S, Száková J, Balík JS (2005) The role of titanium in biomass production and its influence on essential elements' contents in field growing crops. Plant Soil Environ 51:19–25

Trejo-Téllez LI, Gómez-Merino FC (2012) Nutrient solutions for hydroponic systems. In: Asao T (ed) Hydroponics – a standard methodology for plant biological researches. InTech, Rijeka, pp 1–22

Trejo-Téllez LI, Gómez-Merino FC, Gómez-Pérez V, Castro-García FA (2014) Cobalt in postharvest of gladiolus (*Gladiolus grandiflorus* Hort.) Rev Mex Cienc Agríc 9:1575–1587

Trejo-Téllez LI, Gómez-Merino FC, Alcántar-González G (2016) Elementos benéficos: potencialidades y limitantes. In: Alcántar-González G, Trejo-Téllez LI, Gómez-Merino FC (eds) Nutrición de cultivos. Colegio de Postgraduados, Mexico City, pp 59–101

Turra C, Fernandes EAN, Bacchi MA et al (2015) Effects of lanthanum on citrus plant. Int J New Tech Res 1:48–50

Vachirapatama N, Jirakiattikul Y, Dicinoski G et al (2011) Effect of vanadium on plant growth and its accumulation in plant tissues. Songklanakarin J Sci Technol 33:255–261

Wahid PA, Valiathan MS, Kamalam NV et al (2000) Effect of rare earth elements on growth and nutrition of coconut palm and root competition for these elements between the palm and *Calotropis gigantean*. J Plant Nutr 23:329–338. https://doi.org/10.1080/01904160009382019

Wang JF, Liu Z (1999) Effect of vanadium on the growth of soybean seedlings. Plant Soil 216:47–51. https://doi.org/10.1023/A:1004723509113

Wang XD, Ou-yang C, Fan ZR et al (2010) Effects of exogenous silicon on seed germination and antioxidant enzyme activities of *Momordica charantia* under salt stress. J Anim Plant Sci 6:700–708

Wang Q, Ma X, Zhang W et al (2012) The impact of cerium oxide nanoparticles on tomato (*Solanum lycopersicum* L.) and its implications for food safety. Metallomics 4:1105–1112. https://doi.org/10.1039/C2MT20149F

Wang Q, Ebbs SD, Chen Y et al (2013a) Trans-generational impact of cerium oxide nanoparticles on tomato plants. Metallomics 5:753–759. https://doi.org/10.1039/c3mt00033h

Wang H, Wang T, You L et al (2013b) Effects of vanadate supply on plant growth, Cu accumulation, and antioxidant capacities in *Triticum aestivum* L. Environ Geochem Health 35:585–592. https://doi.org/10.1007/s10653-013-9541-z

Wang L, Fan XW, Pan JL et al (2015) Physiological characterization of maize tolerance to low dose of aluminum, highlighted by promoted leaf growth. Planta 242:1391–1403. https://doi.org/10.1007/s00425-015-2376-3

Watanabe T, Jansen S, Osaki M (2005) The beneficial effect of aluminium and the role of citrate in Al accumulation in *Melastoma malabathricum*. New Phytol 165:773–780. https://doi.org/10.1111/j.1469-8137.2004.01261.x

Whitted-Haag B, Kopsell DE, Kopsell DA et al (2014) Foliar silicon and titanium applications influence growth and quality characteristics of annual bedding plants. Open Hortic J 7:6–15. https://doi.org/10.2174/1874840601407010006

Wu M, Wang PY, Sun LG et al (2014) Alleviation of cadmium toxicity by cerium in rice seedlings is related to improved photosynthesis, elevated antioxidant enzymes and decreased oxidative stress. Plant Growth Regul 74:251–260. https://doi.org/10.1007/s10725-014-9916-x

Yan S, Huang X, Zhou Q (2007) Effect of lanthanum (III) on reactive oxygen metabolism of soybean seedlings under supplemental UV-B irradiation. J Rare Earths 25:352–358. https://doi.org/10.1016/S1002-0721(07)60435-9

Yin S, Ze Y, Liu C et al (2009) Cerium relieves the inhibition of nitrogen metabolism of spinach caused by magnesium deficiency. Biol Trace Elem Res 132:247–258. https://doi.org/10.1007/s12011-009-8392-z

Yousuf PY, Ahmad A, Ganie AH et al (2016) Salt stress-induced modulations in the shoot proteome of *Brassica juncea* genotypes. Environ Sci Pollut Res 23:2391. https://doi.org/10.1007/s11356-015-5441-3

Yuan J, Hu M, Zhou Z (2013) Selenium treatment mitigates the effect of lead exposure in *Coleus blumei* Benth. J Environ Anal Toxicol 3:191. https://doi.org/10.4172/2161-0525.1000191

Zeng FL, Shi P, Zhang MF et al (2000) Effect of lanthanum on ion absorption in cucumber seedling leaves. Biol Trace Elem Res 78:265–270. https://doi.org/10.1385/BTER:78:1-3:265

Zhang X, Wang L, Zhou A et al (2016) Alterations in cytosol free calcium in horseradish roots simultaneously exposed to lanthanum (III) and acid rain. Ecotoxicol Environ Saf 126:62–70. https://doi.org/10.1016/j.ecoenv.2015.12.014

Zhu Z, Chen Y, Zhang X et al (2016) Effect of foliar treatment of sodium selenate on postharvest decay and quality of tomato fruits. Sci Hortic 198:304–310. https://doi.org/10.1016/j.scienta.2015.12.002

Plant Adaptation to Stress Conditions: The Case of Glutathione S-Transferases (GSTs)

Evangelia Stavridou, Georgia Voulgari, Irini Bosmali, Evangelia G. Chronopoulou, Luca Lo Cicero, Angela Roberta Lo Piero, Nikolaos E. Labrou, Athanasios Tsaftaris, Irini Nianiou-Obeidat, and Panagiotis Madesis

Abstract Plants, unlike animals, are anchored to one place and, therefore, forced to sustain any environmental condition present. Unfavourable environmental conditions include abiotic (extreme temperatures, water deficits, floods, salinity, light intensities) and biotic (pests, viral, bacterial and fungal diseases) stress factors. Both types of stresses induce the production of reactive oxygen species (ROS), which damage macromolecules such as proteins, lipids, nucleic acids and cell structures like membranes. The effect of each stress factor depends on its intensity. When the stress is severe and the production of ROS is high, it might result to plant death. To avoid such event, plants have developed advanced physiological and chemical defence mechanisms of stress avoidance and/or tolerance, which allow growth only when the environmental conditions are optimum for each species, like in the case of seed dormancy. Plants have also evolved specific enzymatic defence mechanisms, including enzymes like catalase, peroxidase, super oxide dismutase and glutathione transferases. These defence mechanisms help plants either to avoid adverse environmental conditions or to combat their negative effects. A major defence mechanism involves the action of antioxidant enzymes. Glutathione transferases (GSTs) are antioxidant enzymes of great importance for the detoxification of plants from toxic compounds. GSTs have also important involvement in plant

E. Stavridou · G. Voulgari · I. Bosmali · P. Madesis (✉)
Institute of Agrobiotechnology, CERTH, Thessaloniki, Greece
e-mail: pmadesis@certh.gr

E. G. Chronopoulou · N. E. Labrou
Laboratory of Enzyme Technology, Department of Agricultural Biotechnology, Agricultural University of Athens, Athens, Greece

L. L. Cicero · A. R. L. Piero
Dipartimento di ScienzedelleProduzioniAgrarie e Alimentari (DISPA), Università di Catania, Catania, Italy

A. Tsaftaris
Perrotis College, American Farm School, Thessaloniki, Greece

I. Nianiou-Obeidat
Department of Genetics and Plant Breeding, School of Agriculture, Aristotle University of Thessaloniki, Thessaloniki, Greece

S. Vats (ed.), *Biotic and Abiotic Stress Tolerance in Plants*,
https://doi.org/10.1007/978-981-10-9029-5_7

stress tolerance against biotic and abiotic stress tolerance like extreme heat, cold, salt and herbicides.

Keywords Glutathione transferase · GST · Oxidative stress · Herbicide detoxification · Salinity · Water deficit · High and low temperatures · Heavy metal stress · Tolerance mechanisms

1 Introduction

Climatic conditions have never been stable over time; even at a certain given area there are fluctuation and extreme conditions like heat, drought, cold, moisture, etc., the same variability applies also for plant pathogen attacks. Plants are forced to germinate and grow at a certain site as they lack the ability to move. This feature of plants has obvious implications for their tolerance to biotic and abiotic stresses. Thus, plants have developed an arsenal of mechanisms in order to withstand stressful conditions. They have developed physiological mechanisms like avoiding to germinate when the conditions are not favourable or they have developed mechanism like stomatal closure. In addition, they have also developed an array of chemical compounds like anthocyanins, which help the detoxification of plants from free radicals. In addition to the aforesaid mechanisms, plants have also developed a very sophisticated enzymatic mechanism to detoxify the free toxic compounds and especially the oxygen free radicals like superoxide anion radical ($O_2^{\cdot -}$), singlet oxygen ($^1O_2$), hydroxyl radical (OH) and perhydroxyl radical ($HO_2^{\cdot}$) (termed 'reactive oxygen species' (ROS)) (Ahsan et al. 2003). Enzymes involved in ROS detoxification include superoxide dismutase (SOD), catalase (CAT), ascorbate peroxidase (APX), monodehydroascorbate reductase (MDHAR), dehydroascorbate reductase (DHAR) and glutathione reductase (GR) glutathione transferase (GST) (Sharma et al. 2012). GSTs are soluble proteins consisted of two polypeptide subunits. The role of GSTs is to catalyse the reaction between the tripeptide glutathione and a co-substrate, which contains a reactive electrophilic centre; the result of this reaction is the formation of a polar S-glutathionylated reaction product and thus the inactivation of any toxic compounds. These enzymes were reported for the first time in animals in the 1960s as enzymes active in the metabolism and detoxification of drugs. Later in the 1970s, the GSTs were also discovered in maize plants as the enzymes responsible for the detoxification of the chloro-S-triazine atrazine herbicides and thus protecting the crop from herbicide damage (Frear and Swanson 1970). GSTs have received increasing attention not only due to their multiple roles in plant metabolism but mainly for their central role in plant detoxification from xenobiotics. This chapter will emphasize on the GSTs role as part of the defence mechanism in plants against biotic and abiotic stress conditions.

Stress tolerance of plants is of great interest to both basic and applied research. Understanding plant responses and adaptation mechanisms to severe stress conditions is the key to the improvement of economically important crops. Environmental stresses, like temperature extremes, salinity, water deficit, toxic metals, nutrient deficient soils and pathogen attacks, along with anthropogenic factors (herbicide toxicity and pollutants) are among the major constraints regarding plant growth and crop yield worldwide. Because of the climate changes, abiotic stresses are already affecting plant growth and yield, worldwide, on a large percentage of arable lands (Lane and Jarvis 2007). Considering the predictions for 39% increase in population by 2050, the stagnation of agricultural production mainly due to increasing loss from abiotic stresses is now a major concern (Popelka et al. 2009; Mittler and Blumwald 2010). Moreover, it has been estimated that the reduction of crop yield worldwide will become more severe due to the forthcoming climate changes, which are expected to cause acute reduction on average yields of the major staple crop plants by more than 50% over the next years (Boyer 1982; Wang et al. 2003). For instance, the production of winter crops, such as wheat, oats, rye and apples, is anticipated to decrease by approximately 15% over the next 50 years, whereas the production of strawberries will drop up to 32% (Ramakrishna and Ravishankar 2011) simply because of the climate disruption (Pimm 2009).

Biotic and abiotic stresses have forced species to evolve; in this aspect natural selection favoured the fittest individuals to proliferate. Plants have evolved in diverse conditions, which allowed and even forced them to develop an arsenal of adaptive mechanisms that allow the detection of precise environmental changes and are able to respond to multiple stresses, reduce damage, while at the same time retain their precious resources to be used for growth and reproduction (Verslues et al. 2006). Plants are required to alter their physiology, metabolism, gene expression and developmental processes under stress conditions (Rao et al. 2006). These anatomical, morphological, physiological and biochemical adjustments include osmotic adjustment, selective ion uptake and cytoplasmic and cystic compartmentalization (Amini et al. 2007). The effect of abiotic stresses on plant performance is largely analogous of their intensity, which may differ not only for individual organisms but also for different organs of the same organism. Maintaining plant growth, development and productivity requires adaptation to stress conditions and activation of various defence mechanisms in such way that the restoration of the homeostasis and the normal cell functions are achieved (Wang et al. 2003).

Plants in their natural habitat face multiple stresses, certain molecular mechanisms either highly specific or more general control plants stress response, and form a complex regulatory network which involves ROS, small RNAs, maybe methylation alterations, transcription factors and kinase cascades (Atkinson and Urwin 2012). Plant responses to abiotic stresses are mostly polygenic and of complex nature; thus a large number of genes are involved in plants effort to maintain its homeostasis; however, many of these genes have not yet been fully identified (Vij and Tyagi 2007). Tolerance mechanisms occur at cellular and biochemical level, from expression of genes to stress associated changes in the metabolome. Such mechanisms involve the activation of the antioxidant machinery, which may also be impaired by the effect of stress level and related signalling pathways or induction of similar cellular responses, such as stress-related protein synthesis and upregulation of antioxidant enzymes (Baxter et al. 2014).

Reactive oxygen species (ROS) and reactive nitrogen species (RNS) production is induced by abiotic and biotic stresses (Mittler et al. 2011; Fones and Preston 2012; Molassiotis and Fotopoulos 2011), causing changes in the cellular redox environment and, consecutively, modifying enzyme activity and gene regulation. During adverse environmental stimuli, the ROS homeostasis is disturbed causing significant damage to cell structures (Dat et al. 2000; Mittler et al. 2004; Møller et al. 2007). Besides their toxic effect on the cell homeostasis, when in low levels, ROS participate in signalling events which aim to control abiotic stress responses (Mittler 2002; Shao et al. 2008). Enzymatic and non-enzymatic defence systems both have an important role in stabilizing redox levels and in preventing oxidative damage (Bowler et al. 1994; Foyer et al. 1994). The enzymatic defence system of plants involves many different enzymes like ascorbate peroxidases (APX), superoxide dismutases, catalases, GST and glutathione peroxidases (GPX) which catalyse the scavenging of ROS. The activities of GST, GPX and APX largely depend on the availability of glutathione (GSH) and reduced ascorbate (ASA). This availability is secured by enzymes like GR, DHAR and MDHAR, which use NAD(P)H as an electron donor (Roxas et al. 2000). However, the function and the efficiency of the plant antioxidative systems depend mainly on plant type, species and their genetic make-up.

GST (EC 2.5.1.18.) are abundant enzymes encoded by an ancient gene family exhibiting high divergence with multiple functions in stress tolerance such as the detoxification of xenobiotic substrates and thus the prevention of oxidative damage. The ability of GSTs to detoxify herbicides is well studied mainly because of their importance in determining herbicide selectivity (Labrou et al. 2015; Skopelitou et al. 2015; Skopelitou et al. 2017). In fact, GSTs have been found to have multiple roles and functions in multiple cellular processes. GSTs play a role in targeting numerous secondary metabolites, which may be phytotoxic to an appropriate cellular localization (Marrs 1996). Some GSTs probably act in order to protect the cell from oxidative damage by quenching reactive molecules through the enzymatic catalysis of glutathione (GSH) with the xenobiotic (Fig. 1) (McGonigle et al. 2000) and thus decreasing the toxic organic hydroperoxides (Dixon 2010).

GSTs have also a glutathione-dependent peroxidases (GPOX) function which allows them to reduce the oxidative stress products, such as organic hydroxyperoxides (Frova 2003; Basantani and Srivastava 2007), to protect cells from cytotoxicity. GSTs have also a role as carriers and transport of proteins or other molecules implying a wide range of functions, including cellular signalling by binding to diverse metabolites, such as porphyrins, flavonoids, anthocyanins and plant hormones as well as other secondary metabolites (Cummins et al. 2013; Dixon et al. 2011). Other GST functions involve reversible ligand that plays a role in auxin regulation (Smith et al. 2003), protection of plants from the ultraviolet (UV) radiation-induced damage (Liu and Li 2002), regulation of apoptosis (Kampranis et al. 2000), transferring of growth regulators and flavonoids like anthocyanins to the vacuole or plastids (Dixon et al. 2008) and operating as stress signalling proteins (Rentel et al. 2004; Moons 2005).

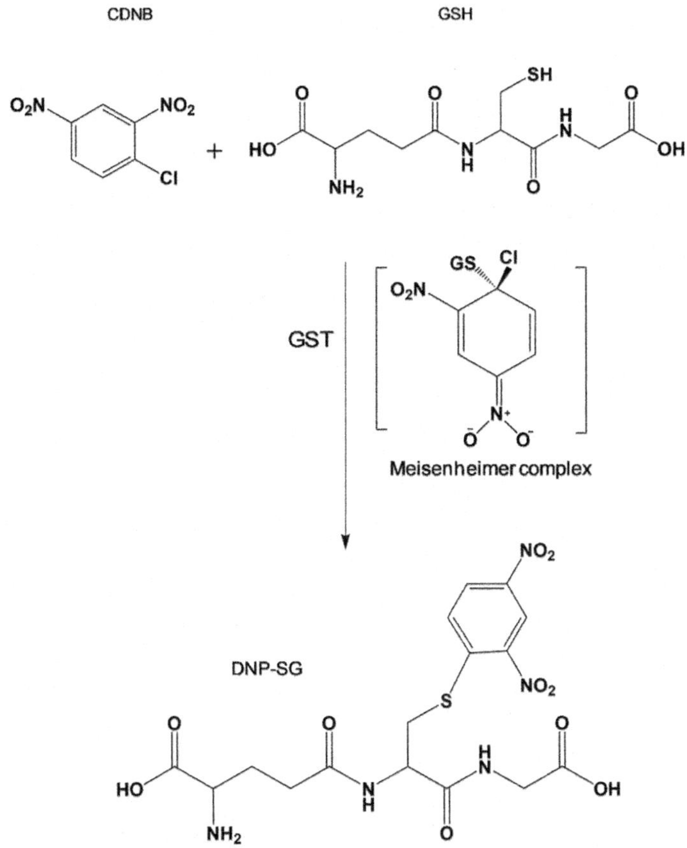

Fig. 1 Example of aromatic substitution reaction catalysed by GSTs. The figure depicts the reaction of GSH with 1-chloro-2,4-dinitrobenzene (CDNB). In this conjugation, a covalent intermediate is formed, the σ-complex (Meisenheimer complex). The aromatic character of CDNB is regained as chloride leaves from the collapsing σ-complex

GSTs and GSHs have evolved in parallel in aerobic organisms, at the same time they are distributed extensively in almost all forms of life. They are found to be expressed on every stage of plant development and in all plant tissues investigated thus far for their presence (McGonigle et al. 2000). The natural glutathione tripeptide (y-glu-cys-gly) is found both in plant and animal cells. It has multiple functions in the cellular metabolism and is a major redox buffer playing a significant role in cell homeostasis. In addition, it is involved in the antioxidative defence system as a substrate for the antioxidant enzymes or it can react directly with the reactive oxygen species (ROS) (Peñuelas 2008). It exists in two forms in the cell, the reduced (GSH) and the oxidized (GSSG). In the reduced form, it is involved in the scavenging of ROS or in the regeneration of the ascorbate during the ascorbate-glutathione cycle (Noctor and Foyer 1998). In plants, it is a major pool of sulphur and the main transport form of reduced sulphur, while it has been shown that regulates sulphur uptake at root level (Herschbach et al. 2000; Lappartient and Touraine 1997). It is a precursor of phytochelatins, with a significant role in the detoxification of heavy metals and especially of cadmium (Cd) (Jozefczak et al. 2014). Its role in plant tolerance against pathogens has also been demonstrated (Parisy et al. 2007). Glutathione along with glutaredoxins (GRXs) are involved in deglutathionylation/glutathionylation reactions which is an important mechanism function in the regulation and redox signalling in plants (Peñuelas 2008).

The elimination of xenobiotics in plant cells initiates with their conjugation to glutathione. As mentioned earlier, the GSTs are the enzymes that perform the catalytic biding of glutathione to external molecules such as xenobiotics. Plant GSTs are classified into six classes, phi, tau, theta, zeta, lambda and glutathione-dependent dehydroascorbate reductase (DHAR), while tau and theta classes are considered as the major GST classes found in plants (Fig. 2), which are involved mainly in the detoxification of xenobiotics (Frova 2006; Cummins et al. 2011).

The soluble GSTs are biologically active as dimers of approximately 23–30 kDa identical subunits forming a dimer with globular shape (Axarli et al. 2009). Each subunit has two distinct sub-active sites, the G site at the N-terminal which is a GSH-binding site and an H site at the C-terminal which is an electrophile substrate binding site (Fig. 3a, b); both G and H sites consist an independent active site of each subunit (Axarli et al. 2017). Plant GST gene families are highly diverse and usually have many members in plants, for example, there have been reported to be 25 in soybean, 42 in maize and 53 in *Arabidopsis* (Dixon et al. 2002; Sappl et al. 2009; Labrou et al. 2015; Skopelitou et al. 2015; Skopelitou et al. 2017).

The expression of GST genes of various classes is either constitutive or tissue and/or developmental stage-specific but also varies as a result of hormones and/or abiotic and biotic stresses influence (Jain et al. 2010; Kumar et al. 2013a). Regarding GSTs compartmentalization, they are considered to be mostly cytoplasmic, yet GST isoforms have also been reported to be present in microsoms, plastids, nucleus and apoplasts (Frova 2003).

Plant adaptation and enhanced stress tolerance are vital for the future of agriculture in a changing environment. Novel technologies have allowed the progress of crop breeding strategies to improve yields under optimal growth conditions. Yet, a better understanding of the biological mechanisms and the genetic basis underpinning stress adaptation is required if we want to enhance plant stress tolerance. Over the past decade, our knowledge on plant adaptation to environmental stresses has grown considerably. The development and application of systems biology and omics approaches used to shed light onto some of the key regulatory pathways involved in plant molecular responses to abiotic stress will accelerate the progress in the field (Cramer et al. 2011).

Considering all the above, GSTs are an important enzyme family for the genetic plant breeding through the development of tolerant crop varieties to multiple stresses, with maximum yield potential. The recent discovery of methods for creating recombinant DNA molecules and horizontal gene transfer from an organism to another has opened new ways to plant breeding. The use of genetic engineering renders the targeted plant breeding possible beyond the limitation of natural

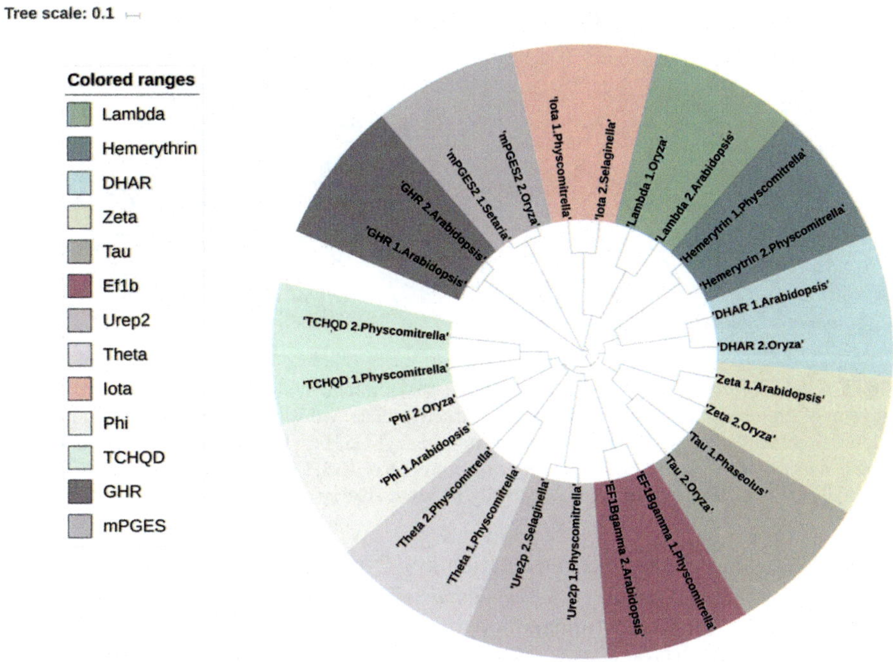

Fig. 2 Circular phylogenetic tree of plant GSTs. The sequences used are those identified in *Arabidopsis thaliana*, *Oryza sativa*, *Phaseolus vulgaris*, *Physcomitrella patens* and *Selaginella moellendorffii* (Lan et al. 2009; Liu et al. 2013; Chronopoulou et al. 2014; Lallement et al. 2014). Sequences were aligned with CLUSTAL Omega (Sievers et al. 2011), and phylogenetic tree was constructed with Geneious 9.1.2 (http://www.geneious.com, Kearse et al. 2012) with UPGMA tree building method. Various classes can be distinguished: phi (GSTF), tau (GSTU), lambda (GSTL), theta (GSTT), dehydroascorbate reductase (DHAR), elongation factor 1Bγ (EF1Bγ), glutathionyl hydroquinone reductase (GHR), hemerythrin (GSTH), iota (GSTI), zeta (GSTZ), microsomal prostaglandin E synthase type 2 (mPGES-2), tetrachloro-hydroquinone dehalogenase (TCHQD) and Ure2p. The accession numbers of proteins that were used for this phylogenetic tree are the following: phi_1. *Arabidopsis thaliana* (CAA72973.1), phi_2. *Oryza sativa* (ABF93846.1); tau_1. *Phaseolus vulgaris* (AEX38000.1), tau_2. *Oryza sativa*(AAQ02687.1); lambda_1. *Oryza sativa* (AAF70831.1), lambda_2. *Arabidopsis thaliana* (NP_191064.1); theta_1. *Physcomitrella patens* (AFZ39142.1), theta_2. *Physcomitrella patens* (AFZ39143.1); DHAR_1. *Arabidopsis thaliana* (AAF98403.1); DHAR_2. *Oryza sativa* (AAL71856.1); EF1Bgamma_1. *Physcomitrella patens* (AFZ39147.1), EF1Bgamma_2. *Arabidopsis thaliana* (BAH56923.1); GHR_1. *Arabidopsis thaliana* (NP_199315), GHR_2. *Arabidopsis thaliana* (NP_001031671.1); hemethrin_1. *Physcomitrella patens* (AFZ39150.1), hemethrin_2. *Physcomitrella patens* (AFZ39151.1); iota_1. *Physcomitrella patens* (AFZ39144.1); iota_2. *Selaginella moellendorffii* (XP_002968645.1); zeta_1. *Arabidopsis thaliana* (AAO60039.1), zeta_2. *Oryza sativa* (ABA96700.2); mPGES2_1. *Setaria italic* (XP_004969028.1), mPGES2_2. *Oryza sativa* (CAH67930.1); TCHQD_1. *Physcomitrella patens* (AFZ39137.1), TCHQD_2. *Physcomitrella patens* (AFZ39138.1); and Ure2p_1. *Physcomitrella patens* (AFZ39145.1), Ure2p_2. *Selaginella moellendorffii* (EFJ21054.1)

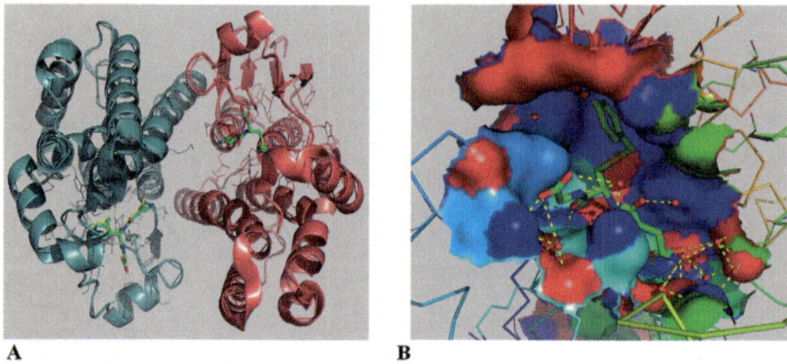

Fig. 3 The structure of GSTU4-4 from *Glycine max* protein model (PDB: 2VO4). (**a**) Ribbon diagram of high resolution GST homodimeric structure. The two subunits are depicted in different colours (light teal and deep salmon). Bound S-nitrobenzyl-GSH (in both subunits) is shown in stick representation. (**b**) Close-up of the binding of S-nitrobenzyl-GSH in the G and H site of GSTU4-4 from *Glycine max*. The figures were created by PyMOL (DeLano 2002)

hybridization and existing genetic variability (Nianiou-Obeidat et al. 2017). The precise physiological roles of GSTs especially in plant stress tolerance is still required to be completely understood, despite the fact that a huge number of GSTs, from many different plant species, has been functional characterized (Labrou et al. 2015; Skopelitou et al. 2015; Skopelitou et al. 2017).

We review in this chapter the impact of abiotic and biotic stresses on plant adaptation, highlighting the GST-mediated defence mechanisms employed by plants during adverse environmental conditions. In the scope of the GST defence line, we address the extent to which plant mechanisms are interconnected in response to quite diverse environmental stresses, describing how the plant utilizes and integrates common signalling and subsequent pathways to cope with adverse environmental conditions.

2 Antioxidant Defence Mechanisms

In plants, photosynthesis and respiration are two vital physiological processes; yet, the photorespiration and mitochondrial respiration result in the production of ROS, like the $O_2^{\cdot -}$, the hydrogen peroxide (H_2O_2), the OH· and the perhydroxyl radical (HO^2) (Gill and Tuteja 2010). ROS generation occurs due to the function of photosystems I and II (PSI and PSII) in the chloroplasts (Asada 2006), complexes I and III and ubiquinone of the mitochondrial electron transport chain (ETC) (Møller et al. 2007), membrane and matrix of the peroxisomes (Mittova et al. 2003; del Río et al. 2006) as well as the apoplast (Hernandez et al. 2001; Hu et al. 2007). Despite being a source of ROS, photorespiration plays also a vital role in readjustment of redox homeostasis under abiotic stress conditions (Voss et al. 2013). Reactive oxygen species (ROS) are also produced as by-products of various metabolic reactions and bioenergetic pathways (Suzuki et al. 2011). When the growth conditions are optimal, ROS production is retained at low levels and serves as signalling molecules which regulate plant responses to environmental stress and also growth and development (Foyer and Noctor 2005; Miller et al. 2010, 2011). Furthermore, acclimation of plants to abiotic stressed occurs through networks of ROS/redox signalling in the chloroplast and mitochondria (Suzuki et al. 2012). However, the redox state of such interactive organelles involved in carbon metabolism and energy balance (NAD(P)H and ATP) can be affected in plants exposed to adverse environmental conditions, and excessive ROS generation and accumulation may lead to cell death (Gill and Tuteja 2010; Miller et al. 2010; Sekmen et al. 2014; Zinta et al. 2014).

Plants prevent the toxic results of free radicals through their efficient and complex enzymatic and non-enzymatic antioxidant defence systems (Perez-Lopez et al. 2009; Sharma et al. 2012; Xu et al. 2015b). The non-enzymatic system consists of low and high molecular weight compounds like ascorbic acid, glutathione, proline, carotenoids, polyamines, phenolic acids, flavonoids, phytohormones and tannins (Foyer and Noctor 2005). The enzymatic antioxidant system includes the following categories: (1) antioxidative enzymes (i.e. the protein uses substrates such as superoxide, H_2O_2 or organic peroxide), (2) proteins implicated in the redox state maintenance (i.e. regeneration of reduced forms of reductants) and (3) proteins that control metabolite signals released at a secondary stage (i.e. conjugases). In the first group, SODs and CAT catalyse dismutation reactions, whereas APXs require reductant in the form of ascorbate. The second category of antioxidative enzymes includes DHARs, GRs and NADPH-generating dehydrogenases, as well as some glutaredoxins and thioredoxins. The third category comprises of enzymes such as glyoxylases, aldo/ketoreductases, cytochrome P450s (CYPs), conjugase-type glutathione transferases (GTs) and glycosyltransferases (Foyer and Noctor 2009).

Scavenging capacity of plants is species specific as well as tissue and developmental stage dependent. The antioxidant responses of cultivars differing in stress tolerance have been investigated in several crops under drought (wheat, Hameed et al. 2011; Gietler et al. 2016; bean, Türkan et al. 2005; rice, Basu et al. 2009; cotton, Deeba et al. 2012; sugarcane, Cia et al. 2012; tomato, Sánchez-Rodríguez et al. 2010), heat (wheat, Gupta et al. 2013; cotton, Mahan and Mauget 2005; Gür et al. 2010; maize and rice, Kumar et al. 2012) and salinity (rice, Mishra et al. 2013; barley, Seckin et al. 2010; tomato, Mittova et al. 2003) single stresses, as well as combinations of drought and heat stress (cotton, Sekmen et al. 2014; tobacco, Rizhsky et al. 2002; wheat, Rampino et al. 2012), salinity and heat stress (tomato, Rivero et al. 2014) and heat and cadmium (rice, Nahakpam and Shah 2011).

According to Jozefczak et al. (2014), exposure to cadmium (Cd) induced a time-dependent differential response of leaves and roots of *Arabidopsis* plants which demonstrated a biphasic GSH-related chelating and antioxidant capacity in roots. As an early response in roots, GSH levels were reduced due to the preferentially allocated GSH to the synthesis of phytochelatin (PC) for Cd chelation. However, after a period of 24 h, secondary reactions initiated with the increase of expression of multiple antioxidative enzymes, including ascorbate (AsA), SOD and CAT, to ensure the efficient removal of Cd-induced ROS. Thus, Cd retention and detoxification in roots along with high contents of thiol in leaves and possibly signalling responses from the roots allowed sufficient time for the leaves to activate their defence mechanisms (Jozefczak et al. 2014). Functional analysis of soybean *GmGSTL1*, a gene upregulated under salt stress, revealed that its protective effect against salinity might be the result of its interactions with the antioxidant flavonoids quercetin and kaempferol, providing evidence that this enzyme might catalyse the reduction of oxidized flavonoids thus restoring their antioxidant function and via this action might contribute to oxidative stress tolerance (Chan and Lam 2017). Therefore, elucidating the molecular pathways that control ROS enzymatic defence system in cells in relation to GSTs during abiotic stresses could provide an integrated approach to enhance crop tolerance to adverse environmental conditions.

3 GST-Mediated Environmental Stress Tolerance

Different environmental stimuli can induce GST expression, entailing abiotic stresses (Gallé et al. 2013), herbicides (Dill et al. 2008), heavy metals (Ezaki et al. 2004), wounding (Reymond 2000), biotic stresses such as pathogen infection and fungal elicitors (Dongli Pei 2012) and also hormones (Nutricati et al. 2006; Wagner et al. 2002; Xu et al. 2002), which are found at different plant developmental stages. The use of genetic engineering techniques allows for efficient management of these response mechanisms through targeted overexpression or suppression of specific genes (Zhang et al. 2004; Umezawa et al. 2006).

3.1 GSTs Conferring Tolerance to Abiotic Stresses

Environmental stresses are among the main factors resulting in crop losses. GSTs have been shown to be involved in many plant–environment interactions and particularly in stress tolerance mechanisms. GSTs and especially members of the tau and phi classes have been shown to be differentially expressed in response to abiotic stress signals (Csiszár et al. 2014). Increased GST expression has been correlated to enhanced stress tolerance, as observed in crops like tomato (Gallé et al. 2009; Sun et al. 2010), wheat (Gallé et al. 2011), barley (Rezaei et al. 2013), cotton (Dong et al. 2016) and rice (Moons 2003). The GSTL class expression profile in rice revealed that there is a tissue and developmental stage regulation, with *OsGSTL2* exhibiting overall the highest expression levels as well as increased tolerance to drought, salinity and cold stress (Kumar et al. 2013b). Several studies have employed genetic engineering to investigate the *in planta* function of specific GSTs under stress stimuli and their mechanistic involvement in abiotic stress tolerance.

Tobacco plants overexpressing the *GsGST* gene from *Glycine soja*, a high salt- and drought-tolerant species, demonstrated dehydration tolerance, while the T_2 seedlings showed higher tolerance to salt and mannitol with significant growth advantages compared to the wild-type plants (Ji et al. 2010). These results emphasize the evidence that *GsGST* could be an efficient target for engineering in order to improve environmental stress tolerance in important crops. In another example, transgenic tobacco plants overexpressing a *Prosopis juliflora* (drought-tolerant tree species) GST (*PjGSTU*1) survived under 15% PEG conditions, which simulates drought stress. Further investigation revealed that the *PjGSTU1* was localized in the chloroplast of transgenic plants, which is correlated with its role in ROS removal (George et al. 2010).

The signalling functions of plant GSTs could be used further in order to enhance plant performance under abiotic stress conditions. Overexpressing a phi class GST from *Arabidopsis* (*AtGST10*) in *Arabidopsis* transgenic plants conferred tolerance to salt and oxidative stress, whereas downregulation of *AtGSTF10* via RNA interference resulted in decreased tolerance to abiotic stress (Ryu et al. 2009). However, silencing of *AtGSTU17* resulted in increased drought and salt stress tolerance through anatomical and physiological changes accompanied with higher ABA and GSH content, suggesting that the suppression of *AtGSTU*17 expression could play a significant role in fine-tuning GSH homeostasis, redox status and stress-responsive genes for the adaptation to environmental signal changes (Chen et al. 2012). Transgenic *Arabidopsis* plants overexpressing *AtGSTU*19 showed enhanced tolerance to salt, drought and methyl viologen stress and increased percentage of seed germination and cotyledon emergence and was correlated with increased proline level and antioxidant enzymes activities, along with decreased lipid peroxidation under stress conditions (Xu et al. 2016).

The regulation of the differential expression of GSTs seems to be under the control of a complex stress-dependent induced mechanism operating in plants. A GST from *Suaeda salsa* has been overexpressed in *Arabidopsis* (Qi et al. 2010) and co-expressed, along with CAT, in rice (Zhao and Zhang 2006). In the first occasion, the specific gene, which was shown to have dual GST/GPX activity (Wang et al. 2002), provided the transgenic plants with enhanced salinity tolerance up to 200 mM NaCl, and despite the 1.5-fold lower photosynthetic rates, the plants exhibited reduced lipid peroxidation and a high metabolic rate (Qi et al. 2010). In the second occasion, the co-expression of *SsGST* with CAT resulted in reduced oxidative damage under salt and paraquat stress conditions. However, the GST activity increased only under the herbicide stress. The enhanced tolerance might be the result of the synergistic effect of the two enzymes and the observed increase in SOD activity (Zhao and Zhang 2006). Another GST with dual GST and glutathione peroxidase activity from *Limonium bicolor* (*LbGST*1), overexpressed in tobacco plants, exhibited higher levels of activity of peroxidase (POD), superoxide dismutase and catalase when compared to wild-type plants, particularly when grown under salt stress. The *LbGST1* was found to be localized in the nucleus, suggesting a possible role in mediating certain physiological pathways or protecting the DNA from oxidative damage (Diao et al. 2011). Furthermore, overexpression of *ThGSTZ1*

Fig. 4 Transgenic tomato plants under salinity stress. Effect of 240 mM NaCl on transgenic line A1 overexpressing the *gmgstu*4 gene compared with the WTCLX3711 (Felina)

Fig. 5 Transgenic tomato plants under osmotic stress. Effect of mannitol on transgenic line A5 overexpressing the *gmgstu*4 gene compared with the WTCLX3711 (Felina)

from *Tamarix hispida* demonstrated enhance tolerance towards drought and salinity stress with simultaneous increase of total GST and GPOX activity and ROS scavenging ability (Yang et al. 2014).

The plant-specific tau class GST (GSTU) genes can be further induced by different abiotic stresses. The expression of *SbGST* gene, isolated from *Salicornia brachiate*, an extreme halophyte, was upregulated under different abiotic stresses (salt, cold, drought) and plant growth regulator treatments (ABA), except salicylic acid treatment. In addition, Jha et al. (2011) showed that *SbGST* gene overexpression in transgenic tobacco plants led to improved seed germination and growth under salinity, suggesting its vital role in abiotic stress tolerance (Jha et al. 2011). Interestingly, overexpression of rice *OsGSTU*4 in *Arabidopsis* conferred salt and oxidative stress tolerance and was accompanied by lower sensitivity to ABA and auxin and at the same time upregulation of pathways of sulphate reduction and phenylpropanoid and flavonoid biosynthesis which are related to defence responses, thus indicating pleiotropic interactions that could merely be explained by the enzymes catalytic functions (Sharma et al. 2014).

In more recent examples, transgenic tobacco plants overexpressing a specific soybean *GmGSTU4* isoenzyme have been created in our laboratory and showed enhanced salt and osmotic stress tolerance (Figs. 4 and 5), while at the same time, the transgenic plants showed significant differences in their metabolome towards maintaining metabolic homeostasis while exhibiting higher concentration of metabolites such as proline and trehalose which have protective role, compared to wild-type plants (Kissoudis et al. 2015a). Such pleiotropic effects might provide explanation to the enhanced tolerance to drought and salt stress that transgenic tobacco plants show when overexpressing a sweet orange *CsGSTU*, despite the absence of in vitro enzyme scavenging activity (Lo Cicero et al. 2015). Transgenic *Arabidopsis* plants overexpressing the *LeGSTU2* gene from tomato exhibited increased tolerance to salt and drought stress which was also linked to changes in metabolites like proline and malondialdehyde and also antioxidative enzymes activities (Xu et al. 2015a).

Transgenic lines of *Medicago sativa* overexpressing the *GsGSTU13* showed ameliorated growth and physiological traits under alkaline stress compared to wild alfalfa plants. Further co-transformation of *GsGSTU13* together with *SCMRP* into two alfalfa cultivars increased the methionine content and enhanced the tolerance of the transgenic plants to alkaline and salt stresses (Jia et al. 2016).

Regarding the function of GSTs under thermal stress, transgenic rice overexpressing a rice zeta (ζ) class GST gene under the control of the ubiquitin promoter showed increased germination and growth at low temperatures, which coincided with increased GST and GPOX activity of rice leaf extracts, even under submergence, thus enabling the direct sowing of rice in cooler regions and reducing the production cost (Takesawa et al. 2002). In rice, the presence of a *OsGSTZ2* allelic variant was correlated with reduced cold tolerance, but the interesting thing here is that this is a naturally occurring with significantly lower isomerase activity (Kim et al. 2011).

Plants overexpressing GSTs seem to have adaptive capability in a fast changing environment. It has been shown that when GST and GPX are overexpressed in tobacco transgenic seedlings exhibited enhanced growth under stressed environment and showed increased GSH-dependent peroxide scavenging activity and changes in GSH and ASH metabolism which resulted in enhanced oxidative stress tolerance through GPOX activity. The transgenic seedlings grew faster than wild type, even under different stress conditions such as chilling, heating and salt stress, and also showed reduced lipid peroxidation (Roxas et al. 2000). Transgenic plants of *Dianthus superbus* overexpressing a *Nicotiana tabacum* GST (NT107) showed enhanced tolerance to high light intensity and increased photosynthetic rates under high light and in drought conditions. In addition, the transgenic plants showed increased copper accumulation. These characteristics can be extremely useful in the impending climate change conditions (Lim et al. 2005). Nevertheless, transgenic cotton seedlings overexpressing the tobacco GST (*Nt107*) failed to show enhanced tolerance to salt, cold or herbicides (atrazine and imazethapyr), even though they exhibited five- to tenfold higher GST activity compared to wild-type plants (Light et al. 2005). In this case, the dosage effect should be taken into consideration since the application of 200 mM NaCl compared to the usual dosage (100–150 mM NaCl) might have a significant impact on stress response and gene regulation. Overexpression of *PpGST* from the fruit of *Pyrus pyrifolia* in tobacco increased T1 transgenic tobacco lines tolerance to oxidative damage caused by NaCl, drought and Cd stresses; yet, the molecular mechanism through which *PpGST* is involved in tolerance to abiotic stresses still remains to be investigated (Liu et al. 2013).

Engineering GSTs directly into the chloroplast is an interesting approach for enhanced plant stress tolerance, since the chloroplast is the organelle where photosynthesis takes place and thus, a place where reactive oxygen species are generated. Le Martret et al. (2011) have expressed GST in tobacco chloroplasts, either alone or in combination with DHAR and glutathione reductase, which led to enhanced salt (200 mM NaCl) and cold (4 °C) tolerance of the transplastomic plants compared to wild type. Although a GST gene has been expressed directly in the chloroplast before (Dixon et al. 2008), there is increasing evidence that overexpression of ROS scavenging enzymes within the chloroplast enhances the plant's ability to tolerate abiotic stress (Le Martret et al. 2011).

The genetic manipulation of agronomical important crops using specific genes conferring high tolerance to abiotic stresses might provide alternative resource for the cultivation of such crops on marginal soils and overcome the world increasing food demand. Transgenic overexpression has provided further insights into the functional mechanism of GSTs towards abiotic stress adaptation. Different experimental approaches and scientific evidence indicate that GSTs contribute significantly in plant acclimatization and tolerance to environmental stresses like salinity, heat, drought and cold stress. Thus, it is of paramount importance to further understand the GST function and regulation under different stress stimuli in order to improve our perception of the underlying mechanisms and development of tolerant crops able to withstand the changing climates.

3.2 GST-Mediated Heavy Metal and Pollutant Tolerance as a Phytoremediation Strategy

Transcriptomics and proteomic studies have indicated an extensive induction of GSTs expression under various heavy metal stress conditions (Alvarez et al. 2009; Lin et al. 2013). GST increased activity was also observed in leaves and roots of *P. sativum* plants when exposed to Cd (Dixit et al. 2001) and in roots of *O. sativa* (Moons 2003) and *Phragmites australis* (Lannelli et al. 2002), while He et al. (2015) identified 17 genes encoding GSTs that were upregulated in rice roots subjected to cadmium (Cd) stress. It is interesting to note here that GSTs could be employed through genetic engineering approaches for the phytoremediation of environmental pollution caused by organic xenobiotics, including herbicides, chemicals used in industry and explosives (Gunning et al. 2014).

Cadmium is considered as one of the most phytotoxic heavy metals due to its accumulation in the soil and its relatively highly mobile in the soil–plant system (Benavides et al. 2005). Transgenic tobacco plants overexpressing a fungal GST from *Trichoderma virens* revealed enhanced tolerance to different concentrations of

Cd compared to wild-type plants, without enhancing its accumulation in the plant biomass (Dixit et al. 2011). This might signify a potential function of GSTs as carriers of cadmium or cadmium-binding substances and their transfer to pumps that exert the heavy metal out of the plant, with additional value in developing Cd-tolerant crops while limiting Cd availability in the food chain (Dixit et al. 2011). On the other hand, some GSTs are responsible for the sequestration of heavy metals in plants, as it was demonstrated on *Dianthus superbus* L. plants, overexpressing an auxin-regulated tobacco glutathione S-transferase (GST) (NT107). The transgenic plants accumulated significantly higher amounts of copper in shoots and roots when compared to wild-type plants and were found to synthesize phytochelatin (PC), which functions by sequestering and detoxifying excess copper ions, suggesting that both GSTs and GSH coordinately function to enhance chelation and sequestration of GSH–chromium complexes into vacuoles (Lim et al. 2005).

The highly specific catalytic activity which GSTs exhibit against targeted chemical compound is necessary in order to maximize the remediation efficiency (Benekos et al. 2010). It was observed before that when the poplar *GSTU51* was overexpressed, it led to the selective higher tolerance to mercury but not cadmium stress. The selectivity of *PatgGSTU51* might be associated with different mechanisms employed to manage different heavy metal toxicities with GSH (Choi et al. 2013). The expression of rice *OsGSTL* genes is highly induced by arsenic stress with demonstrated differential expression of these genes in arsenic sensitive and tolerant genotypes; overexpression of *OsGSTL2* in *Arabidopsis* plants enhanced their tolerance to different heavy metals like chromium, arsenate and cadmium (Kumar et al. 2013b). It was further suggested that the expression of *OsGSTL2* in *Arabidopsis* may enhanced antioxidant system in transgenic lines and/or flavonol levels, which may participate in conferring tolerance towards abiotic stresses (Kumar et al. 2013b). These results are suggestive of GSTs roles in regulating the binding and transport of tolerance-related compounds *in planta* (Dixon et al. 2011), and possibly the flavonols and their derivatives as well as oxidized derivatives of tocopherols (vitamin E) bind tightly to GSTs and are possible substrates for GSTs (Hernandez et al. 2004).

Nanoparticles (NPs) have become widely used in manufacturing and medical processes, and while their toxic effect on animals has started to receive attention, their impacts on the environment and their effects on plant life have yet to be better understood (Cox et al. 2017). Gene expression analysis in *A. thaliana* has shown upregulation of sulphur assimilation, glutathione biosynthesis, GSTs and glutathione reductase genes upon exposure to AgNPs compared to Ag ions which suggests that exposure to silver nanoparticles exacerbated the toxic response of *Arabidopsis* plants with reduction in total chlorophyll and increase in anthocyanin content and lipid peroxidation (Nair and Chung 2014). This increase in anthocyanin might be explained by the role of GSTs as non-enzymatic carriers (ligandins) for intracellular transport that catalyse anthocyanin–GSH conjugates which allow transport into vacuoles by glutathione pump (Marrs 1996).

Regarding aluminium stress tolerance, *Arabidopsis* plants overexpressing a tobacco glutathione S-transferase gene (parB) showed enhanced tolerance compared to the wild-type plants under aluminium and copper stress (Ezaki et al. 2000) exhibiting significantly lower lipid peroxidation and thus oxidative damage caused by Al stress (Ezaki et al. 2001). Later on, two aluminium (Al)-induced genes from *Arabidopsis*, *AtGST1* and *AtGST11*, when overexpressed in *Arabidopsis* showed GST induction by Al treatment as well as by cold stress, heat stress and oxidative damage, suggesting a common induction mechanism in response to abiotic stresses (Ezaki et al. 2004). An interesting result was the possible existence of a deduced signalling system between the root and shoot under Al stress, since the gene expression was observed in the leaf and only the root was exposed to Al stress (Ezaki et al. 2004).

As a part of detoxification mechanism, metalloid arsenic (As) is chelated and sequestered into the vacuoles via sulphur containing compounds such as glutathione (GSH) and phytochelatins (PCs) (Norton et al. 2014). Nevertheless, under limiting sulphur conditions, exposure of plants to As leads to phytotoxic effects (Dixit et al. 2015). In *A. thaliana* accessions, biochemical analysis and expression profiling of the genes responsible for sulphur transport and assimilation as well as metal detoxification and accumulation revealed significantly enhanced sulphur assimilation mechanism, with the tolerant accession demonstrating enhanced level of GSH and increased expression of GSTLs in As and combined limiting sulphur and As stresses (Khare et al. 2017).

Co-expression of GST and CAT1 in tobacco improved resistance of transgenic plants to cadmium and combined cadmium and heat stress compared to wild-type plants. The increased tolerance of transgenic plants might be attributed not only to the high levels of expression of the transgenes but also to the important effect of the coordinated co-expression on the antioxidant system of the ascorbate-glutathione cycle (Zhao et al. 2009). The co-induction of *ZmGST27*, an ABC transporter (*ZmMRP1*) and a glutathione transporter (*ZmGT1*) in maize leaves after treatment by a range of xenobiotics suggests that glutathione transporters are also one of the components in the glutathione conjugation-related plant detoxification system of plants along with GSH and GSTs and also ABC transporters (Pang et al. 2012). Another synergistic effect conferring heavy metal tolerance seems to be implemented by the interaction of GSTs with plant hormones such as brassinosteroids. It was recently shown that application of brassinosteroids induced increased GST activity and GSH expression in tomato plants under cadmium stress compared to plants under As stress alone (Ahammed et al. 2012).

Phytoremediation of land contaminated with inorganic and/or organic pollutants has attracted much attention and research over the last decade (Zhao and McGrath 2009). Transgenic approaches towards enhanced phytoremediation efficiency under situations of contaminants co-occurrence may be achieved through pyramiding GST genes with other components of the cellular detoxification machinery resulting in synergistic effects. Enhanced GST activity may be useful for developing strategies to enhance heavy metal tolerance and to limit their mobility in plants such as rice (Zhang et al. 2013). Nevertheless, further research on the potential mechanisms underlying GSTs functions in relation to co-expression with other components of the detoxification machinery or plant hormone regulators on xenobiotic toxicity and accumulation in the environment should be further pursued. Environmental risk assessments of the phytoextraction methods efficacy in contaminated soils should also be performed under field conditions to establish phytoremediation strategies.

3.3 GSTs Role in Herbicide Detoxification

Plant detoxification mechanism involves a three-phase detoxification system, comprising of specific enzymes which function successively (Powles and Yu 2010). GSTs and glycosyltransferases (GTs) enzymes are mainly involved in phase II of enzymatic detoxification, catalysing the conjugation of the xenobiotics with highly hydrophilic molecules, such as GSH or glucose, thus increasing its hydrophilicity (Cummins et al. 2011). In phase III, this conjugation reaction allows the xenobiotic secretion from the cytoplasm and its compartmentation in the vacuole by specialized ATP-binding cassette transporters localized at the tonoplast or can be secreted via the root tips (Rea 2007).Thus, the GSTs play a key role in plant detoxification system, with the GSH-conjugated xenobiotics becoming permanently non-toxic while at the same time could be further available to other metabolic procedures (Schroeder et al. 2001).

GSTs have been found to be also involved in crop and weed tolerance against herbicides as the GST/GSH system has been regarded as a major player in the detoxification of different herbicides with diverse biochemical mechanism of action. GSTs also contribute in the selectivity between crops and weeds (Cummins et al. 2013). GSTs from different plant species have been found to confer resistance also to other groups of herbicides. GSTs were found to enhance tolerance through protecting from oxidative stress and detoxification of herbicides from the phenoxy group (Bakkali et al. 2007), chloroacetamide group (Deng and Hatzios 2002) and chloroacetanilide group by GSH conjugation (Cho and Kong 2005). In other cereals, including the hexaploid bread wheat (*Triticum aestivum* L.) (Cummins et al. 2003), the GSH conjugation with herbicides, through the enzymatic action of GSTs, was also shown to play an important role in the metabolism and detoxification of selective herbicides, such as the chloroacetamide, dimethenamid, aryloxy-phenoxy-propionate (APP), fenoxaprop-ethyl and the sulphonylurea flupyrsulfuron-methyl. The wheat GSTs involved in the process described above were found to be closely related to the maize GSTUs (Thom et al. 2002).

Unlike other plants, in legumes like soybean and beans, the predominant thiols are the homoglutathione (hGSH, γ-glutamyl-L-cysteinyl-L-alanine) rather than GSH. This difference has direct results in the differential catalytic efficiency of several *Gm*GSTs when using either of the thiol substrates (McGonigle et al. 1997). For instance, soybean isoenzymes *Gm*GSTU1 and *Gm*GST2U6 overexpressed in transgenic tobacco plants with a dual construct (hGSH and GST), conjugate more efficiently diphenyl ether herbicide fomesafen with hGSH rather than solely GSH in order to become resistant (Skipsey et al. 2005). It was further found that GST activity was enhanced when fluazifop-*p*-butyl of the aryloxyphenoxypropionic group of herbicides was applied on *Phaseolus vulgaris*, and three inducible GST isoenzymes were isolated showing high homology with GSTs which belong to phi and tau classes(Chronopoulou et al. 2012).

A major contribution of transgenic technology towards tolerance against herbicides is the development of plants overexpressing GST isoenzymes. Overexpression of the maize *ZmGSTF27* in transgenic wheat (Droog 1997) and *ZmGSTF1* in transgenic tobacco plants (Karavangeli et al. 2005) has resulted in increased tolerance to alachlor compared to non-transgenic plants in terms of root, leaf and vigorous development. However, the *ZmGSTF27*did not provide any tolerance against atrazine or oxyfluorodifen (Milligan et al. 2001)showing the potential of GSTs for developing herbicide-specific resistant plants. Overexpression of *OsGSTL1* of the lambda class in rice enhanced tolerance to chlorsulfuron and glyphosate (Hu et al. 2009; Hu 2014). Another GST isoenzyme (*Gm*GSTU4) of the tau class was induced by fluorodifen and isolated from soybean (Axarli et al. 2009). Later, tobacco plants overexpressing the *GmGSTU4* showed increased tolerance to fluorodifen and oxyfluorfen (200μM) and the chloroacetanilide alachlor (7.5 mg/L) compared to wild-type plants, expressed as reduced electrolyte leakage (Benekos et al. 2010) and thus showed that they might have a protective role to plant cell membrane.

The use of reverse genetics approach has also verified the involvement of GSTs in the herbicide detoxification. The downregulation of the *OsGST* III subunit in rice which is active in the pretilachlor detoxification resulted in reduced tolerance of transformed lines to pretilachlor (Deng et al. 2003), suggesting that the *OsGST* III gene has significant role in the detoxification of pretilachlor and maybe the metabolism of other phenolic compounds. The development of GST isoenzymes with optimal properties (Chronopoulou and Labrou 2009) or increased catalytic activities towards xenobiotics is possible through modifications of GSTs using directed mutagenesis and forced evolution approaches that could result in environmental friendly and effective weed control. Using a forced evolution method, Dixon et al. (2003) mutated maize *ZmGSTU1* and *ZmGSTU2* and recognized seven different enzymes with enhanced detoxifying activity against fluorodifen. One of the mutant enzymes had 29-fold higher activity compared to the parental enzymes, and when expressed in *Arabidopsis thaliana*, the optimized recombinant enzyme conferred enhanced tolerance to fluorodifen, compared to the parental enzymes (Dixon et al. 2003).

A more recent study by Kissoudis et al. (2015b), with a metabolomics perspective, investigated the responses of *GmGSTU4* overexpressing tobacco plants to alachlor in comparison to wild-type plants. The transgenic plants exhibited higher induction rates of abiotic stress-responsive metabolites, accumulation of secondary metabolites and metabolic detoxification by-products suggesting that the increased metabolic capacity of *GmGSTU4* overexpressing plants is accompanied by different metabolic alterations (Kissoudis et al. 2015b). Transgenic tobacco plants overexpressing a sweet orange (*Citrus sinensis*), tau GST (*CsGSTUs*), acquired tolerance to the diphenyl ether herbicide fluorodifen (Lo Cicero et al. 2015). In a sequel study, Lo Cicero et al. (2017) overexpressed the *CsGSTU2* isoform in transgenic tobacco plants which showed enhanced tolerance against alachlor probably due to the high specific conjugation activity of the in vitro expressed *Cs*GSTU2 protein.

GSTs are involved in the detoxification of herbicides, a highly significant function with potential use in agriculture and industry, such as phytoremediation of remaining xenobiotics in the environment towards the development of practices of soil–water protection strategies for the detoxification of herbicide pollutants in agricultural fields (Karavangeli et al. 2005; Lo Cicero et al. 2015, 2017). GSTs could also be the target for further manipulation towards the development of efficient management of GST induced, nontarget site, herbicide resistance in weeds (Kissoudis et al. 2015b). The underlying mechanism of resistance to atrazine in the weed Palmer amaranth was recently shown that is a nontarget site-based tolerance mechanism mediated by GSTs paralleled with nuclear inheritance of the trait exacerbating the difficulty of controlling its spread (Nakka et al. 2017).

Further research is required towards the elucidation of the molecular mechanisms underlined, such as the responsible transcription factors, the mechanism of homo- or heterodimerization for the efficient detoxification of a certain xenobiotics and the induced protective role of GSTs. This knowledge will allow the development of integrated weed management protocols and new breeding strategies for the development of tolerant crops.

4 Future Perspectives

Humanity will face serious challenges in the coming years. The increasing population and greater demand for food and other products from plants such as fibres, fuels, etc. and the new challenges posed due to climate change render it urgent to find novel ways of increasing plant yield with less natural resources, such as water and fossil fuels, and less chemical inputs. It is therefore imperative to intensify our research on plant stress adaptation and tolerance. In this aspect, GSTs should play an important role towards the development of stress tolerant plants. We should mention here that genetic engineering through the novel gene-editing technologies like CRISP/Cas9 system could offer novel innovative solutions in plant breeding for the development of high-yielding varieties, which will be stress tolerant and at the same time will have decreased demands in inputs. However, in order to be able to use such technologies, basic studies in plant adaptation are necessary. The era of omics technologies, genomics, transcriptomics, proteomics, metabolomics, methylomics, as well as the metagenomis studies of microbial communities living in the plant rhizosphere and their interaction with the plant system will contribute significantly into the existing knowledge which will facilitate our conventional and biotechnological breeding efforts.

References

Ahammed GJ, Gao CJ, Ogweno JO, Zhou YH, Xia XJ, Mao WH, Shi K, Yu JQ (2012) Brassinosteroids induce plant tolerance against phenanthrene by enhancing degradation and detoxification in *Solanum lycopersicum* L. Ecotoxicol Environ Saf 80:28–36. https://doi.org/10.1016/j.ecoenv.2012.02.004

Ahsan H, Ali A, Ali R (2003) Oxygen free radicals and systemic autoimmunity. Clin Exp Immunol 131:398–404. https://doi.org/10.1046/j.1365-2249.2003.02104.x

Alvarez S, Berla BM, Sheffield J, Cahoon RE, Jez JM, Hicks LM (2009) Comprehensive analysis of the *Brassica juncea* root proteome in response to cadmium exposure by complementary proteomic approaches. Proteomics 9:2419–2431. https://doi.org/10.1002/pmic.200800478

Amini F, Ehsanpour AA, Hoang QT, Shin JS (2007) Protein pattern changes in tomato under in vitro salt stress. Russ J Plant Physiol 54:464–471. https://doi.org/10.1134/S102144370704005X

Asada K (2006) Production and scavenging of reactive oxygen species in chloroplasts and their functions. Plant Physiol 141:391–396. https://doi.org/10.1104/pp.106.082040

Atkinson NJ, Urwin PE (2012) The interaction of plant biotic and abiotic stresses: from genes to the field. J Exp Bot 63:3523–3544. https://doi.org/10.1093/jxb/err313

Axarli I, Dhavala P, Papageorgiou AC, Labrou NE (2009) Crystallographic and functional characterization of the fluorodifen-inducible glutathione transferase from *Glycine max* reveals an active site topography suited for diphenylether herbicides and a novel L-site. J Mol Biol 385:984–1002. https://doi.org/10.1016/j.jmb.2008.10.084

Axarli I, Muleta AW, Chronopoulou EG, Papageorgiou AC, Labrou NE (2017) Directed evolution of glutathione transferases towards a selective glutathione-binding site and improved oxidative stability. Biochim Biophys Acta – Gen Subj 1861:3416–3428. https://doi.org/10.1016/j.bbagen.2016.09.004

Bakkali Y, Ruiz-Santaella JP, Osuna MD, Wagner J, Fischer AJ, De Prado R (2007) Late watergrass (*Echinochloa phyllopogon*): mechanisms involved in the resistance to fenoxaprop-p-ethyl. J Agric Food Chem 55:4052–4058. https://doi.org/10.1021/jf0624749

Basantani M, Srivastava A (2007) Plant glutathione transferases—a decade falls short. Can J Bot 85:443–456. https://doi.org/10.1139/B07-033

Basu S, Roychoudhury A, Saha PP, Sengupta DN (2009) Differential antioxidative responses of indica rice cultivars to drought stress. Plant Growth Regul 60:51–59. https://doi.org/10.1007/s10725-009-9418-4

Baxter A, Mittler R, Suzuki N (2014) ROS as key players in plant stress signalling. J Exp Bot 65:1229–1240. https://doi.org/10.1093/jxb/ert375

Benavides MP, Gallego SM, Tomaro ML (2005) Cadmium toxicity in plants. Braz J Plant Physiol 17:21–34. https://doi.org/10.1590/S1677-04202005000100000

Benekos K, Kissoudis C, Nianiou-Obeidat I, Labrou N, Madesis P, Kalamaki M, Makris A, Tsaftaris A (2010) Overexpression of a specific soybean GmGSTU4 isoenzyme improves diphenyl ether and chloroacetanilide herbicide tolerance of transgenic tobacco plants. J Biotechnol 150:195–201. https://doi.org/10.1016/j.jbiotec.2010.07.011

Bowler C, Van Camp W, Van Montagu M, Inzé D, Asada K (1994) Superoxide dismutase in plants superoxide dismutase in plants. CRC Crit Rev Plant Sci 13:199–218

Boyer JS (1982) Plant productivity and environment. Science 218:443–448. https://doi.org/10.1126/science.218.4571.443

Chan C, Lam H (2017) A putative lambda class glutathione S -transferase enhances plant survival under salinity stress. Plant Cell Physiol 55:570–579. https://doi.org/10.1093/pcp/pct201

Chen J-H, Jiang H-W, Hsieh E-J, Chen H-Y, Chien C-T, Hsieh H-L, Lin T-P (2012) Drought and salt stress tolerance of an Arabidopsis glutathione S-transferase U17 knockout mutant are attributed to the combined effect of glutathione and abscisic acid. Plant Physiol 158:340–351. https://doi.org/10.1104/pp.111.181875

Cho H-Y, Kong K-H (2005) Molecular cloning, expression, and characterization of a phi-type glutathione S-transferase from Oryza sativa. Pestic Biochem Physiol 83:29–36. https://doi.org/10.1016/j.pestbp.2005.03.005

Choi YI, Noh EW, Kim HJ, Shim D (2013) Overexpression of poplar *GSTU51* confers selective tolerance to both mercury and methyl viologen but not to CDNB or cadmium in transgenic poplars. Plant Biotechnol Rep 7:175–184. https://doi.org/10.1007/s11816-012-0246-z

Chronopoulou EG, Labrou NE (2009) Glutathione transferases: emerging multidisciplinary tools in red and green biotechnology. Recent Pat Biotechnol 3:211–223

Chronopoulou E, Madesis P, Asimakopoulou B, Platis D, Tsaftaris A, Labrou NE (2012) Catalytic and structural diversity of the fluazifop-inducible glutathione transferases from *Phaseolus vulgaris*. Planta 235:1253–1269. https://doi.org/10.1007/s00425-011-1572-z

Chronopoulou E, Madesis P, Tsaftaris A, Labrou NE (2014) Cloning and characterization of a biotic-stress-inducible glutathione transferase from *Phaseolus vulgaris*. Appl Biochem Biotechnol 172:595–609. https://doi.org/10.1007/s12010-013-0509-3

Cia MC, Guimarães ACR, Medici LO, Chabregas SM, Azevedo RA (2012) Antioxidant responses to water deficit by drought-tolerant and -sensitive sugarcane varieties. Ann Appl Biol 161:313–324. https://doi.org/10.1111/j.1744-7348.2012.00575.x

Cox A, Venkatachalam P, Sahi S, Sharma N (2017) Plant physiology and biochemistry reprint of: silver and titanium dioxide nanoparticle toxicity in plants: a review of current research. Plant Physiol Biochem 110:33–49. https://doi.org/10.1016/j.plaphy.2016.08.007

Cramer GR, Urano K, Delrot S, Pezzotti M, Shinozaki K (2011) Effects of abiotic stress on plants: a systems biology perspective. BMC Plant Biol 11:163. https://doi.org/10.1186/1471-2229-11-163

Csiszár J, Horváth E, Váry Z, Gallé Á, Bela K, Brunner S, Tari I (2014) Glutathione transferase supergene family in tomato: salt stress-regulated expression of representative genes from distinct GST classes in plants primed with salicylic acid. Plant Physiol Biochem 78:15–26. https://doi.org/10.1016/j.plaphy.2014.02.010

Cummins I, O'Hagan D, Jablonkai I, Cole DJ, Hehn A, Werck-Reichhart D, Edwards R (2003) Cloning, characterization and regulation of a family of phi class glutathione transferases from wheat. Plant Mol Biol 52:591–603. https://doi.org/10.1023/A:1024858218804

Cummins I, Dixon DP, Freitag-Pohl S, Skipsey M, Edwards R (2011) Multiple roles for plant glutathione transferases in xenobiotic detoxification. Drug Metab Rev 43:266–280. https://doi.org/10.3109/03602532.2011.552910

Cummins I, Wortley DJ, Sabbadin F et al (2013) Key role for a glutathione transferase in multiple-herbicide resistance in grass weeds. Proc Natl Acad Sci U S A 110:5812–5817. https://doi.org/10.1073/pnas.1221179110

Dat J, Vandenabeele S, Vranová E, Van Montagu M, Inzé D, Van Breusegem F (2000) Dual action of the active oxygen species during plant stress responses. Cell Mol Life Sci 57:779–795. https://doi.org/10.1007/s000180050041

Deeba F, Pandey AK, Ranjan S, Mishra A, Singh R, Sharma YK, Shirke PA, Pandey V (2012) Physiological and proteomic responses of cotton (*Gossypium herbaceum* L.) to drought stress. Plant Physiol Biochem 53:6–18. https://doi.org/10.1016/j.plaphy.2012.01.002

DeLano WL (2002) Pymol: an open-source molecular graphics tool. CCP4 Newsl Protein Crystallogr 40:82–92

del Río LA, Sandalio LM, Corpas FJ, Palma JM, Barroso JB (2006) Reactive oxygen species and reactive nitrogen species in peroxisomes. Production, scavenging, and role in cell signaling. Plant Physiol 141:330–335. https://doi.org/10.1104/pp.106.078204

Deng F, Hatzios KK (2002) Purification and characterization of two glutathione S-transferase isozymes from indica-type rice involved in herbicide detoxification. Pestic Biochem Physiol 72:10–13. https://doi.org/10.1006/pest.2001.2580

Deng F, Jelesko J, Cramer CL, Wu J, Hatzios KK (2003) Use of an antisense gene to characterize glutathione S-transferase functions in transformed suspension-cultured rice cells and calli. Pestic Biochem Physiol 75:27–37. https://doi.org/10.1016/S0048-3575(03)00015-4

Diao G, Wang Y, Wang C, Yang C (2011) Cloning and functional characterization of a novel glutathione S-transferase gene from *Limonium bicolor*. Plant Mol Biol Rep 29:77–87. https://doi.org/10.1007/s11105-010-0212-2

Dill GM, Cajacob CA, Padgette SR (2008) Glyphosate-resistant crops: adoption, use and future considerations. Pest Manag Sci 64:326–336. https://doi.org/10.1002/ps.1501

Dixit V, Pandey V, Shyam R (2001) Differential antioxidative responses to cadmium in roots and leaves of pea (*Pisum sativum* L. cv. Azad). J Exp Bot 52:1101–1109. https://doi.org/10.1093/jexbot/52.358.1101

Dixit P, Mukherjee PK, Ramachandran V, Eapen S (2011) Glutathione transferase from Trichoderma virens enhances cadmium tolerance without enhancing its accumulation in transgenic *Nicotiana tabacum*. PLoS One 6:e16360. https://doi.org/10.1371/journal.pone.0016360

Dixit G, Singh AP, Kumar A, Singh PK, Kumar S, Dwivedi S, Trivedi PK, Pandey V, Norton GJ, Dhankher OP, Tripathi RD (2015) Sulfur mediated reduction of arsenic toxicity involves efficient thiol metabolism and the antioxidant defense system in rice. J Hazard Mater 298:241–251. https://doi.org/10.1016/j.jhazmat.2015.06.008

Dixon (2010) Glutathione transferases. In: The Arabidopsis book. American Society of Plant Biologists, p e0131

Dixon DP, Davis BG, Edwards R (2002) Functional divergence in the glutathione transferase superfamily in plants: identification of two classes with putative functions in redox homeostasis in *Arabidopsis thaliana*. J Biol Chem 277:30859–30869. https://doi.org/10.1074/jbc.M202919200

Dixon DP, McEwen AG, Lapthorn AJ, Edwards R (2003) Forced evolution of a herbicide detoxifying glutathione transferase. J Biol Chem 278:23930–23935. https://doi.org/10.1074/jbc.M303620200

Dixon DP, Lapthorn A, Madesis P, Mudd EA, Day A, Edwards R (2008) Binding and glutathione conjugation of porphyrinogens by plant glutathione transferases. J Biol Chem 283:20268–20276. https://doi.org/10.1074/jbc.M802026200

Dixon DP, Sellars JD, Edwards R (2011) The Arabidopsis phi class glutathione transferase *AtGSTF2*: binding and regulation by biologically active heterocyclic ligands. Biochem J 438:63–70. https://doi.org/10.1042/BJ20101884

Dong Y, Li C, Zhang Y, He Q, Daud MK, Chen J, Zhu S (2016) Glutathione S-transferase gene family in *Gossypium raimondii* and *G. arboreum*: comparative genomic study and their expression under salt stress. Front Plant Sci 7:139. https://doi.org/10.3389/fpls.2016.00139

Droog F (1997) Plant glutathione S -transferases, a tale of theta and tau. J Plant Growth Regul 16:95–107

Ezaki B, Gardner RC, Ezaki Y, Matsumoto H (2000) Expression of aluminum-induced genes in transgenic arabidopsis plants can ameliorate aluminum stress and/or oxidative stress. Plant Physiol 122:657–665. https://doi.org/10.1104/pp.122.3.657

Ezaki B, Katsuhara M, Kawamura M, Matsumoto H (2001) Different mechanisms of four aluminum (Al)-resistant transgenes for Al toxicity in Arabidopsis. Plant Physiol 127:918–927. https://doi.org/10.1104/pp.010399.918

Ezaki B, Suzuki M, Motoda H, Kawamura M, Nakashima S (2004) Mechanism of gene expression of Arabidopsis in response to aluminum stress 1. Sci Technol 134:1672–1682. https://doi.org/10.1104/pp.103.037135.plants

Fones H, Preston GM (2012) Reactive oxygen and oxidative stress tolerance in plant pathogenic pseudomonas. FEMS Microbiol Lett 327:1–8. https://doi.org/10.1111/j.1574-6968.2011.02449.x

Foyer CH, Noctor G (2005) Redox homeostasis and antioxidant signaling: a metabolic Interface between stress perception and physiological responses. Plant Cell 17:1866–1875. https://doi.org/10.1105/tpc.105.033589

Foyer CH, Noctor G (2009) Redox regulation in photosynthetic organisms: signaling, acclimation, and practical implications. Antioxid Redox Signal 11:862–905. https://doi.org/10.1089/ars.2008.2177

Foyer CH, Descourvières P, Kunert KJ, Descourvieres P (1994) Protection against oxygen radicals-an important defense mechanism studied in transgenic plants. Plant Cell Environ 17:507–523. https://doi.org/10.1111/j.1365-3040.1994.tb00146.x

Frear DS, Swanson HR (1970) Biosynthesis of S-(4-Ethylamino-6-Isopropylamino-2-S-Triazino) glutathione: partial purification and properties of a glutathione S-transferase from corn. Phytochemistry 9:2123–2132

Frova C (2003) The plant glutathione transferase gene family: genomic structure, functions, expression and evolution. Physiol Plant 119:469–479. https://doi.org/10.1046/j.1399-3054.2003.00183.x

Frova C (2006) Glutathione transferases in the genomics era: new insights and perspectives. Biomol Eng 23:149–169. https://doi.org/10.1016/j.bioeng.2006.05.020

Gallé Á, Csiszár J, Secenji M, Guóth A, Cseuz L, Tari I, Györgyey J, Erdei L (2009) Glutathione transferase activity and expression patterns during grain filling in flag leaves of wheat genotypes differing in drought tolerance: response to water deficit. J Plant Physiol 166:1878–1891. https://doi.org/10.1016/j.jplph.2009.05.016

Gallé Á, Csiszár J, Secenji M, Erdei L, Benyo D, Györgyey J, Tari I (2011) Induction and regulation of glutathione transferases in wheat species exposed to PEG induced osmotic stress. Acta Biol Szeged 55:79–80. https://doi.org/10.1016/j.jprot.2011.05.015

Gallé Á, Csiszár J, Secenji M, Guóth A, Cseuz L, Tari I, Györgyey J, Erdei L (2013) Drought response strategies during grain filling in wheat. J Plant Physiol 170:1389–1399

George S, Venkataraman G, Parida A (2010) A chloroplast-localized and auxin-induced glutathione S-transferase from phreatophyte *Prosopis juliflora* confer drought tolerance on tobacco. J Plant Physiol 167:311–318. https://doi.org/10.1016/j.jplph.2009.09.004

Gietler M, Nykiel M, Zagdańska BM (2016) Changes in the reduction state of ascorbate and glutathione, protein oxidation and hydrolysis leading to the development of dehydration intolerance in *Triticum aestivum* L. seedlings. Plant Growth Regul 79:287–297. https://doi.org/10.1007/s10725-015-0133-z

Gill SS, Tuteja N (2010) Reactive oxygen species and antioxidant machinery in abiotic stress tolerance in crop plants. Plant Physiol Biochem 48:909–930. https://doi.org/10.1016/j.plaphy.2010.08.016

Gunning V, Tzafestas K, Sparrow H, Johnston EJ, Brentnall AS, Potts JR, Rylott EL, Bruce NC (2014) Arabidopsis glutathione transferases U24 and U25 exhibit a range of detoxification activities with the environmental pollutant and explosive, 2,4,6-trinitrotoluene. Plant Physiol 165:854–865. https://doi.org/10.1104/pp.114.237180

Gupta NK, Agarwal S, Agarwal VP, Nathawat NS, Gupta S, Singh G (2013) Effect of short-term heat stress on growth, physiology and antioxidative defence system in wheat seedlings. Acta Physiol Plant 35:1837–1842. https://doi.org/10.1007/s11738-013-1221-1

Gür A, Demirel U, Özden M, Kahraman A, Opur O (2010) Diurnal gradual heat stress affects antioxidant enzymes, proline accumulation and some physiological components in cotton (*Gossypium hirsutum* L.) Afr J Biotechnol 9:1008–1015. https://doi.org/10.5897/AJB09.1590

Hameed A, Bibi N, Akhter J, Iqbal N (2011) Differential changes in antioxidants, proteases, and lipid peroxidation in flag leaves of wheat genotypes under different levels of water deficit conditions. Plant Physiol Biochem 49:178–185. https://doi.org/10.1016/j.plaphy.2010.11.009

He F, Liu Q, Zheng L, Cui Y, Shen Z, Zheng L (2015) RNA-Seq analysis of rice roots reveals the involvement of post-transcriptional regulation in response to cadmium stress. Front Plant Sci 6:1–16. https://doi.org/10.3389/fpls.2015.01136

Hernandez JA, Ferrer MA, Jimenez A, Barcelo AR, Sevilla F (2001) Antioxidant systems and O2.-/H2O2 production in the apoplast of pea leaves. Its relation with salt-induced necrotic lesions in minor veins. Plant Physiol 127:817–831. https://doi.org/10.1104/pp.010188

Hernandez I, Alegre L, Munne-Bosch S (2004) Drought-induced changes in flavonoids and other low molecular weight antioxidants in *Cistus clusii* grown under Mediterranean field conditions. Tree Physiol 24:1303–1311. https://doi.org/10.1093/treephys/24.11.1303

Herschbach C, van Der Zalm E, Schneider A, Jouanin L, De Kok LJ, Rennenberg H (2000) Regulation of sulfur nutrition in wild-type and transgenic poplar overexpressing gamma-glutamylcysteine synthetase in the cytosol as affected by atmospheric H_2S. Plant Physiol 124:461–473. https://doi.org/10.1104/pp.124.1.461

Hu T (2014) A glutathione s-transferase confers herbicide tolerance in rice. Crop Breed Appl Biotechnol 14:76–81. https://doi.org/10.1590/1984-70332014v14n2a14

Hu Y, Burucs Z, von Tucher S, Schmidhalter U (2007) Short-term effects of drought and salinity on mineral nutrient distribution along growing leaves of maize seedlings. Environ Exp Bot 60:268–275. https://doi.org/10.1016/j.envexpbot.2006.11.003

Hu T, Qv X, Xiao G, Huang X (2009) Enhanced tolerance to herbicide of rice plants by over-expression of a glutathione S-transferase. Mol Breed 24:409–418. https://doi.org/10.1007/s11032-009-9302-y

Iannelli MA, Pietrini F, Fiore L, Petrilli L, Massacci A (2002) Antioxidant response to cadmium in *Phragmites australis* plants. Plant Physiol Biochem 40:977–982. https://doi.org/10.1016/S0981-9428(02)01455-9

Jain M, Ghanashyam C, Bhattacharjee A (2010) Comprehensive expression analysis suggests overlapping and specific roles of rice glutathione S-transferase genes during development and stress responses. BMC Genomics 11:73. https://doi.org/10.1186/1471-2164-11-73

Jha B, Sharma A, Mishra A (2011) Expression of SbGSTU (tau class glutathione S-transferase) gene isolated from *Salicornia brachiata* in tobacco for salt tolerance. Mol Biol Rep 38:4823–4832. https://doi.org/10.1007/s11033-010-0625-x

Ji W, Zhu Y, Li Y, Yang L, Zhao X, Cai H, Bai X (2010) Over-expression of a glutathione S-transferase gene, GsGST, from wild soybean (Glycine soja) enhances drought and salt tolerance in transgenic tobacco. Biotechnol Lett 32:1173–1179. https://doi.org/10.1007/s10529-010-0269-x

Jia B, Sun M, Sun X et al (2016) Overexpression of *GsGSTU13* and SCMRP in *Medicago sativa* confers increased salt-alkaline tolerance and methionine content. Physiol Plant 156:176–189. https://doi.org/10.1111/ppl.12350

Jozefczak M, Keunen E, Schat H et al (2014) Differential response of Arabidopsis leaves and roots to cadmium: glutathione-related chelating capacity vs antioxidant capacity. Plant Physiol Biochem 83:1–9. https://doi.org/10.1016/j.plaphy.2014.07.001

Kampranis SC, Damianova R, Atallah M, Toby G, Kondi G, Tsichlis PN, Makris AM (2000) A novel plant glutathione S -transferase/peroxidase suppresses bax lethality in yeast. J Biol Chem 275:29207–29216. https://doi.org/10.1074/jbc.M002359200

Karavangeli M, Labrou NE, Clonis YD, Tsaftaris A (2005) Development of transgenic tobacco plants overexpressing maize glutathione S-transferase I for chloroacetanilide herbicides phytoremediation. Biomol Eng 22:121–128. https://doi.org/10.1016/j.bioeng.2005.03.001

Kearse M, Moir R, Wilson A et al (2012) Geneious basic: an integrated and extendable desktop software platform for the organization and analysis of sequence data. Bioinformatics 28:1647–1649. https://doi.org/10.1093/bioinformatics/bts199

Khare R, Kumar S, Shukla T, Ranjan A, Trivedi PK (2017) Differential sulphur assimilation mechanism regulates response of *Arabidopsis thaliana* natural variation towards arsenic stress under limiting sulphur condition. J Hazard Mater 337:198–207. https://doi.org/10.1016/j.jhazmat.2017.05.009

Kim SD, Lee JH, Hur EH et al (2011) Influence of GST gene polymorphisms on the clearance of intravenous busulfan in adult patients undergoing hematopoietic cell transplantation. Biol Blood Marrow Transplant 17:1222–1230. https://doi.org/10.1016/j.bbmt.2010.12.708

Kissoudis C, Kalloniati C, Flemetakis E, Madesis P, Labrou NE, Tsaftaris A, Nianiou-Obeidat I (2015a) Stress-inducible *GmGSTU4* shapes transgenic tobacco plants metabolome towards increased salinity tolerance. Acta Physiol Plant 37:1–11. https://doi.org/10.1007/s11738-015-1852-5

Kissoudis C, Kalloniati C, Flemetakis E, Madesis P, Labrou NE, Tsaftaris A, Nianiou-Obeidat I (2015b) Maintenance of metabolic homeostasis and induction of cytoprotectants and secondary metabolites in alachlor-treated *GmGSTU4*-overexpressing tobacco plants, as resolved by metabolomics. Plant Biotechnol Rep 9:287–296. https://doi.org/10.1007/s11816-015-0364-5

Kumar S, Gupta D, Nayyar H (2012) Comparative response of maize and rice genotypes to heat stress: status of oxidative stress and antioxidants. Acta Physiol Plant 34:75–86. https://doi.org/10.1007/s11738-011-0806-9

Kumar S, Asif M, Chakrabarty D (2013a) Differential expression of rice lambda class GST gene family members during plant growth, development, and in response to stress conditions. Plant Mol Biol 31:569–580

Kumar S, Asif MH, Chakrabarty D, Tripathi RD, Dubey RS, Trivedi PK (2013b) Expression of a rice lambda class of glutathione S-transferase, OsGSTL2, in Arabidopsis provides tolerance to heavy metal and other abiotic stresses. J Hazard Mater 248–249:228–237. https://doi.org/10.1016/j.jhazmat.2013.01.004

Labrou NE, Papageorgiou AC, Pavli O, Flemetakis E (2015) Plant GSTome: structure and functional role in xenome network and plant stress response. Curr Opin Biotechnol 32:186–194. https://doi.org/10.1016/j.copbio.2014.12.024

Lallement PA, Brouwer B, Keech O, Hecker A, Rouhier N (2014) The still mysterious roles of cysteine-containing glutathione transferases in plants. Front Pharmacol 5:1–22. https://doi.org/10.3389/fphar.2014.00192

Lan T, Yang Z-L, Yang X, Liu Y-J, Wang X-R, Zeng Q-Y (2009) Extensive functional diversification of the Populus glutathione S-transferase supergene family. Plant Cell 21:3749–3766. https://doi.org/10.1105/tpc.109.070219

Lane A, Jarvis A (2007) Changes in climate will modify the geography of crop suitability: agricultural biodiversity can help with adaptation. SAT eJ 4:1–12. https://doi.org/10.3914/ICRISAT.0094

Lappartient AC, Touraine B (1997) Glutathione-mediated regulation of ATP sulfurylase activity, SO_4^{2-} uptake, and oxidative stress response in intact canola roots. Plant Physiol 114:177–183

Le Martret B, Poage M, Shiel K, Nugent GD, Dix PJ (2011) Tobacco chloroplast transformants expressing genes encoding dehydroascorbate reductase, glutathione reductase, and glutathione-S-transferase exhibit altered anti-oxidant metabolism and improved abiotic stress tolerance. Plant Biotechnol J 9:661–673. https://doi.org/10.1111/j.1467-7652.2011.00611.x

Light GG, Mahan JR, Roxas VP, Allen RD (2005) Transgenic cotton (*Gossypium hirsutum* L.) seedlings expressing a tobacco glutathione S-transferase fail to provide improved stress tolerance. Planta 222:346–354. https://doi.org/10.1007/s00425-005-1531-7

Lim JD, Hahn SJ, Yu CY, Chung IM (2005) Expression of the glutathione S-transferase gene (*NT107*) in transgenic *Dianthus superbus*. Plant Cell Tissue Organ Cult 80:277–286. https://doi.org/10.1007/s11240-004-1032-6

Lin CY, Trinh NN, Lin CW, Huang HJ (2013) Transcriptome analysis of phytohormone, transporters and signaling pathways in response to vanadium stress in rice roots. Plant Physiol Biochem 66:98–104. https://doi.org/10.1016/j.plaphy.2013.02.007

Liu XF, Li JY (2002) Characterization of an ultra-violet inducible gene that encodes glutathione S-transferase in Arabidopsis thaliana. Acta Genet Sin 29:458–460

Liu D, Liu Y, Rao J, Wang G, Li H, Ge F, Chen C (2013) Overexpression of the glutathione S-transferase gene from *Pyrus pyrifolia* fruit improves tolerance to abiotic stress in transgenic tobacco plants. Mol Biol 47:515–523. https://doi.org/10.1134/S0026893313040109

Lo Cicero L, Madesis P, Tsaftaris A, Lo Piero AR (2015) Tobacco plants over-expressing the sweet orange tau glutathione transferases (*CsGSTUs*) acquire tolerance to the diphenyl ether herbicide fluorodifen and to salt and drought stresses. Phytochemistry 116:69–77. https://doi.org/10.1016/j.phytochem.2015.03.004

Lo Cicero L, Catara V, Strano CP, Bella P, Madesis P, Lo Piero AR (2017) Over-expression of *CsGSTU* promotes tolerance to the herbicide alachlor and resistance to *Pseudomonas syringae* pv. tabaci in transgenic tobacco. Biol Plant 61:169–177. https://doi.org/10.1007/s10535-016-0659-6

Mahan JR, Mauget SA (2005) Antioxidant metabolism in cotton seedlings exposed to temperature stress in the field. Crop Sci 45:2337–2345. https://doi.org/10.2135/cropsci2005.0106

Marrs KA (1996) The functions and regulation of glutathione S-transferases in plants. Annu Rev Plant Physiol Plant Mol Biol 47:127–158. https://doi.org/10.1146/annurev.arplant.47.1.127

McGonigle B, Lau SMC, O'Keefe DP (1997) Endogenous reactions and substrate specificity of herbicide metabolizing enzymes. In: Regulation of enzymatic systems detoxifying xenobiotics in plants, NATO ASI Series. Kluwer Academic Publishers, Dordrecht, pp 9–18

McGonigle B, Keeler SJ, Cindy Lau S-M, Koeppe MK, O DP (2000) A genomics approach to the comprehensive analysis of the glutathione S-transferase gene family in soybean and maize. Plant Physiol 124:1105–1120. https://doi.org/10.1104/PP.124.3.1105

Miller G, Suzuki N, Ciftci-Yilmaz S, Mittler R (2010) Reactive oxygen species homeostasis and signalling during drought and salinity stresses. Plant Cell Environ 33:453–467. https://doi.org/10.1111/j.1365-3040.2009.02041.x

Milligan AS, Daly A, Parry MAJ, Lazzeri PA, Jepson I, Al H (2001) The expression of a maize glutathione S -transferase gene in transgenic wheat confers herbicide tolerance, both *in planta* and in vitro. Mol Breed 7:301–315

Mishra P, Bhoomika K, Dubey RS (2013) Differential responses of antioxidative defense system to prolonged salinity stress in salt-tolerant and salt-sensitive indica rice (*Oryza sativa* L.) seedlings. Protoplasma 250:3–19. https://doi.org/10.1007/s00709-011-0365-3

Mittler R (2002) Oxidative stress, antioxidants and stress tolerance. Trends Plant Sci 7:405–410. https://doi.org/10.1016/S1360-1385(02)02312-9

Mittler R, Blumwald E (2010) Genetic engineering for modern agriculture: challenges and perspectives. Annu Rev Plant Biol 61:443–462. https://doi.org/10.1146/annurev-arplant-042809-112116

Mittler R, Vanderauwera S, Gollery M, Van Breusegem F (2004) Reactive oxygen gene network of plants. Trends Plant Sci 9:490–498. https://doi.org/10.1016/j.tplants.2004.08.009

Mittler R, Vanderauwera S, Suzuki N, Miller G, Tognetti VB, Vandepoele K, Gollery M, Shulaev V, Van Breusegem F (2011) ROS signaling: the new wave? Trends Plant Sci 16:300–309. https://doi.org/10.1016/j.tplants.2011.03.007

Mittova V, Tal M, Volokita M, Guy M (2003) Up-regulation of the leaf mitochondrial and peroxisomal antioxidative systems in response to salt-induced oxidative stress in the wild salt-tolerant tomato species *Lycopersicon pennellii*. Plant Cell Environ 26:845–856. https://doi.org/10.1046/j.1365-3040.2003.01016.x

Molassiotis A, Fotopoulos V (2011) Oxidative and nitrosative signaling in plants. Plant Signal Behav 6:210–214. https://doi.org/10.4161/psb.6.2.14878

Møller IM, Jensen PE, Hansson A (2007) Oxidative modifications to cellular components in plants. Annu Rev Plant Biol 58:459–481. https://doi.org/10.1146/annurev.arplant.58.032806.103946

Moons A (2003) Osgstu3 and Osgtu4, encoding tau class glutathione S-transferases, are heavy metal- and hypoxic stress-induced and differentially salt stress-responsive in rice roots. FEBS Lett 553:427–432. https://doi.org/10.1016/S0014-5793(03)01077-9

Moons A (2005) Regulatory and functional interactions of plant growth regulators and plant glutathione S-transferases (GSTs). Vitam Horm 72:155–202. https://doi.org/10.1016/S0083-6729(05)72005-7

Nahakpam S, Shah K (2011) Expression of key antioxidant enzymes under combined effect of heat and cadmium toxicity in growing rice seedlings. Plant Growth Regul 63:23–35. https://doi.org/10.1007/s10725-010-9508-3

Nair PMG, Chung IM (2014) Assessment of silver nanoparticle-induced physiological and molecular changes in *Arabidopsis thaliana*. Environ Sci Pollut Res 21:8858–8869. https://doi.org/10.1007/s11356-014-2822-y

Nakka S, Godar AS, Thompson CR, Peterson DE, Jugulam M (2017) Rapid detoxification via glutathione S-transferase (GST) conjugation confers a high level of atrazine resistance in palmer amaranth (*Amaranthus palmeri*). Pest Manag Sci 73:2236–2243. https://doi.org/10.1002/ps.4615

Nianiou-Obeidat I, Madesis P, Kissoudis C, Voulgari G, Chronopoulou E, Tsaftaris A, Labrou NE (2017) Plant glutathione transferase-mediated stress tolerance: functions and biotechnological applications. Plant Cell Rep 36:791–805. https://doi.org/10.1007/s00299-017-2139-7

Noctor G, Foyer CH (1998) Ascorbate and glutathione: keeping active oxygen under control. Annu Rev Plant Physiol Plant Mol Biol 49:249–279. https://doi.org/10.1146/annurev.arplant.49.1.249

Norton GJ, Douglas A, Lahner B et al (2014) Genome wide association mapping of grain arsenic, copper, molybdenum and zinc in rice (*Oryza sativa* L.) grown at four international field sites. PLoS One 9:1–12. https://doi.org/10.1371/journal.pone.0089685

Nutricati E, Miceli A, Blando F, De Bellis L (2006) Characterization of two *Arabidopsis thaliana* glutathione S-transferases. Plant Cell Rep 25:997–1005. https://doi.org/10.1007/s00299-006-0146-1

Pang S, Ran Z, Liu Z, Song X, Duan L, Li X, Wang C (2012) Co-induction of a glutathione-S-transferase, a glutathione transporter and an ABC transporter in maize by xenobiotics. PLoS One 7:1–5. https://doi.org/10.1371/journal.pone.0048085

Parisy V, Poinssot B, Owsianowski L, Buchala A, Glazebrook J, Mauch F (2007) Identification of PAD2 as a γ-glutamylcysteine synthetase highlights the importance of glutathione in disease resistance of Arabidopsis. Plant J 49:159–172. https://doi.org/10.1111/j.1365-313X.2006.02938.x

Pei D (2012) Cloning and expression of a tomato glutathione S- transferase (GST) in *Escherichia coli*. Afr J Biotechnol 11:6402–6408. https://doi.org/10.5897/AJB12.143

Peñuelas J (2008) An increasingly scented world. New Phytol 180(4):735–738

Perez-Lopez U, Robredo A, Lacuesta M, Sgherri C, Mun A, Pe U, Navari-izzo F, Mena-petite A (2009) The oxidative stress caused by salinity in two barley cultivars is mitigated by elevated CO_2 use. Physiol Plant 135:29–42. https://doi.org/10.1111/j.1399-3054.2008.01174.x

Pimm SL (2009) Climate disruption and biodiversity. Curr Biol 19:R595–R601. https://doi.org/10.1016/j.cub.2009.05.055

Popelka M, Tuinstra M, Weil CF (2009) Discovering genes for abiotic stress tolerance in crop plants. In: Jenks MA, Wood AJ (eds) Genes for plant abiotic stress. Blackwell Publishing Ltd, Ames, pp 281–302

Powles SB, Yu Q (2010) Evolution in action: plants resistant to herbicides. Annu Rev Plant Biol 61:317–347. https://doi.org/10.1146/annurev-arplant-042809-112119

Qi YC, Liu WQ, Qiu LY, Zhang SM, Ma L, Zhang H (2010) Overexpression of glutathione S-transferase gene increases salt tolerance of *Arabidopsis*. Russ J Plant Physiol 57:233–240. https://doi.org/10.1134/S102144371002010X

Ramakrishna A, Ravishankar GA (2011) Influence of abiotic stress signals on secondary metabolites in plants. Plant Signal Behav 6:1720–1731. https://doi.org/10.4161/psb.6.11.17613

Rampino P, Mita G, Fasano P, Borrelli GM, Aprile A, Dalessandro G, De Bellis L, Perrotta C (2012) Novel durum wheat genes up-regulated in response to a combination of heat and drought stress. Plant Physiol Biochem 56:72–78. https://doi.org/10.1016/j.plaphy.2012.04.006

Rao KVM, Raghavendra AS, Reddy KJ (2006) Physiology and molecular biology of stress tolerance in plants. Springer Science & Business Media, Netherlands

Rea PA (2007) Plant ATP-binding cassette transporters. Annu Rev Plant Biol 58:347–375. https://doi.org/10.1146/annurev.arplant.57.032905.105406

Rentel MC, Knight MR, Kingdom U (2004) Oxidative stress-induced calcium signaling. Plant Physiol 135:1471–1479. https://doi.org/10.1104/pp.104.042663.1

Reymond P (2000) Differential gene expression in response to mechanical wounding and insect feeding in *Arabidopsis*. Plant Cell 12:707–720. https://doi.org/10.1105/tpc.12.5.707

Rezaei MK, Shobbar ZS, Shahbazi M, Abedini R, Zare S (2013) Glutathione S-transferase (GST) family in barley: identification of members, enzyme activity, and gene expression pattern. J Plant Physiol 170:1277–1284. https://doi.org/10.1016/j.jplph.2013.04.005

Rivero RM, Mestre TC, Mittler R, Rubio F, Garcia-Sanchez F, Martinez V (2014) The combined effect of salinity and heat reveals a specific physiological, biochemical and molecular response in tomato plants. Plant Cell Environ 37:1059–1073. https://doi.org/10.1111/pce.12199

Rizhsky L, Liang H, Mittler R (2002) The combined effect of drought stress and heat shock on gene expression in tobacco. Plant Physiol 130:1143–1151. https://doi.org/10.1104/pp.006858.then

Roxas VP, Lodhi SA, Garrett DK, Mahan JR, Allen RD (2000) Stress tolerance in transgenic tobacco seedlings that overexpress glutathione S-transferase/glutathione peroxidase. Plant Cell Physiol 41:1229–1234. https://doi.org/10.1093/pcp/pcd051

Ryu HY, Kim SY, Park HM, You JY, Kim BH, Lee JS, Nam KH (2009) Modulations of *AtGSTF10* expression induce stress tolerance and BAK1-mediated cell death. Biochem Biophys Res Commun 379:417–422. https://doi.org/10.1016/j.bbrc.2008.11.156

Sánchez-Rodríguez E, Rubio-Wilhelmi MM, Cervilla LM, Blasco B, Rios JJ, Rosales MA, Romero L, Ruiz JM (2010) Genotypic differences in some physiological parameters symptomatic for oxidative stress under moderate drought in tomato plants. Plant Sci 178:30–40. https://doi.org/10.1016/j.plantsci.2009.10.001

Sappl PG, Carroll AJ, Clifton R, Lister R, Whelan J, Harvey Millar A, Singh KB (2009) The Arabidopsis glutathione transferase gene family displays complex stress regulation and co-silencing multiple genes results in altered metabolic sensitivity to oxidative stress. Plant J 58:53–68. https://doi.org/10.1111/j.1365-313X.2008.03761.x

Schroeder JI, Allen GJ, Hugouvieux V, Kwak JM, Waner D (2001) Guard cell signal transduction. Annu Rev Plant Physiol Plant Mol Biol 52:627–658

Seckin B, Turkan I, Sekmen AH, Ozfidan C (2010) The role of antioxidant defense systems at differential salt tolerance of *Hordeum marinum* Huds. (sea barleygrass) and *Hordeum vulgare* L. (cultivated barley). Environ Exp Bot 69:76–85. https://doi.org/10.1016/j.envexpbot.2010.02.013

Sekmen AH, Ozgur R, Uzilday B, Turkan I (2014) Reactive oxygen species scavenging capacities of cotton (*Gossypium hirsutum*) cultivars under combined drought and heat induced oxidative stress. Environ Exp Bot 99:141–149. https://doi.org/10.1016/j.envexpbot.2013.11.010

Shao H-B, Chu L-Y, Jaleel CA, Zhao C-X (2008) Water-deficit stress-induced anatomical changes in higher plants. C R Biol 331:215–225. https://doi.org/10.1016/j.crvi.2008.01.002

Sharma P, Jha AB, Dubey RS, Pessarakli M (2012) Reactive oxygen species, oxidative damage, and antioxidative defense mechanism in plants under stressful conditions. J Bot 2012:1–26. https://doi.org/10.1155/2012/217037

Sharma R, Sahoo A, Devendran R, Jain M (2014) Over-expression of a rice tau class glutathione S-transferase gene improves tolerance to salinity and oxidative stresses in *Arabidopsis*. PLoS One 9:1–11. https://doi.org/10.1371/journal.pone.0092900

Sievers F, Wilm A, Dineen D et al (2011) Fast, scalable generation of high-quality protein multiple sequence alignments using Clustal Omega. Mol Syst Biol 7:539. https://doi.org/10.1038/msb.2011.75

Skipsey M, Cummins I, Andrews CJ, Jepson I, Edwards R (2005) Manipulation of plant tolerance to herbicides through co-ordinated metabolic engineering of a detoxifying glutathione transferase and thiol cosubstrate. Plant Biotechnol J 3:409–420. https://doi.org/10.1111/j.1467-7652.2005.00134.x

Skopelitou K, Muleta AW, Papageorgiou AC, Chronopoulou E, Labrou NE (2015) Catalytic features and crystal structure of a tau class glutathione transferase from *Glycine max* specifically upregulated in response to soybean mosaic virus infections. Biochim Biophys Acta, Proteins Proteomics 1854:166–177. https://doi.org/10.1016/j.bbapap.2014.11.008

Skopelitou K, Muleta AW, Papageorgiou AC, Chronopoulou EG, Pavli O, Flemetakis E, Skaracis GN, Labrou NE (2017) Characterization and functional analysis of a recombinant tau class glutathione transferase *GmGSTU2-2* from *Glycine max*. Int J Biol Macromol 94:802–812

Smith AP, Nourizadeh SD, Peer WA, Xu J, Bandyopadhyay A, Murphy AS, PBG Ã (2003) Arabidopsis AtGSTF2 is regulated by ethylene and auxin, and encodes a glutathione S-transferase that interacts with flavonoids. Plant J 36:433–442. https://doi.org/10.1046/j.1365-313X.2003.01890.x

Sun W, Xu X, Zhu H, Liu A, Liu L, Li J, Hua X (2010) Comparative transcriptomic profiling of a salt-tolerant wild tomato species and a salt-sensitive tomato cultivar. Plant Cell Physiol 51:997–1006. https://doi.org/10.1093/pcp/pcq056

Suzuki N, Miller G, Morales J, Shulaev V, Torres MA, Mittler R (2011) Respiratory burst oxidases: the engines of ROS signaling. Curr Opin Plant Biol 14:691–699. https://doi.org/10.1016/j.pbi.2011.07.014

Suzuki N, Koussevitzky S, Mittler R, Miller G (2012) ROS and redox signalling in the response of plants to abiotic stress. Plant Cell Environ 35:259–270. https://doi.org/10.1111/j.1365-3040.2011.02336.x

Takesawa T, Ito M, Kanzaki H, Kameya N, Nakamura I (2002) Over-expression of glutathione ζ glutathione S-transferase in transgenic rice enhances germination and growth at low temperature. Mol Breed 9:93–101

Thom R, Cummins I, Dixon DP, Edwards R, Cole DJ, Lapthorn AJ (2002) Structure of a tau class glutathione S-transferase from wheat active in herbicide detoxification. Biochemistry 41:7008–7020. https://doi.org/10.1021/bi015964x

Türkan I, Bor M, Özdemir F, Koca H (2005) Differential responses of lipid peroxidation and antioxidants in the leaves of drought-tolerant *P. acutifolius* gray and drought-sensitive *P. vulgaris* L. subjected to polyethylene glycol mediated water stress. Plant Sci 168:223–231. https://doi.org/10.1016/j.plantsci.2004.07.032

Umezawa T, Fujita M, Fujita Y, Yamaguchi-Shinozaki K, Shinozaki K (2006) Engineering drought tolerance in plants: discovering and tailoring genes to unlock the future. Curr Opin Biotechnol 17:113–122. https://doi.org/10.1016/j.copbio.2006.02.002

Verslues PE, Agarwal M, Katiyar-Agarwal S, Zhu J, Zhu J (2006) Methods and concepts in quantifying resistance to drought, salt and freezing, abiotic stresses that affect plant water status. Plant J 45:523–539

Vij S, Tyagi AK (2007) Emerging trends in the functional genomics of the abiotic stress response in crop plants. Plant Biotechnol J 5:361–380

Voss I, Sunil B, Scheibe R, Raghavendra AS (2013) Emerging concept for the role of photorespiration as an important part of abiotic stress response. Plant Biol 15:713–722. https://doi.org/10.1111/j.1438-8677.2012.00710.x

Wagner U, Edwards R, Dixon DP, Mauch F (2002) Probing the diversity of the *Arabidopsis* glutathione S-transferase gene family. Plant Mol Biol 49:515–532. https://doi.org/10.1023/A:1015557300450

Wang LP, Qi YC, Zhao YX, Zhang H (2002) Cloning and sequencing of GST gene of *Suaeda salsa* and its expression characteristics. J Plant Physiol Mol Biol 28:133–136

Wang W, Vinocur B, Altman A (2003) Plant responses to drought, salinity and extreme temperatures: towards genetic engineering for stress tolerance. Planta 218:1–14. https://doi.org/10.1007/s00425-003-1105-5

Xu F, Lagudah ES, Moose SP, Riechers DE (2002) Tandemly duplicated safener-induced glutathione S-transferase genes from *Triticum tauschii* contribute to genome- and organ-specific expression in hexaploid wheat. Society 130:362–373. https://doi.org/10.1104/pp.004796.362

Xu J, Xing XJ, Tian YS, Peng RH, Xue Y, Zhao W, Yao QH, Zhang H (2015a) Transgenic Arabidopsis plants expressing tomato glutathione S-transferase showed enhanced resistance to salt and drought stress. PLoS One 10:1–16. https://doi.org/10.1371/journal.pone.0136960

Xu Z, Jiang Y, Zhou G (2015b) Response and adaptation of photosynthesis, respiration, and antioxidant systems to elevated CO_2 with environmental stress in plants. Front Plant Sci 6:1–17. https://doi.org/10.3389/fpls.2015.00701

Xu J, Tian YS, Xing XJ, Peng RH, Zhu B, Gao JJ, Yao QH (2016) Over-expression of *AtGSTU19* provides tolerance to salt, drought and methyl viologen stresses in *Arabidopsis*. Physiol Plant 156:164–175. https://doi.org/10.1111/ppl.12347

Yang G, Wang Y, Xia D, Gao C, Wang C, Yang C (2014) Overexpression of a GST gene (*ThGSTZ1*) from *Tamarix hispida* improves drought and salinity tolerance by enhancing the ability to scavenge reactive oxygen species. Plant Cell Tissue Organ Cult 117:99–112. https://doi.org/10.1007/s11240-014-0424-5

Zhang C-H, Wu Z-Y, Ju T, Ge Y (2013) Purification and identification of glutathione S-transferase in rice root under cadmium stress. Rice Sci 20(3):173–178

Zhang JZ, Creelman RA, Zhu J-K (2004) From laboratory to field. Using information from Arabidopsis to engineer salt, cold, and drought tolerance in crops. Plant Physiol 135:615–621. https://doi.org/10.1104/pp.104.040295

Zhao FJ, McGrath SP (2009) Biofortification and phytoremediation. Curr Opin Plant Biol 12:373–380. https://doi.org/10.1016/j.pbi.2009.04.005

Zhao F, Zhang H (2006) Salt and paraquat stress tolerance results from co-expression of the *Suaeda salsa* glutathione S-transferase and catalase in transgenic rice. Plant Cell Tissue Organ Cult 86:349–358. https://doi.org/10.1007/s11240-006-9133-z

Zhao F, Liu W, Zhang S (2009) Different responses of plant growth and antioxidant system to the combination of cadmium and heat stress in transgenic and non-transgenic Rice. J Integr Plant Biol 51:942–950. https://doi.org/10.1111/j.1744-7909.2009.00865.x

Zinta G, Abdelgawad H, Domagalska MA et al (2014) Physiological, biochemical, and genome-wide transcriptional analysis reveals that elevated CO_2 mitigates the impact of combined heat wave and drought stress in *Arabidopsis thaliana* at multiple organizational levels. Glob Chang Biol 20:3670–3685. https://doi.org/10.1111/gcb.12626

Phosphite as an Inductor of Adaptive Responses to Stress and Stimulator of Better Plant Performance

Libia Iris Trejo-Téllez and Fernando Carlos Gómez-Merino

Abstract Phosphite (Phi) is emerging as a novel molecule that can be used as a biostimulant to enhance plant performance in limiting environments. In addition, Phi is effective against some pathogenic bacteria, oomycetes, fungi, and nematodes that significantly affect crop production and productivity. As a biostimulant, Phi may improve the yield and quality of a number of important crop species and can induce better performance of plants exposed to abiotic stress factors. In conventional agricultural systems, Phi cannot be used as a nutrient source and hence cannot substitute or complement inorganic phosphate (Pi) fertilizers. Instead, novel genetic engineering approaches are currently allowing its use as an alternative Pi fertilizer and herbicide, although it is not yet widely used on a commercial basis. This innovative biotechnology is addressing the challenges of Pi reserve depletion and multiple herbicide tolerance in weeds. In terms of biostimulation and induction of better plant performance, the beneficial effects of Phi on plant metabolism are more evident in conditions of Pi sufficiency. Additionally, Phi applications are more efficient when properly timed to match plant requirements, which in turn depend on the genotype of the crop plant used, type of soil and climate where plants are grown, cultural practices, as well as the dose, rate, and Phi source to be used. This chapter outlines the recent research advances on the effects of Phi as a potential biostimulator, pesticide, and a dual fertilizer and herbicide in agriculture and discusses potentialities and challenges of its use, especially those related to its utilization as an inductor of adaptive responses to stress in plants.

Keywords Phosphorus · Phosphorous acid · Phi · Biotic stress · Abiotic stress · Biostimulation

L. I. Trejo-Téllez
Colegio de Postgraduados Campus Montecillo, Texcoco, State of Mexico, Mexico

F. C. Gómez-Merino (✉)
Colegio de Postgraduados Campus Córdoba, Amatlán de los Reyes, Veracruz, Mexico
e-mail: fernandg@colpos.mx

S. Vats (ed.), *Biotic and Abiotic Stress Tolerance in Plants*,
https://doi.org/10.1007/978-981-10-9029-5_8

1 Introduction

Plants are continually challenged by both biotic and abiotic stresses that severely reduce their growth and development. Plant responses to these stresses are complex and involve numerous physiological, biochemical, and molecular mechanisms (Rejeb et al. 2014). In turn, those mechanisms may lead to cellular adaptations allowing plants to cope with adverse changes in their environment and thus avoid the detrimental effects of stress agents (Barkla et al. 2013). In the quest for improved abiotic stress tolerance, biostimulants are receiving increasing interest. By definition, a plant biostimulant is any substance or microorganism applied to plants in order to enhance nutrition efficiency, abiotic stress tolerance, and/or crop quality traits (du Jardin 2015). Phosphite (Phi), an isostere of the phosphate (Pi) anion, is emerging as a potential biostimulator since it has been effective against different stress factors and has been proved to enhance crop yield and quality traits (Gómez-Merino and Trejo-Téllez 2015). Furthermore, Phi may play an important role as an indicator of plant defense mechanisms against a number of biotic factors (Lim et al. 2013; Massoud et al. 2012). While Pi is the sole source of phosphorus (P) of vital importance in plant nutrition under conventional agricultural production systems, Phi is an alternative biostimulator and metabolic inductor which is becoming an innovative driving force in modern crop production (Gómez-Merino and Trejo-Téllez 2016; Achary et al. 2017).

Phosphorus (P) is one of the primary macronutrients required by all forms of life on the Earth, making up about 0.2% of a plant's dry weight biomass. This macronutrient is an essential component of biomolecules such as sugar phosphates (i.e., dihydroxyacetone phosphate, glucose-6-phosphate, and fructose-6-phosphate), phospholipids (i.e., phosphatidic acid, phosphatidylethanolamine, and phosphatidylinositol), phosphoproteins (i.e., polymerases, phosphoenolpyruvate carboxylase, and cryptochromes), enzymes, and energy-rich compounds such as adenosine triphosphate (ATP) and nicotinamide adenine dinucleotide phosphate (NADP), as well as the nucleic acids deoxyribonucleic acid (DNA) and ribonucleic acid (RNA) (Gómez-Merino and Trejo-Téllez 2015, 2016; Manna et al. 2016). Thus, P plays a pivotal role in genetic heredity, membrane structure, signal perception and transduction, and metabolism. Importantly, the processes of phosphorylation and dephosphorylation involved in cellular protection, defense responses, gene activation, and metabolism are mediated by P supply. Consequently, P regulates diverse cellular functions for proper plant growth and development, triggering responses to environmental stimuli and stress factors as well (Gómez-Merino and Trejo-Téllez 2015, 2016).

The demand for phosphorus is supplied in the Pi form of P ($H_2PO_4^-$ or HPO_4^{2-}), which is the sole P-containing nutrient important for optimal plant performance. Phosphate is derived from phosphate rock, which is a nonrenewable resource mined from Pi reserves mainly located in China, Morocco, the United States, Russia, Jordan, Egypt, Saudi Arabia, Peru, Israel, Tunisia, and Vietnam. Other important P reserves are located in Algeria, Syria, South Africa, Australia, Peru, Kazakhstan,

India, Senegal, and Mexico (Van Kauwenbergh 2010). Recent studies suggest that the world reserves of phosphate rock might be exhausted within 100–200 years if current consumption is maintained (Cordell et al. 2009; Schröder et al. 2010; Dawson and Hilton 2011; Achary et al. 2017), which represents one of the most critical challenges facing the quest to achieve food security and sustainable development.

In terms of plant biology, in order to guarantee functional metabolic reactions, Pi homeostasis must be kept between 5 and 20 mM in the cytoplasm. Plants absorb P only in its soluble inorganic forms, $H_2PO_4^-$ or HPO_4^{2-}, which occur in the soil between 0.1 and 1 μM (Malboobi et al. 2012, 2014). Because of such a tremendous difference between P supply and demand, this macronutrient becomes critical for plant metabolism. In fact, P is ranked as the second most vital element for plant growth and development, just after nitrogen (N). Hence, P is considered as a major limiting factor in agriculture and food production worldwide (Gómez-Merino and Trejo-Téllez 2016).

Phosphorus is a very reactive element, and once in the soil solution, it rapidly combines with other elements to form compounds with variable oxidation states, the most oxidized being Pi. Because of its chemical properties, Pi is largely immobile in the soil. It reacts with the calcium (Ca) and iron (Fe) present therein, which makes Pi substantially unavailable to be absorbed by plant roots (Achary et al. 2017). Furthermore, some soil microorganisms rapidly convert Pi into organic forms that are not absorbed by plants (Syers et al. 2008). Though P is the 11th most abundant element in the Earth's crust (4×10^{15} mt), only a small part of it (20–50%) is available for plants in the forms of Pi (Schröder et al. 2010). Because of such low Pi availability in agricultural soils, conventional cultural practices commonly apply excessive P fertilizers to crops. However, nearly 80% of P fertilizers applied to crops are lost because of precipitation and adsorption to mineral surfaces or converted to organic forms; in very sandy and organic soils, P leaching can also take place (Lehmann and Schroth 2003; Manna et al. 2016). Thus, the current agronomic practices related to P management increase crop production costs, contribute to the deterioration of soil health, and lead to the eutrophication of rivers, streams, lakes, and oceans (Achary et al. 2017). In addition to the risk of Pi depletion in the near future, modern agricultural production systems have to deal with weeds, and herbicides have become less effective in controlling such unwanted plants (Manna et al. 2016). Therefore, approaches aimed at developing a more efficient use of this finite resource while reducing the environmental impacts related to its excessive use are gaining momentum. Novel agricultural practices and technologies as well as innovative strategies to achieve sustainable use of P can attenuate negative environmental impacts and enhance the long-term supply of this essential nutrient (Syers et al. 2011). One of the novel technologies that is currently gaining attention in this regard is the use of Phi ($H_2PO_3^-$ or HPO_3^{2-}). Recent reports describe the development of Phi-mediated fertilization, weed control, and selectable marker platforms useful in plant genetic transformation approaches with a wide spectrum of applications in agriculture (López-Arredondo and Herrera-Estrella 2012; Manna et al. 2016; Achary et al. 2017). Under controlled conditions (i.e. laboratory and greenhouse), this technology proved to be effective. However, field experiments will be necessary to allow its agronomic use.

Nowadays, Phi is emerging as a novel biostimulant in agriculture, improving yield and quality of crops. Phi also exhibits both direct antibiotic effects on pathogens and inhibition through enhanced plant defense responses. Furthermore, Phi induces diverse mechanisms of tolerance against a number of abiotic stress factors (Gómez-Merino and Trejo-Téllez 2016). Though it has been largely applied as an alternative fertilizer in conventional agricultural systems, its contribution to P nutrition is limited, and it has been the subject of controversy in the technical and scientific worlds.

In this chapter, we outline the recent advances in research concerning the use of Phi as a pesticide, an inductor of plant resistance against pathogens, and a biostimulant that improves yield, harvest quality, and responses to environmental stressors. In addition, we explore the recent development of Phi-mediated fertilization, weed control, and selectable marker platforms with a wide spectrum of applications in biotechnology and agriculture.

2 Chemical Properties of Phi and Its Metabolism in Plants

Orthophosphate or inorganic phosphate (Pi) is the most oxidized form of P found in nature. It has an empirical formula of PO_4^{3-} and a molar mass of 94.97 g mol^{-1}. Its molecule displays a tetrahedral structure, with the P atom located at the center and the oxygen (O) atoms distributed evenly at the points of the structure. Its formal charge is -3. Pi is defined as the conjugate base of the hydrogen phosphate ion (HPO_4^{2-}), which is the conjugate base of the dihydrogen phosphate ion ($H_2PO_4^-$). In turn, the $H_2PO_4^-$ ion is defined as the conjugate base of phosphoric acid (H_3PO_4). The formation of phosphate salts takes place when a positively charged ion attaches to the negatively charged oxygen atoms of the ion (Fig. 1). At neutral pH in the soil solution, the Pi ion is present as a combination of $H_2PO_4^-$ and HPO_4^{2-} ions, with 62% of it as $H_2PO_4^-$ and the remaining 38% as HPO_4^{2-}, while the form $H_2PO_4^-$ is how Pi is normally metabolized in plant cells (McDonald et al. 2001).

Nonetheless, over the past 30 years, Phi has increasingly been used to enhance yield and quality of a number of crop species (Table 1), and just recently, novel approaches based on genetic engineering have developed transgenic plants that can use Phi as a P source, serving not only as an alternative fertilizer but also as a weed control (López-Arredondo and Herrera-Estrella 2012; Manna et al. 2016; Achary et al. 2017). Similar to Pi, the Phi molecule displays a formal charge of -3. It is defined as a conjugate base of the hydrogen phosphite ion (HPO_3^{2-}), which is the conjugate base of the dihydrogen phosphite ion ($H_2PO_3^-$). In turn, the $H_2PO_3^-$ ion is considered the conjugate base of phosphorous acid (H_3PO_3). Phi is therefore defined an isostere of the Pi anion, in which a hydrogen (H) atom replaces one of the O atoms bound to the P atom (Varadarajan et al. 2002; Gómez-Merino and Trejo-Téllez 2015) (Fig. 1).

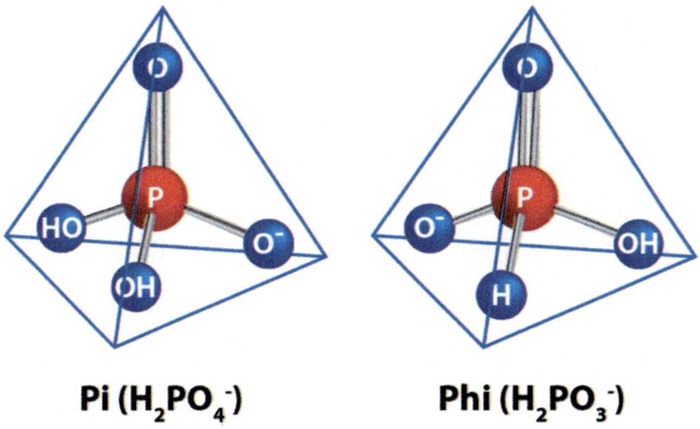

Fig. 1 Tetrahedral molecular geometry of phosphate (Pi, $H_2PO_4^-$) and phosphite (Phi, $H_2PO_3^-$). The Pi ion has one more oxygen (O) atom than the Phi one

Phosphorous acid and phosphonate are alternative names given to phosphite. Nevertheless, the term phosphonate encompasses a wide range of compounds containing carbon-phosphorus bonds like Fosetyl-Al (McDonald et al. 2001; Metcalf and van der Donk 2009). Fosetyl-Al was indeed one of the first trademarks patented in the United States, and when the corresponding patent expired, several companies formulated a series of Phi-containing products with other ions (i.e., Ca, Cu, K, and Na, among others) (Gómez-Merino and Trejo-Téllez 2016).

Although Pi and Phi display similar chemical structures, the lack of an O atom in Phi significantly changes the nature and reactivity of the resultant molecule. Both Pi and Phi display tetrahedral molecular geometry, but because of the structural difference, the charge distribution is distinct in each molecule. Thus, the binding of Pi and Phi to their interacting molecules is influenced not only by the shape but also by the charge distribution of the structure. Consequently, most enzymes involved with phosphoryl transfer reactions easily distinguish Phi from Pi ions (Plaxton and Tran 2011). Nonetheless, some proteins found in higher organisms, including membrane Pi transporters and proteins involved in Pi sensing, may not discriminate between Phi and Pi (McDonald et al. 2001), and Phi may disrupt Pi starvation responses (Varadarajan et al. 2002). Enhanced root growth and root-to-shoot ratio, a hallmark of Pi stress response, is strongly inhibited by Phi. At the molecular level, the expression of Pi starvation-induced genes, including high-affinity Pi transporters (i.e., *LePT1*, *LePT2*, *AtPT1*, and *AtPT2*) and phosphatases (i.e., *LePS2* and *PAP1*), is suppressed by Phi (Varadarajan et al. 2002).

Because plants lack the biochemical mechanisms to metabolize Phi, this ion usually persists within plant tissues for months (Ouimette and Coffey 1990), displaying systemic effects therein. If the amount of Phi applied and the Pi status of the plant are appropriate, Phi may trigger beneficial effects as a positive biostimulator or inductor of defense mechanisms against a number of pathogenic agents, mainly fungi and oomycetes. However, if it is applied at high concentrations

Table 1 Effects of phosphite as a biostimulator or inductor of better plant performance in response to biotic and abiotic stress factors

Plant species	Concentrations and source of Phi used (optimal dosage)[a]	Method of application	Beneficial effects observed (detrimental effects, if any, are indicated in parenthesis)	Reference
Lupinus angustifolius, *Nicotiana tabaccum*, and *Carica papaya*	20 mL per pot of 99% pure anhydrous phosphorous acid	Root drench	Effective control of *Phytophthora cinnamomi*, *P. nicotianae*, and *P. palmivora*	Smillie et al. (1989)
Capsicum annuum	0.0, 0.1, and 1.0 mM Phi as phosphorous acid (optimum, 0.1 and 1.0 mM Phi)	Nutrient solution in hydroponics	Significant reduction of incidence of *Phytophthora capsici* (reduction of plant growth)	Förster et al. (1998)
Zea mays	20% phosphonic acid and its formulations like Aliette (0.25%) and Akomin (0.25%)	Foliar sprays	Reduction in disease incidence and severity caused by *Peronosclerospora sorghi*, and yield increase by up to 73%	Panicker and Gangadharan (1999)
Adenanthos barbiger, *Daviesia decurrens*, and *Xanthorrhoea preissii*	0.0%, 0.2%, 0.5%, and 2% Phi supplied from Fosject 200 containing 200 g L^{-1} Phi as mono- and dipotassium phosphite (optimum, ≥0.5% Phi)	Foliar applications	Efficient control of *Phytophthora cinnamomi* in native plant communities (treatment with 2% phosphite led to the development of severe phytotoxicity symptoms)	Pilbeam et al. (2000)
Vitis vinifera	7 kg ha^{-1} $year^{-1}$ of potassium Phi	Aqueous solutions	Effective control against *Plasmopara viticola* but not against *Oidium tuckeri* and *Pseudopezicula tracheiphila* (the application of Phi led to Phi residues in the wine)	Speiser et al. (2000)
Different plant species from the jarrah forest and northern-	5–20 g L^{-1} Phi	Foliar sprays	Foliar applications of Phi have considerable potential in reducing the impact of *P. cinnamomi* in	Tynan et al. (2001)

(continued)

Table 1 (continued)

Plant species	Concentrations and source of Phi used (optimal dosage)[a]	Method of application	Beneficial effects observed (detrimental effects, if any, are indicated in parenthesis)	Reference
sandplain near Perth, Australia			native plant communities in the short term. In order to maintain adequate control, Phi should be sprayed every 6–12 months, depending on the species and/or plant community	
Banksia grandis, B. hookeriana, Dampiera linearis, Dryandra sessilis, and Hibbertia commutata	0, 10 and 20 mg L^{-1} Phi	Foliar applications	Foliar application of Phi slowed, but did not completely inhibit, colonization of stems by *P. cinnamomi*	Wilkinson et al. (2001)
Banksia brownie	0, 12, 24, and 96 kg ha^{-1} of Foli-R-Fos 400, a 40% solution of mono- and dipotassium phosphite (optimum, 24–96 kg ha^{-1})	Foliar applications	Potent control of *Phytophthora cinnamomi* in the early stages of infection, providing protection to individuals that have avoided infection (less than 10% of the foliage were affected even at the highest rate)	Barrett et al. (2003)
Malus domestica	Potassium phosphite (0, 50, and 500 ppm) was applied as a solution of mono- and dipotassium phosphonate ions (50–500 ppm Phi)	Soaking of fruits in the laboratory and sprays to trees in the field at different bloom stages	Efficient control of moldy core caused by *Alternaria alternata* in apple fruits	Reuveni et al. (2003)
Catharanthus roseus	0.0, 0.1, 0.3, and 0.5 mM Phi applied individually or in combination with	Foliar application	Effective protection against *Phytophthora nicotianae*, similar to foliar	Banko and Hong (2004)

(continued)

Table 1 (continued)

Plant species	Concentrations and source of Phi used (optimal dosage)[a]	Method of application	Beneficial effects observed (detrimental effects, if any, are indicated in parenthesis)	Reference
	Pi. Phi was supplied as Nutri-Grow® PK, 0–28–26 (optimum, 0.5 mM phi)		applications of Aliette fungicide	
Solanum tuberosum	Phosphonic acid (>98% pure) and commercial Fosetyl-Al at different concentrations and rates (optimum, 4 kg ha^{-1} Phi in a single application)	Sprays to plants and tubers	Reduced infection of potato tubers by *Phytophthora infestans*. Healthy tubers removed from treated plants did not develop disease symptoms. By contrast, foliar application of Fosetyl-Al did not significantly reduce tuber blight development	Cooke and Little (2002)
Solanum tuberosum	Phosphorous acid at 0.0, 4.68, 7.49, and 9.37 kg ha^{-1} a.i. (optimum, 7.49 or 9.37 kg ha^{-1} a.i.)	Foliar sprays	Reduced incidence and severity of *Phytophthora infestans* and *P. erythroseptica* (Phi had no effect on the control of *Pythium ultimum*)	Johnson et al. (2004)
Cucumis sativus	Aqueous solutions of various phosphonate formulations (0.0–0.28% a.i. (v/m) were applied (optimum, 0.140% or 0.280% a.i.)	Preplanting amendments or postplanting drench treatments applied as peat-based mix, muck soil, and sandy loam soil	Control of *Pythium* damping-off and disease suppression that increased with the concentration of phosphonate	Abbasi and Lazarovits (2005)
Brassica rapa var. *chinensis*, *B. rapa* var. *perkinensis*, and *B. oleracea* var. *capitata*	Aqueous solutions of various phosphonate formulations (0.00–0.35% a.i. (v/m) were applied (optimum, 0.07% or 0.21% a.i.)	Replanting amendments in some trials or postplanting drench in most trials	Consistent reduction of clubroot (caused by *Plasmodiophora brassicae*) severity when Phi was applied before or after planting	Abbasi and Lazarovits (2006)

(continued)

Table 1 (continued)

Plant species	Concentrations and source of Phi used (optimal dosage)[a]	Method of application	Beneficial effects observed (detrimental effects, if any, are indicated in parenthesis)	Reference
Banksia species and *Eucalyptus marginata*	0, 50, 100, and 200 g L^{-1} Phi, applied as Fosject 200, containing 200 g $H_2(PO_3H)$ L^{-1} present as the mono- and dipotassium phosphite (optimum, 50–100 g L^{-1} Phi)	Stem injection	Practical control of *Phytophthora cinnamomi* and protection of threatened native flora	Shearer et al. (2006)
Fragaria x ananassa	Agri-Fos at 2.34 kg ha^{-1} a.i.	Sprays of plants and fruits	Protection against leather rot of strawberry, caused by *Phytophthora cactorum* for up to 7 days, as well as curative activity of at least 36 h	Rebollar-Alviter et al. (2007)
Citrus spp.	Cannon at 0.0–400 μg mL, with 0.25% final concentration in spray (optimum, 50 μg L_{-1})	Inoculation of fruits in laboratory and sprays in the field	Inhibition of mycelium growth of *Alternaria alternata* pv. *citri*, slightly at 3.12 μg mL^{-1} and completely at 50 μg mL^{-1} (it did not inhibit germination of conidia)	Yogev et al. (2006)
Solanum tuberosum	3 L ha^{-1} Phi to seed tubers and 3, 4.5, or 6 L ha $^{-1}$ as foliar sprays. Phi was supplied as potassium and calcium phosphite (optimum, 3 L ha^{-1})	Sprays to seed tubers and foliar applications to plants	Reduction of disease severity caused by *Phytophthora infestans*, *Fusarium solani*, and *Rhizoctonia solani* in potato seed tubers and foliage, while crop senescence was delayed	Lobato et al. (2008)
Citrus sinensis	Phosphorous acid and Nutriphite (optimum, 0.87 mM	Foliar sprays	Enhanced root resistance to *Phytophthora* root rot of citrus	Orbovic et al. (2008)

(continued)

Table 1 (continued)

Plant species	Concentrations and source of Phi used (optimal dosage)[a]	Method of application	Beneficial effects observed (detrimental effects, if any, are indicated in parenthesis)	Reference
	KH_2PO_4 + 75 μL Nutriphite +0.7 mM of K_2SO_4)		seedlings treated with Phi alone or in combination with Pi. Phi has a direct antifungal effect and elicits effective host defense responses	
Lactuca sativa, *Solanum lycopersicum*, and *Musa paradisiaca*	Different combinations of Pi and Phi: phosphoric acid for Pi and phosphorous acid for Phi (optimum, 50% Pi plus 50% Phi)	Nutrient solution in hydroponics	Increase P content, foliar area, and biomass dry weight (negative effects when a balance between Pi and Phi was not achieved)	Bertsch et al. (2009)
Fragaria x ananassa	Commercial fertilizer Phosfik, a liquid NPK fertilizer at 3:12:15 containing Phi	Soaking and irrigation of plants	Activation of plant defense mechanisms, since fruit ascorbic acid and anthocyanin content increased	Moor et al. (2009)
Lambertia inermis and *Banksia grandis*	0, 24, or 48 ha^{-1} Phi. Phi was provided as Foli-R-Fos 400, which has 400 g L^{-1} phosphorus acid present as mono- and dipotassium phosphite (optimum, 48 kg ha^{-1} Phi)	Foliar sprays	Efficient control of *P. cinnamomi*, which will depend more on plant species composition than soil nutrient status	Shearer and Crane (2009)
Solanum tuberosum	Calcium, potassium, and copper phosphites, with final concentration ranging from 0.0 to 3.82 mg mL^{-1} Phi (optimum, 0.04–0.87 mg mL^{-1} Phi)	Potato slices cultivated in vitro, with different growth media	Reduction of disease symptoms produced by *Phytophthora infestans*, *Fusarium solani*, and *Rhizoctonia solani*	Lobato et al. (2010)

(continued)

Table 1 (continued)

Plant species	Concentrations and source of Phi used (optimal dosage)[a]	Method of application	Beneficial effects observed (detrimental effects, if any, are indicated in parenthesis)	Reference
Malus domestica	Potassium phosphite containing 78% of mono-(KH_2PO_3) and di-(K_2HPO_3)-potassium phosphite salts (optimum, 2 mg mL^{-1})	Growth media *for* in vitro assays; immersion in Phi solution for fruit treatment	Inhibition of mycelial growth and conidial germination of *Penicillium expansum*, reduction of disease incidence, and effective control against natural blue mold infections after 6 months of storage	Amiri and Bompeix (2011)
Zea mays	Potassium phosphite at 25% of total P in the nutrient solution	Nutrient solution in hydroponics	Stimulation of guaiacol peroxidase activity and lignin biosynthesis (inhibition of P uptake, regardless of Pi status)	Ávila et al. (2011)
Fragaria x ananassa	Phosphorous acid at 0%, 20%, 30%, 40%, and 50% Phi (optimum, 20–30%)	Nutrient solution in hydroponics	Increased concentrations of sugars, chlorophylls, and total free amino acids in leaves	Estrada-Ortiz et al. (2011)
Solanum tuberosum	Potassium phosphite applied at 1% of the commercial product, equivalent to 3 L ha^{-1}, 5 mL per plant	Foliar spray	Reduced susceptibility to *Phytophthora infestans*, *Fusarium solani*, and *Erwinia carotovora* infections. Phi induced a systemic defense response by activating a higher synthesis of phytoalexins, chitinases, glucanases, peroxidases, and polyphenol oxidases	Lobato et al. (2011)
Glycine max	Four rates (0, 375, 750, and 1500 g ha^{-1} a.i.) of potassium phosphite, containing 30% P_2O_5 + 20% K_2O (optimum, 750–1500 g ha^{-1} a.i.)	Foliar applications	Combination of two fungicide applications (pyraclostrobin and epoxiconazole) following Phi application significantly improved the control of *Phakopsora*	Silva et al. (2011)

(continued)

Table 1 (continued)

Plant species	Concentrations and source of Phi used (optimal dosage)[a]	Method of application	Beneficial effects observed (detrimental effects, if any, are indicated in parenthesis)	Reference
			pachyrhizi and *Microsphaera diffusa*, yield, and seed weight when compared to a single fungicide application	
Solanum tuberosum	Potassium phosphite was applied at 3 L ha^{-1}	Foliar applications	Reduction of damage growth rate caused by *Phytophthora* ssp. and tended to increase yields	Cicore et al. (2012)
Zea mays	The product Hortifós PK provides potassium phosphite at 1.5 mL L^{-1}	Foliar sprays	Decrease of the population of *Pratylenchus brachyurus* in maize while increasing plant height	Dias-Arieira et al. (2012)
Arabidopsis thaliana	0, 5, 10, 25, 50, 75, and 100 mM Phi as phosphorous acid (optimum, 5–10 mM Phi)	Soil drenching	Induction of resistance to *Hyaloperonospora arabidopsidis*, priming salicylic acid accumulation, transcription, and mobilization of essential components of basal resistance	Massoud et al. (2012)
Solanum tuberosum	Application of either 1.07 kg ha^{-1} potassium phosphite plus a combination of fungicides (carboxin, captan, dimethomorph, and mancozeb) to seed tubers or 3 L ha^{-1} potassium phosphite to seed tubers and crops	Foliar and seed tuber treatments	Induction of defense responses against *Fusarium solani* in tuber periderm and cortex, with higher pectin accumulation and increased content and/or activity of polygalacturonase, proteinase inhibitor, and chitinase	Olivieri et al. (2012)

(continued)

Table 1 (continued)

Plant species	Concentrations and source of Phi used (optimal dosage)[a]	Method of application	Beneficial effects observed (detrimental effects, if any, are indicated in parenthesis)	Reference
Phaseolus vulgaris	Phosphorous acid at 0, 16, 32, 64, 128, 256, and 512 μM Phi and 2 levels of Pi: 80 and 800 μM, corresponding to Pi-starved and Pi-sufficient plants, respectively (optimum, up to 0. 32 μM Phi)	Nutrient solution in hydroponics	In Pi-starved plants, catalase activity was considerably higher when Phi was applied at low levels (medium and high Phi levels reduced substantially the activity of this enzyme, while growth and grain yield were negatively affected)	Ávila et al. (2013)
Fragaria x ananassa	0, 20, 30, 40, and 50% Phi supplied as phosphonic acid (optimum, 20–30% Phi)	Nutrient solution in hydroponics	Activation of plant defense mechanisms by producing a higher concentration of anthocyanins, while yield and fruit size were slightly improved	Estrada-Ortiz et al. (2013)
Glycine max	Manganese phosphate, at a concentration of 20 ml L^{-1} of the commercial product	Spray onto the shoots	Significant reduction of *Meloidogyne javanica* eggs per gram of roots	Puerari et al. (2013)
Solanum tuberosum	40 mL 1.25% (v/v) Proalexin, corresponding to 36 mM Phi	Foliar spray	Changes of the transcriptome and secretome, leading to long-lasting resistance against *Phytophthora infestans*. Transcripts associated with defense, wounding, and oxidative stress constituted the core of the responses, along with adjustments in primary metabolism and cell wall-related processes	Burra et al. (2014)

(continued)

Table 1 (continued)

Plant species	Concentrations and source of Phi used (optimal dosage)[a]	Method of application	Beneficial effects observed (detrimental effects, if any, are indicated in parenthesis)	Reference
Arabidopsis thaliana	0.0, 0.5, 2.5, and 20.0 mM Phi as phosphorous acid (optimum, up to 2.5 mM Phi)	Nutrient solution in hydroponics	Induction of resistance to *Phytophthora cinnamomi*	Eshraghi et al. (2014)
Solanum tuberosum	Afital K, applied at 3000 cm^3 ha^{-1}	Sprays to seed tubers	Reduction of the period between planting and emergence, while leaf area, dry matter, and indigenous mycorrhizal colonization were increased	Tambascio et al. (2014)
Malus domestica	Phosphojet, providing mono- and dipotassium salts of phosphorous acid 45.8%, applied 2 × 22.5 mL per tree	Trunk injection	Significant reduction of blossom and shoot blight symptoms caused by *Erwinia amylovora* and induction of *PR-1*, *PR-2*, and *PR-8* genes, indicating a possible activation of systemic acquired resistance (SAR)	Aćimović et al. (2015)
Banksia grandis and *Eucalyptus marginata* trees	Phosphite (Fosject 200) was injected at 1 mL cm^{-1} of stem circumference. Soluble powder implants of phosphite (Phoscap and Medicap MD) were applied at 10 cm intervals	Stem injections and novel implants	Both liquid phosphite injections and novel phosphite implants are effective at controlling lesion extension in *B. grandis* and *E. marginata*, caused by *P. cinnamomi*	Scott et al. (2015)
Solanum tuberosum	The product Afital Potassium Phosphite was applied at 5 mL per plant, equivalent to 3 L ha^{-1}	Foliar application	Induction of UV-B stress tolerance, by increasing chlorophyll content and expression of the *psbA* gene, as well	Oyarburo et al. (2015)

(continued)

Table 1 (continued)

Plant species	Concentrations and source of Phi used (optimal dosage)[a]	Method of application	Beneficial effects observed (detrimental effects, if any, are indicated in parenthesis)	Reference
	i.a. The dose utilized was 1% (v/v) of the commercial product		as the accumulation of glucanases and chitinases, preventing oxidative stress, decreasing the accumulation of hydrogen peroxide, enhancing guaiacol peroxidase (POD) and superoxide dismutase (SOD) activities, and inducing the transcription rate of a gene involved in flavonoid synthesis	
Glycine max	Manganese phosphite at a dosage of 200 mL per 100 kg of seeds	Immersion of seeds in Phi solution	The combination of the isolate *Bacillus subtilis* 54 with manganese phosphite caused 82% of control of *Macrophomina phaseolina* (Phi applied singly improved the pathogen performance)	Simonetti et al. (2015)
Lactuca sativa and *Beta vulgaris* var. *cicla*	Phosphorous acid, providing 0, 0.25, and 0.50 mM Phi (optimum, 0.25 mM Phi)	Nutrient solution in hydroponics	Induction of chlorophyll contents and increase in P concentration	Estrada-Ortiz et al. (2016)
Coffea arabica	Manganese phosphite at 5.0 mL L^{-1}	Foliar sprays	Control the severity of rust (caused by *Hemileia vastatrix*) in 63% while inducing defense responses with increased transcription of the genes *POX*, *CAT*, *GLU*, and *PAL* in non-inoculated	Monteiro et al. (2016)

(continued)

Table 1 (continued)

Plant species	Concentrations and source of Phi used (optimal dosage)[a]	Method of application	Beneficial effects observed (detrimental effects, if any, are indicated in parenthesis)	Reference
			plants and increased activity of APX, SOD, and PPO enzymes in plants inoculated with rust and in non-inoculated plants	
Pinus spp. and *Pseudotsuga menziesii*	Phi at 0%, 1%, and 4%	Culture medium in vitro and foliar sprays in the field	Effective inhibition of *Fusarium circinatum* mycelial growth in a dose-dependent manner	Cerqueira et al. (2017)

[a]Commercial products are mentioned for academic purposes and are not recommended over similar products in this chapter

and the Pi status of the plant is deficient, Phi may accumulate and cause detrimental effects (Loera-Quezada et al. 2015). Negative effect derived from a non-appropriate use of Phi includes the repression of genes involved in Pi-starvation responses (Ticconi et al. 2001; Varadarajan et al. 2002), which disrupt P nutrition. By inhibiting Pi uptake in a competitive manner, Phi alters the homeostasis of phosphorus within the plant (Kobayashi et al. 2006; Danova-Alt et al. 2008; Berkowitz et al. 2013). Importantly, Phi uptake is significantly hindered by Pi (Pratt et al. 2009; Jost et al. 2015). When Phi is assimilated by the plant, it accumulates, especially in sink tissues (Nartvaranant et al. 2004; Jost et al. 2015).

The three O atoms in the Phi molecule give this anion increased mobility in plant tissues through both the xylem and the phloem, so that it can be successfully applied at low concentrations throughout the plant in order to induce beneficial effects. Such high mobility allows Ph*i* to be absorbed and translocated within the plant more readily than Pi (Ratjen and Gerendas 2009; Jost et al. 2015). Conversely, commercial P fertilizers are usually solid, have low solubility in water, react strongly with the soil matrix, and are more prone to be adsorbed to soil particles than Phi. These facts render Pi largely immobile in the soil and only a small fraction of it is available to the plant, eroding over time within the soil solution. Importantly, commercial Phi-containing products have higher concentrations of P ($\geq$40% P) than traditional Pi-based fertilizer (~30% P). Moreover, Phi-containing salts exhibit a higher solubility than that of their analogous Pi-containing ones, which render Phi uptake by leaf and root a more efficient process (Gómez-Merino and Trejo-Téllez 2016). Phi may trigger hormesis, which is a biphasic dose response characterized by a low-dose stimulation or beneficial effect and a high-dose inhibitory or toxic effect (Mattson

2008). Therefore, the application of Phi-containing compounds must be tightly regulated, since excessive applications of Phi may cause toxic or deleterious effects to plants (Gómez-Merino and Trejo-Téllez 2015, 2016).

Phosphate transporters (PHT) are the primary proteins involved in the absorption of both Pi and Phi by plants (Jost et al. 2015; Sun et al. 2017; Wang et al. 2017), although these transporters are fundamentally involved in the Pi acquisition (Guest and Grant 1991; Ullrich-Eberius et al. 1981), and their role in Phi uptake is secondary (d'Arcy-Lameta and Bompeix 1991; Danova-Alt et al. 2008; Jost et al. 2015). PHT proteins are distributed throughout the plant, and consequently Pi and Phi can be absorbed by the leaves through foliar sprays or by the roots as soil drenches and through irrigation water, nutrient solution, or growth medium. Since Phi is highly water soluble (Jost et al. 2015), its mobility within the plant is more rapid than Pi (Ratjen and Gerendas 2009). Furthermore, because in nature plants lack the mechanisms to metabolize Phi, it remains relatively stable and is not significantly oxidized within the plant cells, and thus its effects are usually long lasting (Lovatt and Mikkelsen 2006). According to Adams and Conrad (1953), the approximate half-life for Phi oxidation to Pi in soil is approximately 12–16 weeks. Moreover, foliar applications are not as long lasting as injections. For instance, the effects of foliar sprays may last some weeks, whereas stem injections may protect plants for at least 4 years (Shearer and Fairman 2007a). Nevertheless, applications of Phi to the soil facilitate the contact of this ion with microorganisms, which mediate the oxidation of Phi to Pi. Consequently, after metabolization of Phi by soil microorganisms, Phi can become available to the plant as a P nutrient, albeit this conversion may take some months or even years to be completed (McDonald et al. 2001).

The acquisition of Phi by plant cells depends on pH, while Pi is an antagonist to Phi (Ouimette and Coffey 1990; Hanrahan et al. 2005). Once within the plant, Phi shows systemic effects and is highly stable, while it can be mobilized via vascular tissues throughout the whole plant. This mobility is important when using Phi as a biostimulant, since it favors its transport to all tissues in the plant (Smillie et al. 1989; Brunings et al. 2005). Mobility of Phi in both xylem and phloem is carried out by PHT proteins, in a similar manner to Pi (Ouimette and Coffey 1989).

To date a number of PHT enzymes have been identified and characterized in some crop species, including apple (*Malus domestica*), barley (*Hordeum vulgare*), maize (*Zea mays*), potato (*Solanum tuberosum*), rice (*Oryza sativa*), soybean (*Glycine max*), and tomato (*Solanum lycopersicum*) (López-Arredondo et al. 2014; Sun et al. 2017; Teng et al. 2017; Wang et al. 2017). The functional characterization of such transporters has revealed a significant divergence among genotypes. The distinct reported affinities and subcellular localizations of PHT proteins may reflect diverse functional roles such as uptake from the soil or the nutrient solution, as well as translocation and remobilization of stored Pi within the plant (Nussaume et al. 2011; Ceasar et al. 2014).

When using Phi in conventional agriculture, a sufficient supply of Pi will ensure an efficient use of low to moderate Phi concentrations without showing detrimental effects in the plant (Thao and Yamakawa 2009). However, since Phi displays hormetic effects on plant physiology, beneficial responses of Phi would depend on sufficient Pi supply to the plant, the dose of Phi applied, and the genetic background of the genotypes evaluated, among other factors. Indeed, both the cellular metabolic state and the reserves of Pi within the cell significantly influence the subcellular localization of Phi in plants (Martinoia et al. 2000), which may also alter the interactions between Phi and Pi signaling processes, as well as plant performance under different Pi states (i.e., nutrient preloading, sufficient supply, deprivation, or resupply) (Danova-Alt et al. 2008; Gómez-Merino and Trejo-Téllez 2015).

A summary of the effects of Phi on plant response to biotic and abiotic stress factors, or as a biostimulator that improves some yield and quality traits, is presented in the Table 1.

3 The Role of Phi Against Biotic Stressors

It has been widely proven that Phi can act as an efficient pesticide against various species of phytopathogenic organisms, including nematodes, fungi, oomycetes, and bacteria (Chase 1993; Smillie et al. 1989; Deliopoulos et al. 2010; Hofgaard et al. 2010; Dias-Arieira et al. 2013; Percival and Banks 2014; Puerari et al. 2015). This is because Phi stimulates a broad-spectrum resistance against pathogens (Jost et al. 2015), playing an important role as a priming molecule of plant defense responses (Machinandiarena et al. 2012; Massoud et al. 2012; Dalio et al. 2014).

In the case of pathogenic bacteria, the application of either 1.0 or 0.67% (v/v) potassium Phi inhibited the growth of *Streptomyces scabies* in potato by approximately 80% and 60%, respectively (Lobato et al. 2010). When potassium phosphite is applied to the leaves of potato plants grown in the field, harvested tubers display a reduced susceptibility to *Erwinia carotovora* infections, demonstrating that this Phi-salt triggers a systemic defense response (Lobato et al. 2011). In apple, the application of Phi significantly reduced blue mold incidence caused by *Penicillium expansum* in wounded and inoculated fruits (Amiri and Bompeix 2011), while the injection of Phi proved to be effective in controlling fire blight caused by *Erwinia amylovora* in apple trees (Aćimović et al. 2015).

Phi can prime plants for a faster and stronger defense response against a number of fungi and oomycetes including the genera *Phytophthora*, *Fusarium*, and *Rhizoctonia*, among others (Smillie et al. 1989; Förster et al. 1998; Pilbeam et al. 2000; Machinandiarena et al. 2012; Alexandersson et al. 2016). In this chapter we recognize that oomycetes are within the kingdom Protoctista rather than the kingdom Fungi. Nonetheless, we will use the terms fungicide or fungistatic to include activity against members of either group.

According to Förster et al. (1998), the occurrence of *Phytophthora capsici* was significantly reduced in pepper plants (*Capsicum annuum*) grown hydroponically in the presence of Phi, while the foliar application of Phi reduced late blight infection and tuber blight caused by *Phytophthora infestans* in potato (Andreu et al. 2006; Kromann et al. 2012). In strawberry (*Fragaria* × *ananassa*) foliar applications of Phi to plants and fruits rendered efficient protection against leather rot caused by

Phytophthora cactorum for up to 7 days, as well as curative activity of at least 36 h (Rebollar-Alviter et al. 2007).Similarly, foliar applications of Phi conferred reduced incidence and severity of *P. infestans* and *P. erythroseptica*, though Phi had no effect on the control of *Pythium ultimum* (Johnson et al. 2004). In cucumber (*Cucumis sativus*), Phi efficiently controlled *Pythium* damping-off and triggered disease suppression that increased with the concentration of Phi (Abbasi and Lazarovits 2005). In grape (*Vitis vinifera*), Speiser et al. (2000) reported effective control of Phi against *Plasmopara viticola*, but not against *Oidium tuckeri* and *Pseudopezicula tracheiphila*. In this study, the application of Phi led to phosphonate residues in the wine, which nonetheless were being considered toxicologically harmless.

According to Reuveni et al. (2003), Phi applications to apple fruits or trees confer an efficient control of moldy core caused by *Alternaria alternata*. In bok choy (*Brassica rapa* var. *chinensis*), chinese cabbage (*B. rapa* var. *perkinensis*), and cabbage (*B. oleracea* var. *capitata*), the application of Phi conferred consistent reduction of the severity of clubroot caused by *Plasmodiophora brassicae* (Abbasi and Lazarovits 2006).

With the combination of foliar field applications and postharvest applications of Phi, potato tubers were efficiently protected against pink rot caused by *Phytophthora erythroseptica* during storage (Miller et al. 2006; Taylor et al. 2011). Similarly, the postharvest application of Phi efficiently controlled the spread of potato tuber blight during storage (Lobato et al. 2011; Johnson 2010). Moreover, the susceptibility of tubers to *P. infestans* was decreased when plants were treated with foliar applications of Phi (Cooke and Little 2002; Liljeroth et al. 2016), while the application of Phi to the tubers once harvested effectively controlled this oomycete during their storage (Lobato et al. 2008). Furthermore, combined applications of the pesticide mancozeb with Phi provided potato tubers with greater protection against blight caused by *P. infestans* than the pesticide alone (Cooke and Little 2002). Similarly, the most efficient control against potato late blight under field conditions was observed by combining Phi with the broad-spectrum nonsystemic fungicide chlorothalonil (Liljeroth et al. 2016).

In vinca (*Catharanthus roseus*), foliar applications of 0.5 mM Phi at 3–6-day intervals effectively protected plants against *Phytophthora nicotianae*, as done by the foliar applications of 3 g L^{-1} Aliette at 14-day intervals (Banko and Hong 2004). In *Banksia brownii*, Phi conferred a potent control of *Phytophthora cinnamomi* in the early stages of infection, also providing protection to individuals that had avoided infection (Barrett et al. 2003). In *Banksia grandis*, *B. coccinea*, and *Eucalyptus marginata*, stem injection of Phi offered a practical control of this pathogen and rendered an effective protection of threatened native flora against this pathogen (Shearer et al. 2006). In addition, the foliar application of Phi delayed and reduced the rate of mortality of four *Banksia* species infected with *P. cinnamomi* as well (Shearer and Fairman 2007b; Shearer and Crane 2009). In species of the genus *Lambertia*, Shearer and Crane (2012) observed variations within genotypes regarding the efficacy of Phi spray applied to control this oomycete, and Eshraghi et al. (2014) demonstrated that Phi induces resistance to this pathogen in *Arabidopsis thaliana*, which has also been proved for *Banksia grandis* and *Eucalyptus marginata*

(Scott et al. 2015). In *Pinus* spp. and *Pseudotsuga menziesii*, Cerqueira et al. (2017) reported that Phi effectively inhibits *Fusarium circinatum* mycelial growth in a dose-dependent manner.

In maize, Panicker and Gangadharan (1999) found a significant reduction in disease incidence and severity caused by *Peronosclerospora sorghi*, while yield increased by up to 73%. In soybean, Silva et al. (2011) reported that Phi decreased the downy mildew damage caused by the fungus *Peronospora manshurica*, whereas Simonetti et al. (2015) observed for the first time the efficient control of *Macrophomina phaseolina* using combined treatment with plant growth-promoting rhizobacteria (PGPR) and Phi in soybean grown under greenhouse conditions. Hence, Phi may represent a good control to inhibit the spread of this pathogen in crop plants (Shafique et al. 2016), since this is a cosmopolitan fungus causing an important number of diseases, including charcoal rot, stalk rot, and dry root rot/stem canker, in more than 500 plant species, including vegetables, fruits, and potatoes (Khan 2007).

Some Phi-containing products found in the market can act efficiently as fungicides (i.e., potassium Phi), while others have been shown to be fungistatic (i.e., copper Phi and calcium Phi) (Lobato et al. 2010). Importantly, Phi has proved to be effective against *Phytophthora infestans*, *Fusarium solani*, and *Rhizoctonia solani*, which represent some of the most destructive pathogens that drastically reduce plant yields and quality.

Plant-parasitic nematodes are of great economic importance. In maize, the application of potassium phosphite was effective in the control of *Pratylenchus brachyurus* (Dias-Arieira et al. 2012), probably due to the capacity of the Phi to stimulate plant defense mechanisms leading to an enhanced synthesis of phytoalexins (Dercks and Creasy 1989). Moreover, manganese Phi was effective against *Meloidogyne javanica* in soybean, reducing the number of eggs per gram of root when applied 7 days before the inoculation of nematodes in pest-resistant cultivar MG/BR 46 Conquista (Puerari et al. 2013). Likewise, Oka et al. (2007) found that the application of potassium Phi in the shoots of wheat (*Triticum aestivum*) and oat (*Avena sativa*) plants effectively controlled *Heterodera avenae* and *Meloidogyne marylandi*, confirming the capability of Phi to stimulate phytoalexin synthesis in Phi-treated plants (Dercks and Creasy 1989). Because nematodes are very common in some vegetables and potatoes, Phi is an effective means to control such pathogens in agriculture.

Based on the aforementioned, Phi represents an efficient agricultural input to protect crops against pathogenic organisms by inducing vital defense mechanisms, thus acting as a plant resistance inducer (PRI) of paramount significance for novel plant protection approaches (Alexandersson et al. 2016). Defense responses triggered by Phi include the accumulation of phytoalexins, while lignification of the cell wall is also common. Importantly, hypersensitive cell death may also be induced by Phi, thus avoiding the proliferation of infected cells. Lytic enzymes produced by the plant in response to Phi may also contribute to pathogen control.

The effect of Phi on pathogen control depends on application time, cultivar evaluated, location, and disease incidence and severity (Cicore et al. 2012). Because plants can acquire Phi and translocate it to different organs via both xylem and phloem, this oxyanion is particularly flexible and can be applied to the plant in different ways, including foliar sprays, fertigation, trunk sprays, trunk injections, trunk paints, and in-furrow and soil drenches. Nevertheless, application of Phi to the soil surface must be practiced only when the area under the tree canopy is weed-free and application can be followed with adequate irrigation area coverage to move Phi into the root zone. If these conditions for application are not met, an extended residence time of the Phi in the soil may risk microbial conversion of Phi to Pi and loss the efficacy for pathogen control (Graham 2011). The application method would finally depend on the crop-pathogen interactions, but foliar applications are more common than the other methods (Kiirika et al. 2013; Deliopoulos et al. 2010; Alexandersson et al. 2016).

To date a number of mechanisms explaining how Phi functions in plants have been postulated. It appears that Phi triggers complex processes against pathogens, comprising both direct (inhibition of reproduction or decreases in the development rate) and indirect effects (immediate and robust induction of plant defense mechanisms) (Smillie et al. 1989; Grant et al. 1990; Guest and Bompeix 1990; Guest and Grant 1991; Jackson et al. 2000; Hardy et al. 2001; Brunings et al. 2005; Daniel and Guest 2005; Deliopoulos et al. 2010). Though it was believed that such mechanisms implicated in the prophylactic effects of Phi may have a limited effect on the development of pathogen resistance to Phi (Landschoot and Cook 2005), a naturally occurring *P. cinnamomi* strain resistant to Fosetyl-Al has been reported (Grant et al. 1990). Furthermore, Dobrowolski et al. (2008) observed selection for isolates of *P. cinnamomi* less sensitive to Phi from areas of prolonged Phi applications. Moreover, insensitivity to Phi was reported in isolates of the lettuce downy mildew pathogen *Bremia lactucae* (Brown et al. 2004). However, how Phi signals are primarily recognized within the plant and how signal perception and transduction mechanisms triggered by Phi are adjusted to prime defense responses still need to be precisely explored (Schothorst et al. 2013; Jost et al. 2015).

4 Biostimulant Effects of Phi in Response to Abiotic Stress Factors

Phosphite may also induce tolerance mechanisms against a number of abiotic stressors. In maize, the replacement of 1/4 of Pi by Phi stimulated guaiacol peroxidase activity and lignin biosynthesis (Ávila et al. 2011). However, Phi decreased the biomass production of the plants grown under low Pi supply, while no effect was observed in the plants grown under adequate Pi supply (Ávila et al. 2011). In Phi-pretreated potato leaves exposed to UV stress, Phi increased chlorophyll content and expression of the *psbA* gene, which encodes the PSII reaction center protein D1 (Oyarburo et al. 2015). In potato, Phi has also been proved to prevent oxidative stress driven by UV-B, thus mediating the UV-B stress tolerance. In plants exposed to abiotic stressors, Phi may also increase the abundance of proteins related to cell wall formation as a tolerance mechanisms. Therefore, Phi roles as an inductor of resistance or tolerance responses are not restricted to plant defense mechanisms against pathogens, but also responses to abiotic stress and primary metabolism have been proved to be altered in Phi-treated plants.

5 Biostimulant Effects of Phi on Yield and Quality of Crops

The biostimulant effects of Phi on crop plants have been well documented, resulting in improved production and yield quality (Gómez-Merino and Trejo-Téllez 2015, 2016).

One of the first reports summarizing the biostimulant effects of Phi was compiled by Rickard (2000). Phi was shown to increase the production and yield quality of potato, pepper, onion (*Allium cepa*), and celery (*Apium graveolens*). As well, soil or foliar applications of Phi improved the quality of peaches (*Prunus persica*) and oranges (*Citrus sinensis*) when applied as a sole P source. The results were attributed to a possible conversion of Phi to Pi by microorganisms in the soil or leaves, though this idea is subject to considerable debate. In fact, studies on black mustard (*Brassica nigra*) and rapeseed (*Brassica napus*) (Carswell et al. 1996, 1997), as well as pepper and tomato (Förster et al. 1998; Varadarajan et al. 2002), demonstrated that Phi cannot be used as a proper fertilizer since it does not provide phosphorus nutrition. However, Phi applied to the soil will not get in contact with the roots immediately, and hence it may be converted to Pi by some soil microbial communities. In contrast, Phi is taken up rapidly by leaves, especially if an adjuvant is used. If plants are not supplied with sufficient Pi, or Phi applications are excessive, Phi causes deleterious effects including reduced growth and toxicity. For instance, Bertsch et al. (2009) found that applying Pi plus Phi (50% as H_3PO_4 and 50% as H_3PO_3) in the nutrient solution to lettuce (*Lactuca sativa*), tomato, and banana (*Musa paradisiaca*) in hydroponics increased P content, foliar area, and biomass dry weight in the whole plant. When foliar treatments using 100% of P as Phi were applied to those crops, severe decrease in plant growth were observed, with visible foliage damage and root deterioration (Bertsch et al. 2009). According to Estrada-Ortiz et al. (2016), Phi differentially affects chard and lettuce metabolism in hydroponics; when applied in concentrations lower than 0.25 mM in sufficient Pi conditions, Phi may induce positive responses, including increased biomass production and nutrient contents.

In potatoes and tomatoes, Lovatt and Mikkelsen (2006) found that Phi can enhance floral intensity and increase yield and quality of final products. Such responses were attributed to the effect of Phi on sugar metabolism, changes in phytohormones and secondary metabolites, as well as induction of the shikimic acid pathway. This pathway is of paramount importance since the aromatic amino acids phenylalanine, tyrosine, and tryptophan derive from it. Apart from being essential components of protein synthesis, these amino acids also serve as precursors for a wide range of secondary metabolites, including pigments, alkaloids, hormones, and cell wall components, which are important for plant growth as well as for human health and nutrition (Maeda and Dudareva 2012; Tzin and Galili 2010).

Phi has also been demonstrated to enhance pectin content in both periderm and cortex tissue of potato tubers (Olivieri et al. 2012). Furthermore, Phi induced defense responses associated with a higher content and activity of polygalacturonase, chitinase, and proteinase inhibitor in Phi-treated potato plants. Increased contents of phytoalexin and chitinase, as well as improved peroxidase and polyphenol oxidase activities, have been found in tubers produced by potato plants treated with foliar applications of potassium Phi (Lobato et al. 2011). Likewise, reduced time to emergence after planting and enhanced leaf area and dry matter of potato plants were detected after potassium Phi treatments (Tambascio et al. 2014). In addition, seed tubers increased mycorrhizal colonization after Phi application, a phenomenon also reported in other plant species such as *Agonis flexuosa* (Howard et al. 2000).

The quality of fruit may also be improved by Phi applications in different horticultural crops. For instance, in P-deficient citrus (*Citrus sinensis*) seedlings, the foliar application of potassium Phi and calcium Pi induced similar responses, while Phi applications stimulated plant growth (Lovatt 1990). As well, orange trees receiving foliar applications of potassium Phi produced significantly more large-sized fruit with high commercial value, and their soluble solid (SS) contents and the ratio of SS to acids also increased (Lovatt 1998, 1999). Likewise, Valencia orange trees treated with foliar applications of Phi had an increased number of flowers, fruit set and yield, and total SS in fruits (Albrigo 1999). Furthermore, Phi moderately enhanced Pi absorption by mycorrhizas colonizing citrus roots and stimulated root colonization by the symbiotic fungi (Graham and Drouillard 1999; Graham 2011). In citrus and avocado (*Persea americana*) trees treated with foliar applications of Phi, Lovatt and Mikkelsen (2006) observed increased floral intensity, enhanced yield, and improved fruit quality in terms of total SS and anthocyanin concentrations (Lovatt and Mikkelsen 2006). Similarly, the foliar application of Phi in citrus trees increased yield while improving SS content, acidity, and yield of navel oranges (*C. sinensis*). Phi foliar sprays may also improve the quality of stone fruits (*Prunus* spp.). In peach trees, the application of Phi increases sugars and SS contents. Similarly, the application of Phi produced greater firmness in red raspberry (*Rubus idaeus*) (Rickard 2000).

Strawberry plants irrigated with Phi-containing solutions improved fruit quality by stimulating anthocyanins and ascorbic acid biosynthesis (Moor et al. 2009), which is consistent with the results reported by Estrada-Ortiz et al. (2013). Anthocyanins act as light attenuators with the potential to mitigate photooxidative injury in leaves, both by shielding chloroplasts from excess high-energy quanta and by scavenging reactive oxygen species (Ticconi et al. 2001; Neill and Gould 2003). The potent antioxidant properties of these metabolites are of great importance for plant physiology and human health (Lo Piero 2015). Furthermore, Phi may increase the concentrations of sugars, chlorophylls, and total free amino acids in leaves (Estrada-Ortiz et al. 2011), as well the content of total soluble sugars, degrees Brix, and fruit firmness (Estrada-Ortiz et al. 2012).

The synthesis of antioxidant enzymes and metabolites has been proved to be stimulated by Phi. For instance, in Pi-starved common bean plants (*Phaseolus vulgaris*), catalase activity was considerably higher when Phi was applied at low levels. Conversely, medium and high Phi levels substantially reduced the activity of this enzyme, whereas growth and grain yield were negatively affected (Ávila et al. 2013).

Phi is most effective when the rate and the application are properly timed to match the needs of the crop, which depend on the plant genotype (Lovatt and Mikkelsen 2006), phenological stages, and environmental conditions. Moreover, considering that Phi displays hormetic effects, its applications must be strictly supervised to avoid plant damage as a consequence of the toxicity it may cause (Gómez-Merino and Trejo-Téllez 2015, 2016).

6 Phi Uses in Dual Fertilization and Weed Control Systems

In nature, plants are not capable of using Phi as a proper phosphorus nutrient. However, recent advances in genetic engineering have developed a dual P fertilizer and herbicide system (López-Arredondo and Herrera-Estrella 2012; Manna et al. 2016; Achary et al. 2017). Such a system consists of transgenic plants harboring a *phosphite oxidoreductase* (*ptxD*) bacterial gene, which allows plants to utilize Phi as an alternative P source. Transgenic plants overexpressing the *ptxD* gene require up to 50% less Pi supply when provided with Phi to accomplish comparable production levels to those achieved by the same plants using Pi as a sole P source (López-Arredondo and Herrera-Estrella 2012). Furthermore, transgenic plants supplied with Phi and grown in competition with weeds accumulate two to ten times greater biomass than when fertilized with Pi. This innovative use of Phi (as a dual fertilizer and herbicide) could be of paramount importance for agricultural sustainability and food security, since this approach may decrease the excessive use of P fertilizers and minimize both eutrophication in water bodies and the development of herbicide resistance (López-Arredondo and Herrera-Estrella 2012; Manna et al. 2016). However, the novel use of Phi as fertilizer would imply a much larger input than its current use as a biostimulant or plant defense inductor against pathogens. Even if applied at 50% of the total P, its high mobility in the soil may cause accumulation in aquifers with generalized pollution, and potential negative impacts on different ecosystems may occur. Furthermore, this novel approach would require the use of genetically modified (GM) plants that cannot be cultivated in many countries. In general, public opinion is currently inclined against GM organisms. Consequently, the use of GM crops has a number of technical and political restrictions, which have to be considered when outlining programs aimed at using GM approaches related to the employment of Phi as an herbicide and fertilizer. Moreover, the effects of Phi on human health are unknown, and the European Union (EU) has lowered the maximum residual level (MRL) of Phi in agricultural products. Despite the fact that under experimental conditions this technology has proved to be effective in reducing not only phosphorus fertilizer use but also the growth of the tested weeds, field trials with a variety of soils and climates are required to validate its commercial implementation. So far, no commercial transgenic plant harboring a recombinant ptxD protein is available in the international market (Gómez-Merino and Trejo-Téllez 2016).

7 Phi Use as a Selectable Marker

López-Arredondo and Herrera-Estrella (2013) reported for the first time the development of a biotechnological system aimed at selecting transgenic plants under in vitro and greenhouse conditions based on Phi metabolism. Subsequently, Kanda et al. (2014) reported the application of a bacterial *ptxD* gene as a novel dominant selection marker for *Saccharomyces cerevisiae*, potentially applicable on an industrial scale. Recently, Nahampun et al. (2016) developed a system using Phi as an effective selectable marker for *Agrobacterium*-mediated plant transformation. Therefore, novel technologies for Phi application are under development, and new avenues for the usage of Phi are foreseen, which could be of great significance for the development of future agriculture based on biotechnologies (Gómez-Merino and Trejo-Téllez 2016).

8 Phosphite in the Market

The list of commercial Phi-containing products accessible in the global market comprises a considerable number of trademarks. These products are composed of alkali salts (i.e., Al^{3+}-, Ca^{2+}-, Cu^{2+}-, K^{+}-, Na^{+}-, NH_4^{+}-, and Mg^{2+}-Phi, among others) of phosphorous acid. Although Phi does not contribute to P nutrition in plants under natural conditions (i.e., non-GM crops), agrochemical companies still market Phi as a fertilizer, rather than as a pesticide. This is especially remunerative for those companies, as they avoid spending significant time and money on registering an agricultural pesticide (Gómez-Merino and Trejo-Téllez 2015, 2016). Nevertheless, in 2013, the EU changed the designation of Phi-containing compounds as both fertilizers and pesticides to only pesticides. This evolution is currently affecting the global trade of agri-foods that have been treated with Phi and definitely will impact the future use of Phi in agriculture all over the world. Importantly, on January 1, 2016, the EU MRL for Fosetyl-Al for several fruits and vegetables reverted back to the detection level set at 2 mg kg^{-1}, from 75 mg kg^{-1} set before 2015 (EU 2016; USDA 2016). Imports of fruits and other commodities that use Fosetyl-Al or other phosphonate inputs will likely be threatened by the return to the default MRL. Hence, farmers and growers using Phi to cultivate and export their agricultural products to the EU should analyze the EU's MRLs for Phi-containing products to ensure compliance (Gómez-Merino and Trejo-Téllez 2015, 2016).

Because of the recent widespread use of Phi in agriculture, several environmental and human health concerns have arisen. For instance, the excessive use of Phi as a pesticide can result in genetic resistance of pathogenic species. Indeed, Guest and Grant (1991) reported a naturally occurring Fosetyl-Al resistant isolate of *Phytophthora cinnamomi*. Furthermore, by using chemical mutagenesis, two Phi-resistant *Phytophthora* strains have been produced (Fenn and Coffey 1984). In addition, microorganisms that are capable of using Phi as a P source may be subject to a strong selective pressure due to the regular use of Phi in crops. In turn, a significant selective pressure against organisms incapable of utilizing Phi as a source of P may also occur. As a result, these changes could have adverse effects on the ecosystem as a whole. Since some results on the effect of Phi on symbiotic microbes (i.e., mycorrhizal fungi and nitrogen-fixing bacteria) are controversial (Despatie et al. 1989; Sukarno et al. 1993), further research into this area would be beneficial. It is well documented that Phi disrupts plant metabolism, especially under suboptimal P supply (Carswell et al. 1996, 1997; Förster et al. 1998; Varadarajan and Raghothama 2000). Hence, its use must be performed under strict control. Furthermore, one must also take into account regulatory issues aimed at guaranteeing not only the efficacy of the product but also its harmlessness to human or animal health when present at concentrations that may be found in food products or ecosystems. The effectiveness of Phi as a pesticide relies on its stability and high mobility within the plant, often ending up in edible tissues or products. Consequently, there is a dire need to record and analyze Phi concentrations found in foods produced by Phi-treated crops and to guarantee that long-term consumption of these products poses no hazard to the final consumers (McDonald et al. 2001; Gómez-Merino and Trejo-Téllez 2016). To determine Phi contents in agri-food products, different methodologies have been implemented, including ion chromatography (IC) (Smillie et al. 1988), nuclear magnetic resonance spectroscopy (NMR) (Niere et al. 1990), two-dimensional ion chromatography system coupled with capillary ion chromatography (2D-CIC) (Pech et al. 2009), and suppressed ion chromatography coupled with electrospray mass spectrometry (IC-ESI-MS) (Qui et al. 2013), among others.

9 Conclusions and Final Considerations

Current agricultural production systems are using Phi as a biostimulant in order to enhance nutrition efficiency, yield, crop quality, and abiotic stress tolerance. Moreover, Phi is also used as an efficient inductor of resistance or defense mechanisms against a number of pathogenic agents. Just recently, an innovative genetic engineering system for the effective development of Phi-mediated dual fertilization and weed control has been created, which is allowing the use of Phi as a potential fertilizer and herbicide. Nonetheless, in nature, plants lack the mechanisms to exploit Phi as a proper P fertilizer, and its use can cause deleterious effects to plant cells if its administration is not properly managed.

Since Phi displays hormetic effects, it may promote positive responses when applied at low dosages, though when applied at high levels, negative effects leading to cell damage or death can be observed. Hence, its application must be controlled and supervised to ensure better responses in non-biotech crops. In this chapter, we have provided evidence that Phi can be utilized as biostimulant, antibiotic, and plant resistance inducer. As a plant biostimulator, Phi may activate a number of molecular, biochemical, and physiological mechanisms leading to the induction of plant tolerance responses to abiotic stress factors and to the improvement of crop production and productivity parameters (Gómez-Merino and Trejo-Téllez 2016).

To ensure high efficiency in the use of Phi and avoidance of negative effects, the Pi status of the plant has to be first taken into consideration. Furthermore, the rate and dosage of Phi to be applied must be properly timed to fulfill crop requirements (Lovatt 2013), which in turn would depend on the genetic background of the crops, climate and soil conditions, cultural practices, chemical source, and concentrations of Phi to be applied (Gómez-Merino and Trejo-Téllez 2015, 2016).

With the avenue of modern tools provided by the omic sciences, it is now possible to explore how and to what extent Phi alters molecular processes that trigger defense responses in either wild or genetically engineered crop plants. With more than 270 plant genome sequencing projects either completed or currently under way (Gómez-Merino et al. 2015; http://www.ncbi.nlm.nih.gov/genome/browse/), better understanding of the molecular mechanisms of P use efficiency can be achieved. In biotech crops engineered to use Phi, this isostere of the Pi anion may allow the growth of crops in soils with low Pi availability while addressing the problems of P depletion and herbicide resistance. However, there are various issues to be considered regarding the widespread use of Phi in agriculture, including the development of Phi resistance in pathogens, the effect of Phi on soil microflora, and the possible threat to public health (Gómez-Merino and Trejo-Téllez 2016). Therefore, there is an obvious need to study and document all these phenomena in the near future.

In conclusion, Phi can stimulate positive effects in crop plants if it is properly combined with Pi. In conventional agronomic systems (i.e., non-biotech crops), Phi may efficiently serve as a biostimulator that enhances yield, quality, and plant performance under stress conditions. Postharvest quality of fruits may also be improved by Phi. Furthermore, it can be employed in the control of pathogenic organisms including bacteria, oomycetes, fungi, and nematodes. Thanks to its particular mobility throughout the whole plant, Phi can be applied in different ways, including not only foliar sprays and hydroponic systems but also via fertigation, trunk sprays, injections, paints, and in-furrow and soil drenches. Determining the right method of application for different agricultural species and cultivars remains an unmet challenge. Environmental impacts and new trends in international agri-food markets regarding the allowed residual levels of Phi must be taken into consideration and their compliance ensured.

References

Abbasi PA, Lazarovits G (2005) Effects of AG3 phosphonate formulations on incidence and severity of *Pythium* damping-off of cucumber seedlings under growth room, microplot, and field conditions. Can J Plant Pathol 27:420–429

Abbasi PA, Lazarovits G (2006) Effect of soil application of AG3 phosphonate on the severity of clubroot of bok choy and cabbage caused by *Plasmodiophora brassicae*. Plant Dis 90:1517–1522

Achary VMM, Ram B, Manna M, Datta D, Bhatt A, Reddy MK et al (2017) Phosphite: a novel P-fertilizer for weed management and pathogen control. Plant Biotechnol J 15:1493–1408. https://doi.org/10.1111/pbi.12803

Aćimović SG, Zeng Q, Mcghee GC, Sundin GW, Wise JC (2015) Control of fire blight (*Erwinia amylovora*) on apple trees with trunk-injected plant resistance inducers and antibiotics and assessment of induction of pathogenesis-related protein genes. Front Plant Sci 6:16. https://doi.org/10.3389/fpls.2015.00016

Adams F, Conrad JP (1953) Transition of phosphite to phosphate in soils. Soil Sci 75:361–371

Albrigo LG (1999) Effects of foliar applications of urea or Nutriphite on flowering and yields of Valencia orange trees. Proc Fla State Hort Soc 112:1–4

Alexandersson E, Mulugeta T, Lankinen A, Liljeroth E, Andreasson E (2016) Plant resistance inducers against pathogens in Solanaceae species-from molecular mechanisms to field application. Int J Mol Sci 17:1673. https://doi.org/10.3390/ijms17101673

Amiri A, Bompeix G (2011) Control of *Penicillium expansum* with potassium phosphite and heat treatment. Crop Prot 30:222–227. https://doi.org/10.1016/j.cropro.2010.10.010

Andreu AB, Guevara MG, Wolski EA (2006) Enhancement of natural disease resistance in potatoes by chemicals. Pest Manag Sci 62:162–170. https://doi.org/10.1002/ps.1142

Ávila FW, Faquin V, Araújo JL, Marques DJ, Lobato AKS, Ramos SJ et al (2011) Phosphite supply affects phosphorus nutrition and biochemical responses in maize plants. Austr J Crop Sci 5:646–653

Ávila FW, Faquin V, Lobato AKS, Ávila PA, Marques DJ, Guedes EMS et al (2013) Effect of phosphite supply in nutrient solution on yield, phosphorus nutrition and enzymatic behavior in common bean (*Phaseolus vulgaris* L.) plants. Austr J Crop Sci 7:713–722

Banko TJ, Hong CX (2004) Evaluation of nutrient phosphite for the control of *Phytophthora* shoot blight on annual vinca. J Environ Hortic 22:41–44

Barkla BJ, Vera-Estrella V, Pantoja O (2013) Progress and challenges for abiotic stress proteomics of crop plants. Proteomics 13:131801–131815. https://doi.org/10.1002/pmic.201200401

Barrett SR, Shearer BL, Hardy GESJ (2003) The efficacy of phosphite applied after inoculation on the colonisation of *Banksia brownii* stems by *Phytophthora cinnamomi*. Australas Plant Pathol 32:1–7

Berkowitz O, Jost R, Kollehn DO, Fenske R, Finnegan PM, O'Brien PA et al (2013) Acclimation responses of *Arabidopsis thaliana* to sustained phosphite treatments. J Exp Bot 64:1731–1743. https://doi.org/10.1093/jxb/ert037

Bertsch F, Ramírez F, Henríquez C (2009) Evaluación del fosfito como fuente fertilizante de fósforo vía radical y foliar. Agron Costarric 33:249–265

Brown S, Koike ST, Ochoa OE, Laemmlen F, Michelmore RW (2004) Insensitivity to the fungicide fosetyl-aluminum in California isolates of the lettuce downy mildew pathogen, *Bremia lactucae*. Plant Dis 88:502–508

Brunings AM, Datnoff LE, Simonne EH (2005) Phosphorous acid and phosphoric acid: when all P sources are not equal, Publication HS1010. University of Florida, Gainesville. http://edis.ifas.ufl.edu/HS254. Accessed 10 Sept 2017

Burra DD, Berkowitz O, Hedley PE, Morris J, Resjö S, Levander F, Liljeroth E, Andreasson E, Alexandersson E (2014) Phosphite-induced changes of the transcriptome and secretome in Solanum tuberosum leading to resistance against Phytophthora infestans. BMC Plant Biol 14:254. https://doi.org/10.1186/s12870-014-0254-y

Carswell MC, Grant BR, Theodorou ME, Harris J, Niere JO, Plaxton WC (1996) The fungicide phosphonate disrupts the phosphate starvation response in *Brassica nigra* seedlings. Plant Physiol 110:105–110. https://doi.org/10.1104/pp.110.1.105

Carswell MC, Grant BR, Plaxton WC (1997) Disruption of the phosphate-starvation response of oilseed rape suspension cells by the fungicide phosphonate. Planta 203:67–74. https://doi.org/10.1007/s00050166

Ceasar SA, Hodge A, Baker A, Baldwin SA (2014) Phosphate concentration and arbuscular mycorrhizal colonisation influence the growth, yield and expression of twelve *PHT1*family phosphate transporters in foxtail millet (*Setaria italica*). PLoS One 9:e108459. https://doi.org/10.1371/journal.pone.0108459

Cerqueira A, Alves A, Berenguer H, Correia B, Gómez-Cadenas A, Diez JJ, Monteiro P, Pinto G (2017) Phosphite shifts physiological and hormonal profile of Monterey pine and delays *Fusarium circinatum* progression. Plant Physiol Biochem 114:88–99. https://doi.org/10.1016/j.plaphy.2017.02.020

Chase AR (1993) Efficiency of fosetyl-Al for control of some bacterial diseases on ornamentals. Plant Dis 77:771–776

Cicore PL, Suarez PA, Andreu AB (2012) Phosphite effect on late blight control and physiological parameters in commercial potato (*Solanum tuberosum* L.) in Argentina. Pest Technol 6:27–31

Cooke LR, Little G (2002) The effect of foliar application of phosphonate formulations on the susceptibility of potato tubers to late blight. Pest Manag Sci 58:17–25. https://doi.org/10.1002/ps.408

Cordell D, Dranger JO, White S (2009) The story of phosphorus: global food security and food for thought. Glob Environ Chang 19:292–305. https://doi.org/10.1016/j.gloenvcha.2008.10.009

D'Arcy-Lameta A, Bompeix G (1991) Systemic transport of tritiated phosphonate in tomato plantlets (*Lycopersicon esculentum* mill). Pest Manag Sci 32:7–14. https://doi.org/10.1002/ps.2780320103

Dalio RJD, Fleischmann F, Humez M, Osswald W (2014) Phosphite protects *Fagus sylvatica* seedlings towards *Phytophthora plurivora* via local toxicity, priming and facilitation of pathogen recognition. PLoS One 9:e87860. https://doi.org/10.1371/journal.pone.0087860

Daniel R, Guest D (2005) Defence responses induced by potassium phosphonate in *Phytophthora palmivora*-challenged *Arabidopsis thaliana*. Physiol Mol Plant Pathol 67:194e201. https://doi.org/10.1016/j.pmpp.2006.01.003

Danova-Alt R, Dijkema C, Waard PDE, Köch M (2008) Transport and compartmentation of phosphite in higher plant cells. Kinetic and nuclear magnetic resonance studies. Plant Cell Environ 31:1510–1521. https://doi.org/10.1111/j.1365-3040.2008.01861.x

Dawson CJ, Hilton J (2011) Fertiliser availability in a resource-limited world: production and recycling of nitrogen and phosphorus. Food Policy 36:S14–S22

Deliopoulos T, Kettlewell PS, Hare MC (2010) Fungal disease suppression by inorganic salts: a review. Crop Prot 29:1059–1075. https://doi.org/10.1016/j.cropro.2010.05.011

Dercks W, Creasy LL (1989) Influence of fosetyl-Al on phytoalexin accumulation in the *Plasmopara viticola*-grapevine interaction. Physiol Mol Plant Pathol 34:203–213. https://doi.org/10.1016/0885-5765(89)90044-1

Despatie S, Furlan V, Fortin JA (1989) Effects of successive applications of fosetyl-Al on growth of *Allium cepa* L. associated with endomycorrhizal fungi. Plant Soil 113:175–180. https://doi.org/10.1007/BF02280178

Dias-Arieira CR, Marini PM, Fontana LF, Roldi M, Benetoli da Silva TR (2012) Effect of *Azospirillum brasilense*, Stimulate® and potassium phosphite to control *Pratylenchus brachyurus* in soybean and maize. Nematropica 42:170–175

Dias-Arieira CR, Santana-Gomes SM, Puerari HH, Fontana LF, Ribeiro LM, Mattei D (2013) Induced resistance in the nematodes control. Afr J Agric Res 8:2312–2318. https://doi.org/10.5897/AJARx12.012

Dobrowolski MP, Shearer BL, Colquhoun IJ, O'Brien PA, Hardy GESJ (2008) Selection for decreased sensitivity to phosphite in *Phytophthora cinnamomi* with prolonged use of fungicide. Plant Pathol 57:928–936. https://doi.org/10.1111/j.1365-3059.2008.01883.x

du Jardin P (2015) Plant biostimulants: definition, concept, main categories and regulation. Sci Hortic 196:3–14. https://doi.org/10.1016/j.scienta.2015.09.0210304-42

Eshraghi L, Anderson JP, Aryamanesh N, McComb JA, Shearer B, Hardy GESJ (2014) Suppression of the auxin response pathway enhances susceptibility to *Phytophthora cinnamomi* while phosphite-mediated resistance stimulates the auxin signalling pathway. BMC Plant Biol 14:68. https://doi.org/10.1186/1471-2229-14-68

Estrada-Ortiz E, Trejo-Téllez LI, Gómez-Merino FC, Núñez-Escobar R, Sandoval-Villa M (2011) Biochemical responses in strawberry plants supplying phosphorus in the form of phosphite. Rev Chapingo Ser Hort 17:129–138

Estrada-Ortiz E, Trejo-Téllez LI, Gómez-Merino FC, Núñez-Escobar R, Sandoval-Villa M (2012) Phosphite on growth and fruit quality in strawberry. Acta Hortic 947:277–282. https://doi.org/10.17660/ActaHortic.2012.947.35

Estrada-Ortiz E, Trejo-Téllez LI, Gómez-Merino FC, Núñez-Escobar R, Sandoval-Villa M (2013) The effects of phosphite on strawberry yield and fruit quality. J Soil Sci Plant Nutr 13:612–620. https://doi.org/10.4067/S0718-95162013005000049

Estrada-Ortiz E, Trejo-Téllez LI, Gómez-Merino FC, Silva-Rojas HV, Castillo-González AM, Avitia-García E (2016) Physiological responses of chard and lettuce to phosphite supply in nutrient solution. J Agric Sci Technol 18:1079–1090

EU (2016) Commission Regulation (EU) 2016/75 of 21 January 2016 amending. Annex III to Regulation (EC) No 396/2005 of the European Parliament and of the Council as regards maximum residue levels for fosetyl in or on certain products. Official Journal of the European Union 23.1.2016.http://faolex.fao.org/docs/pdf/eur151668.pdf. Accessed 8 Sept 2017

Fenn ME, Coffey MD (1984) Antifungal activity of fosetyl-Al and phosphorous acid. Phytopathology 74:606–611

Förster H, Adaskaveg JE, Kim DH, Stanghellini ME (1998) Effect of phosphite on tomato and pepper plants and on susceptibility of pepper to *Phytophthora* root and crown rot in hydroponic culture. Plant Dis 82:1165–1170. https://doi.org/10.1094/PDIS.1998.82.10.1165

Gómez-Merino FC, Trejo-Téllez LI (2015) Biostimulant activity of phosphite in horticulture. Sci Hortic 196:82–90. https://doi.org/10.1016/j.scienta.2015.09.035

Gómez-Merino FC, Trejo-Téllez LI (2016) Conventional and novel uses of phosphite in horticulture: potentialities and challenges. Italus Hortus 23:1–13

Gómez-Merino FC, Trejo-Téllez LI, Alarcón A (2015) Plant and microbe genomics and beyond: potential for developing a novel molecular plant nutrition approach. Acta Physiol Plant 37:208. https://doi.org/10.1007/s11738-015-1952-2

Graham JH (2011) Phosphite for control of *Phytophthora* diseases in citrus: model for management of *Phytophthora* species on forest trees? New Zeal J For Sci 41S:S49–S56

Graham JH, Drouillard DL (1999) Phosphite and phosphate uniquely affect root carbohydrate pools, root exudation and activity of citrus mycorrhizas. Second international symposium on the dynamics of physiological processes in Woody Roots, Nancy, France

Grant BR, Dunstan RH, Griffith JM, Niere JO, Smillie RH (1990) The mechanism of phosphonic (phosphorous) acid action in *Phytophthora*. Australas Plant Pathol 19:115–121

Guest DI, Bompeix G (1990) The complex mode of action of phosphonates. Australas Plant Pathol 19:113–115

Guest D, Grant BR (1991) The complex action of phosphonates as antifungal agents. Biol Rev 66:159–187. https://doi.org/10.1111/j.1469-185X.1991.tb01139.x

Hanrahan G, Salmassi TM, Khachikian CS, Foster KL (2005) Reduced inorganic phosphorus in the natural environment: significance, speciation and determination. Talanta 66:415–444. https://doi.org/10.1016/j.talanta.2004.10.004

Hardy GESJ, Barrett SR, Shearer BL (2001) The future of phosphite as a fungicide to control the soilborne plant pathogen *Phytophthora cinnamomi* in natural ecosystems. Australas Plant Pathol 30:133–139

Hofgaard IS, Ergon A, Henriksen B, Tronsmo AM (2010) The effect of potential resistance inducers on development of *Microdochium majus* and *Fusarium culmorum* in winter wheat. Eur J Plant Pathol 128:269–281

Howard K, Dell B, Hardy GESJ (2000) Phosphite and mycorrhizal formation in seedlings of three Australian Myrtaceae. Aust J Bot 48:725–729. https://doi.org/10.1071/BT00007

Jackson TJ, Burgess T, Colquhoun I, Hardy GESJ (2000) Action of the fungicide phosphite on *Eucalyptus marginata* inoculated with *Phytophthora cinnamomi*. Plant Pathol 49:147–154. https://doi.org/10.1046/j.1365-3059.2000.00422.x

Johnson SB (2010) Postharvest applications of phosphorous acid for control of *Phytophthora infestans* on potatoes. In: Proceedings of Twelfth EuroBlight Workshop, Arras, France, 3–6 May 2010

Johnson DA, Inglis DA, Miller JS (2004) Control of potato tuber rots caused by oomycetes with foliar applications of phosphorous acid. Plant Dis 88:1153–1159

Jost R, Pharmawati M, Lapis-Gaza HR, Rossig C, Berkowitz O, Lambers H et al (2015) Differentiating phosphate-dependent and phosphate-independent systemic phosphate-starvation response networks in *Arabidopsis thaliana* through the application of phosphite. J Exp Bot 66:2501–2514. https://doi.org/10.1093/jxb/erv025

Kanda K, Ishida T, Hirota R, Ono S, Motomura K, Ikeda T et al (2014) Application of a phosphite dehydrogenase gene as a novel dominant selection marker for yeasts. J Biotechnol 182–183:68–73. https://doi.org/10.1016/j.jbiotec.2014.04.012

Khan SN (2007) *Macrophomina phaseolina* as causal agent or charcoal rot of sunflower. Mycopathologia 5:111–118

Kiirika LM, Stahl F, Wydra K (2013) Phenotypic and molecular characterization of resistance induction by single and combined application of chitosan and silicon in tomato against *Ralstonia solanacearum*. Physiol Mol Plant Pathol 81:1–12. https://doi.org/10.1016/j.pmpp.2012.11.002

Kobayashi K, Masuda T, Takamiya K, Ohta H (2006) Membrane lipid alteration during phosphate starvation is regulated by phosphate signaling and auxin/cytokinin cross-talk. Plant J 47:238–248. https://doi.org/10.1111/j.1365-313X.2006.02778.x

Kromann P, Pérez WG, Taipe A, Schulte-Geldermann E, Sharma BP, Andrade-Piedra JL et al (2012) Use of phosphonate to manage foliar potato late blight in developing countries. Plant Dis 96:1008–1015. https://doi.org/10.1094/PDIS-12-11-1029-RE

Landschoot P, Cook J (2005) Understanding the phosphonate products. Department of Crop and Soil Sciences, The Pennsylvania State University, University Park. http://turfgrassmanagement.psu.edu/pdf/understanding_the_phosphonate_products.pdf. Accessed 2 Sept 2017

Lehmann J, Schroth G (2003) Nutrient leaching. In: Schroth G, Sinclair FL (eds) Trees, crops, and soil fertility. CAB International Publishing, Wallingford, pp 151–166

Liljeroth E, Lankinen Å, Wiik L (2016) Potassium phosphite combined with reduced doses of fungicides provides efficient protection against potato late blight in large-scale field trials. Crop Prot 86:42–55. https://doi.org/10.1016/j.cropro.2016.04.003

Lim S, Borza T, Peters RD, Coffin RH, Al-Mughrabi KI, Pinto DM et al (2013) Proteomics analysis suggests broad functional changes in potato leaves triggered by phosphites and a complex indirect mode of action against *Phytophthora infestans*. J Proteome 93:207–223. https://doi.org/10.1016/j.jprot.2013.03.010

Lo Piero AR (2015) The state of art on biosynthesis of anthocyanins and its regulation in pigmented sweet oranges [(*Citrus sinensis*) L. Osbeck]. J Agric Food Chem 63:4031–4041. https://doi.org/10.1021/acs.jafc.5b01123

Lobato MC, Olivieri FP, Gonzalez-Altamiranda E, Wolski EA, Daleo GR, Caldiz DO et al (2008) Phosphite compounds reduce disease severity in potato seed tubers and foliage. Eur J Plant Pathol 122:349–358

Lobato MC, Olivieri FP, Daleo GR, Andreu AB (2010) Antimicrobial activity of phosphites against different potato pathogens. J Plant Dis Prot 117:102–109

Lobato MC, Machinandiarena MF, Tambascio C, Dosio GAA, Caldiz DO, Daleo GR et al (2011) Effect of foliar applications of phosphite on post-harvest potato tubers. Eur J Plant Pathol 130:155–163. https://doi.org/10.1007/s10658-011-9741-2

Loera-Quezada MM, Leyva-González MA, López-Arredondo D, Herrera-Estrella L (2015) Phosphite cannot be used as a phosphorus source but is non-toxic for microalgae. Plant Sci 231:124–130. https://doi.org/10.1016/j.plantsci.2014.11.015

López-Arredondo DL, Herrera-Estrella L (2012) Engineering phosphorus metabolism in plants to produce a dual fertilization and weed control system. Nat Biotechnol 30:889–893. https://doi.org/10.1038/nbt.2346

López-Arredondo DL, Herrera-Estrella L (2013) A novel dominant selectable system for the selection of transgenic plants under in vitro and greenhouse conditions based on phosphite metabolism. Plant Biotechnol J 11:516–525. https://doi.org/10.1111/pbi.12063

López-Arredondo DL, Leyva-González MA, González-Morales SI, López-Bucio J, Herrera-Estrella L (2014) Phosphate nutrition: improving low-phosphate tolerance in crops. Annu Rev Plant Biol 65:95–123. https://doi.org/10.1146/annurev-arplant-050213-035949

Lovatt CJ (1990) Foliar phosphorus fertilization of citrus by foliar application of phosphite. In: Summary of citrus research. Citrus Research Advisory Committee, University of California, Riverside, pp 25–26

Lovatt CJ (1998) Managing yield with foliar fertilization. Calif Citrograph 84:8–13

Lovatt CJ (1999) Timing citrus and avocado foliar nutrient applications to increase fruit set and size. HortTechnology 9:607–612

Lovatt CJ (2013) Properly timing foliar-applied fertilizers increases efficacy: a review and update on timing foliar nutrient applications to citrus and avocado. HortTechnology 23:536–541

Lovatt CJ, Mikkelsen RL (2006) Phosphite fertilizers: what are they? Can you use them? What can they do? Better Crops 90:11–13

Machinandiarena MF, Lobato MC, Feldman ML, Daleo GR, Andreu AB (2012) Potassium phosphite primes defense responses in potato against *Phytophthora infestans*. J Plant Physiol 169:1417–1424. https://doi.org/10.1016/j.jplph.2012.05.005

Maeda H, Dudareva N (2012) The shikimate pathway and aromatic amino acid biosynthesis in plants. Annu Rev Plant Biol 63:73–105. https://doi.org/10.1146/annurev-arplant-042811-105439

Malboobi MA, Samaeian R, Sabet MS, Lohrasebi T (2012) Plant phosphate nutrition and environmental challenges. In: Dhal NK (ed) Plant science. InTech, Rijeka, pp 3–4. https://doi.org/10.5772/53424

Malboobi MA, Zamani K, Lohrasebi T, Sarikhani MR, Samaian A, Sabet MS (2014) Phosphate: the silent challenge. Progr Biol Sci 4:1–32

Manna M, Achary VMM, Islam T, Agrawa PK, Reddy MK (2016) The development of a phosphite mediated fertilization and weed control system for rice. Sci Rep 6:24941. https://doi.org/10.1038/srep24941

Martinoia E, Massonneau A, Frangne N (2000) Transport processes of solutes across the vacuolar membrane of higher plants. Plant Cell Physiol 41:1175–1186

Massoud K, Barchietto T, Le Rudulier T, Pallandre L, Didierlaurent L, Garmier M et al (2012) Dissecting phosphite-induced priming in *Arabidopsis* infected with *Hyaloperonospora arabidopsidis*. Plant Physiol 159:286–298. https://doi.org/10.1104/pp.112.194647

Mattson MP (2008) Hormesis defined. Ageing Res Rev 7:1–7. https://doi.org/10.1016/j.arr.2007.08.007

McDonald AE, Grant BR, Plaxton WC (2001) Phosphite (phosphorous acid): its relevance in the environment and agriculture and influence on plant phosphate starvation response. J Plant Nutr 24:1505–1519. https://doi.org/10.1081/PLN-100106017

Metcalf WW, van der Donk WA (2009) Biosynthesis of phosphonic and phosphinic acid natural products. Annu Rev Biochem 78:65–94. https://doi.org/10.1146/annurev.biochem.78.091707.100215

Miller JS, Olsen N, Woodell L, Porter LD, Clayson S (2006) Post-harvest applications of zoxamide and phosphite for control of potato tuber rots caused by oomycetes at harvest. Am J Potato Res 83:269–278. https://doi.org/10.1007/BF02872163

Monteiro ACA, de Resende MLV, Valente TCT, Ribeiro Junior PM, Pereira VF, da Costa JR et al (2016) Manganese phosphite in coffee defence against *Hemileia vastatrix*, the coffee rust fungus. Biochem Mol Analyses J Phytopathol 164:1043–1053. https://doi.org/10.1111/jph.12525

Moor U, Põldma P, Tõnutare T, Karp K, Starast M, Vool E (2009) Effect of phosphite fertilization on growth, yield and fruit composition of strawberries. Sci Hortic 119:264–269. https://doi.org/10.1016/j.scienta.2008.08.005

Nahampun HN, López-Arredondo D, Xu X, Herrera-Estrella L, Wang K (2016) Assessment of *ptxD* gene as an alternative selectable marker for *Agrobacterium*-mediated maize transformation. Plant Cell Rep 35:1121–1132. https://doi.org/10.1007/s00299-016-1942-x

Nartvaranant P, Hamill S, Leonardi J, Whiley AW, Subhadrabandhu S (2004) Seasonal effects of foliar application of phosphonate on phosphonate translocation: in vitro pollen viability and pollen germination in 'Hass' avocado (*Persea americana* mill.) J Hortic Sci Biotechnol 79:91–96. https://doi.org/10.1080/14620316.2004.11511741

Neill SO, Gould KS (2003) Anthocyanins in leaves: light attenuators or antioxidants? Funct Plant Biol 30:865–873

Niere JO, Grifth JM, Grant BR (1990) ^{31}P-NMR studies on the effect of phosphite on *Phytophthora palmivora*. J Gen Microbiol 136:147–156

Nussaume L, Kanno S, Javot H, Marin E, Pochon N, Ayadi A et al (2011) Phosphate import in plants: focus on the PHT1 transporters. Front Plant Sci 2:83. https://doi.org/10.3389/fpls.2011.00083

Oka Y, Tkachi N, Mor M (2007) Phosphite inhibits development of the nematode *Heterodera avenae* and *Meloidogyne marylandi* in cereals. Nematology 97:396–404. https://doi.org/10.1094/PHYTO-97-4-0396

Olivieri FP, Feldman ML, Machinandiarena MF, Lobato MC, Caldiz DO, Dalio GR et al (2012) Phosphite applications induce molecular modifications in potato tuber periderm and cortex that enhance resistance to pathogens. Crop Prot 32:1–6. https://doi.org/10.1016/j.cropro.2011.08.025

Orbovic V, Syvertsen JP, Bright D, van Clief DL, Graham JH (2008) Citrus seedling growth and susceptibility to root rot as affected by phosphite and phosphate. J Plant Nutr 31:774–787. https://doi.org/10.1080/01904160801928448

Ouimette DG, Coffey MD (1989) Phosphonate levels in avocado (*Persea americana*) seedlings and soil following treatment with fosetyl-Al or potassium phosphonate. Plant Dis 73:212–215. https://doi.org/10.1094/PD-73-0212

Ouimette DG, Coffey MD (1990) Symplastic entry and phloem translocation of phosphonate. Pestic Biochem Physiol 38:18–25. https://doi.org/10.1016/0048-3575(90)90143-P

Oyarburo NS, Machinandiarena MF, Feldman ML, Daleo GR, Andreu AB, Olivieri FP (2015) Potassium phosphite increases tolerance to UV-B in potato. Plant Physiol Biochem 88:1–8. https://doi.org/10.1016/j.plaphy.2015.01.003

Panicker S, Gangadharan K (1999) Controlling downy mildew of maize caused by *Peronosclerospora sorghi* by foliar sprays of phosphoric acid compounds. Crop Prot 18:115–118

Pech H, Henry A, Khachikian CS, Salmassi TM, Hanrahan G, Foster KL (2009) Detection of geothermal phosphite using high performance liquid chromatography. Environ Sci Technol 43:7671–7675

Percival GC, Banks JM (2014) Evaluation of plant defence activators for the potential control of *Pseudomonas syringae* pv. *aesculi*. Arboricultural J 36:76–88. https://doi.org/10.1080/03071375.2014.921396

Pilbeam RA, Colquhoun IJ, Shearer BL, Hardy GESJ (2000) Phosphite concentration: its effect on phytotoxicity symptoms and colonisation by *Phytophthora cinnamomi* in three understorey species of *Eucalyptus marginata* forest. Australas Plant Pathol 29:86–95

Plaxton WC, Tran HT (2011) Metabolic adaptations of phosphate-starved plants. Plant Physiol 156:1006–1015. https://doi.org/10.1104/pp.111.17528

Pratt J, Boisson AM, Gout E, Bligny R, Douce R, Aubert S (2009) Phosphate (Pi) starvation effect on the cytosolic Pi concentration and Pi exchanges across the tonoplast in plant cells. An in vivo ^{31}P-NMR study using methyl-phosphonate as a Pi analogue. Plant Physiol 151:1646–1657. https://doi.org/10.1104/pp.109.144626

Puerari HH, Dias-Arieira CR, Tavares-Silva CA, Arieira JO, Biela F, Poletine JP (2013) Ecolife® and manganese phosphite in the control of *Meloidogyne javanica* and the development of soybean cultivars susceptible and resistant to the nematode. Nematropica 43:105–112

Puerari HH, Dias-Arieira CR, Cardoso MR, Hernandes I, Brito ODC (2015) Resistance inducers in the control of root lesion nematodes in resistant and susceptible cultivars of maize. Phytoparasitica 43:383–389. https://doi.org/10.1007/s12600-014-0447-9

Qui HM, Geng JJ, HAN C, Ren HQ (2013) Determination of phosphite, phosphate, glyphosate and aminomethylphosphonic acid by two-dimensional ion chromatography system coupled with capillary ion chromatography. Chin J Anal Chem 41:1910–1914. https://doi.org/10.1016/S1872-2040(13)60700-8

Ratjen AM, Gerendas J (2009) A critical assessment of the suitability of phosphite as a source of phosphorus. J Plant Nutr Soil Sci 172:821–828. https://doi.org/10.1002/jpln.200800287

Rebollar-Alviter A, Madden LV, Ellis MA (2007) Pre- and post-infection activity of azoxystrobin, pyraclostrobin, mefenoxam, and phosphite against leather rot of strawberry, caused by *Phytophthora cactorum*. Plant Dis 91:559–564

Rejeb IB, Pastor V, Mauch-Mani B (2014) Plant responses to simultaneous biotic and abiotic stress: molecular mechanisms. Plants (Basel) 3:458–475. https://doi.org/10.3390/plants3040458

Reuveni M, Sheglov D, Cohen Y (2003) Control of moldy-core decay in apple fruits by β-aminobutyric acids and potassium phosphites. Plant Dis 87:933–936

Rickard DA (2000) Review of phosphorus acid and its salts as fertilizer materials. J Plant Nutr 23:161–180. https://doi.org/10.1080/01904160009382006

Schothorst J, Kankipati HN, Conrad M, Samyn DR, van Zeebroeck G, Popova Y et al (2013) Yeast nutrient transceptors provide novel insight in the functionality of membrane transporters. Curr Genet 59:197–206. https://doi.org/10.1007/s00294-013-0413-y

Schröder JJ, Cordell D, Smit AL, Rosemarin A (2010) Sustainable use of phosphorus. Plant Research International. Report No. 457. Wageningen, The Netherlands. 140 p. http://ec.europa.eu/environment/natres/pdf/sustainable_use_phosphorus.pdf. Accessed 8 Sept 2017

Scott PM, Barber PA, Hardy GESJ (2015) Novel phosphite and nutrient application to control *Phytophthora cinnamomi* disease. Australasian Plant Pathol 44:431–436. https://doi.org/10.1007/s13313-015-0365-4

Shafique HA, Viqar Sultana V, Ehteshamul-Haque S, Athar M (2016) Management of soil-borne diseases of organic vegetables. J Plant Prot Res 56:221–230. https://doi.org/10.1515/jppr-2016-0043

Shearer BL, Crane CE (2009) Influence of site and rate of low-volume aerial phosphite spray on lesion development of *Phytophthora cinnamomi* and phosphite persistence in *Lambertia inermis* var. inermis and *Banksia grandis*. Australas Plant Pathol 38:288–304

Shearer BL, Crane CE (2012) Variation within the genus *Lambertia* in efficacy of low-volume aerial phosphite spray for control of *Phytophthora cinnamomi*. Australas Plant Pathol 41:47–57

Shearer BL, Fairman RG (2007a) A stem injection of phosphite protects *Banksia* species and *Eucalyptus marginata* from *Phytophthora cinnamomi* for at least four years. Australasian Plant Pathol 36:78–86

Shearer BL, Fairman RG (2007b) Application of phosphite in a high-volume foliar spray delays and reduces the rate of mortality of four *Banksia* species infected with *Phytophthora cinnamomi*. Australas Plant Pathol 36:358–368

Shearer BL, Fairman RG, Grant MJ (2006) Effective concentration of phosphite in controlling *Phytophthora cinnamomi* following stem injection of *Banksia* species and *Eucalyptus marginata*. For Path 36:119–135

Silva OC, Santos HAA, Dalla Pria M, May-De Mio LL (2011) Potassium phosphite for control of downy mildew of soybean. Crop Prot 30:598–604. https://doi.org/10.1016/j.cropro.2011.02.015

Simonetti E, Viso NP, Montecchia M, Zilli C, Balestrasse K, Carmona M (2015) Evaluation of native bacteria and manganese phosphite for alternative control of charcoal root rot of soybean. Microbiol Res 180:40–48. https://doi.org/10.1016/j.micres.2015.07.004

Smillie RH, Grant BR, Cribbes RL (1988) Determination of phosphate and phosphite in plant material by gas chromatography-mass spectrometry and ion chromatography. J Chromat A 455:253–261. https://doi.org/10.1016/S0021-9673(01)82123-3

Smillie RH, Grant BR, Guest D (1989) The mode of action of phosphite. Evidence for both direct and indirect modes of action on three *Phytophthora* spp. in plants. Phytopathology 79:921–926

Speiser B, Berner A, Haseli A, Tamm L (2000) Control of downy mildew of grapevine with potassium phosphonate: effectivity and phosphonate residues in wine. Biol. Agric Hortic 17:305–312

Sukarno N, Smith SE, Scott E (1993) The effect of fungicides on vesicular-arbuscular mycorrhizal symbiosis. I. The effects on vesicular-arbuscular mycorrhizal fungi and plant growth. New Phytol 125:139–147. https://doi.org/10.1111/j.1469-8137.1993.tb03872.x

Sun T, Li M, Shao Y, Yu L, Ma F (2017) Comprehensive genomic identification and expression analysis of the phosphate transporter (PHT) gene family in apple. Front Plant Sci 8:426. https://doi.org/10.3389/fpls.2017.00426

Syers JK, Johnston AE, Curtin D (2008) Efficiency of soil and fertilizer phosphorus use. Reconciling changing concepts of soil phosphorus behavior with agronomic information, FAO Fertilizer and Plant Nutrition Bulletin No. 18. FAO, Rome, p 110

Syers JK, Bekunda M, Cordell D, Corman J, Johnston J, Rosemarin A et al (2011) UNEP year book 2011. P and Food Production. http://www.unep.org/yearbook/2011/pdfs/P_and_food_productioin.pdf. Accessed 10 Sept 2017

Tambascio C, Covacevich F, Lobato MC, de Lasa C, Caldiz D, Dosio G et al (2014) The application of K phosphites to seed tubers enhanced emergence, early growth and mycorrhizal colonization in potato (*Solanum tuberosum*). Am J Plant Sci 5:132–137. https://doi.org/10.4236/ajps.2014.51017

Taylor RJ, Pasche JS, Gudmestad NC (2011) Effect of application method and rate on residual efficacy of mefenoxam and phosphorous acid fungicides in the control of pink rot of potato. Plant Dis 95:997–1006. https://doi.org/10.1094/PDIS-09-10-0694

Teng W, Zhao YY, Zhao XQ, He X, Ma WY, Deng Y et al (2017) Genome-wide identification, characterization, and expression analysis of PHT1 phosphate transporters in wheat. Front Plant Sci 8:543. https://doi.org/10.3389/fpls.2017.00543

Thao HTB, Yamakawa T (2009) Phosphite (phosphorous acid): fungicide, fertilizer or bio-stimulator. Soil Sci Plant Nutr 55:228–243. https://doi.org/10.1111/j.1747-0765.2009.00365

Ticconi CA, Delatorre CA, Abel S (2001) Attenuation of phosphate starvation responses by phosphite in *Arabidopsis*. Plant Physiol 127:963–972. https://doi.org/10.1104/pp.010396

Tynan KM, Wilkinson CJ, Holmes JM, Dell B, Colquhoun IJ, McComb JA et al (2001) The long-term ability of phosphite to control *Phytophthora cinnamomi* in two native plant communities of Western Australia. Aust J Bot 46:761–770

Tzin V, Galili G (2010) The biosynthetic pathways for shikimate and aromatic amino acids in *Arabidopsis thaliana*. In: The arabidopsis book, vol 8. The American Society of Plant Biologists, Washington, DC, p e0132. https://doi.org/10.1199/tab.0132

Ullrich-Eberius CI, Novacky A, Fischer E, Lüttge U (1981) Relationship between energy-dependent phosphate uptake and the electrical membrane potential in *Lemna gibba* G1. Plant Physiol 67:797–801. https://doi.org/10.1104/pp.67.4.797

USDA (2016) New EU MRLfor Fosetyl takes effect. USDA Foreign Agricultural Service. Global Agricultural Information Network. http://www.usda-eu.org/reports/. Accessed 10 Sept 2017

Van Kauwenbergh SJ (2010) World phosphate rock reserves and resources. The International Fertilizer Development Center. Muscle Shoals, AL, USA. http://pdf.usaid.gov/pdf_docs/Pnadw835.PDF. Accessed 10 Sept 2017

Varadarajan D, Raghothama KG (2000) Phosphite (HPO_3^{2-}): a structural analog of phosphate ($H_2PO_4^-$) suppresses phosphate starvation induced molecular responses in tomato. In: Proceedings of annual meeting. American Society of Plant Physiologists 2000, San Diego, CA, 15–19 July 2000

Varadarajan DK, Karthikeyan AS, Matilda PD, Raghothama KG (2002) Phosphite, an analog of phosphate, suppresses the coordinated expression of genes under phosphate starvation. Plant Physiol 129:1232–1240. https://doi.org/10.1104/pp.010835

Wang DLS, Jiang P, Li Y (2017) Roles, regulation, and agricultural application of plant phosphate transporters. Front Plant Sci 8:817. https://doi.org/10.3389/fpls.2017.00817

Wilkinson CJ, Holmes JM, Tynan KM, Colquhoun IJ, McComb JA, Hardy GESJ et al (2001) Ability of phosphite applied in a glasshouse trial to control *Phytophthora cinnamomi* in five plant species native to Western Australia. Australas Plant Pathol 30:343–351. https://doi.org/10.1071/AP01055

Yogev E, Sadowsky A, Solel Z, Oren Y, Orbach Y (2006) The performance of potassium phosphite for controlling *Alternaria* brown spot of citrus fruit. J Plant Dis Prot 113:207–213. https://doi.org/10.1007/BF03356182

Nitric Oxide and Reactive Oxygen Species Interactions in Plant Tolerance and Adaptation to Stress Factors

Renata Bączek-Kwinta

Abstract The research on the regulatory role of reactive oxygen species (ROS) and reactive nitrogen species (RNS) in plants' life indisputably proved the involvement of these compounds in numerous life processes, including developmental and stress ones. Generation of both ROS and RNS occurs concomitantly, leading to some specific plant responses, and each group of compound interacts with the other one, which involves complexity and is sometimes difficult to understand and study. For this reason, the chapter will integrate the papers on biotic and abiotic stress response and provides an overview of the molecular mechanism of:

- ROS/RNS signalling
- The phenotypic response
- The perspective of use ROS and RNS in biotechnology and food production

Keywords Hydrogen peroxide · ROS · RNS · Phytohormones · Plans stress response · ROS and RNS signalling · Shelf life of fruits

1 Introduction

Nitric oxide (NO) and its derivatives, as well as reactive oxygen species (ROS), are molecules of differentiated half-life generated during various physiological processes in both plants and animals. In a surplus, they can be toxic, but as different cellular mechanisms partake in their control, they are involved in many metabolic pathways. In plants, their role in development and stress response has been studied for over 20 years. Recently, it is known that they act in concert, regulating signalling processes. Hence, their mode of action at the molecular level as well as some possibilities to transfer this knowledge into practice will be described in this chapter.

R. Bączek-Kwinta (✉)
Faculty of Agriculture and Economics, Department of Plant Physiology
University of Agriculture in Cracow, Cracow, Poland
e-mail: rrbaczek@cyf-kr.edu.pl

S. Vats (ed.), *Biotic and Abiotic Stress Tolerance in Plants*,
https://doi.org/10.1007/978-981-10-9029-5_9

2 Regulatory Roles of ROS

ROS, including superoxide anion O_2*^-, hydrogen peroxide H_2O_2, hydroxyl radical *OH and singlet oxygen $^1O_2$, are formed as by-products in many physiological processes of plants (Fig. 1). The main sources of O_2*^- as primary ROS are the electron transport chains in chloroplasts (Foyer and Shigeoka 2011) and mitochondria (Foyer and Noctor 2005; Vanlerberghe 2013), while the rich sources of H_2O_2 are the enzymes operating in peroxisomes and glyoxysomes (Corpas 2015). The Fenton reaction between O_2*^- and H_2O_2 in the presence of the ions of transition metals such as Zn, Fe or Cu leads to *OH formation. *OH is considered the most toxic ROS, but it is noteworthy to mention that it is involved in non-enzymic scission of polysaccharides and cell wall loosening (Schopfer et al. 2002; Vreeburg and Fry 2005). $^1O_2$ is usually linked with the chloroplasts and light phase of photosynthesis, where it is formed as a result of "normal" triplet oxygen excitation by the excited chlorophyll within the photosynthetic antennae (Asada 1994a, b). However, the studies of Mor et al. (2014) revealed that it can be emanated in the dark from other cellular compartments such as mitochondria, peroxisomes and the nucleus, which implies the new approach to the physiological role of this molecule.

ROS partake in cell wall expansion (Schopfer et al. 2002), stomatal closure (Pei et al. 2000; Zhang et al. 2001), embryogenesis (Żur et al. 2014) and other numerous processes at the tissue level. They stimulate both seed germination and ageing (Fath et al. 2001; Bailly 2004), senescence of different tissues and organs, as well as ripening and abscission of fruits (Leshem 1988; Bartoli et al. 1997; Yang et al. 2015). Their role in stress sensing, sensitivity and response has been described in many papers (Politycka 1996; Bączek-Kwinta et al. 2005; Vellosillo et al. 2010; Kreslavski et al. 2012; Mor et al. 2014). The interorganellar communication and the newest concept of "plant intelligence" also comprises the involvement of ROS (Karpinski and Szechynska-Hebda 2010; Kopczewski and Kuzniak 2013).

The regulatory role of ROS is possible as long as and they remain under control of antioxidants, because their excessive amounts in relation to the repairing processes result in irreversible damage. However, redox pathways and cycles, regulated by the number and concentrations of various antioxidants, give ROS the attitudes of

Fig. 1 Examples of O_2*^- visualization using histochemical method with nitroblue tetrazolium in germinating lupine seeds and in tomato leaflet. Violet and navy-blue areas reveal the sites of O_2*^- generation. No stress factor was involved (Photos by the author)

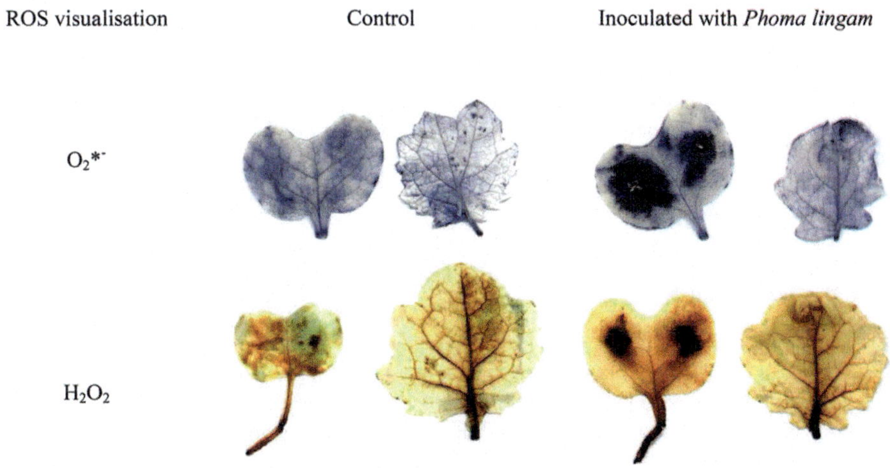

Fig. 2 $O_2{}^{*-}$ (navy-blue) and H_2O_2 (bronze areas) accumulation in cotyledons and leaves of winter rape at 48 h after inoculation with spores of the fungal pathogen *Phoma lingam* (According to Hura et al. 2014)

secondary messengers and metabolic regulators. On the molecular level, the physiologically active ROS is H_2O_2, inducing gene expression and modulating signalling proteins, such as kinases, phosphatases, calcium channels and transcription factors (Kreslavski et al. 2012; Kopczewski and Kuzniak 2013). In some cases, the excess of ROS is eligible. The so-called oxidative burst, namely, the rapid generation of O_2*, and H_2O_2 in consequence, occurs in the apoplast as a common response to pathogens, wounding, but also heat, UV and ozone (Wang et al. 2013). ROS may kill plant pathogens during the hypersensitive response (HR) of a host plant, strengthen the cell wall, but also act as signals in different forms of immunity (Vellosillo et al. 2010; Hura et al. 2014). An example of elevated $O_2{}^{*-}$ and H_2O_2 levels as a result of infection is given in Fig. 2.

3 Generation and Key Reactions of RNS and Their Regulatory Role in Plants

Nitric oxide was of interest to biologists for many years due to its involvement in air pollution, being a constituent of acid rain, and one of the factors depleting the ozone layer as well (Driscoll 1997; Portmann et al. 2012). Moreover, it was identified as an important signal in the human vascular system functioning, and then its role in plant defence responses against bacterial pathogens was established (Ignarro et al. 1987; Palmer et al. 1987; Noritake et al. 1996). In addition, since the enzymes generating NO were identified in mammals, it became obvious that this compound and its

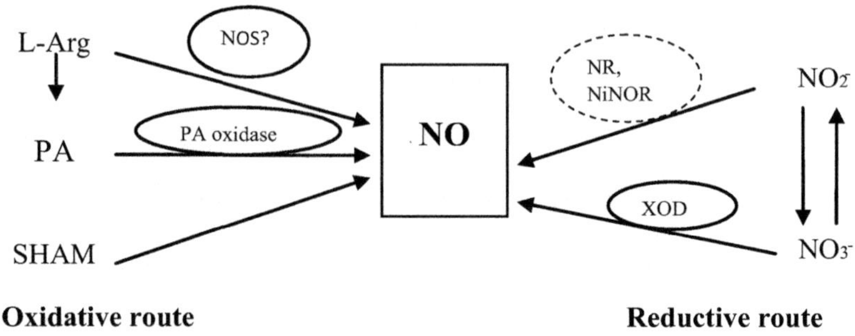

Fig. 3 NO synthesis routes in plants. Metabolites: *L-Arg* L-arginine, *PA* polyamines, *SHAM* salicylhydroxamic acid, NO_2^- nitrite, NO_3^- nitrate, enzymes: *NOS* NO synthase, *NR* nitrate reductase, *NiNOR* nitrite:NO reductase, *XOD* xanthine dehydrogenase. The reduction of NO_2^- to NO may occur either enzymatically or nonenzymatically (Based on Corpas et al. 2011; Bellin et al. 2013 and other references given in the text)

derivatives must play the physiological role in various organisms (Alderton et al. 2001).

The half-life of NO is 6 s, and the reactivity of the molecule is high. Upon the specific chemical circumstances, it exists in three forms: radical NO*, anionic NO^- and cationic NO^+ (Wendehenne et al. 2001). In the presence of oxygen, it creates dioxide NO_2*, whereas in water solutions is subjected to oxidation, forming nitrite NO_2, while the reaction with the superoxide leads to the formation of biologically active peroxynitrite ($ONOO^-$). NO reacts also with the hydroxyl radical producing HNO_2 (Tuteja et al. 2004).

Generation of NO occurs via two pathways, the oxidative and reductive ones, and depends on the oxygen availability and pH (Corpas et al. 2011; Gupta et al. 2011) (Fig. 3). The presence of NO synthase (NOS) in plants has not been proved, although some experiments suggested that its existence is possible (Corpas et al. 2011). However, the research of Jeandroz et al. (2016) on over 1000 species of plants and algae suggests that land plants, instead of generating NO with evolutionarily conserved NOS enzymes, have evolved finely regulated nitrate assimilation and reduction processes to synthesize NO through a mechanism different than that in animals. Hence, one potential source of NO in the oxidative route may be polyamines (PA), but this mechanism is unknown, and another one is salicylhydroxamate (SHAM) (Rümer et al. 2009). In the reductive pathway, NO can be formed from the nitrite, either enzymatically by nitrate reductase or nitrite:NO reductase (EC 1.7.99.4 and EC 1.7.2.1, respectively) or nonenzymatically or from the nitrate by the xanthine oxidoreductase (Moreau et al. 2010). The last enzyme is well-known and even used in some biochemical tests as a generator of O_2^{*-}. It can also create NO, but the direction of the reaction depends on what form of the enzyme prevails, the oxidase (XO, EC 1.1.3.22) or, under anoxic conditions, dehydrogenase (XOD, EC 1.1.1.204) (Wang et al. 2010; Gupta et al. 2011).

Irrespectively from such controversies, the involvement of NO in plant life was experimentally demonstrated 20 years ago, and its partake in the key plant life

Table 1 Literature examples on the role of protein S-nitrosylation in plant response to stress, developmental processes and hormonal regulation

Stress/process	Plant species	Literature
Desiccation tolerance	*Antiaris toxicaria*	Bai et al. (2011)
Thermotolerance	*Arabidopsis thaliana*	Lindermayr et al. (2005)
Photomorphogenesis		Lee et al. (2008)
Auxin signalling		Lozano-Juste and Leon (2011)
Regulation of metacaspase activity		Terrile et al. (2012)
		Belenghi et al. (2007)
Floral transition		He et al. (2004)
Rubisco activity	*Brassica juncea*	Abat and Deswal (2009)
Rubisco activity	*Kalanchoe pinnata*	Abat et al. (2008)
Leaf cell death	*Oryza sativa*	Lin et al. (2012)
Salinity	*Pisum sativum*	Camejo et al. (2013)
Biotic stress	*Pelargonium peltatum*	Floryszak-Wieczorek et al. (2007)
Seed germination, de-etiolation, hypocotyl elongation	*Lactuca sativa*	Beligni and Lamattina (2000)
Programmed cell death of seed aleurone cells	*Hordeum sativum*	Beligni et al. (2002)
Adventitious root formation induced by indole-3-acetic acid (IAA)	*Cucumis sativus*	Pagnussat et al. (2003)
Root nodule signalling	*Medicago truncatula*	Boscari et al. (2013), Silva and Carvalho (2013)

processes and stress responses has been gradually revealed (Noritake et al. 1996; Delledonne et al. 1998; Durner et al. 1998). An overview of some examples is given in Table 1. As it can be noticed, the list begins with germination and growth/developmental processes, through plant movement, stomatal opening, blossoming, ending with fruit ripening and senescence (Beligni and Lamattina 2000; Hu et al. 2005; García-Mata and Lamattina 2003; He et al. 2004; Manjunatha et al. 2010; Leshem et al. 1998). It is worth recalling that these processes are regulated by ROS, and ROS/RNS interplay will be discussed later.

Similarly to ROS, the most important sources of NO are mitochondria, chloroplasts and peroxisomes (Galatro et al. 2013). To plants, the most important processes involving NO are nitrosylation and nitration. Nitrosylation of proteins is triggered by the specific form of NO, peroxynitrite $ONOO^-$, and, similarly to phosphorylation, is considered the factor regulating of many plant proteins, because it is fast and reversible (Malik et al. 2011; Lamotte et al. 2015). Cysteine (Cys) residues are prone to nitrosylation, and it is noteworthy that the 90% of the plant proteome consists such residues. It must be kept in mind, however, that nitrosylation occurs only when in the vicinity of the cysteine acid amino acids and some specific motifs occur (Hess et al. 2005). The signalling mode of NO action involves also protein modification by binding to critical Fe-S centres and heme groups (Besson-Bard et al. 2008).

The tripeptide glutathione (γ-Glu-Cys-Gly, GSH) is well-known as an antioxidant and one of the redox regulators. As can be predicted, it also subjects to nitrosylation due to the presence of Cys in its amino acid chain (Malik et al. 2011; Lamotte et al. 2015). Both redox regulation and NO transport and mode of action play an important role in plant stress. Interestingly, NO can be transported in form of S-nitrosothiols (SNOs), among which the nitrosoglutathione (GSNO) is important (Corpas et al. 2013). GSNO catabolism is mediated by the ubiquitous GSNO reductase proteins (GSNOR), upregulating 99 and downregulating 170 genes, which was demonstrated on *Arabidopsis*. Thirty percent of these genes are involved in biotic stress; some of them are associated with developmental processes (stem and trichome branching) and the others with iron metabolism; hence they are the key genes for the whole plant functioning (Xu et al. 2013).

Some examples of the study on the involvement of protein S-nitrosylation in signal transduction, intracellular homeostasis and stress response are given in Table 1. Nitrosylation plays also an important role in the regulation of metacaspases, the proteolytic enzymes taking part in the programmed cell death (PCD), which was revealed in the case of *Arabidopsis thaliana* metacaspase 9 (AtMC9, Belenhgi et al. 2007) (see also Sect. 4).

In contrast, nitration of the amino acid tyrosine, a type of a post-translational modification, is often considered the pathological process (Saito et al. 2006; Valderrama et al. 2007; Romero-Puertas et al. 2008; Leterrier et al. 2012; Chaki et al. 2013), but its regulatory role also has been described (Besson-Bard et al. 2008; Silva and Carvalho 2013). For example, in root nodules of *Medicago truncatula*, one of the key nitrogen assimilatory enzyme glutamine synthetase (GS) is post-translationally regulated by tyrosine nitration. Probably the inactivation of GS by NO is connected to the inhibition of the key microbial enzyme necessary for N_2 fixation, nitrogenase, and is related to metabolite channelling to boost the nodule antioxidant defences (Silva and Carvalho 2013).

4 ROS/RNS Interactions and Their Physiological Roles

As it was mentioned, ROS and RNS are often generated concomitantly, and the dynamic balance between them leads to the specific reactions, beginning with the molecular level, ending with the phenotypic response. First, as it was already mentioned, NO and $O_2{}^{*-}$ can be generated by the same enzyme, xanthine oxidoreductase (Wang et al. 2010; Gupta et al. 2011). Second, the salicylhydroxamate (SHAM), the substrate for NO synthesis, suppresses the mitochondrial source of $O_2{}^{*-}$, alternative oxidase (Rümer et al. 2009). Third, NO can scavenge ROS, but its amount is under control of a specific antioxidant glutathione (GSH, mentioned earlier), and NO triggers plant response either similar to this caused by ROS or the opposite one. Moreover, both ROS and RNS are involved in hormonal metabolism, which makes the molecular network more complicated than it might seem on the basis of the role of a ROS or RNS acting individually. Hence, the results of many studies on ROS or RNS "themselves" carry the load of the second group unless we use a specific inhibitor of ROS or RNS.

4.1 NO Scavenges ROS and Stimulates Antioxidative Enzymes in Abiotic Stresses

NO reacts with $O_2{}^{*-}$ directly and indirectly, by increased superoxide dismutase (SOD) activity (Chen et al. 2015). It also diminishes the amount of H_2O_2 via the increment of catalase (CAT), non-specific peroxidase (POX) and ascorbate peroxidase (APX), which was proved for different kinds of plant stress. For example, the use of NO in a form of sodium nitroprusside (SNP) in maize diminishes the oxidative stress triggered by the iron deficiency. The mechanism involves a direct interplay with ROS or by increasing activities of H_2O_2-scavenging enzymes such as CAT, POX and APX (Sun et al. 2007). The protective effects of SNP added into controlled-release fertilizer or sprayed on leaves were also obtained on peanut growing on calcareous soil, which implied iron deficiency stress, too (Zhang et al. 2012).

SNP partially alleviated UV-B-induced impairment in photosynthesis and the oxidative damage to the thylakoid membrane by increased activities of SODs, APXs and CATs and a decrease in H_2O_2 content as a result. To test whether there was an effect of NO, a specific NO scavenger (potassium salt of 2-(4-carboxyphenyl)-4,4,5,5-tetramethylimidazoline-1-oxyl-3-oxide) was used. This arrested the protective effects, providing the protective role of NO (Shi et al. 2005).

Similar protective effects were obtained for cucumber roots in oxidative stress induced by salt, for soybean in aluminium stress as well as for wheat seedlings and water plant *Pistia stratiotes* in arsenic stress (Cai et al. 2011; Hasanuzzaman and Fujita 2013; Farnese et al. 2017). In the case of mercury toxicity, it was found that the activities of antioxidant enzymes were not enhanced by SNP, but $O_2{}^{*-}$ was scavenged directly (Chen et al. 2015). In the same paper, it was established that both absorption and translocation of mercury in rice roots can be reduced directly, and it was obtained also in the experiment on *Brassica napus* treated with nickel (Kazemi et al. 2010).

NO can positively affect the immobilization of heavy metals in the root tissue by increased pectin and hemicelluloses content root cell walls; hence the aerial parts are protected from cadmium (Xiong et al. 2009). It is known that some metals, e.g. zinc, can enhance oxidative stress via the Fenton reaction (Eq. 1), whereas cadmium is not involved in such catalysis. On the other hand, some plant species can tolerate and even accumulate large amounts of metals (Shah et al. 2010; Solanki and Dhankhar 2011; Kusznierewicz et al. 2012). Hence, it could be interesting to elucidate whether the type of NO-mediated response depends on the metal and its concentration, plant species or their interaction. Some examples can be found in Sahay and Gupta (2017) and references therein.

$$H_2O_2 + Me^{+n} \rightarrow Me^{+n+1} + OH + HO^{\bullet} \quad (1)$$

Equation 1: Fenton reaction generating hydroxyl radical $HO^{\bullet}$. Me – a metal ion, usually of Fe, Cu or Zn

4.2 What Is the Role of ROS/RNS in Cadmium Toxicity?

NO donors might alleviate cadmium toxicity by direct ROS scavenging or antioxidant enzymes activation (Kopyra and Gwóźdź 2003). On the other hand, S-nitrosylation of metal-binding proteins called phytochelatins trigger Cd toxicity, and it can be manifested as programmed cell death (PCD) induction (Arasimowicz-Jelonek et al. 2011). An explanation of such conflicting data may be a result of other interactions, probably with phytohormones (Xu et al. 2010), but this example is only a top of the iceberg of the issue.

4.3 NO Protects Plants from ROS-Generating Herbicide

Another example of ROS/RNS interaction is the herbicide resistance. *Arabidopsis* plants treated with NO donors, SNP and GSNO (S-nitrosoglutathione), displayed resistance to paraquat (1,1′-dimethyl-4,4′-bipyridinium dichloride). The positive impact of NO was proven using *Arabidopsis* mutant *paraquat resistant2-1* (*par2-1*) (Chen et al. 2009).

4.4 The Role of GSH in NO Storage and Release: Nitrosoglutathione

In the last case described, the role of GSNO was raised, which is an important issue in the dissemination of the NO-metabolizing pathways. GSNO is formed by the covalent addition of a NO molecule to a cysteine thiol. It was already mentioned in the Sect. 2 that the donor for the cysteine is often glutathione (GSH), the tripeptide which is extremely important for the antioxidant purposes (Foyer et al. 1997). However, while nitrosylated, it acts as the major mobile reservoir of biologically active NO affecting the equilibrium between GSNO and *S*-nitrosylated proteins. Protein nitrosylation is reversible, and thus the protein can be activated or deactivated this way (Belenhgi et al. 2007; Kato et al. 2013).

Nitrosothiols are considered the key factors in plant defence against pathogens and regulate various developmental processes (Feechan et al. 2005; Leterrier et al. 2012; Bellin et al. 2013; Xu et al. 2013), so it is obvious that the issue is of interests of many researchers. However, according to Leterrier et al. (2012), a reliable method is necessary to detect and quantify the GSNO in plant tissues, allowing to establish the direct relationship between the substrate and enzyme. The precise subcellular location is also necessary, as well as the regulatory mechanism.

4.5 ROS/RNS in Biotic Stress and Senescence

The common denominator of biotic stress and senescence is the programmed cell death (PCD) resembling animal apoptosis. PCD occurs in the case of so-called hypersensitive response (HR), characterized by the cell death in the sites surrounding infected cells. It comprises membrane dysfunction, formation of lytic vacuoles, chromatin condensation and enzymatic DNA cleavage (Van Doorn 2011). In the case of both pathogen recognition and senescence programme, both NO and ROS are generated.

The cross-talk of ROS and RNS in PCD is depicted in Fig. 4. O_2^{*-} is formed as a result of a specific plant NADPH oxidases called RBOH (respiratory burst oxidase homologues) activity. Then it is scavenged by SOD producing H_2O_2. The H_2O_2 pool is under control of NO via the catalases (CAT) and peroxiredoxins (PRX). The mechanism involves signal transduction via mitogen-activated protein kinases (MAPKs), phosphatases and cyclic guanosine monophosphate (cGMP) by NO. However, NO suppresses senescence-associated genes (SAGs), which are, in turn, activated by H_2O_2. The cross-talk of NO and H_2O_2, even more complex, as seen in Fig. 4, leads to leaf cell death in the site of pathogen invasion (Delledonne et al. 2001; Wang et al. 2013).

4.6 ROS, RNS and Phytohormones: Practical Implications

As it was mentioned earlier, the interplay of both groups of molecules with specific hormones is also important. NO attenuates auxin transport and/or prevents oxidative stress, which was proved, e.g. for rice seedlings treated with mercury (Chen et al. 2015). Triiodobenzoic acid (TIBA) auxin, added to the plant medium, caused root growth, which was attenuated when SNP was used, but stunted while TIBA inhibitor was implemented. On the other hand, in the same experiment, NO scavenged O_2^{*-}, because the level of SOD activity was not influenced (Chen et al. 2015). This individual case illustrates the complexity of the interplay between hormones and redox system, which was revealed in numerous papers (Barna et al. 2012; Bartoli et al. 2013; Zhou et al. 2014; Khan et al. 2015; Herrera-Vásquez et al. 2015). If there

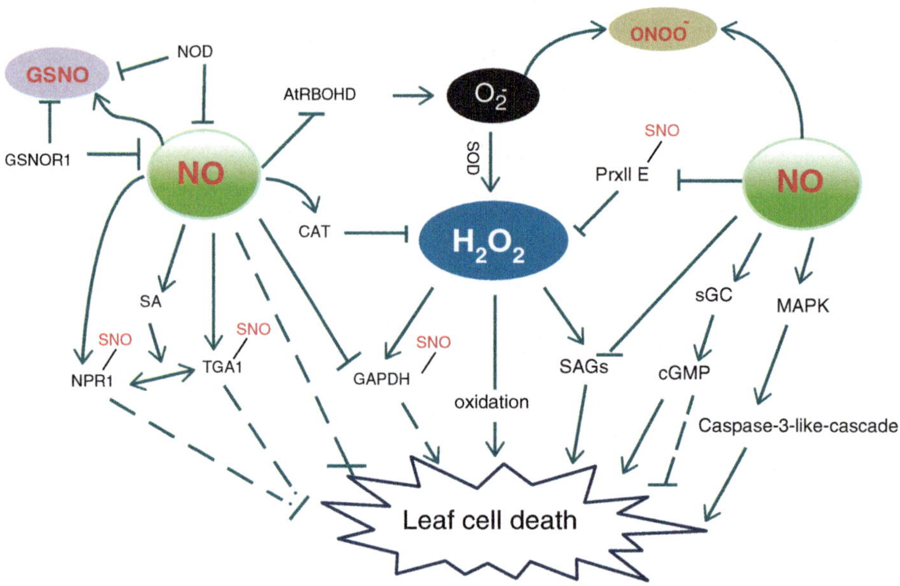

Fig. 4 Cross-talk of RNS and ROS in leaf cell death. *AtRBOHD* NADPH oxidase, *GAPDH* glyceraldehyde3-phosphatedehydrogenase, *GSNO* *S*-nitrosoglutathione, *GSNOR1* *S*-nitrosoglutathionereductase1, *NPR1* non-expression of pathogenesis-related protein1, *TGA1* TGACG motif-binding factor1, *NR* nitrate reductase, *SAGs* senescence-associated genes, *ONOO*$^-$ peroxynitrite, *PrxIIE* peroxiredoxin IIE, *NOD* NO-degrading dioxygenase, *sGC* soluble guanylate cyclase, *MAPK* mitogen-activated protein kinase, *SOD* superoxide dismutase, *CAT* catalase, *cGMP* cyclic guanosine monophosphate, *sGC* soluble guanylate cyclase, *SNO* S-nitrosothiol (Adapted from Wang et al. 2013)

is such mutual influence between redox buffers and NO, some other interactions between the specific hormones and growth factors must be involved, too.

Such complex and well-known example is the interaction of different molecules involved in stomatal closure (Fig. 5). For many years it has been known that abscisic acid (ABA) phytohormone is the factor responsible for the process, but the precise mode of its action has been gradually elucidated since 2005, and now it is certain that ABA needs H_2O_2 for this (Desikan et al. 2005). However, H_2O_2 generation is preceded by $O_2{}^{*-}$ formation by "NADPH oxidases," and both ABA and ethylene are involved, too (Desikan et al. 2006; Grefen et al. 2008). Moreover, ABA can also induce NO synthesis (Bright et al. 2006) and interacts with jasmonic acid in order to stimulate stomatal closure (Daszkowska-Golec and Szarejko 2013). In addition, as small molecules such as ascorbic acid are often responsible for the redox balance of a cell compartment, alterations in reduced and oxidized ascorbate forms may influence stomatal closure. To make things more complicated, ascorbate accumulation is dependent on ethylene, which was exemplified on the transgenic plants (Chen and Gallie 2004).

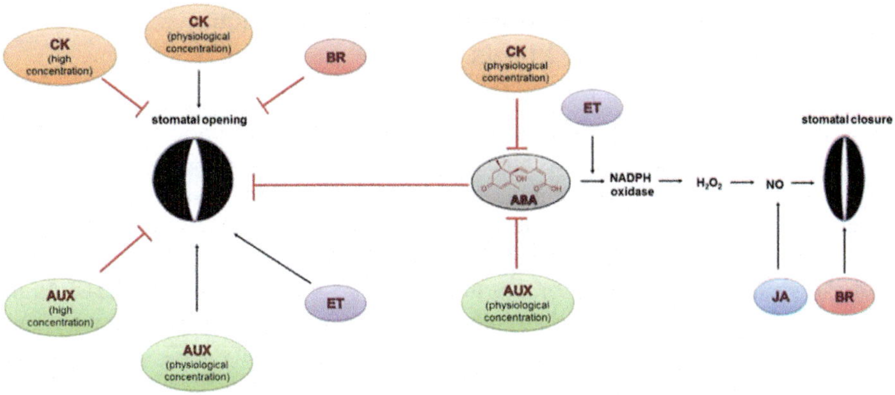

Fig. 5 Hormonal cross-talk in the regulation of stomatal closure and opening during water stress. *ABA* abscisic acid, *AUX* auxins, *BR* brassinosteroids, *CK* cytokinins, *JA* jasmonates, *ET* ethylene. (Adapted from Daszkowska-Golec and Szarejko 2013)

An interesting example of RNS/ROS/hormones interaction, and of practical use, is the impact of NO on ripening control. Some so-called climacteric fruits are prone to ethylene, which stimulates respiration and senescence. The process is natural, but for the marketers, it is very unfavourable, because it shortens the shelf-life of fruit crops and reduces the profit; hence various methods have been developed to delay either ethylene emission or the susceptibility of fruits to ethylene (Brody et al. 2001; Farneti et al. 2015). The first attempt of applying NO in its gaseous form revealed the usefulness of the method for kiwi fruits and strawberry (Leshem et al. 1998), and then it was confirmed by the teams of Zhu and Zhou (2007) and Zhu et al. (2008) and finally applied to mango stored at 5 °C, which resulted in the delay of fruit ripening and diminished chilling injury (Zaharah and Singh 2011). Similarly, the use of NO donors, diethylenetriamine/nitric oxide adduct (DETA/NO) and SNP extended the shelf-life of plums (Zhang et al. 2007), as well as apples and banana products (Pristijono et al. 2008; Huque et al. 2013; Cheng et al. 2009). Apart from fruits, the senescence of some vegetables, namely, broccoli, green bean and bok choy (a type of Chinese cabbage), was also delayed by DETA/NO (Soegiarto and Wills 2004). The same refers to flowers, for example, these of *Dianthus caryophyllus* (carnations), very popular in the Anglo-Saxon countries (Bowyer et al. 2003), but also other species belonging to different botanical taxa (Badiyan et al. 2004). Other examples were described in a review by Manjunatha et al. (2010). However, the proper formulation of NO is important due to the high reactivity of the molecule; hence its widespread commercial use is not considered yet.

5 New Prospectives: Nitro-Fatty Acids and Their Physiological Role in Plants

Nitric oxide binds to unsaturated fatty acids, forming nitro-fatty acids (NO_2-FAs). As they participate in cardiovascular regulation and diseases as well as in the inflammatory processes of endothelium, their physiological role was firstly reported in animal and human biology (Villacorta et al. 2015; Ambrozova et al. 2016). It is well known that plant tissues and organs possess mores unsaturated fatty acids than animal ones; hence the interest of researchers has been turned towards NO_2-FAs in plants, first in terms of the Mediterranean diet known for its anti-inflammatory and anti-hypertensive effects, rich in the sources of unsaturated fatty acids from oils, fish or dairy products as well as nitrates and nitrites from leafy vegetables (Mata-Pérez et al. 2017 and references therein). From the point of view of a plant physiologist and biochemist, it is noteworthy that nitro-linolenic acid (NO_2-Ln) was detected in seeds and both roots and leaves of seedlings of pea and rice and in their mitochondria and peroxisomes to the amounts of these established in animal tissues (Mata-Pérez et al. 2016).

High amounts of NO_2-Ln were found at the beginning of plant development, especially in seeds and 14-day-old seedlings, and then it declined (Mata-Pérez et al. 2016). As it has been mentioned earlier, the involvement of NO_2-Ln in plant development is confirmed (Table 1). This means the possible impact of NO_2-Ln as NO-releasing factor mediating S-nitrosylation mechanism of key transcription factors regulating seed germination and seedling growth (Albertos et al. 2015). Moreover, the physiological impact of NO_2-Ln should be regarded in relation to the involvement of NO in plant stress response and signalling described by Mata-Perez et al. (2016, 2017), who revealed high induction of ascorbate peroxidase 2 (APX2) gene. Such findings open new possibilities for further basic and applied research.

6 Conclusion

As it can be deduced from the results of many papers cited in the text, the studies on ROS/RNS interactions are difficult, but many issues can be resolved, too, by teams of plant physiologists, biochemists and molecular biologists. Some practical uses seem to be promising due to the protective impact of NO for plants subjected to environmental stresses. This implies the possibility of the extension of fruits, vegetables and flowers shelf-life, although high reactivity of NO encounters difficulties in its practical use. From the "pure" scientific point of view, some interactions between RNS, ROS and phytohormones on the molecular level, as well as nitro-fatty acids mode of action, should be intensively studied to reveal the genes and possible signalling processes involved in this interaction, in different plants and their organs, as well as in various environmental conditions.

References

Abat JK, Deswal R (2009) Differential modulation of S-nitrosoproteome of *Brassica juncea* by low temperature: change in S-nitrosylation of Rubisco is responsible for the inactivation of its carboxylase activity. Proteomics 9:4368–4380

Abat JK, Mattoo AK, Deswal R (2008) S-nitrosylated proteins of a medicinal CAM plant *Kalanchoe pinnata*–ribulose-1, 5-bisphosphate carboxylase/oxygenase activity targeted for inhibition. FEBS J 275:2862–2872

Albertos P, Romero-Puertas MC, Tatematsu K et al (2015) S-nitrosylation triggers ABI5 degradation to promote seed germination and seedling growth. Nat Commun 6:8669. https://doi.org/10.1038/ncomms9669

Alderton WK, Cooper CE, Knowles RG (2001) Nitric oxide synthases: structure, function and inhibition. Biochem J 357:593–615

Ambrozova G, Martiskova H, Koudelka A et al (2016) Nitro-oleic acid modulates classical and regulatory activation of macrophages and their involvement in pro-fibrotic responses. Free Radic Biol Med 90:252–260

Arasimowicz-Jelonek M, Floryszak-Wieczorek J, Gwóźdź E (2011) The messenger of nitric oxide in cadmium-challenged plants. Plant Sci 181:612–620

Asada K (1994a) Mechanisms for scavenging reactive molecules generated in chloroplasts under light stress. In: Baker NR, Bowyer JR (eds) Photoinhibition of photosynthesis: from molecular mechanisms to the field. Bios Scientific Publishers, Oxford, pp 129–142

Asada K (1994b) Production and action of active oxygen species in photosynthetic tissues. In: Foyer CH, Mullineaux PM (eds) Causes of photooxidative stress and amelioration of defense system in plants. CRC Press, Boca Raton, pp 77–104

Bączek-Kwinta R, Miszalski Z, Niewiadomska E (2005) Physiological role of reactive oxygen species in chill-sensitive plants. Phyton – Annales Rei Botanicae 45:25–37

Badiyan D, Wills RBH, Bowyer MC (2004) Use of a nitric oxide donor compound to extend the vase life of cut flowers. Hortscience 39:1371–1372

Bai X, Yang L, Tian M et al (2011) Nitric oxide enhances desiccation tolerance of recalcitrant *Antiaris toxicaria* seeds via protein S-nitrosylation and carbonylation. PLoS One 6(6):e20714. https://doi.org/10.1371/journal.pone.0020714

Bailly C (2004) Active oxygen species and antioxidants in seed biology. Seed Sci Res 14:93–107

Barna B, Fodor J, Harrach BD et al (2012) The Janus face of reactive oxygen species in resistance and susceptibility of plants to necrotrophic and biotrophic pathogens. Plant Physiol Biochem 59:37–43

Bartoli CG, Simontacchi M, Montaldi ER et al (1997) Oxidants and antioxidants during aging of chrysanthemum petals. Plant Sci 129:157–165

Bartoli CC, Casalongué C, Simintacchi M et al (2013) Interactions between hormone and redox signaling pathways in the control of growth and cross tolerance to stress. Environ Exp Bot 94:73–88

Belenhgi B, Romero-Puertas M-C, Vercammen D et al (2007) Metacaspase activity of *Arabidopsis thaliana* is regulated by nitrosylation of a critical cysteine residue. J Biol Chem 282:1352–1358

Beligni MV, Lamattina L (2000) Nitric oxide stimulates seed germination and de-etiolation, and inhibits hypocotyl elongation, three light-inducible responses in plants. Planta 210:215–221

Beligni MV, Fath A, Bethke PC, Lamattina L, Jones RL (2002) Nitric oxide acts as an antioxidant and delays programmed cell death in barley aleurone layers. Plant Physiol 129(4):1642–1650

Bellin D, Asai S, Delledonne M et al (2013) Nitric oxide as a mediator for defense responses. MPMI 26:271–277

Besson-Bard A, Pugin A, Wendehenne D (2008) New insights into nitric oxide signalling in plants. Annu Rev Plant Biol 59:21–39

Boscari A, Del Giudice J, Ferrarini A et al (2013) Expression dynamics of the *Medicago truncatula* transcriptome during the symbiotic interaction with *Sinorhizobium meliloti*: which role for nitric oxide? Plant Physiol 161:425–439

Bowyer MC, Wills RBH, Badiyan D et al (2003) Extending the postharvest life of carnations with nitric oxide comparison of fumigation and in vivo delivery. Postharvest Biol Technol 30:281–286

Bright J, Desikan R, Hancock JT et al (2006) ABA-induced NO generation and stomatal closure in *Arabidopsis* are dependent on H_2O_2 synthesis. Plant J 45:113–122

Brody AL, Strupinsky EL, Kline LR (eds) (2001) Active packaging for food applications. CRC Press, Boca Raton, pp 112–120

Cai MZ, Zhang SN, Wang FM et al (2011) Protective effect of exogenously applied nitric oxide on aluminum-induced oxidative stress in soybean plants. Russ J Plant Physiol 58:791–779

Camejo D, Romero-Puertas MC, Rodríguez-Serrano M et al (2013) Salinity-induced changes in S-nitrosylation of pea mitochondrial proteins. J Proteome 79:87–99

Chaki M, Carreras A, Lopez-Jaramillo J et al (2013) Tyrosine nitration provokes inhibition of sunflower carbonic anhydrase (beta-CA) activity under high temperature stress. Nitric Oxide 29:30–33

Chen Z, Gallie DR (2004) The ascorbic acid redox state controls guard cell signalling and stomatal movement. Plant Cell 16:1143–1162

Chen R, Sun S, Wang C et al (2009) The *Arabidopsis* PARAQUATRESISTANT2 gene encodes an S-nitrosoglutathione reductase that is a key regulator of cell death. Cell Res 19:1377–1387

Chen Z, Zhang L, Zhu C (2015) Exogenous nitric oxide mediates alleviation of mercury toxicity by promoting auxin transport in roots or preventing oxidative stress in leaves of rice seedlings. Acta Physiol Plant 37:197. https://doi.org/10.1007/s11738-015-1931-7

Cheng G, Yang E, Lu W et al (2009) Effect of nitric oxide on ethylene synthesis and softening of banana fruit slice during ripening. J Agric Food Chem 57:5799–5804

Corpas FJ (2015) What is the role of hydrogen peroxide in plant peroxisomes? Plant Biol J 17:1099–1103

Corpas FJ, Leterrier M, Valderrama R et al (2011) Nitric oxide imbalance provokes a nitrosative response in plants under abiotic stress. Plant Sci 181:604–611

Corpas FJ, Alché JD, Barroso JB (2013) Current overview of S-nitrosoglutathione (GSNO) in higher plants. Front Plant Sci 4:126. https://doi.org/10.3389/fpls.2013.00126

Daszkowska-Golec A, Szarejko I (2013) Open or close the gate – stomata action under the control of phytohormones in drought stress conditions. Front Plant Sci 4:138. https://doi.org/10.3389/fpls.2013.00138

Delledonne M, Xia Y, Dixon RA et al (1998) Nitric oxide functions as a signal in plant disease resistance. Nature 394:585–588

Delledonne M, Zeier J, Marocco A et al (2001) Signal interactions between nitric oxide and reactive oxygen intermediates in the plant hypersensitive disease response. Proc Natl Acad Sci U S A 98:13454–13459

Desikan R, Hancock JT, Bright J et al (2005) A role for ETR1 in hydrogen peroxide signaling in stomatal guard cells. Plant Physiol 137:831–834

Desikan R, Last K, Harrett-Williams R et al (2006) Ethylene-induced stomatal closure in *Arabidopsis* occurs via AtrbohF-mediated hydrogen peroxide synthesis. Plant J 47:907–916

Driscoll JA (1997) Acid rain demonstration: the formation of nitrogen oxides as a by-product of high-temperature flames in connection with internal combustion engines. J Chem Educ 74:1424. https://doi.org/10.1021/ed074p1424

Durner J, Wendehenne D, Klessig DF (1998) Defense gene induction in tobacco by nitric oxide, cyclic GMP, and cyclic ADPribose. Proc Natl Acad Sci U S A 95:10328–11033

Farnese FS, Oliveira JA, Paiva EAS et al (2017) The involvement of nitric oxide in integration of plant physiological and ultrastructural adjustments in response to arsenic. Front Plant Sci 8:516. https://doi.org/10.3389/fpls.2017.00516

Farneti B, Khomenko J, Cappellin L et al (2015) Dynamic volatile organic compound fingerprinting of apple fruit during processing. LWT Food Sci Technol 63:21–28

Fath A, Bethke PC, Jones RL (2001) Enzymes that scavenge reactive oxygen species are down-regulated prior to gibberellic acid-induced programmed cell death in barley aleurone. Plant Physiol 126:156–166

Feechan A, Kwon E, Yun B-W et al (2005) A central role for S-nitrosothiols in plant disease resistance. Proc Natl Acad Sci 102:8054–8059

Floryszak-Wieczorek J, Arasimowicz M, Milczarek G et al (2007) Only an early nitric oxide burst and the following wave of secondary nitric oxide generation enhanced effective defence responses of pelargonium to a necrotrophic pathogen. New Phytol 175:718–730

Foyer CH, Noctor G (2005) Redox homeostasis and antioxidant signaling: a metabolic interface between stress perception and physiological responses. Plant Cell 17:1866–1875

Foyer CH, Shigeoka S (2011) Understanding oxidative stress and antioxidant functions to enhance photosynthesis. Plant Physiol 155:93–100

Foyer CH, Lopez-Delgado H, Dat JF et al (1997) Hydrogen peroxide- and glutathione-associated mechanism of acclimatory stress tolerance and signalling. Physiol Plant 100:241–254

Galatro A, Puntarulo S, Guiamet JJ et al (2013) Chloroplast functionality has a positive effect on nitric oxide level in soybean cotyledons. Plant Physiol Biochem 66:26–33

García-Mata C, Lamattina L (2003) Abscisic acid, nitric oxide and stomatal closure – is nitrate reductase one of the missing links? Trends Plant Sci 1:20–26

Grefen C, Städele K, Ruzicka K et al (2008) Subcellular localization and in vivo interactions of the *Arabidopsis thaliana* ethylene receptor family members. Mol Plant 1:308–320

Gupta KJ, Fernie AR, Kaiser WM et al (2011) On the origins of nitric oxide. Trends Plant Sci 16:160–168

Hasanuzzaman M, Fujita M (2013) Exogenous sodium nitroprusside alleviates arsenic-induced oxidative stress in wheat *(Triticum aestivum* L.) seedlings by enhancing antioxidant defense and glyoxalase system. Ecotoxicology 22:584–596

He YK, Tang RH, Yi H et al (2004) Nitric oxide represses the *Arabidopsis* floral transition. Science 305:1968–1971

Herrera-Vásquez A, Salinas P, Holuigue L (2015) Salicylic acid and reactive oxygen species interplay in the transcriptional control of defense genes expression. Front Plant Sci 5:171. https://doi.org/10.3389/fpls.2015.00171

Hess DT, Matsumoto A, Kim S et al (2005) Protein S-nitrosylation: purview and parameters. Nat Rev Mol Cell Biol 6:150–166

Hu X, Neill SJ, Tang Z et al (2005) Nitric oxide mediates gravitropic bending in soybean roots. Plant Physiol 137:663–670

Huque R, Wills RBH, Pristijono P et al (2013) Effect of nitric oxide (NO) and associated control treatments on the metabolism of fresh-cut apple slices in relation to development of surface browning. Postharvest Biol Technol 78:16–23

Hura K, Hura T, Bączek-Kwinta R et al (2014) Induction of defense mechanisms in seedlings of oilseed winter rape inoculated with *Phoma lingam* (*Leptosphaeria maculans*). Phytoparasitica 42:145–154

Ignarro IJ, Buga GM, Wood KS et al (1987) Endothelium-derived relaxing factor produced and released from artery and vein is nitric oxide. Proc Natl Acad Sci U S A 84:9265–9269

Jeandroz S, Wipf D, Stuehr DJ et al (2016) Occurrence, structure, and evolution of nitric oxide synthase–like proteins in the plant kingdom. Sci Signal 9(417):re 2. https://doi.org/10.1126/scisignal.aad4403

Karpinski S, Szechynska-Hebda M (2010) Secret life of plants. From memory to intelligence. Plant Signal Behav 5:1391–1394

Kato H, Takemoto D, Kawakita K (2013) Proteomic analysis of S-nitrosylated proteins in potato plant. Physiol Plant 148:371–386

Kazemi N, Khavari-Nejad RA, Fahimi H et al (2010) Effects of exogenous salicylic acid and nitric oxide on lipid peroxidation and antioxidant enzyme activities in leaves of *Brassica napus* L. under nickel stress. Sci Hortic 126:402–407

Khan MI, Fatma M, Per TS et al (2015) Salicylic acid-induced abiotic stress tolerance and underlying mechanisms in plants. Front Plant Sci 6:462. https://doi.org/10.3389/fpls.2015.00462

Kopczewski T, Kuzniak E (2013) Redox signals as a language of interorganellar communication in plant cells. Centr Eur J Biol 8:1153–1183

Kopyra M, Gwóźdź E (2003) Nitric oxide stimulates seeds germination and counteracts the inhibitory effect of heavy metal and salinity on root growth of *Lupinus luteus*. Plant Physiol Biochem 441:1011–1017

Kreslavski VD, Los DA, Allakhverdiev SI et al (2012) Signaling role of reactive oxygen species in plants under stress. Russ J Plant Physiol 59:141–154

Kusznierewicz B, Bączek-Kwinta R, Bartoszek A et al (2012) The dose-dependent influence of zinc and cadmium contamination of soil on their uptake and glucosinolate content in white cabbage (*Brassica oleracea var. capitata* f. alba). Environ Toxicol Chem 31:2482–2489

Lamotte O, Bertoldo JB, Besson-Bard A et al (2015) Protein S-nitrosylation: specificity and identification strategies in plants. Front Chem 2:114. https://doi.org/10.3389/fchem.2014.00114

Lee U, Wie C, Fernandez BO et al (2008) Modulation of nitrosative stress by S-nitrosoglutathione reductase is critical for themotolerance and plant growth in *Arabidopsis*. Plant Cell 20:786–802. https://doi.org/10.1105/tpc.107.052647

Leshem YY (1988) Plant senescence processes and free radicals. Free Rad Biol Med 5:39–49

Leshem YY, Wills RBH, Ku VVV (1998) Evidence for the function of the free radical gas nitric oxide (NO) as an endogenous maturation and senescence regulating factor in higher plants. Plant Physiol Biochem 36:825–833

Leterrier M, Airaki M, Palma JM et al (2012) Arsenic triggers the nitric oxide (NO) and *S*-nitrosoglutathione (GSNO) metabolism in *Arabidopsis*. Environ Pollut 166:136–143

Lin A, Wang Y, Tang J et al (2012) Nitric oxide and protein S-nitrosylation are integral to hydrogen peroxide-induced leaf cell death in rice. Plant Physiol 158:451–464

Lindermayr C, Saalbach G, Durner J (2005) Proteomic identification of S-nitrosylated proteins in *Arabidopsis*. Plant Physiol 137:921–930

Lozano-Juste J, Leon J (2011) Nitric oxide regulates DELLA content and PIF expression to promote photomorphogenesis in *Arabidopsis*. Plant Physiol 156:1410–1423

Malik SI, Hussain A, Yun BW et al (2011) GSNOR-mediated de-nitrosylation in the plant defence response. Plant Sci 181:540–544

Manjunatha G, Lokesh V, Neelwarne B (2010) Nitric oxide in fruit ripening: trends and opportunities. Biotechnol Adv 28:489–499

Mata-Pérez C, Sánchez-Calvo B, Padilla-Serrano MN et al (2016) Nitro-fatty acids in plant signaling: nitro-linolenic acid induces the molecular chaperone network in *Arabidopsis*. Plant Physiol 170:686–670

Mata-Pérez C, Sánchez-Calvo B, Padilla MN et al (2017) Nitro-fatty acids in plant signaling: new key mediators of nitric oxide metabolism. Redox Biol 11:554–561. https://doi.org/10.1016/j.redox.2017.01.002

Mor A, Koh E, Weiner L et al (2014) Singlet oxygen signatures are detected independent of light in chloroplast in response to multiple stresses. Plant Physiol 165:249–261

Moreau M, Lindermayr C, Durner J et al (2010) NO synthesis and signalling in plants – where do we stand? Physiol Plant 138:372–383

Noritake T, Kawakita K, Doke N (1996) Nitric oxide induces phytoalexin accumulation in potato tuber tissues. Plant Cell Physiol 37:113–116

Pagnussat GC, Lanteri ML, Lamattina L (2003) Nitric oxide and cyclic GMP are messengers in the indole acetic acid-induced adventitious rooting process. Plant Physiol 132:1241–1248

Palmer RMJ, Ferrige AG, Moncada S (1987) Nitric oxide release accounts for the biological activity of endothelium-derived relaxing factor. Nature 327:524–526

Pei Z-M, Murata Y, Benning G et al (2000) Calcium channels activated by hydrogen peroxide mediate abscisic acid signalling in guard cells. Nature 406:731–734

Politycka B (1996) Peroxidase activity and lipid peroxidation in roots of cucumber seedlings influenced by derivatives of cinnamic and benzoic acids. Acta Physiol Plant 18:365–370
Portmann RW, Daniel JS, Ravishankara AS (2012) Stratospheric ozone depletion due to nitrous oxide: influences of other gases. Philos Trans R Soc Lond B Biol Sci 367:1256–1264
Pristijono P, Wills RBH, Golding JB (2008) Use of the nitric oxide-donor compound, diethylenetriamine-nitric oxide (DETANO), as an inhibitor of browning in apple slices. J Hortic Sci Biotechnol 83:555–558
Romero-Puertas MC, Campostrini N, Mattè A et al (2008) Proteomic analysis of S-nitrosylated proteins in *Arabidopsis thaliana* undergoing hypersensitive response. Proteomics 8:1459–1469
Rümer S, Kapuganti JG, Kaiser W (2009) Oxidation of hydroxylamines to NO by plant cells. Plant Signal Behav 4:853–855
Sahay S, Gupta M (2017) An update on nitric oxide and its benign role in plant responses under metal stress. Nitric Oxide 67:39–52. https://doi.org/10.1016/j.niox.2017.04.011
Saito S, Yamamoto-Katou A, Yoshioka H et al (2006) Peroxynitrite generation and tyrosine nitration in defense responses in tobacco BY-2 cells. Plant Cell Physiol 47:689–697
Schopfer P, Liszkay A, Bechtold M et al (2002) Evidence that hydroxyl radicals mediate auxin-induced extension growth. Planta 214:821–828
Shah FR, Ahmad N, Masood KR et al (eds) (2010) Plant adaptation and phytoremediation. Springer, New York, pp 71–97
Shi S, Wang G, Wang Y et al (2005) Protective effect of nitric oxide against oxidative stress under ultraviolet-⍰ radiation. Nitric Oxide 13:1–9
Silva L, Carvalho H (2013) Possible role of glutamine synthetase in the NO signaling response in root nodules by contributing to the antioxidant defenses. Front Plant Sci 4:372. https://doi.org/10.3389/fpls.2013.00372
Soegiarto L, Wills RBH (2004) Short term fumigation with nitric oxide gas in air to extend the postharvest life of broccoli, green bean, and bok choy. HortTechnology 14:538–540
Solanki R, Dhankhar R (2011) Biochemical changes and adaptive strategies of plants under heavy metal stress. Biologia 66:195–204
Sun B, Jing Y, Chen K et al (2007) Protective effect of nitric oxide on iron deficiency-induced oxidative stress in maize (*Zea mays*). J Plant Physiol 164:536–543
Terrile MC, París R, Calderón-Villalobos LI et al (2012) Nitric oxide influences auxin signaling through S-nitrosylation of the *Arabidopsis* TRANSPORT INHIBITOR RESPONSE 1 auxin receptor. Plant J 70:492–500
Tuteja N, Chandra M, Tuteja R et al (2004) Nitric oxide as a unique bioactive signaling messenger in physiology and pathophysiology. J Biomed Biotechnol 4:227–237
Valderrama R, Corpas FJ, Carreras A et al (2007) Nitrosative stress in plants. FEBS Lett 581:453–461
Van Doorn VG (2011) Classes of programmed cell death in plants, compared to those in animals. J Exp Bot 62:4749–4761
Vanlerberghe GC (2013) Alternative oxidase: amitochondrial respiratory pathway to maintain metabolic and signaling homeostasis during abiotic and biotic stress in plants. Int J Mol Sci 14:6805–6847
Vellosillo T, Vicente J, Kulasekaran S et al (2010) Emerging complexity in reactive oxygen species production and signaling during the response of plants to pathogens. Plant Physiol 154:444–448
Villacorta L, Gao Z, Schopfer FJ et al (2015) Nitro-fatty acids in cardiovascular regulation and diseases: characteristics and molecular mechanisms. Front Biosci (Landmark Ed) 21:873–889
Vreeburg AM, Fry SC (2005) Reactive oxygen species in cell walls. In: Smirnoff N (ed) Antioxidants and reactive oxygen species in plants. Blackwell Publishing, Oxford, pp 197–214
Wang BL, Tang XY, Cheng LY et al (2010) Nitric oxide is involved in phosphorus deficiency-induced cluster-root development and citrate exudation in white lupin. New Phytol 187:1112–1123

Wang Y, Loake GJ, Chu C (2013) Cross-talk of nitric oxide and reactive oxygen species in plant programmed cell death. Front Plant Sci 4:314. https://doi.org/10.3389/fpls.2013.00314

Wendehenne D, Pugin A, Klessig DF et al (2001) Nitric oxide: comparative synthesis and signaling in animal and plant cells. Trends Plant Sci 6:177–183

Xiong J, An L, Lu H et al (2009) Exogenous nitric oxide enhances cadmium tolerance of rice by increasing pectin and hemicelluloses content in root cell wall. Planta 230:755–765

Xu J, Wang W, Yin H et al (2010) Exogenous nitric oxide improves antioxidative capacity and induces auxin degradation in roots of *Medicago truncatula* seedlings under cadmium stress. Plant Soil 326:321–330

Xu S, Guerra D, Lee U et al (2013) S-nitrosoglutathione reductases are low-copy number, cysteine-rich proteins in plants that control multiple developmental and defense responses in *Arabidopsis*. Front Plant Sci 4:430. https://doi.org/10.3389/fpls.2013.00430

Yang Z, Zhong X, Fan Y et al (2015) Burst of reactive oxygen species in pedicel-mediated fruit abscission after carbohydrate supply was cut off in longan (*Dimocarpus longan*). Front Plant Sci 6:360. https://doi.org/10.3389/fpls.2015.00360

Zaharah SS, Singh Z (2011) Postharvest nitric oxide fumigation alleviates chilling injury, delays fruit ripening and maintains quality in cold-stored 'Kensington Pride' mango. Postharvest Biol Technol 60:202–210

Zhang X, Zhang L, Dong F et al (2001) Hydrogen peroxide is involved in abscisic acid-induced stomatal closure in *Vicia faba*. Plant Physiol 126:1438–1448

Zhang DD, Cheng GP, Li J et al (2007) Effect of nitric oxide on disorder development and quality maintenance of plum stored at low temperature. ISHS Acta Hortic 804:549–554

Zhang XW, Dong YJ, Qiu XK et al (2012) Exogenous nitric oxide alleviates iron-deficiency chlorosis in peanut growing on calcareous soil. Plant Soil Environ 58:111–120

Zhou J, Wang J, Li X et al (2014) H_2O_2 mediates the crosstalk of brassinosteroid and abscisic acid in tomato responses to heat and oxidative stress. J Exp Bot 65:4371–4383

Zhu S, Zhou E (2007) Effect of nitric oxide on ethylene production in strawberry fruit during storage. Food Chem 100:1517–1522

Zhu S, Lina S, Mengchen L et al (2008) Effect of nitric oxide on reactive oxygen species and antioxidant enzymes in kiwi fruit during storage. J Sci Food Agric 88:2324–2331

Żur I, Dubas E, Krzewska M et al (2014) Antioxidant activity and ROS tolerance in triticale (x *Triticosecale* Wittm.) anthers affect the efficiency of microspore embryogenesis. Plant Cell Tissue Organ Cult 119:79–97

Involvement of Reactive Species of Oxygen and Nitrogen in Triggering Programmed Cell Death in Plants

Vineet Kumar Maurya, Dhananjay Kumar, Chandramani Pathak, and Budhi Sagar Tiwari

Abstract Programmed cell death (PCD) is a multifaceted process involved in cell number control, removal of diseased or inflamated cells and maintaining homeostasis between dying and newborn cells. It is a fine regimented process under strict genetic control. The process is not only associated with developmental programs of plants and animals but also observed during extreme fluctuations in environmental factors, as well as during noncompatible biotic interactions. In particular, stress-induced PCD in plants has been hypothesized as one of the survival strategies. At the regulatory stages of the process, amplification of cellular reactive species of oxygen (ROS) and nitrogen (RNS) acts as key signalling events for execution of PCD. Although roles of ROS and RNS in execution of PCD have been well studied independently, information about cross-talks between ROS and RNS are limited. In this chapter, efforts have been made to compile the available information regarding involvement of ROS, RNS and their cross-talk during the execution of PCD in plants.

Keywords ROS · RNS · Signalling · PCD · Cross-talk · Abiotic stress · Chloroplasts · Mitochondria

1 Introduction

Every surviving organism in the universe is destined to die, thus signifying that life and death are the two wheels of a cart. Due to natural urge of longer and better life, continuous efforts have been made to decipher bio-physiochemical events of life, both in plants and animals. Compared to life, death seems an unimportant

V. K. Maurya
Department of Microbiology, H.N.B. Garhwal University, Srinagar, Uttarakhand, India

D. Kumar
Department of Botany, H.N.B. Garhwal University, Srinagar, Uttarakhand, India

C. Pathak · B. S. Tiwari (✉)
Indian Institute of Advanced Research, Gandhinagar, Gujarat, India
e-mail: bstiwari@iar.ac.in

S. Vats (ed.), *Biotic and Abiotic Stress Tolerance in Plants*,
https://doi.org/10.1007/978-981-10-9029-5_10

phenomenon, about which no one would have been interested in earlier days of human civilization. But in the dusk of nineteenth century, some scholars aroused interest in death while deciphering the physiological events behind it, so that these events could be prevented and death either could be averted or postponed. Walter Flemming was the pioneer to describe the morphological features of cell death in 1885. Death, which seems to be simple and unimportant process, is actually a complex and useful process for life. Death at the cellular level could be highly ordered and genetically programmed, called 'programmed cell death' (PCD) (Mocarski et al. 2014; Daneva et al. 2016) or 'apoptosis' (a common term used in animal system) (Hochreiter-Hufford and Ravichandran 2013; Poon et al. 2014; Croce and Reed 2016). In a multicellular organism, death starts at cellular level and eventually leads to either a well-developed multicellular body or collapse of whole organism, depending upon stages of life. Living organisms are continuously exposed to a variety of stress factors which can jeopardize their survival, unless properly encountered. While animals can simply evade stressors by escaping, plants can't opt this strategy, on account of being sessile. But to protect themselves from stresses, plants have developed many complex strategies like programmed cell death, necrosis, autophagy, etc. PCD is an active, genetically controlled process, which is initiated to remove damaged and unwanted tissues, thereby ensuring the survival and proper development of the organism. In multicellular organism PCD plays an important role in maintaining homeostasis between dying and newly generated cells, removal of diseased or damaged cells and development of organs (Van Breusegem and Dat 2006; Mocarski et al. 2014; Huysmans et al. 2017). Besides being an integral component of organism's developmental program (Fuchs and Steller 2011; Huang et al. 2016; Liebthal and Dietz 2017), PCD plays a crucial role under perturbed environmental conditions, caused by biotic and abiotic stresses. When the intensity of a stress factor is moderately high, plants employ the induction of PCD as one of the survival mechanisms. Under such conditions PCD is considered as an adaptive strategy (Petrov et al. 2015; Huysmans et al. 2017). During induction of PCD, levels of reactive oxygen species (ROS) rise in cytoplasm, which are utilized as mediators of the stress signal. In the present decade besides ROS, role of reactive nitrogen species (RNS) has also been discovered in PCD induction (Airaki et al. 2012; Mittler 2017). Developmental PCD is controlled by genetic makeup of plants and under stress-free environmental conditions has nothing to do with crop yield. Unlike developmental PCD, stress-induced PCD has potential of affecting crop yield significantly; hence it is of fundamental biological importance for agriculture scientists. Therefore, a lot of studies have been conducted on decoding the mechanisms leading to both instigation and control of PCD in plants, under unfavourable environmental conditions. In this chapter we have discussed about PCD in plants with special emphasis on role of ROS and RNS in it.

2 PCD in Plants

Like animals, plants do have innately instituted ability to selectively eliminate unwanted and targeted cells by a well-organized, multi-step process involving various signalling molecules and enzymes (Ellis et al. 1991; Van Hautegem et al. 2015; Kazmierczak et al. 2017). *In planta* as well, cells which are targeted and destined to die, are killed systematically without posing any harmful effect on its neighbouring cells (Jacobsen et al. 1997; Fuchs and Steller 2015; Ingram 2017). PCD plays a significant role during the development of plant; at one end, execution of PCD is essential for successful accomplishment of growth and development under non-stress conditions, while on the other hand, different biotic and abiotic stress factors cause induction of PCD. Processes involving PCD for growth and developments include formation and growth of embryo, dissolving of aleurone layer at the time of seed germination in monocot seeds, degeneration of tapetum layer in anthers, development and structural modifications of tracheary elements in xylem (Nakashima et al. 2000; Van Durme and Nowack 2016), formation of trichomes, pollen self-incompatibility, differentiation of parenchymatous tissue in aquatic plants and reshaping of leaf structure (Gunawardena et al. 2004; Maizel 2015), floral organ abscission and senescence of leaves (Gechev et al. 2006) (Fig. 1). A fine balance between death of cells through PCD and new cells birth by mitosis determines the actual growth rate of plant (Van Breusegem and Dat 2006). Various types of stresses which cause PCD to occur are biotic stresses like interaction with incompatible pathogens and abiotic stress like extreme pH, temperature, salinity, heat, drought and oxidative conditions (De Storme and Geelen 2014) (Fig. 1). Crop production is considerably affected by stress-induced PCD, and therefore this area needs special concern in agricultural research (Gregersen et al. 2013). Decline in global food production is a big concern of agriculture sector. Factors responsible for the decline are climate change, crop loss by pathogens, multiple abiotic stresses and shrinkage of cultivable land due to anthropogenic activities. To cope with food crisis in future, which might be caused due to global population burst, boost in food production is urgently required. Since PCD plays significant role in combating biotic and abiotic stresses, focused studies on stress-induced PCD, targeting stress-caused yield loss, are required (Cominelli et al. 2013; Rosenzweig et al. 2014; Kumar et al. 2016).

Nowadays diverse forms of PCD are known; most common among them is apoptosis, which is quite common in animals. Apoptosis involves some unique signatures such as blebbing of plasma membrane, nuclear fragmentation, activation of downstream caspases and fragmentation of cell into smaller parts called apoptotic bodies. Phagocytes engulf the apoptotic bodies and hydrolyse them enzymatically with the help of lysosomes. Phagocytic engulfment is a key feature of apoptosis which prevent local inflammations caused by immunogenic activity against dead cell contents. Apoptosis is absent in plant kingdom probably because of lack of phagocytes and presence of firm cell wall (Doorn et al. 2011). In plants another form of cell death called PCD is widely reported. PCD shows some similarity with apoptosis,

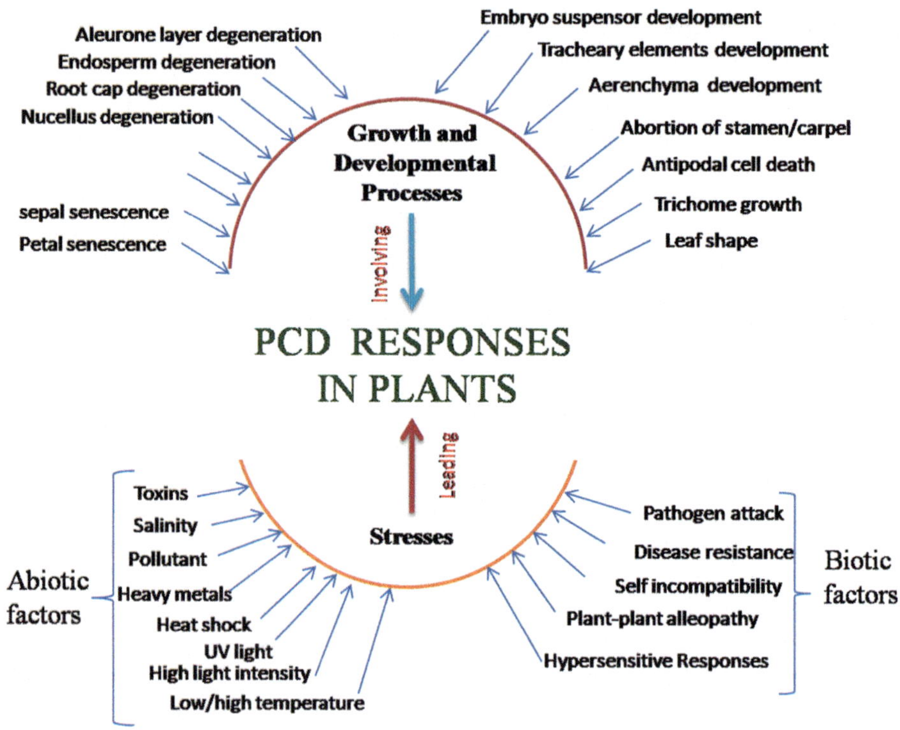

Fig. 1 A diagrammatic listing of PCD responses in plants

viz. shrinkage of cytoplasm, chromosomal condensation and fragmentation of nuclear DNA. In some cases formation of apoptotic bodies is also reported though these apoptotic bodies are not engulfed by phagocytes and remain confined inside the cell wall (Daneva et al. 2016). Based on morphological features, two main types of PCD, namely, vacuolar cell death and necroptosis, have been observed in plants. The former type is responsible for reshaping of leaves and differentiation of vascular tissues via comprising various features like formation of lytic vacuoles, rearrangement of cytoskeleton, rupturing of tonoplast and subsequent degradation of cell organelles, leaving ghost cell wall behind (Muntz 2007; Jones 2001; Minina et al. 2014), while the latter occurs during biotic and abiotic stress and entails features like rupturing of the plasma membrane at early stages of cell death, shrinking of the protoplast and lack of lytic vacuole formation (Majno and Joris 1995; Kroemer et al. 2009; Galluzzi et al. 2017). Besides these two classes of PCD, there are several other developmental incidents, which involve PCD/PCD-like phenomenon. These incidents are (1) xylogenesis, which occurs during xylem development; (2) hypersensitive responses (HR), which initiates after attack of biotrophic pathogens; (3) cell death response during seed germination in cereals; (4) during self-incompatibility phenomenon; and (5) in suspensor cells during embryo development.

Table 1 Different types of ROS in plant system (Halliwell and Gutteridge 2007)

Free radicals	Nonradicals
Superoxide ($O_2^{-\bullet}$)	Hydrogen peroxide (H_2O_2)
Hydroperoxyl (HO_2)	Singlet oxygen ($^1O_2$ or $^1\Delta g$)
Hydroxyl group ($HO^\bullet$)	Ozone (O_3)
Peroxyl ($ROO^\bullet$)	Hypochlorous acid (HOCl)
Alkoxyl ($RO^\bullet$)	Peroxynitrite ($ONOO^-$)

3 Reactive Oxygen Species (ROS) and Their Role in Establishment of PCD in Plants

3.1 Chemical Properties of Reactive Oxygen Species (ROS)

Oxygen is the life-supporting gas for aerobic life, but the same oxygen can be toxic to them in form of reactive oxygen species. Concept of oxygen toxicity in the form ROS was proposed by Rebeca Gerschman (Gerschman et al. 1954). ROS is a wide term, covering all radical and nonradical form of active oxygen (Table 1). Chemically ROS are highly reactive intermediates, formed during reduction of O_2 and undesired by-products of oxygen metabolism. ROS are harmful because they can oxidize various biological molecules and therefore can cause serious damage to various components of the cell (Bailly 2004).

3.2 ROS Generation, Scavenging and Oxidative Signalling in Plants

In case of plants, ROS may arise from electron transport chain of two different organelles: mitochondria and chloroplast. Generation of ROS through enzymatic process, just because of the activity of oxidases and peroxidases, is also known (Mittler et al. 2011). Singlet oxygen ($^1O_2$) and hydrogen peroxide (H_2O_2) are the two most studied ROS. Compared to other ROS, these duos have longer half-lives and entrenched detection methods (Foyer and Noctor 2009). Of the several ROS, functions and production mechanisms have been reported only for a few like H_2O_2, $O_2^{-\bullet}$,$^1O_2$, $ROO^\bullet$, etc.; similar information about other ROS are very scanty. This may be due to absence of suitable detection methods and experimental procedure for other ROS. Among various ROS, only H_2O_2 and $O_2^{-\bullet}$ have been studied for their role during stress and other developmental processes, by different research groups. H_2O_2, the most stable ROS due to its longest half-life, is able to migrate from its site of synthesis to the adjacent compartments, including neighbouring cells and work as juxtacrine signalling molecule. H_2O_2 is generated via two chemical reactions (1) reaction between $^1O_2$ and $O_2^{-\bullet}$ and (2) spontaneous dismutation of $O_2^{-\bullet}$ (Foyer and Noctor 2009). Apart from these chemical reactions, there are various enzymatic sources of H_2O_2 production in plants. These enzymatic sources are acyl-CoA oxidase and glycolate oxidase, which participate in β-oxidation of lipids and photorespiration, respectively. Another most studied ROS is $O_2^{-\bullet}$, which can interact with other ROS like H_2O_2 and forms $^{\bullet}OH$. Plants don't have any mechanism to detoxify $^{\bullet}OH$, so they try to minimize or prevent the formation of $^{\bullet}OH$. $O_2^{-\bullet}$ play an important step in cross-talk between ROS and RNS; it can react with nitric oxide ($NO^{\bullet}$ will be represented as NO, hereafter in this chapter) and forms peroxynitrite ($ONOO^-$); this peroxynitrite is further protonated to form peroxynitrous acid (ONOOH), which has strong affinity towards electron and hence acts as a strong oxidizing agent (Nath et al. 2017).

In order to protect the plants from ROS-induced oxidative damage, antioxidant system comprised of antioxidant enzymes and antioxidant organic molecules evolved inside them (Buchanan et al. 2002; Khan and Khan 2017). These antioxidants have ability to scavenge excess of ROS, produced during metabolic processes, and save plant cell and biomolecules from detrimental effects. As another defence mechanism, plants transport the excess of ROS into the vacuoles for detoxification (Gould et al. 2002; Gautam et al. 2017). Flavonoids, a good amount of which are found in vacuoles, are potent antioxidants and can scavenge hydrogen peroxide and other ROS effectively (Edreva 2005; Tsuda et al. 2000). Inner surface of vacuolar membrane is rich in ascorbate, glutathione and peroxidases. Antioxidant system of plants is a vast network of metabolites and biomolecules capable of ROS production and scavenging. This dual nature of antioxidant biomolecules favoured the selection and evolution of ROS as signalling molecules by nature. Thus, evolution of antioxidant system helped ROS to play dual role. Due to antioxidant system, ROS, which were initially considered only harmful, became important signalling molecules thereby playing significant roles in plant growth and development as also in combating environmental stresses. Particularly, the signalling role of ROS warrants reconsidering the word 'oxidative stress' by replacing it with 'oxidative signalling'. Generally local antioxidant system of plants scavenges ROS immediately after their production, but under excessive ROS burst, this local antioxidant system proves to be insufficient in maintaining ROS level and downstream cascades of ROS signalling initiate.

3.3 *ROS Signalling During PCD Execution*

Studies conducted to explore the factors responsible for PCD suggested that the level of ROS increases during PCD. This particular finding intensified the research work to elucidate the exact role of ROS during the establishment of PCD. Although the signalling roles of ROS in growth and development, stress responses and PCD have been studied for animals, the same things are still in very juvenile phase insofar plants are considered.

ROS are continuously generated in living system, and all the ROS-producing organelles harbour antioxidant systems to maintain a nontoxic level of ROS. During stress and other physiological processes, there are fluctuations in ROS level. Plant cells are capable of perceiving these minute fluctuations and respond accordingly, depending upon the stress intensity and developmental conditions. These responses include either metabolic changes leading to stress tolerance or induction of PCD (Gechev et al. 2002; Das and Roychoudhury 2014). Depending upon its concentration, same ROS can affect different biological responses. So in order to initiate different biological changes, the concentration of ROS must be kept under strict control. Receptor of different ROS must be equally sensitive for sensing the fluctuations in ROS level, so that the required signalling process could be initiated. At varying concentrations, same ROS is able to control different processes; the signalling response of various ROSs depends on many factors which includes (1) the chemical identity of ROS, (2) duration of signal, (3) intensity of the signal, (4) site of ROS production (Gechev et al. 2006; Queval et al. 2007; Choudhury et al. 2017), (5) growth stage of plant, (6) previous stress encounters, (7) plant hormones, (8) reactive nitrogen species (RNS) and lipid messengers. The convoluted interaction of these factors with others determines the final outcome of ROS signalling (Kwak et al. 2006; Zaninotto et al. 2006; Choudhury et al. 2017). Information about specific ROS receptor is unavailable yet, and only downstream components of ROS signalling, specially of H_2O_2 which transfer signal for control of PCD, have been discovered till date. Protein kinases, phosphatases and transcription factors have been identified as downstream components of ROS signalling, leading to commencement of PCD. Most of the downstream components of ROS signalling, discovered till date, are plant specific, having no any close animal homologue, indicating the genetic difference between plant and animal PCD. The presence of plant-specific proteases and nucleases, involved in execution of some forms of PCD in plants, affirms this concept.

3.3.1 H_2O_2 Signalling During PCD Initiation

Ability to be transported from the site of its synthesis to target site is the paramount feature of any signalling molecule, and its concentration must be kept under control through its storage, transport and scavenging. H_2O_2 can move to long distance from its site production and even cross biological membranes with the help of aquaporins especially dedicated to H_2O_2 transport (Bienert et al. 2006; Henzler and Steudle 2000; Choudhury et al. 2017). Through aquaporin-mediated transmembrane transport, local concentrations of H_2O_2 are adjusted for execution of biological responses. This method is used during cross-compartment communication of H_2O_2, where cytosolic burst of H_2O_2 causes disturbance in antioxidant system of chloroplast by inhibiting chloroplastic APX (ascorbate peroxidase), without involving cytosolic APX (Davletova et al. 2005a). Cells are supposed of having storage sites for H_2O_2 and other ROS like $^1O_2$, $O_2^{-\bullet}$ and $^\bullet OH$. Peroxisomes are shown to act as a sink for H_2O_2, with the help of catalase present in it.

At the very first instance, an increase in the level of H_2O_2 send signals for alteration in the concentration of Na^+, K^+ and Ca^{++}. Signal of H_2O_2 is further amplified by modulation in Ca^{++} concentration, in stress-dependent manner. Modulation in Ca^{++} concentration affects downstream Ca^{++}-interacting proteins such as calmodulins and calcium-dependent protein kinases, which helps in over-amplification of the H_2O_2 signal. Besides calcium-dependent protein kinases and mitogen-activated protein kinases (MAPKs) also help in the downstream transmission of H_2O_2, through their large network. The vast network of MAPKs helps in the integration of different stress signals by providing different interaction points, where downstream messengers of different stress signal meet and diverge (Ichimura et al. 2002; Petrov et al. 2015). MAPK kinase kinase (MEKK), MEKK1, *Arabidopsis thaliana* MAPK (AtMPK3), AtMPK6, OXI1 (oxidative signal-inducible-1), OMTK1 (oxidative stress-activated MAP triple-kinase 1) and nucleotide diphosphate kinases (NDKs) are few examples of kinases, which have been identified as a component of H_2O_2 signalling network by different research groups. Asai et al. (2002) and Liu et al. (2007) demonstrated the role of MAPKs cascades in relaying H_2O_2 signal. They demonstrated that stress-generated H_2O_2 burst in chloroplast or H_2O_2 burst in response to pathogen attack leads to PCD with the help of MAPKs. Nakagami et al. (2006) showed proteasome-dependent regulation of MAPK kinase kinase-1 (MEKK1) by H_2O_2 under different stresses. During its downstream signalling, MEKK1 interacts and activates MPK4 and WRKY53. WRKY53 is a transcription factor, which induces PCD during senescence, without requiring other downstream stress signals (Miao et al. 2007). Other MEKK, which is under inducible control of H_2O_2, is ANP1, which regulates expression of H_2O_2-induced gene, by activating two downstream MAPKs, AtMPK3 and AtMPK6 (Kovtun et al. 2000). For full activity of AtMPK3 and AtMPK6, OXI1 (oxidative signal-inducible-1), a serine/threonine kinase, is required. Expression of OXI1 is also upregulated by H_2O_2 and other abiotic stresses. Experiments with *A. thaliana* by Rentel et al. (2004) confirmed the role of OXI1 as one of the essential components of *A. thaliana* H_2O_2 signalling network. *A. thaliana* plants having *Oxi1* mutants exhibited abnormal root hair growth, and their susceptibility towards pathogen increased (Rentel et al. 2004). GST6 and HSP18.2 are the two genes, whose expression is upregulated by H_2O_2 production; besides these duos a complete heat shock regulon is also upregulated by H_2O_2 production. Upregulation of these genes provide protection against oxidative stresses and can be used as a consistent marker for H_2O_2 production (Gechev and Hille 2005; Vanderauwera et al. 2005). OMTK1 is also a H_2O_2-inducible kinase discovered by Nakagami et al. (2004) in alfalfa. Unlike OXI1, which is induced by H_2O_2 and abiotic stresses, OMTK1 is induced by H_2O_2 only. MMK3, a downstream MAPK, is the target of OMITK1 and can be activated by OMITK1 and ethylene both (Nakagami et al. 2004). NDKs play critical role in providing resistance towards cold and salt stress in *A. thaliana* by reducing H_2O_2 accumulation in cells; thus NDKs act as negative regulators of the H_2O_2 signalling network. Miller and Mittler (2006) hypothesized that heat shock proteins can also be used as possible H_2O_2 production markers.

Through an intricate H_2O_2 signalling network, the stress signal eventually reaches to transcription factors. These transcription factors are ROS specific and include LSD1, LOL1, Zat11, Zat12, WRKY523, WRKY57 and heat shock transcription factors (Dietrich et al. 1997; Epple et al. 2003; Gechev and Hille 2005; Miao et al. 2004; Vanderauwera et al. 2005). LSD1 and LOL1 are zinc finger proteins, while

Zat11 and Zat12 are zinc finger transcription factors, WRKY53 is a senescence-specific transcription factor, and WRKY57 is a ROS-inducible transcription factor. Upregulation of Zat12 expression infers protection against oxidative and electromagnetic radiation stresses, while its downregulation makes plants sensitive to H_2O_2-induced oxidative stress (Rizhsky et al. 2004; Davletova et al. 2005b). These stress-specific transcription factors finally activate the components of H_2O_2-induced cell death network.

Under mild stress conditions, increase in H_2O_2 level acts as a cell defender by providing signals for triggering stress acclimation. This was demonstrated by increasing H_2O_2 concentration artificially, which resulted in increased tolerance against high salt, light, cold, heat and oxidative stresses (Karpinski et al. 1999; Lopez-Delgado et al. 1998). Exactly the opposite effect of H_2O_2 was shown by two research groups; these groups demonstrated that during pathogenic infection, H_2O_2 initiates PCD in the cells surrounding hypersensitive reactions (HR) sites. This is a protective mechanism by which systemic acquired resistance is triggered in distant tissues (Alvarez et al. 1998; Torres et al. 2005). Thus, paradoxically, H_2O_2-triggered cell death is vital for the normal growth and development of plants and for proper response to the changing and challenging environmental conditions (Gechev et al. 2006; Petrov et al. 2015).

3.3.2 $^1O_2$ Signalling During PCD Initiation

Besides H_2O_2, other ROS such as singlet oxygen and superoxide radicals also have the potential of initiating PCD in plants (Dat et al. 2003; Op Den et al. 2003; Vranova et al. 2002; Laloi and Havaux 2015). $^{\bullet}OH$ converts lipid molecules into lipid oxides via peroxidation. These lipid oxides are capable of inducing PCD either alone or with combination of other ROS (Montillet et al. 2005; Mueller 2004). Singlet oxygen also induces peroxidation of lipids, but instead of inducing PCD, it only mediates general stress responses. Thus, it is assumed that the final outcome of different ROS-based signalling depends upon their chemical identity and amount of ROS produced during stress (Gechev et al. 2002). As a universal phenomenon, low level of $O_2^{-\bullet}$ and H_2O_2 provides protection against abiotic and oxidative stresses, and high concentrations start PCD, while the tremendously high ROS concentrations might cause necrosis (Vranova et al. 2002; Op den Camp et al. 2003; Montillet et al. 2005; Van Breusegem and Dat 2006).

4 Reactive Nitrogen Species (RNS) and Their Role in PCD

Like ROS, RNS work as signalling molecules during extreme environmental conditions induced abiotic stresses (Corpas et al. 2013a; Khan et al. 2014; Yu et al. 2014) and biotic stress, such as attack of pathogen (Bellin et al. 2013; Trapet et al. 2015; Yu et al. 2014). Under biotic and abiotic stress conditions, rapid amplification

Table 2 Different types of RNS

Free radicals	Nonradicals
Nitric oxide ($NO^{\bullet}$ or NO) **(+2)**	Nitrous acid (HNO_2) **(+3)**
Nitric dioxide ($NO_2^{\bullet}$) **(+4)**	Nitrosonium cation (NO^+) **(+3)**
Nitrate radical ($NO_3^{\bullet}$) **(+5)**	Nitrous anhydride nitrite (NO_2^-) **(+3)**
	Nitroxyl anion (NO^-) **(+1)**
	Peroxynitrite ($ONOO^-$) **(+5)**
	Dinitrogen trioxide (N_2O_3) **(+3)**
	Dinitrogen tetraoxide (N_2O_4) **(+4)**
	S-nitrosoglutathione (GSNO)

Oxidation states of nitrogen is given in parenthesis
Source: Dhawan (2014)

of both ROS and RNS, within the cell, has been experimentally demonstrated (Mittler et al. 2011; Airaki et al. 2012; Sandalio and Foyer, 2015; Baxter et al. 2014; Khan et al. 2014; Yu et al. 2014), and depending upon the intensity of stress, plants either adopt protective mechanisms or undergo PCD.

4.1 *Different Types of RNS*

Reactive nitrogen species (RNS) are a wide term covering various radicals and nonradical forms of oxides of nitrogen which are generated upon reaction of NO• with $O^{2\bullet-}$ and RO•. All the RNS have oxidation states of nitrogen ranging from +5 to −3 (Table 2).

4.2 *RNS Generation, Scavenging and Signalling in Plants*

Plants produce different types of RNS, but the pathways of RNS generation have not yet been identified in them. Among all RNS, studies have been focused mainly on nitric oxide (NO) during recent years. Most of the NO is produced by nitric oxide synthase (NOS) enzyme, in animals (Moncada et al. 1991; Ignarro 2000). NOS oxidizes L-arginine into NO and citrulline and involves FAD, FMN, tetrahydrobiopterin (BH_4), Ca^{2+} and calmodulin for this conversion (Knowles and Moncada 1994; Alderton et al. 2001). NOS or any NOS homologous gene was not reported in *A. thaliana* genome, but NOS activities have been demonstrated in many plants (Barroso et al. 1999; Corpas et al. 2004; del Rio et al. 2004). Besides NOS, presence of other enzymatic and non-enzymatic systems for NO and RNS generation systems in plants has been suggested by many research groups (Del Rio 2015; Hancock 2012; Wilson et al. 2008; Mur et al. 2012; del Río et al. 2014). Enzymatic sources of NO production in plants include nitrate reductase, xanthine oxidoreductase, peroxidase, cytochrome P450, hemeproteins and other NOS analogous enzymes, while the non-enzymatic source is reduction of exogenous NO_2^- low pH, inside apoplast. Besides enzymatic and non-enzymatic sources, plasma membrane, mitochondria, chloroplast and peroxisomes also produce NO in plants (Corpas et al. 2009). NO synthesis in peroxisomes of plants was first reported by (Corpas et al. 2004). Gupta and Kaiser (2010) demonstrated NO generation in mitochondria, while Jasid et al. (2006) demonstrated it in chloroplasts (Chamizo-Ampudia et al. 2017).

Peroxynitrite ($ONOO^-$), which is more reactive RNS than NO, is produced after reaction of NO with $O_2^{\bullet -}$. Peroxynitrate strongly nitrates tyrosine residue of target proteins, and upon reaction with CO_2, it can give rise to other RNS like nitric dioxide ($NO_2^{\bullet}$), NO_2^- and NO_3. *S*-nitrosoglutathione (GSNO) is another RNS which is produced in peroxisomes by reaction of NO with reduced glutathione.

4.3 RNS Signalling During PCD Execution

NO acts as an intercellular and intracellular signalling molecule during various developmental and physiological processes of plants. Involvement of NO in biotic and abiotic stresses responses has been demonstrated by various research groups (Leitner et al. 2009; Gaupels et al. 2011; Mur et al. 2013; Locato et al. 2016). NO also plays a crucial role in germination, pollen tube growth, cell wall lignifications, root growth, Establishment of legume-rhizobium symbiosis, flowering and ripening of fruits. NO controls all these functions by activating secondary messengers and by up-/downregulation of gene involved in them (Besson-Bard et al. 2008; Gaupels et al. 2011).

Presence of unpaired electron makes NO a highly reactive molecule, and after reaction with oxygen, it produces different reactive nitrogen intermediate (RNI), having different reducing states. These RNI include nitroxyl anion (NO^-) and positively charged nitrosonium ion (NO^+) (Gow and Ischiropoulos 2001). NO has a high affinity towards hydrophobic molecules and hence can easily move across biological membranes, while its solubility in water is very poor. Interaction of NO with various biomolecules and ROS propagates downstream signalling of NO (Weremczuk et al. 2017).

NO S-nitrosylated cysteine residue, which directly modulates enzymes and ion channels, is involved in signalling cascade. S-nitrosylation-induced PTM (post-translational modifications) inhibits the activity of peroxisomal catalase and glycolate oxidase. These two enzymes are part of the cellular antioxidant system and help in maintaining cellular level of H_2O_2 and other ROS, thus regulating their downstream signalling processes (Ortega-Galisteo et al. 2012). Nitration of tyrosine residue, directly by $ONOO^-$ and indirectly by NO, causes irreversible protein/enzyme inactivation. Inhibition of NADH-dependent hydroxypyruvate reductase by tyrosine nitration was demonstrated by proteomic studies of isolated pea leaf peroxisomes (Corpas et al. 2013b). $ONOO^-$ causes nitration of tyrosine residues which inhibits phosphorylation of tyrosine. Tyrosine phosphorylation is an important step for downstream signalling; thus, $ONOO^-$ seems to play an important role in controlling RNS-mediated signalling. Plant cells maintain a basal nitration level of tyrosine residues for some regulatory activities, but peroxynitrite-induced nitration of plant proteins induces nitrosative damage of plant cells. Under the conditions favouring high level of NO, along with ROS, plant cells opt for PCD and modulate their signalling accordingly.

5 Possible Cross-Talk Between ROS and RNS

The role of ROS and RNS as signalling molecules is well established; both share common sites of production and are involved in common signalling and physiological processes. These facts indicate that there must be a cross-talking between ROS and RNS during execution of various biological processes (Fig. 2). Fine tuning of biological processes in responses to the altered ROS level is NO mediated. NO interacts with lipid molecules and plant hormones, which further activate secondary

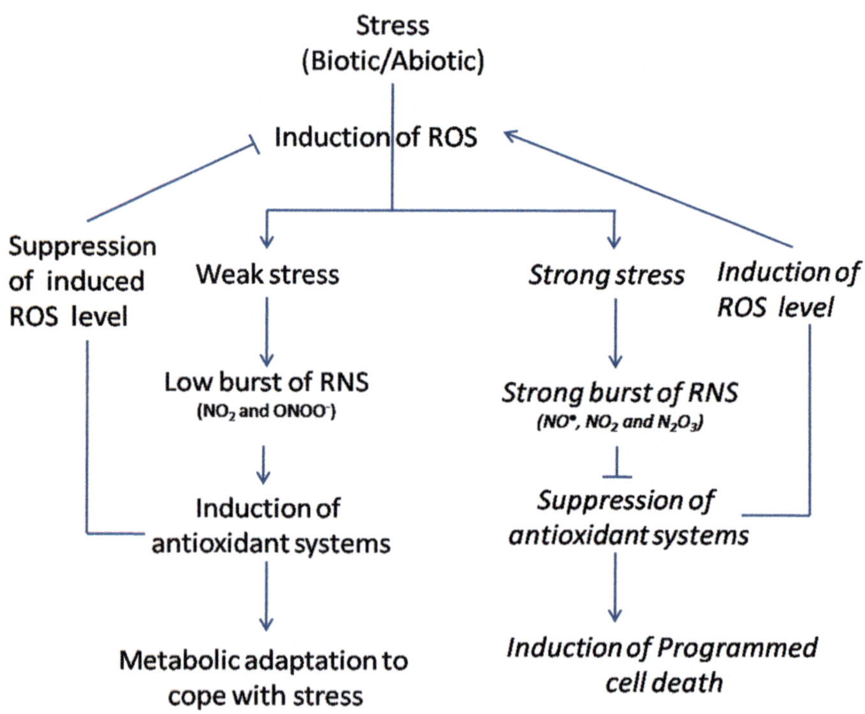

Fig. 2 Stress signals and their fates in the light of the differential expression of ROS and RNS

messengers, leading to up-/down regulation-related transcription factors (Kwak et al. 2006; Zaninotto et al. 2006).

Possible cross-talks between ROS and RNS depend upon the feasibility of chemical reactions among them. Experiments prove that ROS and RNS both control the production and scavenging of each other. H_2O_2 is supposed to react with NO, but this reaction seems unfavourable due to the relative stability of NO. Chemically favourable reaction is binding of NO with $O_2^{-\bullet}$, which produces $ONOO^-$ (peroxynitrite; Radi 2013). $ONOO^-$ is less toxic than ROS and further yields NO_2, NO and N_2O_3 during its degradation into $NO_2^\bullet$ and $NO_3^\bullet$. At neutral pH $ONOO^-$ and ONOOH form peroxynitrates (O_2NOO^- and O_2NOOH), which decay into $NO_2^\bullet$ and O_2, as well as $^1O_2$ and NO. $ONOO^-$ reacts with CO_2 also and forms NO_3^-, CO_3^- and $NO_2^\bullet$. Peroxynitrite is a powerful oxidizing and nitrating agent, whose occurrence has been reported in plant peroxisomes (Corpas and Barroso 2014). Conversion of $O_2^{-\bullet}$ into H_2O_2 by dismutation is also unfavourable one. Under high O_2 concentration, NO reacts with O_2 and gives rise to $NO_2^\bullet$, which further produces N_2O_3 if NO production continues. Depending on the environmental conditions of neighbouring cells, interaction between NO and ROS forms various intermediate products, which help the cell to respond according to altered environmental conditions. In presence of oxygen molecule, NO forms S-nitrosoglutathione (GSNO) by S-nitrosylation of glutathione (GSH). GSNO acts as a transportable sink of NO, and it is found to be present in different plant species (Ortega-Galisteo et al. 2012; Barroso et al. 2013; Corpas et al. 2013b; Xu et al. 2013; Kubienová et al. 2014; Yu et al. 2014). GSNO and peroxynitrite, formed after reaction of NO with GSH and $O_2^{-\bullet}$, respectively, have strong S-nitrosylation and the nitration activity. Both of these molecules change the activity of various enzymes by causing PTMs of proteins, through nitration and S-nitrosylation. PTMs generated by GSNO and peroxynitrite induced are reported in plants under natural as well as stressful conditions (Romero-Puertas et al. 2013; Corpas et al. 2013b).

Concepts of interaction between ROS and RNS came after the experiments of two research groups. The first group demonstrated HR (hypersensitive response)-induced cell death in soybean cells when they were infected with *Pseudomonas syringae* pv. glycinea, an avirulent strain of *P. syringae* (Levine et al. 1994, 1996). They noticed that H_2O_2 accumulation during infection was responsible for HR-induced cell death. To confirm that cell death was due to H_2O_2 accumulation, they treated the same cells with H_2O_2 scavenger, diphenylene iodonium (DPI), which inhibits NADPH oxidase, and observed that cell death was prevented. When same cells were treated with mili-molar concentration of H_2O_2 along with Ca $^{++}$, HR-induced cell death was restored. Thus, they confirmed the role of H_2O_2 in HR-induced cell death (Levine et al. 1994, 1996). Delledonne et al. (1998) demonstrated that like H_2O_2, nitric oxide was also capable of inducing HR-PCD in soybean cells and NO scavengers or NOS inhibitors could regress the phenomenon. Artificial NO donors like SNP restore the HR-induced cell death efficiently in conjunction with ROS. This NO-mediated HR-induced cell death can be prevented by applying ROS scavengers like DPI or catalase. Unlike NO donors, ROS donors alone were unable to kill soybean cells. They could induce cell death only in association with

NO donors (Delledonne et al. 1998). Same results were observed with tobacco BY-2 cells. When both NO donor and H_2O_2 donors were given together to tobacco cells, they caused inactivation of antioxidant systems and subsequently induced PCD in BY-2 cells of tobacco (De Pinto et al. 2002). Interestingly, the results were different when NO donor and H_2O_2 donor were provided individually; therefore, it was established that NO and ROS coordinate each other during cell death signalling. Generally very high concentration of ROS/RNS induces the further synthesis and accumulation of ROS/RNS to induce PCD, while the low concentration of RNS, especially NO, acts as a protector by lowering the ROS level and thus prevents oxidative damage of cell. Beligni et al. (2002) demonstrated a protective role of NO in delaying PCD of aleurone cells. They found that NO counteracts accumulation of ROS in aleurone cells and thus saves them from PCD. Among different ROS and RNS species, only role of H_2O_2, $O_2^{-\bullet}$, NO and $ONOO^-$ has been well studied; there are sparse reports on interaction among other ROS and RNS.

The role of some ROS, $O_2^{-\bullet}$, $^1O_2$ and H_2O_2 in induction of PCD, has been shown by gene manipulation studies, where increased biological concentration of either of these ROS leads to PCD. It was suggested that ROS, especially H_2O_2, induces the synthesis of NO in some parts of plants. When roots of mutant *A. thaliana*, defective in NO accumulations, were treated H_2O_2 donor, they showed H_2O_2-induced synthesis and accumulation of NO (Wang et al. 2010, 2013). H_2O_2-induced synthesis of NO was reconfirmed by Lum et al. (2002) in guard cells of *Phaseolus aureus* leaves and BY-2 cells of tobacco by De Pinto et al. (2006). Although induction of NO biosynthesis through H_2O_2 is well proved, components of signalling network and enzymes involved in NO synthesis are not well characterized yet. There is a possibility that instead of accelerating the synthesis of NO, H_2O_2 may have an inhibitory effect on S-nitrosoglutathione reductase (GSNOR), which is a NO scavenger, to increase NO concentration, but this possibility has not been investigated yet (Gaupels et al. 2008). Increase in H_2O_2 concentration not only induces NO synthesis, but other ROS are also affected, because increased level of NO regulates the production and degradation of ROS. This indicates the presence of an intricate feedback mechanism of regulation between ROS and RNS. ROS are required for the channelling of NO into PCD. Murgia et al. (2004) demonstrated that in the absence of H_2O_2, caused by overexpression of H_2O_2-scavenging enzyme APX in *A. thaliana* mutants, resistance of plants towards NO-induced PCD was increased.

In the above-described experiments, the effect of H_2O_2 on NO concentration was described; however, during vice versa experiments, NO showed the dual effect on H_2O_2 and other ROS production. Experiments showing positive effects of NO concentration on H_2O_2 accumulation were demonstrated using tobacco transgenic plants (35S::nNOS). These mutant plants were capable of NO overproduction and confirmed that accumulation of NO increases the level of H_2O_2. High concentration of NO causes inhibition of catalase in these plants, which results in an increase of H_2O_2 level and subsequent growth reduction as compared to normal plants (Chun et al. 2012). Under high concentration of both NO and H_2O_2, the salicylic acid level increased, spontaneous lesions developed and pathogen-related genes expression were upregulated in 35S::nNOS mutants.

Yun et al. (2011) demonstrated a negative effect of NO concentration on H_2O_2 accumulation. They showed that NO can limit ROS accumulation by inhibiting the ROS-producing enzyme NADPH oxidase. NO S-nitrosylates the cys-890 residue of NADPH oxidase isoform AtRBOHD and inactivates it. Inactivated RBOHD is further incapable of producing ROS; thus, level of H_2O_2 decreases. Besides inactivating NADPH isoforms, NO also enhance the activity of other antioxidant enzymes like catalase (CAT), superoxide dismutase (SOD), ascorbate peroxidase (APX) and glutathione reductase (GR), which scavenge ROS. NO was also shown to inhibit antioxidant system in study of Clark et al. (2000). Different RNS species affect antioxidant enzymes in different manner, by bringing their oxidation/S-nitrosylation/nitrosation/nitration. Negative regulation of ROS by NO enhances abiotic stress tolerance in plants by inhibiting ROS accumulation. Plants treated with artificial NO donors show decreased ROS accumulation resulting in increased abiotic stress tolerance. Scavenging of ROS by NO might be one reason of the negative effects of NO on ROS accumulation. Hence, it can be said that NO bioactivity has its role in both increase and decrease of antioxidant enzyme activities and ROS levels.

Direct evidence of interaction of NO with $O_2^{-\bullet}$ came from the study of Tewari et al. (2013). They showed formation of $ONOO^-$ after NO accumulation and detected it with the help of aminophenyl fluorescein (APF). As stated earlier, $ONOO^-$ is comparatively less toxic than ROS, and most of the ROS ultimately converts to $ONOO^-$. Interaction of NO with H_2O_2 and $O_2^{-\bullet}$ controls plant growth and development by modulating Ca^{2+}level and activities of calcium-dependent protein kinases, cGMP and MAPKs (Neill et al. 2008). Same type of interaction regulates HR-caused cell death during pathogen attack (Delledonne et al. 2001). Researches show that ROS are general stress signals, while NO signalling varies according to stress conditions. Thus, NO cannot be considered as a general stress signal. It can be hypothesized that NO brings specificity to ROS-generated general stress signals, either alone or in association with ROS (Del Rio 2015).

6 Conclusion and Perspectives

Programmed cell death is a unique program for the demise of the targeted cells, which is necessary for survival of the organism. It is observed across the kingdom, but very limited reports are available on stepwise molecular events involved in the process operative in planta. Involvement of ROS and RNS in the signalling and execution events of PCD is universally accepted, and a clear-cut picture of PCD execution events mediated by ROS/RNS in animal system has been proposed. Contrary to this, scattered evidences of the involvement of ROS-/RNS-mediated PCD are available in plants. Some recent studies showed the cross talks between different reactive species, but the exact mechanism of their interaction is not well elucidated till date. Thus, it warrants proper attention of the workers in upcoming days.

References

Airaki M, Leterrier M, Mateos RM, Valderrama R, Chaki M, Barroso JB, del Rio LA, Palma JM, Corpas FJ (2012) Metabolism of reactive oxygen species and reactive nitrogen species in pepper (*Capsicum annuum* L.) plants under low temperature stress. Plant Cell Physiol 35:281–295

Alderton WK, Cooper CE, Knowles RG (2001) Nitric oxide synthases: structure, function and inhibition. Biochem J 357:593–615

Alvarez ME, Pennell RI, Meijer PJ, Ishikawa A, Dixon RA, Lamb C (1998) Reactive oxygen intermediates mediate a systemic signal network in the establishment of plant immunity. Cell 92:773–784

Asai T, Tena G, Plotnikova J, Willmann MR, Chiu WL, Gomez L, Boller T, Ausubel FM, Sheen J (2002) MAP kinase signalling cascade in *Arabidopsis* innate immunity. Nature 415:977–983

Bailly C (2004) Active oxygen species and antioxidants in seed biology. Seed Sci Res 14:93–107

Barroso JB, Corpas FJ, Carreras A, Sandalio LM, Valderrama R, Palma JM, Lupianez JA, del Rio LA (1999) Localization of nitric-oxide synthase in plant peroxisomes. J Biol Chem 274:36729–36733

Barroso JB, Valderrama R, Corpas FJ (2013) Immunolocalization of S-nitroso-glutathione reductase and tyrosine nitration in pea leaf organelles. Acta Physiol Plant 35:2635–2640

Baxter A, Mittler R, Suzuki N (2014) ROS as key players in plant stress signalling. J Exp Bot 65:1229–1240

Beligni MV, Fath A, Bethke PC, Lamattina L, Jones RL (2002) Nitric oxide acts as an anti-oxidant and delays programmed cell death in barley aleurone layers. Plant Physiol 129:1642–1650

Bellin D, Asai S, Delledonne M, Yoshioka H (2013) Nitric oxide as a mediator for defense responses. Mol Plant-Microbe Interact 26:271–277

Besson-Bard A, Pugin A, Wendehenne D (2008) New insights into nitric oxide signaling in plants. Ann Rev Plant Biol 59:21–39

Bienert GP, Schjoerring JK, Jahn TP (2006) Membrane transport of hydrogen peroxide. Biochim Biophys Acta 1758:994–1003

Buchanan B, Gruissem W, Jones R (2002) Biochemistry and molecular biology of plants. Wiley, Hoboken

Chamizo-Ampudia A, Sanz-Luque E, Llamas A, Galvan A, Fernandez E (2017) Nitrate reductase regulates plant nitric oxide homeostasis. Trends Plant Sci 22:p163–p174

Choudhury FK, Rivero RM, Blumwald E, Mittler R (2017) Reactive oxygen species, abiotic stress and stress combination. Plant J 90:856–867

Chun HJ, Park HC, Koo SC, Lee JH, Park CY, Choi MS et al (2012) Constitutive expression of mammalian nitric oxide synthase in tobacco plants triggers disease resistance to pathogens. Mol Cells 34:463–471

Clark D, Durner J, Navarre DA, Klessig DF (2000) Nitric oxide inhibition of tobacco catalase and ascorbate peroxidase. Mol Plant-Microbe Interact 13:1380–1384

Cominelli E, Conti L, Tonelli C, Galbiati M (2013) Challenges and perspectives to improve crop drought and salinity tolerance. Nat Biotechnol 30:355–361

Corpas FJ, Barroso JB (2014) Peroxynitrite ($ONOO^-$) is endogenously produced in *Arabidopsis* peroxisomes and is overproduced under cadmium stress. Ann Bot 113:87–96

Corpas FJ, Barroso JB, Leon AM, Carreras A, Quiros M, Palma JM, Sandalio LM, del Rio LA (2004) Peroxisomes as a source of nitric oxide. In: Magalhaes JR, Singh RP, Passos LP (eds) Nitric oxide signaling in plants. Studium Press, LLC, Houston, pp 111–129

Corpas FJ, Palma JM, del Río LA, Barroso JB (2009) Evidence supporting the existence of L-arginine-dependent nitric oxide synthase activity in plants. New Phytol 184:9–14

Corpas FJ, Alche JD, Barroso JB (2013a) Current overview of S-nitrosoglutathione (GSNO) in higher plants. Front Plant Sci 4:126

Corpas FJ, del Río LA, Barroso JB (2013b) Protein tyrosine nitration in higher plants under natural and stress conditions. Front Plant Sci 4:29

Croce CM, Reed JC (2016) Finally, an apoptosis-targeting therapeutic for cancer. Cancer Res 76:5914–5920

Daneva A, Gao Z, Van Durme M, Nowack MK (2016) Functions and regulation of programmed cell death in plant development. Annu Rev Cell Dev Biol 32:441–468

Das K, Roychoudhury A (2014) Reactive oxygen species (ROS) and response of antioxidants as ROS-scavengers during environmental stress in plants. Front Environ Sci 2:1–3

Dat JF, Pellinen R, Beeckman T, van de Cotte B, Langebartels C, Kangasjarvi J, Inze D, Breusegem VF (2003) Changes in hydrogen peroxide homeostasis trigger an active cell death process in tobacco. Plant J 33:621–632

Davletova S, Rizhsky L, Liang HJ, Zhong SQ, Oliver DJ, Coutu J, Shulaev V, Schlauch K, Mittler R (2005a) Cytosolic ascorbate peroxidase 1 is a central component of the reactive oxygen gene network of *Arabidopsis*. Plant Cell 17:268–281

Davletova S, Schlauch K, Coutu J, Mittler R (2005b) The zinc-finger protein Zat12 plays a central role in reactive oxygen and abiotic stress signaling in *Arabidopsis*. Plant Physiol 139:847–856

de Pinto MC, Tommasi F, de Gara L (2002) Changes in the antioxidant systems as part of the signaling pathway responsible for the programmed cell death activated by nitric oxide and reactive oxygen species in Tobacco Bright-Yellow 2 cells. Plant Physiol 130:698–708

de Pinto MC, Paradiso A, Leonetti P, de Gara L (2006) Hydrogen peroxide, nitric oxide and cytosolic ascorbate peroxidise at the crossroad between defence and cell death. Plant J 48:784–795

De Storme N, Geelen D (2014) The impact of environmental stress on male reproductive development in plants: biological processes and molecular mechanisms. Plant Cell Environ 37:1–8

Del Rio LA (2015) ROS and RNS in plant physiology: an overview. J Exp Bot 66:2827–2837

del Rio LA, Corpasa FJ, Barroso JB (2004) Nitric oxide and nitric oxide synthase activity in plants. Phytochemistry 65:783–792

del Rio LA, Corpas FJ, Barroso JB, López-Huertas E, Palma JM (2014) Function of peroxisomes as a cellular source of nitric oxide and other reactive nitrogen species. In: Nasir Khan M, Mobin M, Mohammad F, Corpas FJ (eds) Nitric oxide in plants: metabolism and role in stress physiology. Springer-Verlag, Berlin/Heidelberg, pp 33–55

Delledonne M, Xia YJ, Dixon RA, Lamb C (1998) Nitric oxide functions as a signal in plant disease resistance. Nature 394:585–588

Delledonne M, Zeier J, Marocco A, Lamb C (2001) Signal interactions between nitric oxide and reactive oxygen intermediates in the plant hypersensitive disease resistance response. Proc Natl Acad Sci U S A 98:13454–13459

Dhawan V (2014) Reactive oxygen and nitrogen species: general considerations. In: Ganguly NK et al (eds) Studies on respiratory disorders, oxidative stress in applied basic research and clinical practice. Springer, New York, pp 27–47. https://doi.org/10.1007/978-1-4939-0497-6_2

Dietrich RA, Richberg MH, Schmidt R, Dean C, Dangl JL (1997) A novel zinc finger protein is encoded by the *Arabidopsis* LSD1 gene and functions as a negative regulator of plant cell death. Cell 88:685–694

Doorn WG, Beers EP, Dang JL, Franklin-Tong VE, Gallois P, Hara-Nishimura I et al (2011) Morphological classification of plant cell deaths. Cell Death Differ 18:1241–1246

Edreva A (2005) The importance of non-photosynthetic pigments and cinnamic acid derivatives in photoprotection. Agric Ecosyst Environ 106:135–146

Ellis RE, Yuan J, Horvitz HR (1991) Mechanisms and functions of cell death. Ann Rev Cell Biol 7:663–698

Epple P, Mack AA, Morris VRF, Dangl JL (2003) Antagonistic control of oxidative stress-induced cell death in *Arabidopsis* by two related, plant-specific zinc finger proteins. Proc Natl Acad Sci U S A 100:6831–6836

Foyer CH, Noctor G (2009) Redox regulation in photosynthetic organisms: signaling, acclimation, and practical implications. Antioxid Redox Signal 11:861–905

Fuchs Y, Steller H (2011) Programmed cell death in animal development and disease. Cell 147:742–758

Fuchs Y, Steller H (2015) Live to die another way: modes of programmed cell death and the signals emanating from dying cells. Nat Rev Mol Cell Biol 16:329

Galluzzi L, Kepp O, Chan FK, Kroemer G (2017) Necroptosis: mechanisms and relevance to disease. Annu Rev Pathol 12:103–130

Gaupels F, Furch AC, Will T, Mur LA, Kogel KH, Van-Bel AJ (2008) Nitric oxide generation in *Vicia faba* phloem cells reveals them to be sensitive detectors as well as possible systemic transducers of stress signals. New Phytol 178:634–646

Gaupels F, Kuruthukulangarakoola GT, Durner J (2011) Upstream and downstream signals of nitric oxide in pathogen defence. Curr Opin Plant Biol 14:707–714

Gautam V, Kaur R, Kohli SK, Verma V, Kaur P, Singh R et al (2017) ROS compartmentalization in plant cells under abiotic stress condition. In: Khan MIR, Khan NA (eds) Reactive oxygen species and antioxidant systems in Plants: role and regulation under abiotic stress. Springer, Singapore, pp 89–114

Gechev TS, Hille J (2005) Hydrogen peroxide as a signal controlling plant programmed cell death. J Cell Biol 168:17–20

Gechev TS, Gadjev I, Van Breusegem F, Inze D, Dukiandjiev S, Toneva V, Minkov I (2002) Hydrogen peroxide protects tobacco from oxidative stress by inducing a set of antioxidant enzymes. Cell Mol Life Sci 59:708–714

Gechev TS, Breusegem VF, Stone JM, Denev I, Laloi C (2006) Reactive oxygen species as signals that modulate plant stress responses and programmed cell death. BioEssays 28:1091–1101

Gerschman R, Gilbert DL, Nye SW, Dwyer P, Fenn WO (1954) Oxygen poisoning and X-irradiation: a mechanism in common. Science 119:623–626

Gould KS, McKelvie J, Markham KR (2002) Do anthocyanins function as antioxidants in leaves? Imaging of H_2O_2 in red and green leaves after mechanical injury. Plant Cell Environ 25:1261–1269

Gow AJ, Ischiropoulos H (2001) Nitric oxide chemistry and cellular signaling. J Cell Physiol 187:277–282

Gregersen PL, Culetic A, Boschian L, Krupinska K (2013) Plant senescence and crop productivity. Plant Mol Biol 82:603–622

Gunawardena AHL, Greenwood JS, Dengler NG (2004) Programmed cell death remodels lace plant leaf shape during development. Plant Cell 16:60–73

Gupta KJ, Kaiser WM (2010) Production and scavenging of nitric oxide by barley root mitochondria. Plant Cell Physiol 51:576–584

Halliwell B, Gutteridge JMC (2007) Free radicals in biology and medicine. Oxford University Press, Oxford

Hancock JT (2012) NO synthase? Generation of nitric oxide in plants. Period Biol 114:19–24

Henzler T, Steudle E (2000) Transport and metabolic degradation of hydrogen peroxide in *Chara corallina*: model calculations and measurements with the pressure probe suggest transport of H_2O_2 across water channels. J Exp Bot 51:2053–2066

Hochreiter-Hufford A, Ravichandran KS (2013) Clearing the dead: apoptotic cell sensing, recognition, engulfment, and digestion. Cold Spring Harb Perspect Biol 5:a008748

Huang S, Van Aken O, Schwarzländer M, Belt K, Millar AH (2016) The roles of mitochondrial reactive oxygen species in cellular signaling and stress response in plants. Plant Physiol 171:1551–1559

Huysmans M, Lema S, Coll NS, Nowack MK (2017) Dying two deaths—programmed cell death regulation in development and disease. Curr Opin Plant Biol 35:37–44

Ichimura K, Shinozaki K, Tena G, Sheen J, Henry Y, Champion A et al (2002) Mitogen activated protein kinase cascades in plants: a new nomenclature. Trends Plant Sci 7:301–308

Ignarro LJ (2000) Nitric oxide. Biology and pathobiology. Academic, San Diego

Ingram GC (2017) Dying to live: cell elimination as a developmental strategy in angiosperm seeds. J Exp Bot 68:785–796

Jacobsen MD, Weil M, Raff MC (1997) Programmed cell death in animal development. Cell 88:347–354

Jasid S, Simontacchi M, Bartoli CG, Puntarulo S (2006) Chloroplasts as a nitric oxide cellular source. Effect of reactive nitrogen species on chloroplastic lipids and proteins. Plant Physiol 142:1246–1255

Jones AM (2001) Programmed cell death in development and defense. Plant Physiol 125:95–97

Karpinski S, Reynolds H, Karpinska B, Wingsle G, Creissen G, Mullineaux P (1999) Systemic signaling and acclimation in response to excess excitation energy in *Arabidopsis*. Science 284:654–657

Kazmierczak A, Doniak M, Bernat P (2017) Membrane-related hallmarks of kinetin-induced PCD of root cortex cells. Plant Cell Rep 36:343–353

Khan MI, Khan NA (2017) Reactive oxygen species and antioxidant systems in plants: role and regulation under abiotic stress. Springer, Singapore. https://doi.org/10.1007/978-981-10-5254-5

Khan NM, Mobin M, Mohammad F, Corpas FJ (eds) (2014) Nitric oxide in plants: metabolism and role in stress physiology. Springer, Berlin

Knowles RG, Moncada S (1994) Nitric oxide synthases in mammals. Biochem J 298:249–258

Kovtun Y, Chiu WL, Tena G, Sheen J (2000) Functional analysis of oxidative stress-activated mitogen-activated protein kinase cascade in plants. Proc Natl Acad Sci U S A 97:2940–2945

Kroemer G, Galluzzi L, Vandenabeele P, Abrams J, Alnemri ES, Baehrecke EH et al (2009) Classification of cell death. Recommendations of the nomenclature committee on cell death. Cell Death Differ 16:3–11

Kubienová L, Ticha T, Jahnova J, Luhova L, Mieslerova B, Petrivalsky M (2014) Effect of abiotic stress stimuli on S-nitrosoglutathione reductase in plants. Planta 239:139–146

Kumar SR, Mohanapriya G, Sathishkumar R (2016) Abiotic stress-induced redox changes and programmed cell death in plants- a path to survival or death? In: Gupta DK, Palma JM, Corpas FJ (eds) Redox state as a central regulator of plant-cell stress responses. Springer, Cham, pp 233–252

Kwak JM, Nguyen V, Schroeder JI (2006) The role of reactive oxygen species in hormonal responses. Plant Physiol 141:323–329

Laloi C, Havaux M (2015) Key players of singlet oxygen-induced cell death in plants. Front Plant Sci 6:1–9

Leitner M, Vandelle E, Gaupels F, Bellin D, Delledonne M (2009) NO signals in the haze: nitric oxide signalling in plant defence. Curr Opin Plant Biol 12:451–458

Levine A, Tenhaken R, Dixon R, Lamb C (1994) H_2O_2 from the oxidative burst orchestrates the plant hypersensitive disease resistance response. Cell 79:583–593

Levine A, Pennell RI, Alvarez ME, Palmer R, Lamb C (1996) Calcium mediated apoptosis in a plant hypersensitive disease resistance response. Curr Biol 6:427–437

Liebthal M, Dietz KJ (2017) The fundamental role of reactive oxygen species in plant stress response. Methods Mol Biol 1631:23–39

Liu YD, Ren DT, Pike S, Pallardy S, Gassmann W, Zhang SQ (2007) Chloroplast-generated reactive oxygen species are involved in hypersensitive response-like cell death mediated by a mitogen-activated protein kinase cascade. Plant J 51:941–954

Locato V, Paradiso A, Sabetta W, De Gara L, de Pinto MC (2016) Chapter nine-nitric oxide and reactive oxygen species in PCD signaling. Adv Bot Res 77:165–192

Lopez-Delgado H, Dat J, Foyer CH, Scott IM (1998) Induction of thermo tolerance in potato microplants by acetylsalicylic acid and H_2O_2. J Exp Bot 49:713–720

Lum HK, Butt YK, Lo SC (2002) Hydrogen peroxide induces a rapid production of nitric oxide in mung bean (*Phaseolus aureus*). Nitric Oxide 6:205–213

Maizel A (2015) A view to a kill: markers for developmentally regulated cell death in plants. Plant Physiol 169:2341–2341

Majno G, Joris I (1995) Apoptosis, oncosis, and necrosis. An overview of cell death. Am J Pathol 146:3–15

Miao Y, Laun T, Zimmermann P, Zentgraf U (2004) Targets of the WRKY53 transcription factor and its role during leaf senescence in *Arabidopsis*. Plant Mol Biol 55:853–867

Miao Y, Laun TM, Smykowski A, Zentgraf U (2007) *Arabidopsis* MEKK1 can take a short cut: it can directly interact with senescence-related WRKY53 transcription factor on the protein level and can bind to its promoter. Plant Mol Biol 65:63–76

Miller G, Mittler R (2006) Could heat shock transcription factors function as hydrogen peroxide sensors in plants? Ann Bot 98:279–288

Minina EA, Smertenko AP, Bozhkov PV (2014) Vacuolar cell death in plants: metacaspase releases the brakes on autophagy. Autophagy 10:928–929

Mittler R (2017) ROS are good. Trends Plant Sci 22:11–19

Mittler R, Vanderauwera S, Suzuki N, Miller G, Tognetti VB, Vandepoele K et al (2011) ROS signaling: the new wave? Trends Plant Sci 16:300–309

Mocarski ES, Kaiser WJ, Livingston-Rosanoff D, Upton JW, Daley-Bauer LP (2014) True grit: programmed necrosis in antiviral host defense, inflammation, and immunogenicity. J Immunol 192:2019–2026

Moncada S, Palmer RMJ, Higgs EA (1991) Nitric oxide: physiology, pathophysiology and pharmacology. Pharmacol Rev 43:109–142

Montillet JL, Chamnongpol S, Rusterucci C, Dat J, van de Cotte B, Agnel JP et al (2005) Fatty acid hydroperoxides and H_2O_2 in the execution of hypersensitive cell death in tobacco leaves. Plant Physiol 138:1516–1526

Mueller MJ (2004) Archetype signals in plants: the phytoprostanes. Curr Opin Plant Biol 7:441–448

Muntz K (2007) Protein dynamics and proteolysis in plant vacuoles. J Exp Bot 58:2391–2407

Mur LAJ, Mandon J, Persijn S, Crisstescu SM, Moshkov IE, Novikova GV, Hall MA, Harren FJM, Hebelstrup KM, Gupta KJ (2012) Nitric oxide in plants: an assessment of the current state of knowledge. AoB Plants 5:pls 052

Mur LA, Mandon J, Persijn S, Cristescu SM, Moshkov IE, Novikova GV et al (2013) Nitric oxide in plants: an assessment of the current state of knowledge. AoB Plants 5:pls052

Murgia I, Tarantino D, Vannini C, Bracale M, Carravieri S, Soave C (2004) *Arabidopsis thaliana* plants over-expressing thylakoidal ascorbate peroxidase show increased resistance to paraquat-induced photooxidative stress and to nitric oxide-induced cell death. Plant J 38:940–953

Nakagami H, Kiegerl S, Hirt H (2004) OMTK1, a novel MAPKKK, channels oxidative stress signaling through direct MAPK interaction. J Biol Chem 279:26959–26966

Nakagami H, Soukupova H, Schikora A, Zarsky V, Hirt H (2006) A mitogen activated protein kinase kinase kinase mediates reactive oxygen species homeostasis in *Arabidopsis*. J Biol Chem 281:38697–38704

Nakashima J, Takabe K, Fujita M, Fukuda H (2000) Autolysis during in-vitro tracheary element differentiation: formation and location of the perforation. Plant Cell Physiol 41:1267–1271

Nath M, Bhatt D, Prasad R, Tuteja N (2017) Reactive Oxygen Species (ROS) metabolism and signaling in plant-mycorrhizal association under biotic and abiotic stress conditions. In: Varma A, Prasad R, Tuteja N (eds) Mycorrhiza-eco-physiology, secondary metabolites, nanomaterials. Springer, Cham, pp 223–232

Neill SJ, Bright J, Desikan R, Hancock JT, Harrison J, Wilson I (2008) Nitric oxide evolution and perception. J Exp Bot 59:25–35

op Den Camp RGL, Przybyla D, Ochsenbein C, Laloi C, Kim C, Danon A (2003) Rapid induction of distinct stress responses after the release of singlet oxygen in *Arabidopsis*. Plant Cell 15:2320–2332

Ortega-Galisteo AP, Rodríguez-Serrano M, Pazmiño DM, Gupta DK, Sandalio LM, Romero-Puertas MC (2012) S-Nitrosylated proteins in pea (*Pisum sativum* L.) leaf peroxisomes: changes under abiotic stress. J Exp Bot 63:2089–2103

Petrov V, Hille J, Mueller-Roeber B, Gechev TS (2015) ROS-mediated abiotic stress-induced programmed cell death in plants. Front Plant Sci 205:6. https://doi.org/10.3389/fpls.2015.00069

Poon IK, Lucas CD, Rossi AG, Ravichandran KS (2014) Apoptotic cell clearance: basic biology and therapeutic potential. Nat Rev Immunol 14:166–180

Queval G, Issakidis-Bourguet E, Hoeberichts FA, Vandorpe M, Gakiere B, Vanacker H et al (2007) Conditional oxidative stress responses in the *Arabidopsis* photorespiratory mutant cat2 demonstrate that redox state is a key modulator of day length-dependent gene expression, and define photoperiod as a crucial factor in the regulation of H2O2-induced cell death. Plant J 52:640–657

Radi R (2013) Protein tyrosine nitration: biochemical mechanisms and structural basis of functional effects. Acc Chem Res 46:550–559

Rentel MC, Lecourieux D, Ouaked F, Usher SL, Petersen L, Okamoto H et al (2004) OXI1 kinase is necessary for oxidative burst-mediated signalling in *Arabidopsis*. Nature 427:858–861

Rizhsky L, Davletova S, Liang HJ, Mittler R (2004) The zinc finger protein Zat12 is required for cytosolic ascorbate peroxidase 1 expression during oxidative stress in *Arabidopsis*. J Biol Chem 279:11736–11743

Romero-Puertas MC, Rodríguez-Serrano M, Sandalio LM (2013) Protein S-nytrosylation in plants under abiotic stress: an overview. Front Plant Sci 4:373

Rosenzweig C, Elliott J, Deryng D, Ruane AC, Muller C, Arneth A et al (2014) Assessing agricultural risks of climate change in the 21st century in a global gridded crop model intercomparison. Proc Natl Acad Sci U S A 111:3268–3273

Sandalio LM, Foyer CH (2015) Unravelling the reactive oxygen and reactive nitrogen signalling networks in plants. J Exp Bot 66:2825–2826

Tewari RK, Prommer J, Watanabe M (2013) Endogenous nitric oxide generation in protoplast chloroplasts. Plant Cell Rep 32:31–44

Torres MA, Jones JDG, Dangl JL (2005) Pathogen-induced, NADPH oxidase derived reactive oxygen intermediates suppress spread of cell death in *Arabidopsis thaliana*. Nat Genet 37:1130–1134

Trapet P, Kulik K, Lamotte O, Jeandroz S, Bourque S, Nicolas-Frances V et al (2015) NO signaling in plant immunity: a tale of messengers. Phytochemistry 112:72–79

Tsuda T, Kato Y, Osawa T (2000) Mechanism for the peroxynitrite scavenging activity by anthocyanins. FEBS Lett 484:207–210

Van Breusegem F, Dat J (2006) Reactive oxygen species in plant cell death. Plant Physiol 141:384–390

Van Durme M, Nowack MK (2016) Mechanisms of developmentally controlled cell death in plants. Curr Opin Plant Biol 29:29–37

Van Hautegem VT, Waters AJ, Goodrich J, Nowack MJ (2015) Only in dying, life: programmed cell death during plant development. Trends Plant Sci 20:102–113

Vanderauwera S, Zimmermann P, Rombauts S, Vandenabeele S, Langebartels C, Gruissem W et al (2005) Genome-wide analysis of hydrogen peroxide-regulated gene expression in *Arabidopsis* reveals a high light-induced transcriptional cluster involved in anthocyanin biosynthesis. Plant Physiol 139:806–821

Vranova E, Atichartpongkul S, Villarroel R, Montagu VM, Inze D, Camp VW (2002) Comprehensive analysis of gene expression in *Nicotiana tabacum* leaves acclimated to oxidative stress. Proc Natl Acad Sci U S A 99:10870–10875

Wang Y, Ries A, Wu K, Yang A, Crawford NM (2010) The *Arabidopsis* prohibitin gene PHB3 functions in nitric oxide-mediated responses and in hydrogen peroxide-induced nitric oxide accumulation. Plant Cell 22:249–259

Wang Y, Loake GJ, Chu C (2013) Cross-talk of nitric oxide and reactive oxygen species in plant programed cell death. Front Plant Sci 4:314

Weremczuk A, Ruszczyńska A, Bulska E, Antosiewicz DM (2017) NO-dependent programmed cell death is involved in the formation of Zn-related lesions in tobacco leaves. Metallomics 9:924–935

Wilson ID, Neill SJ, Hancock JT (2008) Nitric oxide synthesis and signaling in plants. Plant Cell Environ 31:622–631

Xu S, Guerra D, Lee U, Vierling E (2013) S-nitrosoglutathione reductases are low-copy number, cysteine-rich proteins in plants that control multiple developmental and defense responses in *Arabidopsis*. Front Plant Sci 4:430

Yu M, Lamattina L, Spoel SH, Loake GJ (2014) Nitric oxide function in plant biology: a redox cue in deconvolution. New Phytol 202:1142–1156

Yun BW, Feechan A, Yin M, Saidi NB, Bihan LT, Yu M et al (2011) S-nitrosylation of NADPH oxidase regulates cell death in plant immunity. Nature 478:264–268

Zaninotto F, La Camera S, Polverari A, Delledonne M (2006) Cross talk between reactive nitrogen and oxygen species during the hypersensitive disease resistance response. Plant Physiol 141:379–383

Progress and Prospects in *Capsicum* Breeding for Biotic and Abiotic Stresses

Sushil Satish Chhapekar, Vandana Jaiswal, Ilyas Ahmad, Rashmi Gaur, and Nirala Ramchiary

Abstract The genus *Capsicum* (chili), one of the important Solanaceae crop plants, is grown widely for producing vegetables and spices and for extraction of the coloring agent. Chili fruits contain a vast number of metabolites that are crucial for human health, viz., carotenoids (provitamin A), vitamin E, vitamins C, flavonoids, and capsaicinoids (destroy free radicals). However, *Capsicum* production is highly affected by biotic and abiotic stresses and, thus, needs urgent attention of *Capsicum* researchers/breeders. Abiotic stresses mainly include drought, heat, cold, and salinity, while major biotic stresses comprise of root, stem, leaf, and fruit rots; leaf spot, viral, and powdery mildew diseases; and diseases caused by nematodes. Several studies identifying/mapping QTLs/genes conferring resistance/tolerance to major biotic and abiotic stresses have been reported. The global initiative to collect and share and systematic evaluation of phenotypes of *Capsicum* genetic materials for abiotic and biotic stress resistances/tolerances would greatly enhance the understanding of genetic mechanism regulating those traits, thereby helping in sustainable production to meet the worldwide demand and increase the income of the farmers. Furthermore, the introduction of high-throughput next-generation sequencing (NGS) technologies to sequence genomes and transcriptomes within a short period of time with comparatively cheaper cost would be helpful to decipher the genome structure and function of genes.

Keywords *Capsicum* · Plant stress · Gene mapping · Breeding

1 Introduction

The Solanaceae family consists of approx. 2500 flowering plant species in 102 genera and considered to be the third most significantly important plant family after grasses and legumes. The Solanaceae family includes nutritionally rich crop plants

S. S. Chhapekar · V. Jaiswal · I. Ahmad · R. Gaur · N. Ramchiary (✉)
Translational and Evolutionary Genomics Lab, School of Life Sciences, Jawaharlal Nehru University, New Delhi, India

S. Vats (ed.), *Biotic and Abiotic Stress Tolerance in Plants*,
https://doi.org/10.1007/978-981-10-9029-5_11

such as tomato (*Solanum lycopersicun*), potato (*S. tuberosum*), eggplant (*S. melongena*), and *Capsicum* or pepper (*Capsicum annuum*) and is consumed worldwide. Chili pepper or *Capsicum*, one of the important Solanaceae crops, grown for producing vegetables and spices worldwide, has been reported to be originated in Central and South America (Bosland 1996; Perry et al. 2007). Of the total reported 38 species of the genus *Capsicum*, *C. annuum*, *C. baccatum*, *C. chinense*, *C. frutescens*, *C. assamicum*, and *C. pubescens* are cultivated. The world's annual production in 2013 for dry chili was 3.446 million tons (area harvested 2 million ha) and fresh green chili was 31.11 million tons (area harvested 2.6 million ha, FAOSTAT 2015). India is at the top with respect to production (25% of the global production), consumption, and exportation of the chili crop across the world. Chili fruits contain a vast number of metabolites that are crucial for human health, viz., carotenoids (provitamin A), vitamin E, vitamins C, flavonoids, and capsaicinoids (important for destroying free radicals, Maga 1975; Simonne et al. 1997). Due to high medicinal properties of capsaicinoids complex (causing pungency to fruits) and other biochemical components, they are also widely used in traditional medicines. Several reports have been published where capsaicin was used as a pain-relieving agent in various chemotherapy or radiation therapies and reducing cancer (Berger et al. 1996). Apart from this, chili fruits are also used for extracting natural colors.

Like other crops, *Capsicum* production is also highly affected by biotic and abiotic stresses. Due to the sessile nature, plants develop different mechanism to survive in adverse conditions (Abuqamar et al. 2009). During biotic stresses, plant cells experience and show hypersensitive reaction, cell death, defense activation, ion-flux change, etc. Study reports indicate that biotic resistance might be triggered by abiotic stresses, and in few cases response to biotic and abiotic stresses has been suggested to be controlled by common set of genes (Cheong et al. 2002; Fujita et al. 2006). It is widely known in plants that different hormone signaling pathways such as salicylic acid (SA), jasmonic acid (JA), etc. are activated in response to stresses (Garcia et al. 2016). Besides, during stresses, plant may undergo other physiological changes like triggering of the kinase cascades, activation of ion channels, and accumulation of reactive oxygen species (Fujita et al. 2006; Fraire-Velzquez et al. 2011). Transcriptome analysis in *Capsicum* suggested that several genes are highly expressed in response to multiple stresses (both biotic and abiotic, Lee and Choi 2013). Recently, *Pea DNA Helicase 45 (PDH45)* has been identified to be involved in multiple abiotic stress tolerance in pepper (Shivakumara et al. 2017).

Despite the success of conventional breeding in development of tolerant/resistance varieties against abiotic and biotic stresses, and achieving high productivity, the identification of gene(s) governing those traits is limited. Furthermore, conventional breeding is time consuming, influenced by environments and selection at early stage cannot be done. The development and use of molecular markers and quantitative trait loci (QTLs) mapping techniques coupled with the newly developed next-generation sequencing (NGS) and genotyping technologies could identify a vast number of genes/QTL associated with economically important traits which could be further used in systematic breeding program. Furthermore, the availability of the whole genome sequence data of *Capsicum annuum* has greatly enhanced the *Capsicum* breeding program. In this chapter, an attempt has been made to compile the research progress made till date toward the development and breeding of *Capsicum* species against the biotic and abiotic stress resistances/tolerances from classical breeding to the recent use of large-scale transcriptome and genome sequencing technologies.

2 Conventional *Capsicum* Breeding for Development of Stress Resistance Varieties and Hybrids

Conventional breeding involves crossing of parents with contrasting phenotypes and selection of individuals from the segregating population for desired traits. Several public-funded research institutes and private companies around the world have developed high-yielding varieties with desired plant types, high nutritional components, resistance to biotic and abiotic stresses, and of specific durations. Further, several hybrids are being developed for various agronomically important traits such as pungency, color, yield, virus resistance, fruit size and shape, disease tolerance/ resistance, and other quality traits. In India too, several research institutes and companies have developed hybrids in chili pepper to increase the yield. However, many private companies do not publish the details of the hybrids. Some selected commercially released chili pepper varieties and popular hybrids with their characteristics features are listed in Table 1. However, we regret for not being able to list all the varieties and hybrids due to lack of publicly accessible information. In India, screening of *Capsicum* germplasm against major pests could identify several tolerant lines for biotic stresses (Reddy and Reddy 2010; Pandravada et al. 2010; Babu et al. 2011; Kaur et al. 2011; Kumar et al. 2011; Mondal et al. 2013; Reddy et al. 2014a, b; Banerjee et al. 2014). Some of them are Bhut Jolokia, PBC80, LLS, Breck-1, Breck-2, and Jaun (*Colletotrichum* spp. resistant); AVPP0102, PBC66, PBC67, PBC384, PBC385, PBC535, and MC-4 (bacterial wilt resistant); GKC29, PI201234, and IC364063 (*Phytophthora* blight resistant); and BS35, GKC29, and Bhut Jolokia (ChiCLV resistant; Reddy et al. 2014b).

3 Molecular Breeding for Resistance/Tolerance Against Stresses in *Capsicum*

The systematic molecular breeding started with the introduction of DNA markers in late 1980s. Further, continuous efforts to develop simple, robust, less expensive, and high-throughput molecular markers by various researchers around the world resulted in the development of different markers and their genotyping techniques from early generation to present-day next-generation markers (Table 2). The following sections

Table 1 List of popular pepper hybrids/varieties developed and commercialized by public institutes and private seed companies

Serial No.	Name of hybrids/varieties	Developed by	Country of release or commercialized	Characters	References
1	Kashi Surkh, Hybrid-Kashi Early	Indian Institute of Vegetable Research, Varanasi, India	India	Fruits are long (10–12 cm), 1.0–1.2 cm girth, highly attractive, yellowish-green and turn bright red at maturity, pungent with wrinkled surface	Checklist of commercial varieties of vegetables by Dr. Gorakh Singh, Horticulture Commissioner,Govt of India (http://agricoop.nic.in/book792012.pdf)
	Kashi Anmol		India (Punjab, Uttar Pradesh, Bihar, and Jharkhand)	Plants are dwarf in height (60–70 cm) with nodal pigmentation on stem and bear green attractive pendant fruits. First picking starts from 55 days after transplanting	Checklist of commercial varieties of vegetables by Dr. Gorakh Singh, Horticulture Commissioner,Govt of India (http://agricoop.nic.in/book792012.pdf)
2	California Wonder	India Institute of Agricultural Research, New Delhi, India	All over India	The fruits are smooth, heavy, thick-walled, and deep green. The flesh is very thick, sweet, and fine-flavored, turning bright crimson at maturity. The variety has very high market acceptability for its shape, size, and color	http://ztmbpd.iari.res.in/?q=capsicum
3	Arka Sweta, Arka Hybrid	India Institute of Horticulture Research, India	India	High-yielding chili F_1 hybrid developed by using MS line	http://www.iihr.res.in/content/chilli-arka-sweta
	Arka Mohini			The plant has 3–4 lobed dark green blocky fruits with thick fleshed, turns to red at maturity	http://www.iihr.res.in/content/capsicum-arka-mohini
	Arka Basant			Thick-fleshed, 2–3 lobed conical fruits with average fruit weight 50–80 g. The fruits are erect, cream-colored, and turn orange-red at ripening	http://www.iihr.res.in/content/capsicum-arka-basant

	Arka Gaurav			Indeterminate plant habit with green foliage. Thick-fleshed, 3–4 lobed green blocky fruits and will turn to orange-yellow at ripening	http://www.iihr.res.in/content/capsicum-arka-gaurav
	Arka Meghana	India Institute of Horticulture Research, India	India (Punjab, Uttar Pradesh, Gujarat, Rajasthan, Haryana, Delhi, Karnataka, Tamil Nadu Bihar, Odisha, Jharkhand, Chhattisgarh, Arunachal Pradesh, and Kerala)	Widely cultivated in various parts of India, high-yielding chili, field tolerant to viruses and sucking pest	http://www.iihr.res.in/content/chilli-arka-meghana
	Arka Lohit	India Institute of Horticulture Research, India	India (Chhattisgarh, Odisha, Arunachal Pradesh, M.P., Maharashtra, Karnataka, Tamil Nadu, and Kerala)	Widely cultivated, highly pungent, suitable for rainfed and irrigated areas. The fruits are dark green turns to dark red at maturity	www.iihr.res.in/content/chilli-arka-lohit
	Arka Harita	India Institute of Horticulture Research, India	India (Karnataka, Tamil Nadu, and Kerala)	Fruits with length 10 cm and width 1 cm. They are dark green in color and turn to red at maturity. Resistant to powdery mildew and viruses	www.iihr.res.in/content/chilli-arka-harita
	Arka Suphal	India Institute of Horticulture Research, India	India (M.P. and Maharashtra)	Fruits are straight, smooth with pointed tip, 7–9 cm long, green-color fruit changing to deep red at maturity, resistant to powdery mildew and field tolerant to viruses	http://www.iihr.res.in/content/chilli-arka-suphal-pmr-57
4	Pusa Jwala, Pusa Deepti	India Institute of Agricultural Research, India	All over India	Pusa Jwala is the most popular variety grown all over India and used in most popular spicy foods of India. It is widely	http://ztmbpd.iari.res.in/?q=capsicum

(continued)

Table 1 (continued)

Serial No.	Name of hybrids/ varieties	Developed by	Country of release or commercialized	Characters	References
				cultivated due to their favorable characteristics. The fruit is yellowish-green and turns red till maturity, first picking in 60–70 days after transplanting	
	Pusa Sadabahar			Widely cultivated throughout in India, highly pungent, fruits are 6–8 cm long, resistant to CMV, TMV, and leaf curl complex, first picking in 75–80 days after transplanting	Checklist of commercial varieties of vegetables by Dr. Gorakh Singh, Horticulture Commissioner, Govt of India (http://agricoop.nic.in/book792012.pdf)
	Gujarat Vegetable Chilli-101 (GVC-101)	Anand Agricultural University, Anand, Gujarat	India (Gujarat)	The fruits of this variety are green in color with smooth surface having good luster. The fruits are straight in shape with pointed blossom end	http://www.aau.in/crop-varieties?crop_group=4&crop=6742
	Gujarat Vegetable Chilli-121 (GVC-121)			Pungent fruits of dark green color with good luster. Fruits are straight in shape with pointed blossom end	http://www.aau.in/crop-varieties?crop_group=4&crop=6742
	Gujarat Vegetable Chilli-111 (GVC-111)			The fruits are green in color with elongated straight shape and pointed blossom end. There are slightly wrinkled fruit surface with more luster	http://www.aau.in/crop-varieties?crop_group=4&crop=6742
	Vegetable Non Pungent Chilli-131 (AVNPC-131)			The fruits are nonpungent, elongated, straight, compact, short pedicel length, smooth and shining surface with pointed blossom end	http://www.aau.in/crop-varieties?crop_group=4&crop=6742

	Gujarat Anand Vegetable Chilli-112 (GAVC-112)			The pungent variety, with elongated straight and light green color fruit. They have semi wrinkled surface on fruit	http://www.aau.in/crop-varieties?crop_group=4&crop=6742
	Gujarat Anand Vegetable Chilli hybrid-1 (GAVCH-1)			The light green color fruits are elongated straight with pointed blossom end having semi wrinkle surface	http://www.aau.in/crop-varieties?crop_group=4&crop=6742
	Phule Jyoti	Mahatma Phule Krishi Vidyapeeth, Rahuri, Maharashtra	India (Maharashtra)	Cultivated primarily in kharif season. The plants are tall in nature. Fruits are smooth and medium long. Resistant to thrips, mites, and wilt	http://agricoop.nic.in/book792012.pdf
	Phule Mukta			The dark green-colored fruits are small and smooth. The variety is suitable for kharif season and resistant to powdery mildew and fusarium wilt	http://agricoop.nic.in/book792012.pdf
	Hisar Shakti	CCS Haryana Agricultural University, Hisar	India (Haryana)	Highly pungent variety, the fruits are pointed, thin, medium long, upright in clusters, tolerant to leaf curl mosaic virus, fruit rot, and powdery mildew	http://www.hau.ernet.in/research/varreleased.php
	Hisar Vijay			Fruits having blackish tinge on fruit surface under low temperature, also tolerant to TMV, leaf curl, fruit rot	http://www.hau.ernet.in/research/varreleased.php

(continued)

Table 1 (continued)

Serial No.	Name of hybrids/ varieties	Developed by	Country of release or commercialized	Characters	References
5	Tejaswini, MPH-55, MPH-58, MPH-59	Mahyco	All over India	Very hot chili	http://www.mahyco.com/products/2/29/54/vegetable-crops/chillies-hybrid/mhp-1-tejaswini
6	F_1 Hybrid Coral, F_1 Hybrid Dara	Clover Seeds, China	China	NA	http://avrdc.org/download/publications/crop-guides/peppers/Pepper_Chronica%20Hort_Vol%2053%20No%203%202013.pdf
7	VNR38, VNR108, VNR174, (Rani)	VNR Seeds, India	All over India	Intermediate achari variety, good for processing, suitable for rainy season	http://www.vnrseeds.com/showvariety.php?productid=7&varietyid=18
8	VNR200,	VNR Seeds, India	All over India	Medium early achari type pre-sowing period variety (ageti), fruits in rainy weather also	http://www.vnrseeds.com/showvariety.php?productid=7&varietyid=19
9	VNR332	VNR Seeds, India	All over India	High dual-purpose variety, high yield potential	http://www.vnrseeds.com/
10	F_1 Forever	Tropicasem, Senegal, Africa	Sub-Saharan Africa countries	NA	http://avrdc.org/download/publications/crop-guides/peppers/Pepper_Chronica%20Hort_Vol%2053%20No%203%202013.pdf
11	Remington, F1	Alpha Seeds, South Africa	Sub-Saharan Africa countries	NA	http://avrdc.org/download/publications/crop-guides/peppers/Pepper_Chronica%20Hort_Vol%2053%20No%203%202013.pdf

12	F1 TSS AVRDC No.4	Suntech Seeds, Taiwan	Taiwan	NA	http://avrdc.org/download/publications/crop-guides/peppers/Pepper_Chronica%20Hort_Vol%2053%20No%203%202013.pdf
13	F 1 TSS AVRDC No.2	Yung Shan Seeds, Taiwan	Taiwan	NA	http://avrdc.org/download/publications/crop-guides/peppers/Pepper_Chronica%20Hort_Vol%2053%20No%203%202013.pdf
14	Yun Pepper No.2	Horticulture Research Institute, YAAS, China	China	NA	http://avrdc.org/download/publications/crop-guides/peppers/Pepper_Chronica%20Hort_Vol%2053%20No%203%202013.pdf
15	Yun High Pungency No.1	Horticulture Research Institute, YAAS, China	China	NA	http://avrdc.org/download/publications/crop-guides/peppers/Pepper_Chronica%20Hort_Vol%2053%20No%203%202013.pdf
16	F1 Hsing AVRDC No.3 (sweet pepper)	Suntech Seeds, Taiwan	Taiwan	NA	http://avrdc.org/download/publications/crop-guides/peppers/Pepper_Chronica%20Hort_Vol%2053%20No%203%202013.pdf
17	Super F1,	East-West Seeds, Thailand	Sri Lanka	Thai type with long thin chilies and are hot	http://www.eastwestseed.com/international/en/products/detail.php?SECTION_ID=108&ELEMENT_ID=2722
18	Muria F1	East-West Seeds, Thailand	Sri Lanka	Easy fruit setting, resulting high yields of uniform fruits	http://www.eastwestseed.com/international/en/products/

(continued)

Table 1 (continued)

Serial No.	Name of hybrids/ varieties	Developed by	Country of release or commercialized	Characters	References
					detail.php?SECTION_ ID=108&ELEMENT_ ID=662
19	Ulka F1, Masaya 315, Yuvraj IN	East-West Seeds, India	All over India	Plant is vigorous, strong branches and pungent	http://www.eastwestseed.com/ international/en/products/
20	Hybrid	Indus Seeds, India	All over India	NA	http://avrdc.org/download/ publications/
21	HOE-808, HOE-888, HOE-818	Hoechest Seeds, Germany	All over India	NA	http://avrdc.org/download/ publications/
22	NS-1101, NS-1420,	Namdhari Seeds, India	All over India	Fruits are thin walled, straight, short (7–8 cm), and very highly pungent and high level of tolerance to viruses	http://www.namdhariseeds. com/products/chilli/ns-1101-f1, http://www. namdhariseeds.com/compare/ Chilli/238
23	NS-1701	Namdhari Seeds, India	All over India	Fruits of this hybrid are very highly pungent and the plants exhibit high level of tolerance to viruses	http://www.namdhariseeds. com/products/chilli/ns-1701
24	ARCH-006, ARCH-001, ARCH-226, ARCH-228, ARCH-236	Ankur Seeds, India	All over India	NA	http://avrdc.org/download/ publications/
25	BSS-141, BSS-131, BSS-273, Gayatri	Beejo Sheetal Seeds, India	All over India	NA	http://avrdc.org/download/ publications/
26	Champion	Seoul Seeds	South Korea	NA	http://avrdc.org/download/ publications/

27	Delhi hot, Hot green, Skyline	Seminis, India	India, South Korea	NA	http://avrdc.org/download/publications/
	K1	Tamil nadu Agricultural University, Coimbatore, Tamil Nadu	India (Tamil Nadu)	The plants are bushy, and fruits are green in color turns to dark red at ripening, suitable for rainfed condition	http://agritech.tnau.ac.in/horticulture/horti_TNAU_varieties_vegetablecrops.html#chill
	K2			The plants are tall in nature, and fruits are long, red in color at ripening with high seed content	http://agritech.tnau.ac.in/horticulture/horti_TNAU_varieties_vegetablecrops.html#chill
	MDU 1			The variety is suitable for tropical parts of this region. The fruits are in cluster of 4–9; at maturity, the purple color of fruits fades to pale pink	http://agritech.tnau.ac.in/horticulture/horti_TNAU_varieties_vegetablecrops.html#chill
	CO1			High capsaicin content and fruits are long, red in color	http://agritech.tnau.ac.in/horticulture/horti_TNAU_varieties_vegetablecrops.html#chill
	CO3			The dark green-colored fruits are long, slender. They are more suitable for export	http://agritech.tnau.ac.in/horticulture/horti_TNAU_varieties_vegetablecrops.html#chill
	CO4			The fruits are less pungent and dark green color. They are generally suitable for dense planting	http://agritech.tnau.ac.in/horticulture/horti_TNAU_varieties_vegetablecrops.html#chill
	PKM1			Pungent variety and suitable for irrigated cultivation	http://agritech.tnau.ac.in/horticulture/horti_TNAU_varieties_vegetablecrops.html#chill
	PMK1			The plants are medium tall with spreading branches and	http://agritech.tnau.ac.in/horticulture/horti_TNAU_varieties_vegetablecrops.html#chill

(continued)

Table 1 (continued)

Serial No.	Name of hybrids/ varieties	Developed by	Country of release or commercialized	Characters	References
				suitable for semi-dry conditions	
	TNAU Chilli Hybrid CO 1			The fruits are elongated with 10–12 cm long; they are partial resistance to fruit rot disease	http://agritech.tnau.ac.in/horticulture/horti_TNAU_varieties_vegetablecrops.html#chill
	Pant C-1	Govind Ballabh Pant University of Agriculture and Technology Pantnagar, Uttarakhand	All over India	Highly pungent upright fruits, small in size with partial tolerance to mosaic and leaf curl virus	http://www.gbpuat.ac.in/research/ReleasedVarieties.pdf
	Utkal Ava (BC-14-2)	Orissa University of Agriculture and Technology Bhubaneswar, Orissa	India (Orissa, Maharashtra and Gujarat)	Cultivated all around the year, Fruits are erect, straight, medium length, and turns to deep red at ripening	http://www.ouat.ac.in/CropVarietiesReleasedOUAT.aspx and http://dacnet.nic.in/farmer/new/dac/SeedCharacteristics.asp?SCod=04&DCod=0&BCod=0&varietycd=H0703015

NA Not available

Table 2 Reports of candidate genes, QTLs, and linkage maps used in pepper translational research

Population/cultivar	Candidate genes/QTLs	Traits	Marker type	No. of marker	References
NuMex R Naky (*C. annuum*) X PI159234 (*C. chinense*)	–	–	RFLP	192	Prince et al. (1993)
Doux des lands (*C. annuum*) X PI159234 (*C. chinense*)	–	--	RFLP	85	Tanksley et al. (1988)
NuMex R Naky (*C. annuum*) X PI159234 (*C. chinense*)	–	–	RAPD, AFLP, RFLP,	1007	Livingstone et al. (1999)
Maor (*C. annuum*) X Perennial (*C. annuum*)	–	–	RAPD, RFLP	177	Ben-Chaim et al. (2001)
TF68 (*C. annuum*) X Haba-nero (*C. chinense*)	–	–	RFLP, AFLP	580	Kang et al. (2001)
Yo lo Wonder (*C. annuum*) X Criollo de Morelos334 (*C. annuum*)	–	–	RAPD, RFLP, AFLP	208	Lefebvre et al. (2002)
NMCA-30036 (*C. chinense*)	*Pun 1*, *AT3*	Capsaicinoid biosynthesis	–	–	Blum et al. (2002) and Stewart et al. (2007)
Capsicum sp.	*Pal, Ca4h, Comt*, 3-ketoacyl-ACP synthase (*Kas*), putative aminotransferase (*pAmt*), *Acl, Fat*, SB2-149 and SB1-158 clones (Similarity to *pAmt* and *Kas)*	Capsaicinoid biosynthesis	–	–	Curry et al. (1999), Kim et al. (2001), and Aluru et al. (2003)
H3(*C. annuum*) X Vania (*C. annuum*)	–	–	RAPD, AFLP, RFLP,	543	Lefebvre et al. (2002)
Perennial (*C. annuum*) X Yo lo Wonder (*C. annuum*)	–	–	RAPD, AFLP, RFLP	630	Lefebvre et al. (2002)

(continued)

Table 2 (continued)

Population/cultivar	Candidate genes/QTLs	Traits	Marker type	No. of marker	References
Maor (*C. annuum*) X BG 2816 (*C. frutescens*)	–	–	RFLP	92	Rao et al. (2003)
NuMex R Naky (*C. annuum*) X BG 2816 (*C. frutescens*)	*BCAT* and *3A2*	Capsaicinoid biosynthesis	RFLP, AFLP	728	Ben-Chaim et al. (2006)
	3.1, 4.2, 4.13, 4.14, 4.15, 4.16, 5.4, 6.8, 7.3, 10.2, 10.3, 11.8 (12 QTLs)	Capsaicinoid biosynthesis	–	–	Yarnes et al. (2013)
	2.6, 2.7, 2.9, 6.4, 9.3, 9.4, 9.6, 9.8	Fruit lobedness	–	–	Yarnes et al. (2013)
Manganji (*C. annuum*) X Tongari (*C. annuum*)	–	–	RFLP, AFLP	728	Ben-Chaim et al. (2006)
NA	*Pun1*, *CCR*, *KAS* and *HCT*	Capsaicinoid biosynthesis	–	–	Reddy et al. (2014b)
TF68 (*C. annuum*) X Habanero (*C. chinense*)	–	–	SSR, RFLP	180 (SSR) + 63 (RFLP)	Yi et al. (2006)
Integrated Maps: Two interspecific TF68 X Habanero (*C. annuum*) And Two intraspecific CM334 (*C. annuum*) X Chilsungcho (*C. annuum*)	–	–	RFLP, SSR, CAPS, AFLP, WRKY, rRAMP	805 (interspecific) + 745 (intraspecific)	Lee et al. (2008)
NuMex R Naky (*C. annuum*) X BG 2816 (*C. frutescens*)	–	–	COSII	587	Wu et al. (2009)
California Wonder X LCA235 of *C. annuum*	*Ofw.iivr-2.1*, *Otfw.iivr-2.1*, *Otofw.iivr-2.1*, *Onfp.iivr-2.1*,	Plant height, number of fruits per plant, ten fruits weight, fruit length, fruit width, total	SSR, SCAR, RAPD	290 (SSR), 30 (RAPD) and 9 (SCAR)	Dwivedi et al. (2013)

	Opt.iivr-2.1, *Otofw.iivr-3.1* and *Ofl.iivr.3.4*	fruit weight and pericarp thickness			
RN (*C. annuum*) X CA4 (*C. chinense*)	–	–	IBP, COSII, eSNP	512	Park et al. (2014)
Bei A3×B702 (*C. annuum* L. cultivars)	–	–	SNPs	7657	Qin et al. (2014)
BA3 X B702 (*C. annuum* cultivars)	–	–	InDel	251	Li et al. (2015b)
BA3 (*C. annuum*) X YNXML (*C. frutescens*)	–	–	InDel, SSR	224	Tan et al. (2015)
Capsicum sp.	Phytoene synthase (*PSy*), phytoene desaturase (*Pds*), and capsanthin, capsorubin synthase (*Ccs*) genes, *LcyB*, *CrtZ*-2	Fruit color	–	–	Ha et al. (2007), Kim et al. (2010), and Rodriguez-Uribe et al. (2012)
C. annuum	*CaOvate*	Fruit shape	–	–	Tsaballa et al. (2011)
C. annuum	*cl*, *c2* and *y*	Fruit color	–	–	Hurtado-Hernandez and Smith (1985)
Maor (*C. annuum*)X BG 2816 (*C. frutescens*)	*Yld8.1* and *yld1.1*(total 58QTLs were detected)	Fruit weight, fruit diameter, fruit length, fruit shape, pericarp width, yield, fruit number, flowering, maturity, seed weight	RFLP	92	Rao et al. (2003)
C. annum XC. *chinense*	*fs10.1*	Fruit shape	–	–	Ben-Chaim et al. (2003), Borovsky and Paran (2011)
C. chinense X *C. frutescens*	*fw*2.1, *fw4.1* and *fw4*.2	Fruit weight	RFLP	NA	Zygier et al. (2005)
Yolo Wonder X Criollo de Morelos 334 of *C. annuum*	–	–	AFLP, SSRs, RFLPs, SSAP, STS	489	Barchi et al. (2007)

(continued)

Table 2 (continued)

Population/cultivar	Candidate genes/QTLs	Traits	Marker type	No. of marker	References
C. annuum	Methionine sulfoxide reductase (*B2*)	Plant defense			Oh et al. (2010)
California Wonder X LS2341 (JP187992) (*C. annuum* cultivars)	–	–	SNPs	6281	Kim et al. (2014a, b, c)
K9-11XMZC-180 (*C. annuum* cultivars)	–	–	IBP, COSII, eSNP	512	Park et al. (2014)
Dempsey X Perennial (*C. annuum* cultivars)			SNPs	7657	Qin et al. (2014)
C. annuum	*CaBtf3*	Plant defense	–	–	Huh et al. (2012a)
C. annuum	*PmsM1*-CAPS	Genetic male sterility	–	–	Lee et al. (2010a)
C. annuum	*ms1, ms3*	Genetic male sterility	–	–	Lee et al. (2010a, b, c)
Capsicum sp.	*ms8*	Nuclear male sterility	–	–	
C. annuum	*CaMF2, CaMF1*	Male sterility	–	–	Chen et al. (2011, 2012)
C. annuum	*Rf*	Cytoplasmic male sterility	–	–	Lee et al. (2008) and Ma et al. (2013)
C. annuum	*Bs2, Bs3, Bs1, Bs4 bs5* and *bs6*, gdr	*Xanthomonas campestris* pv. *vesicatoria (Xcv)*	–	–	Tai et al. (1999), Pierre et al. (2000), Jones et al. (2002), (2004), and Csillery et al. (2004)
C. annuum	*CaPO2*	*Xcv*	–	–	Choi et al. (2007)
C. annuum	*CaLOX1*	Resistance to *Pseudomonas syringae*pv*tomato*, *Hyaloperonospora arabidopsidis*, and *Alternaria brassicicola*	–	–	Hwang et al. (2010)
C. annuum cv "Bukang"	*Cmr1*	Cucumber mosaic virus (CMV)	–	–	Kang et al. (2010)
C. annuum	*pvr2*	Potyvirus	–	–	Murphy et al. (1998) and Kang et al. (2005)

C. annuum	*pvr2 and pvr6*	Pepper veinal mottle virus	–	–	Ruffel et al. (2006), Hwang et al. (2009), and Rubio et al. (2009)
C. chinense	*Tsw*	Tomato spotted wilt virus	–	–	Moury et al. (2000)
C. annuum	*CaWRKYb, CaWRKYd, CaBtf3*	Tobacco mosaic virus	–	–	Lim et al. (2011) and Huh et al. (2012a, b)
CM33 (*C. annuum*)	*CaRGA2*	*Phytophthora capsici*	–	–	Zhang et al. (2013)
C. annuum	*CaMsrB2*	Plant defense	–	–	Oh et al. (2010)
NMCA10399	*Ipcr*	*Phytophthora capsici*	–	–	Reeves et al. (2013)
Bangchang (*C. annuum*) × PBC932 (*C. chinense*) and PBC80 × CA1316 of *C. baccatum*	*RA932g*and*RA932r* (anthracnose resistances), *RA80rP2, RA80rP3.1, RA80rP3.2* and *RA80rHP1, RA80rHP2*(disease resistance)	Anthracnose resistances, disease resistance	SNPs	PBC932-derived map (214SNPs), PBC80-derived map (403 SNPs)	Mahasuk et al. (2016)
C. annuum	*Me* and *N*-genes	Root-knot nematodes	–	–	Thies and Fery (2000, 2002) and Djian-Caporalino et al. (2001, 2007)
C. annuum	*CaPF1*	Cold, drought	–	–	Yi et al. (2004) and Kim et al. (2004)
NA	*CaRAV1*	Osmotic stress	–	–	Sohn et al. 2006 and Lee et al. (2010d)
C. annuum	*Ca-DREBLP1*	Dehydration and salt stress	–	–	Hong et al. (2005)
C. annuum	*CaXTH3*	Water deficit and salt stress	–	–	Cho et al. (2006)
C. annuum	*CaBI-1*	Heat, water deficit, and salt stress	–	–	Isbat et al. (2009)
C. annuum	*Sod*	Oxidative and drought stress	–	–	Chatzidimitriadou et al. (2009)
C. annuum	*CaELIP*	Heavy metal stress	–	–	Tao et al. (2011)

(continued)

Table 2 (continued)

Population/cultivar	Candidate genes/QTLs	Traits	Marker type	No. of marker	References
C. annuum	*CaPOD*	Salt and drought stress	–	–	Wang et al. (2013)
C. annuum	*CAPIP2*, *CaPO2*, *CaPMEI1*	Biotic and abiotic stress	–	–	Lee et al. (2006), An et al. (2008), and Choi and Hwang (2012)
C. annuum	*CaBZ1*	Salt stress, osmotic stress, and abscisic acid (ABA)	–	–	Moon et al. (2015)

NA Not Available

give a brief description of the development of different molecular markers, their use in generation of linkage/genetic maps, detection and identification of QTL/genes, and development of gene-specific/linked molecular markers tightly linked to morphological, biochemical, and biotic and abiotic stress resistance traits in *Capsicum* species.

3.1 Molecular Technologies and Genetic Advancement in **Capsicum**

Genetic mapping in *Capsicum* species using molecular markers started around three decades back (Reviewed by Ramchiary et al. 2013). In summary, the first genetic map was constructed with 85 restriction fragment length polymorphic (RFLP) markers (Tanksley et al. 1988) followed by construction of genetic maps with advanced markers like simple sequence repeats (SSRs), expressed sequence tag-derived simple sequence repeats (EST-SSRs), amplified fragment length polymorphism (AFLP), cleaved amplified polymorphic sequence (CAPS), single-nucleotide polymorphism (SNP), etc., thereby increasing the density and robustness of the map with advancement of time since 1980s (Prince et al. 1993; Livingstone et al. 1999; Kang et al. 2001; Ben-Chaim et al. 2006; Portis et al. 2007; Lee et al. 2008a; Mimura et al. 2012; Sugita et al. 2013; Cheng et al. 2016). Furthermore, the advent of next-generation sequencing technologies enabled researchers to develop a large number of more advanced markers like SNPs and Indels including SSRs. In *Capsicum*, using whole genome and transcriptome sequences, significant number of markers have been developed (Lu et al. 2012; Kim et al. 2014b; Qin et al. 2014; Li et al. 2015a; Tan et al. 2015). Gongora-Castillo et al. (2012) developed a comprehensive database known as *Capsicum* transcriptome DB. The database was created using a total of 83,116 ESTs generated from 30 cDNA libraries (derived from leaf, flower, stem, root, and fruit tissues) and 1,838,567 reads generated from virus-infected and uninfected leaves of pepper plant by pyrosequencing approach. Further, Nicolai et al. (2012) carried out transcriptome sequencing with Roche 454 pyrosequencer and assembled 23,748 contigs with 60,370 singletons. Later, a pepper GeneChip array (Affymetrix) was developed by using 30,815 unigenes for identification of polymorphism and expression analysis (Hill et al. 2013). In a separate study, a high-throughput transcriptome profiling performed in two *C. annuum* varieties (Ahn et al. 2013) generated 279,221 and 316,357 sequenced reads and 9701 and 12,741 potential SNPs. Further, using these sequence information, a total of 2067 and 2494 potential SSR motifs were also identified (Ahn et al. 2014).

A draft genome sequence of hot pepper (*C. annuum*) cv. "CM-334" was reported by Kim et al. (2014b). The divergence and genetic variation were estimated among CM-334, and three other *Capsicum* genomes including two cultivars, i.e., perennial and Dempsey, and one wild species (*C. chinense* PI159236) revealed 0.35% (10.9

million SNPs), 0.39% (11.9 million SNPs), and 1.85% (56.6 million SNPs) divergence. Using PGA annotation pipeline, 34,903 protein-coding genes were identified which were highly similar to 34,771 genes in tomato (Tomato Genome Consortium 2012) and 39,031 genes in the potato genome (Potato Genome Sequencing Consortium 2011). In the same year, Qin et al. (2014) sequenced the whole genome of two pepper plants, first is of cultivated Zunla-1 (*C. annuum* L.) and other is of wild chiltepin (*C. annuum* var. glabriusculum) pepper. This study helped to elucidate the evolution and domestication process of *Capsicum*. The reads were assembled into scaffolds of 3.48 Gb and 3.35 Gb size, for Zunla-1 and wild chiltepin, respectively. In this study, about 34,476 protein-coding genes were identified and mapped to pepper genome. The gene annotation study resulted in the identification of several candidate genes associated with various traits/characters. Apart from this, the study reported about 6527 long noncoding RNAs (lncRNAs), of which 5976 are intergenic, and 222 are intron-overlapping lncRNAs. The study reported the identification of a total of 5581 phased siRNAs and 176 miRNAs, of which 35 were pepper-specific miRNAs. Furthermore, through comparative genome analysis among cultivated, wild pepper and 20 re-sequenced genotypes, they identified several potentially key genes associated with fruit development and Solanaceae evolution.

3.2 Genes/QTLs for Abiotic Stresses

To struggle with abiotic challenges, plants are evolving with complex mechanism of perception and reactions. Abiotic stresses are recognized by many signaling cascades which later activates ion channels, kinase cascades, and sometimes producing reactive oxygen species and by some other means such as accumulation of hormones, i.e., salicylic acid (SA), ethylene (ET), jasmonic acid (JA), and abscisic acid (ABA) (Fujita et al. 2006).

3.2.1 Heat Stress

Heat stress is one of the major limiting factors for plants all over the world. As a consequence of climate change and global warming, heat stress has an increasingly negative impact on crop growth, survival, and overall productivity. Heat shock proteins (HSPs), also known as molecular chaperons, are present in all organisms and are important for maintaining and restoring the homeostasis of proteins. Heat shock proteins are named according to their molecular weight such as Hsp101, Hsp70, Hsp90, Hsp60, Hsp40, and small heat shock protein (sHsp) (Swindell et al. 2007; Li et al. 2015a). Apart from heat shock proteins, the role of several other heat-responsive genes which are mainly transcription factors has been investigated (Wahid et al. 2012). Since complex mechanism is involved toward the development of heat stress resistance in plants, developing thermotolerant crops seem to be a difficult task.

In *Capsicum*, several genes/QTLs have been identified for tolerance against heat stress. The optimal temperature range for growing pepper is 20–30 °C; temperature higher than this range can affect pollination and fertilization significantly (Guo et al. 2014). QTLs associated with heat tolerance in plants have been reviewed in detail by Driedonks et al. (2016). Using the HiSeq technology, Li et al. (2015a) identified 3799 and 4010 differentially expressed genes (DEGs) in *Capsicum annuum* resistant 'R597' (CaR) and *Capsicum annuumm* susceptible 'S590'(CaS). Gene ontology (GO) and Kyoto Encyclopedia of Genes and Genomes (KEGG) analyses determined that the identified DEGs were involved in heat shock protein, heat shock transcription factors, hormone, as well as calcium and kinase signaling. Such studies will improve our understanding toward the molecular mechanism underlying heat stress response and help in breeding for heat stress-tolerant *Capsicum* varieties. Some of the ethylene-responsive element binding proteins have been found to be upregulated by heat stress. Binding protein (BiP) is also contributing to abiotic stress by reducing ROS accumulation and by increasing the water retention ability, by enhancing UTR pathway, and also by expression of stress-related genes. The gene *CaBiP* was also found to contribute toward tolerance of abiotic stress by similar fashion (Wang et al. 2017).

Association of endophytic fungus with plants enhanced the tolerance against heat stress. Khan et al. (2013) found that *P. resedanum* infestation led to increase in the plant growth during heat stress. Plants having infestation were found to have enhanced the survival rate and produce higher number of leaves as compared to control plants with or without stress excursion. Inoculation with *P. resedanum*, the electrolytic leakage, and lipid peroxidation reduced significantly, thereby minimizing the negative effect of heat stress. Many other endophytes found to help plants to survive in harsh conditions by many ways, such as gibberellins production abilities of *P. funiculosum* LHL06 and *Exophiala* sp. LHL08 extended greater benefits to the host cucumber and soybean plants during salinity and heat stress.

Garruna-Hernandez et al. (2014) reported that the increase in diurnal air temperature raised both stomatal conductance and transpiration rate in habanero pepper (*Capsicum chinense*) causing an increase in temperature deficit (air temperature–leaf temperature). Leaf temperature decreased by 5 °C, allowing a higher CO_2 assimilation rate in plants at diurnal maximum air temperature (40 °C), and showed that the thermal optimum range in a tropical crop such as Habanero pepper is between 30 and 35 °C (leaf temperature, not air temperature).

Heat stress and drought stress often occur simultaneously (Wollenweber et al. 2003) and can induce cellular damage due to the accumulation of ROS (Lei et al. 2006; Wang et al. 2009; Hu et al. 2010). Hu et al. (2010) investigated combination of drought and heat stress in two *C.annuum* genotypes, one tolerant and the other susceptible. There was no difference in relative water content (RWC), electrolyte leakage (EL), and Fv /Fm when both drought-tolerant and drought-sensitive plants are grown at control and heat stress, but the combination of both heat and drought stresses caused higher decrease in RWC and Fv/Fm and increased in EL as compared to drought stress alone.

3.2.2 Drought Stress

Increasing drought is becoming one of the biggest challenges or concern for agricultural productivity worldwide. Drought stress in plants causes disturbance of water flow through the xylem. Interrupting the flow of water causes a decrease in cell turgor pressure. Decrease in turgor affects the process of mitosis, cell elongation, and expansion, as a result of which plant growth and development are severely affected and crop production decreases. Armita et al. (2017) observed decrease of chlorophyll content and increase of secondary metabolites (carotenoid) in *C. frutescens* that have been subjected to drought stress. They also observed that the drought condition caused a long-term effect on the disruption of the metabolic activity in plants. *C. chinense* plants with higher secondary metabolites accumulation and higher antioxidant activities showed more resistance to drought compared to *C. annuum* and *C. frutescens. Capsicum* plants are more susceptible at the vegetative stage to drought than the flowering or fruiting stages. Therefore, *capsicum* plants need more attention at vegetative stage than flowering or fruiting stages (Okunlola et al. 2017). Plant growth-promoting rhizobacteria (PGPR) have also been found to increase resistance and adaptation of plants to drought stresses and have got the potential role in solving future food security issues which are arising due to various stresses. Induced systemic tolerance (IST) term was coined for physical and chemical changes induced by microorganisms in plants which results in enhanced tolerance to drought stresses (Vurukonda et al. 2016).

Higher pungent cultivars of *Capsicum* have high water retention capacity and less affected to drought condition (Phimchan and Techawongstien 2012). Hong and Kim (2005) performed expression analysis of *Ca-DREBLP1* (dehydration-responsive element binding-factor-like protein 1) gene and found that the gene has significantly higher expression in dehydration and salt stress conditions. The expression of *CaGLIP1* (GDSL-type pepper lipase gene) in *Arabidopsis* resulted in marked tolerance to drought and glucose stress. Tomato chloroplast-localized Cu/Zn SOD protein encoded by *sod* gene introduced in pepper was found to improve the regeneration efficiency of pepper explants along with exhibiting tolerance to different oxidative and drought conditions (Chatzidimitriadou et al. 2009). The expression of a cold-stress gene *CaBZ1* (coding for a bZIP family of transcription factor) was found to be induced by abiotic stresses like salt stress, osmotic stress, and abscisic acid (ABA) (Moon et al. 2015).

Silicon, a nonessential element for plants, helps plants to resist drought condition by accumulating more chlorophyll (Matichenkov and Calvert 2002). Under drought stress, plant accumulated proline, which acts as an osmolyte and contributed to stabilize the subcellular structures, scavenges free radicals, and maintained cellular redox potential homeostasis (Hayat et al. 2012).

3.2.3 Cold Stress

Naturally, plants lose water when they find themselves in freezing situation by osmosis. Cold stress is a major and serious threat to plant sustainability which leads to loss in crop yield. Many physiological changes occur in response to cold stress, such as stunting of seedling, yellowing of leaf, poor germination, etc. (Yadav 2010). Abiotic Stress including cold, drought, high salinity, and freezing damage have been shown to be induced by similar mechanisms, most notably, dehydration or water stress. The *late embryogenesis abundant* (LEA) gene from barley has been shown to be effective in increasing cold tolerance when introduced to rice plants (Xu et al. 1996). Perennial plants are not always cold tolerant; in summer they are also sensitive to freezing when exposed to cold condition; adaptation against this change is known as acclimation (Warren et al. 1996).

Many structural and transcriptional factor-encoding genes which get induced by cold stress have been identified in *Capsicum* species including *EREBP* (*CaEREBP-C1 to C4*), *WRKY* (*CaWRKY1*), and *bZIP* (*CaBZ1*) genes. All WRKY proteins harbor either one or two WRKY domains. All four of the *CaEREBP* are induced by cold in *Capsicum* species (Hwang et al. 2005). Brassinosteroids (BRs), a type of polyhydroxysteriod, have been shown to play important role in many processes at different developmental stages of plant like rice, arabidopsis, and maize (Herting and Fock 2002; Wu et al. 2008). Very less work has been done on brassinosteroids in *Capsicum* species. Li et al. (2016) studied the molecular basis of 24-epibrassinolide (EBR) during a chilling stress response and found EBR enhances salicylic acid and jasmonic acid and suppresses the ethylene biosynthesis pathway in cold stress.

3.2.4 Salinity Stress

In the past few decades, salinity has become one of the major limiting factors for growth and productivity of plants since most of the crop plants are glycophytes which cannot withstand salinity (Gupta and Huang 2014). More than 20% of world's land is affected by salinity and still continuing. Each year about 1.5 million hectares of land gets damaged due to enhanced level of salt in soil (Munns and Tester 2008; Neto et al. 2004). Plants are effected by salinity in many ways which includes water stress, ion toxicity, oxidative stress, reduction of cell division, and elongation (Neto et al. 2004; Zhu 2007; Munns and Tester 2008; Qados and Amira 2011).

Water use efficiency (WUE) and nitrogen use efficiency (NUE) were found to decrease drastically by salt stress in pepper plant (Huez-Lopez et al. 2011). On the other hand, WUE increased significantly by decreasing the amount of irrigation water (Al-Harbi et al. 2014). The higher level of irrigation water significantly mitigated the deleterious salinity effects. Excess accumulation of Na^+ and Cl^- ions in the cells inhibits the growth regulator synthesis which are required for cell differentiation. High ion concentration dehydrates and poisons the cells. Al-Hattab et al. (2015) developed salt-stressed chili which might be salt tolerant under field conditions with high content of capsaicin.

Based on gene ontology annotation from *Arabidopsis*, Sanchita (2016) predicted the functions of eight unknown genes responsive to salt stress in *C. annuum*. Out of eight genes, three were found to be upregulated and five were downregulated under stress condition. Further, Zhang et al. (2001) showed that the overexpression of *TaNHX2* in pepper helped to sustain salt stress. Plants having *TaNHX2* had enhanced the level of proline, chlorophyll, RWC, SOD, and APX activities and reduced H_2O_2 levels in chili indicating the involvement of *TaNHX2* in salt stress. Proline is considered as an important osmolyte and osmoprotectant compound when exposed to salt stress. Studies have been shown that upregulation of *osmotin* gene enhances the tolerance against abiotic stresses including salt stress (Maurya et al. 2014).

Many studies have been done to understand the salt-tolerant/resistance mechanism in pepper plants. Maurya et al. (2014) found upregulation of many salt stress-responsive genes such as *CaDREBLP1*, *CaRMaIH1*, *CaKR1*, and *CaOSM1* in *Capsicum*; but genes like *CaPROX1* and *CaPIP2* were downregulated in salt stress. Dehydrins (DHNs) plays an important role in abiotic stress tolerance in plants. Jing et al. (2016) identified seven DHN genes in *C. annuum*, which were divided into two classes based on their highly conserved domains (YnSKn- and SKn-type). *CaDHN3* (YSK2)-silenced pepper plants showed lower resistance to abiotic stresses (cold, salt, and mannitol) than the control plants (TRV2:00), thereby suggesting the involvement of *CaDHN3* gene in abiotic stress. At optimal N level in soil water increased the yield in *Capsicum* plants as compared to when plants are grown on excess nitrogen level in soil water. However, the increase of N levels in soil water causes the salinity stress in pepper (Semiz et al. 2014).

3.3 Genes/QTLs for Biotic Stresses

The *Capsicum* crops get affected by several plant pathogens, which cause major economic loss to farmers. Diseases like root, stem, leaf, and fruit rots caused by *Phytophthora capsici*; leaf spot disease caused by bacteria *Xanthomonas campestris* (Pernezny et al. 2003); and several viral diseases caused by Tobamovirus (Tobacco mosaic virus, TMV, and Tomato mosaic virus ToMV), Topovirus (Tomato spotted wilt virus, TSWV), Cucumovirus (Cucumber mosaic virus, CMV), and Potyviruses (potato virus Y, PVY; tobacco etch virus,TEV; and pepper mottle virus, PepMoV) are major diseases damaging *Capsicum* cultivation. The uses of pesticides, besides causing environmental hazards, have not been so effective in controlling the vectors that transmit virus diseases. Therefore, for sustainable *Capsicum* production, breeding for biotic stress resistance is not the only alternative but also urgently required. The conventional plant breeding through phenotypic selection could develop some resistant *Capsicum* varieties; however, the phenotypic selection for disease resistance is labor intensive, requires lots of time, and is costly. Further, the tests for different disease resistance sometimes are not possible on the same plant. Therefore, several researchers (as highlighted below) used molecular markers and high-throughput genome and transcriptome sequencing to identify genomic regions harboring the disease resistance gene/QTLs in the *Capsicum* genome.

3.3.1 Bacterial Diseases

Bacterial spot disease caused by *Xanthomonas campestris* pv. *vesicatoria* (*Xcv*) is a major problem worldwide (Jones and Stall 1998). The AFLP markers associated with resistance loci, *Bs2*, and *Bs3*, were identified and mapped (Tai et al. 1999; Pierre et al. 2000). Subsequently, four major (*Bs1* to *Bs4*) and three minor genes (*bs5* and *bs6*, gdr) were identified (Jones et al. 2002, 2004; Csillery et al. 2004). The combined effect of *bs5* and *bs6* genes showing complete resistance against P6 race has been reported (Vallejos et al. 2010). Furthermore, the functional codominant marker PR-Bs3 was developed from the promoter of *Bs3* gene (Romer et al. 2010). PR-Bs3 is considered to be the perfect marker for the selection of resistant alleles of *Bs3* gene since it is tightly linked to the gene which was validated in F_2 population and diverse germplasm.

The *C. annuum* peroxidase gene, *CaPO2*, gave significant resistance against *Xanthomonas campestris* pv. *vesicatoria* (*Xcv*) in *Arabidopsis* while the silencing of this gene showed susceptible phenotypes in *Capsicum* plants (Choi et al. 2007). The overexpression of *CaLOX1* gene, encoding nine specific lipoxygenase, extracted from *Xcv*-infected chili leaves, provided tolerance against *Pseudomonas syringae* pv. *tomato*, *Alternaria brassicicola*, and *Hyaloperonospora arabidopsidis* (Hwang and Hwang 2010). Besides playing a role in biotic stress tolerance, *CaLOX1* gene is also found to have major role in giving tolerance against different abiotic stresses such as osmotic, drought, and high salinity stresses (Lim et al. 2015). Therefore, this gene is suggested to be very important, which could be used for the development of biotic and abiotic stress-tolerant *Capsicum* cultivar.

The loss of function of *Capsicum* gene *CaMLO2* showed wide-spectrum resistance against powdery mildew diseases, and silencing of this gene enhanced the resistance against *Xanthomonas campestris* (Kim and Hwang 2012; Zheng et al. 2013). The overexpression of this gene in *Arabidopsis* plant showed susceptibility for *P. syringae* pv. *tomato* and *H. arabidopsidis*. *CaMLO2* interact with *CaCaM1* and causes hypersensitive cell death. Kim et al. (2014c) reported that *CaMLO2* helps in translocation of *CaCaM1* to plasma membrane from cytoplasm, which ultimately suppresses resistance mechanism (Kim et al. 2014c). Silencing of *CaMLO2* gene resulted into significant increase of ROS, cell death, and defense response against *Xcv*, followed by *CaCaM1*, *CaPR1* (PR-1), and *CaPO2* (peroxidase) induction.

Phytophthora capsici, considered as an oomycete, is among the most destructive pathogens, which hamper potential yield of pepper worldwide (Quirin et al. 2005; Bosland 2008). Several markers including SCAR, SNAP, RAPD, CAPS, etc. were found tightly associated with resistant genotypes (Quirin et al. 2005; Kim et al. 2008; Hong Truong et al. 2013). Oh et al. (2010) observed that silencing of *CaMsrB2* gene in pepper plants increased ROS production that accelerated cell death. A *P. capsici*-tolerant gene, known as *CaRGA2*, was isolated from *C. annuum,* cv. CM-334 and characterized by qRT-PCR and VIGS (Zhang et al. 2013). Silencing *CaRGA2* gene by VIGS resulted in significant decrease in the resistance to *P. capsici*. In a pepper variety NMCA10399, a *Ipcr* (disease resistance inhibitor) gene, which exhibits resistance only to *P. capsici*, has been identified by Reeves et al. (2013). The RNA-Seq analysis of a resistant line "PI 201234" (*C. annuum*) could identify 1220 differentially expressed genes (480 upregulated and 740 downregulated) (Wang et al. 2015). Based on the gene annotation study, out of these a total of 211 genes were found to be involved in defense response. Seven genes were responsible for cell wall modification, symptom development, and phytohormone signaling pathways and phytoalexin biosynthesis, which might be playing a crucial role in the prevention of infection caused by exogenous pathogens. It is predicted that the damage cost of *Phytophthora* blight to chili yield in worldwide is more than $100 million annually (Bosland 2008). To control this pathogen, Thabuis et al. (2004) identified 18 QTLs by using three *C. annuum* populations and transferred this resistance factors from CM-334 into a bell pepper genetic background. It was observed that one major locus out of the 18 QTLs were common in all the three populations. In a separate study, Ogundiwin et al. (2005) identified 16 chromosomal parts consisting of single or group of resistant QTLs for root rot and/or foliar blight. An OpD04 RAPD marker was identified, which was tightly linked to *P. capsici*-resistant line, which was further converted to SCAR marker and later mapped to *P. capsici*-resistant *Phyto 5.2* locus of the chromosome 5 (Quirin et al. 2005). Kim et al. (2008) detected three QTLs for damping off and four QTLs for root rot by using an intraspecific F_2 population. Further, from BAC clones, they developed single-nucleotide-amplified polymorphic marker (SNAP), CAPS markers, and two SSRs for *P. capsici*-resistant genotype selection. Liu et al. (2014) identified a major QTL on chromosome 5 for *P. capsici* resistance using F_2 and RIL populations through bulk segregant analysis. Phyto5SAR marker was found to be single dominant marker associated with resistance. Further two putative candidate NBS-LRR genes and one SAR8.2A gene were also identified in the same scaffold where Phyto5SAR was present. NBS-specific marker (Phyto5NBS1) showed 90% accuracy in prediction of resistance, when tested on 100 F_1 commercial cultivars, suggesting that Phyto5NBS1 may prove important marker for selection of resistant genotype. Xu et al. (2016) identified single dominant locus PhR10 for Phytophthora root rot (PRR) resistance using three mapping populations (two backcross populations and one F_2 population). This gene has been mapped on 16.39 Mb interval at the end of long arm of chromosome 10. Using 10 SSR markers (residing in the interval), the locus was fine mapped to about 2.57 Mb. Thirty one genes were identified in the interval that were associated with disease resistance. For fruit rot resistance, QTL interval mapping has also been conducted in *Capsicum*, and additive and epistatic QTLs have been identified using RILs (Naegele et al. 2014). Recently, CaHDZ27, a Homeodomain-Leucine Zipper I Protein, has been reported to regulate *Ralstonia solanacearum* resistance positively in pepper (Mou et al. 2017). Virus-induced gene silencing of this gene downregulated other defense-related genes like *CaHIR1*, *CaACO1*, *CaPR1*, *CaPR4*, *CaPO2*, and *CaBPR1* and significantly decreases the plant resistance against *R. solanacearum*.

3.3.2 Fungal Diseases

Powdery mildew is the major fungal disease of the *Capsicum* and chiefly caused by *Leveillula taurica*. Lefebvre et al. (2003) mapped a resistant QTL using DH population derived from a cross of H3 (resistant) x Vania (susceptible) genotypes. They detected seven genomic regions/QTLs with additive QTLs and epistatic interactions in natural and artificial conditions. Comparative mapping revealed orthology/sytheny of chromosomal region harboring resistance genes for *L. taurica* and *Oidium lycopersicum* in both tomato and pepper. Further, they found a strong resemblance with detected QTLs for virus resistance and fruit color character in *Capsicum*.

Anthracnose fruit rot, an agronomically important disease is caused by *Colletotrichum gloeosporioides* and *C. capsici*. Voorips et al. (2004) through QTL mapping reported the inheritance of tolerance to these two pathogens in an F_2 population obtained from *C. chinense* (resistant) x *C. annuum* (susceptible) cross. The study revealed one major resistant QTL for *C. capsici* and *C. gloeosporioides* and three minor resistant QTLs against *C. gloeosporioides*. Kim et al. (2010b) identified a total of 18 resistant QTLs for *C. baccatum* out of which two were major and 16 were minor QTLs, respectively. Further, Ying et al. (2015) identified main effect QTLs on chromosome 5 for *C. baccatum* resistance in matured green and matured red fruit stages using 385 markers (SSR, InDel, and CAPS) and backcross population. Besides, they also identified four minor QTLs only at green matured stage and suggested that there may be different gene for resistance at different developmental stages.

In an another study, resistant allele against *Colletotrichum acutatum* was successfully intergressed into the susceptible genotype from two resistant lines (PR1 and PR2) using SCAR-Indel and SSR-HpmsE032 markers (Suwor et al. 2017). Selection efficiency was observed as 65% and 77%, respectively.

3.3.3 Virus Diseases

Viruses are another destructive pathogens, which adversely affect *Capsicum* yield and productivity. Several QTLs have been reported against potyviruses, TMV, and CMV which are listed below. CMV has a broad host range and infects members of the Solanaceae family. It has a large insect vector range because of which complete control is difficult. Initially, several researchers reported the detection of QTLs conferring partial resistance against CMV (Pochard et al. 1983; Caranta et al. 1997, 2002; Ben-Chaim et al. 2001). Caranta et al. (1997) reported mapping of three resistant QTLs using double haploid populations explaining together about 57% of the phenotypic variation. Ben-Chaim et al. (2001) identified four major

QTLs providing effective resistance against CMV. Interestingly, the QTL controlling about 16–33% of the observed phenotypic variation (*cmv11.1*) was found linked to the L locus conferring resistance to TMV. It was observed that the QTL which is controlling up to 16–33% of the observed phenotype was associated with the L locus harboring resistance against TMV. In a separate study, Yao et al. (2013) identified an inbred line named BJ0747-1-3-1-1, which conferred tolerance to a CMV isolate (CMVHB). They identified two major QTLs which together explained a total of 55% of the total phenotypic variation. Kang et al. (2010) identified a single major gene known as *Cmr1* (cucumber mosaic resistance 1) from *C. annuum* cv "Bukang" against Korean and FNY strains. The gene showed synteny with ToMV-resistance locus (*Tm*-1) in tomatoes. Kim et al. (2014a) performed full-genome sequencing of 45 CMV isolates. All the known CMV isolate infecting chili plants were found to belong to subgroup I. When this analysis was performed by using RNA1 sequences, the CMV isolates from pepper were divided into three subgroups, while the same falls under two subgroups when analyzed by using RNA2 or RNA3 sequences.

The *Capsicum* crop yield is also affected by PVY. Caranta et al. (1997) using doubled haploid progeny detected 11 resistant QTLs for 2 PVY isolates and 2 potyviruses E. After comparison, it was observed that few QTLs were observed very close to the *pvr2* and *pvr6* loci. Further, CAPS marker, linked to *Pvr4* resistance gene, was developed which showed tolerance to the three pathotypes of PVY and PepMoV.

Potyviruses, which includes pepper veinal mottle virus (PVMV), tobacco etch virus (TEV), chili veinal mottle virus (ChiVMV), PVY, and PepMoV, hamper chili production (Green and Kim 1991). The first characterized gene against potyvirus was *pvr1* (Murphy et al. 1998; Kang et al. 2005). Later, the gene was predicted as eukaryotic translation initiation factor, *4E*(*eIF4E*), and by the alignment of 228 amino acid sequence, only two amino acids difference between the resistant and susceptible genotype was observed (Ruffel et al. 2002). Several studies reported that concurrent and double mutations in *eIF4E* (*pvr2*) and *eIF* (iso) *4E9* (*pvr6*) are essential for preventing the pepper plant from infection of PVMV (Ruffel et al. 2006; Hwang et al. 2009; Rubio et al. 2009). Lee et al. (2013) developed a new genetically dominant resistant line of chili pepper for ChiVMV. For the first time, Banerjee et al. (2014) reported the occurrence of ChiVMV in Naga chili (*C. chinense*) in the North-Eastern region of India (Meghalaya) using various techniques such as mechanical transmission assay, TEM, RT-PCR, and sequence analysis. Further, in various studies, markers such as CAPS, SCAR, SNP, etc. linked to resistance gene for PVY, PepMoV, and PMMoV were identified and could be useful in breeding programs (Arnedo-Andres et al. 2002; Matsunaga et al. 2003; Rubio et al. 2008). Gao et al. (2014) performed a genome sequencing of a novel recombinant isolate of potato virus Y (PVMV-HN) from pepper in mainland China and showed that genome of PVMV-HN is 9793 nucleotides (nts) long excluding the poly (A) tail. It shared 98–99% identity with two other PVMV isolates from Ghana and Taiwan. Resistant pepper (*C. annuum*) landraces, having the ability to change the resistance-breaking capacity by PVY of the *pvr2* locus that encodes the eIF4E, were identified. Moodley et al. (2014) for the first time sequenced the whole genome of PVY isolate (JVW-186) from the pepper. Two ORFs were identified, one for viral poly protein at position 186 and other for frameshift translated protein P3N-PIPO at position 2915 in the genome. *Pvr4* is another potyvirus resistance genes identified in pepper. Caranta et al. (1999) identified AFLP markers associated with potyvirus resistance through bulk segregant analysis. They also converted the closest AFLP marker into CAPS marker for breeding purpose and characterized using *Capsicum* breeding lines. Recently, 5000 SNVs (single-nucleotide variants) have been developed that are tightly linked to *Pvr4* locus (Devran et al. 2015) which are within the interval containing nucleotide binding site – leucine-rich repeat-type disease resistance genes. For utilization of this locus in *Capsicum* breeding, some SNVs are converted into PCR-based marker and also validated using F_2 population (Devran et al. 2015).

Several studies reported that the resistance against TMV is regulated by several alleles of L gene (Lefebvre et al. 1995; Ben-Chaim et al. 2001; Tomita et al. 2008). Transcription factors like *CaWRKYb* (Lim et al. 2011), *CaWRKYd* (Huh et al. 2012b), and *CaBtf3* (Huh et al. 2012a) have been characterized that play a significant role in defense-related mechanism against TMV infection in chili. Hernan et al. (2013) silenced *non-expresser of pathogenesis-related gene 1* (*NPR1*) and found a significant increase in the level of viral symptoms in both *N. benthamiana* and *C. annuum* plants.

Boiteux et al. (1993) reported two *C. chinense* lines ("CNPH 275" and "PI 159236") that were resistant against TSWV-BsB but not toward another isolate, i.e., TSWV-SP. Further, Moury et al. (2000) reported a single major gene, *Tsw*, responsible for tolerance against TSWV, which was mapped along with four RAPD markers tightly linked to *Tsw*. Margaria et al. (2007) identified resistance-breaking (RB) TSWV strains occurring naturally and with further analysis showed that local necrotic response is insufficient for resistance in *Tsw* gene containing peppers. Later, it was found that during hypersensitivity, apoptosis in *C. chinense* plants is caused by the N protein of TSWV (Lovato et al. 2008).

The *Capsicum* chlorosis virus (CCV) has been identified as new pathogen in different parts of the world like Australia and South East Asia that affect the *Capsicum* yield. The transcriptome analysis was performed to understand the molecular mechanism of resistance against CCV, and a total of 2484 genes have been identified that were differentially expressed between resistant and susceptible genotypes (Widana Gamage et al. 2016). Functional annotation revealed that most of these genes were associated with pathogenesis, cell death and hormone-mediated signaling pathways and enzymes for defense-related pathways.

Silvar and Garcia-Gonzalez (2017) screened 180 *Capsicum* germplasm representing worldwide diverse genotypes for major biotic stresses with the help of ten molecular markers. It has been found that ~30% of the genotypes had resistant allele for the loci *Pvr4*, *Phyto*.5.2, and *Cmr1*. For potyvirus resistance, South American genotypes were found to be desirable. They suggested that *C. chinense* may prove important resourse in *Capsicum* breeding, since >80% disease-resistant alleles were found belonging to this particular species.

3.3.4 Nematode Diseases

Root-knot nematodes (RKN, *Meloidogyne* species) are harmful pests, triggering severe damages to several Solanaceae crops. Many genes in *C. annuum* involved in gene-for-gene interactions, showing resistance against RKN, have been reported (Djian-Caporalino et al. 1999, 2001). Thies and Fery (2000, 2002) first characterized the nematode-resistant gene *N*. In an another study, it was found that genes *Me4*, *Mech1*, and *Mech2* located in the orthologous genomic regions of solanaceous crops are specific for some *Meloidogyne* species or its populations (Djian-Caporalino et al. 2001, 2007). Further, Chen et al. (2007) characterized *CaMi* gene (a nematode-resistant gene) in tomato and found enhanced resistance in comparison to untransformed susceptible plants. Fazari et al. (2012) developed PCR-based markers associated with *Me* and *N*-genes which could be beneficial in pepper breeding programs.

3.4 Gene/QTLs for Important Agronomic Traits Including Pungency

In *Capsicum*, a number of QTL mapping studies have been conducted for many economically important agronomic traits. For the number of primary leaves (Nle), Tan et al. (2015) identified one major QTL on chromosome 2 harboring gene that is homolog of *Arabidopsis CLF* gene using interspecific cross. QTLs for Nle have also been identified in different studies (Barchi et al. 2009; Mimura et al. 2010; Alimi et al. 2013; Zong 2013; Duan 2014). For male sterility traits, important for hybrid chili breeding, QTLs are reported on chromosomes 6, 5, and 2 (Wang et al. 2004). Several important genes related to male sterility traits are reported in *Capsicum* such as *Rf* (Lee et al. 2008b; Ma et al. 2013), *ms1* and *ms2* (Lee et al. 2010a, b, c; Jeong et al. 2017), *ms8* (Bartoszewski et al. 2012), and *CaMF1* and *CaMF2* (Chen et al. 2011, 2012). Transcriptome analysis of sterile and fertile lines revealed differentially expressed genes that are associated with pollen development (Chen et al. 2015). *C. baccatum* gained attention only during last couple of years and used for QTL mapping for 11 agronomic traits including plant height, days to flowering, days to fruiting, crown diameter, fruits per plant, fruit weight, fruit length, fruit diameter, fruit pulp thickness, soluble solids, and fruit dry weight (Moulin et al. 2015).

Wide diversity of fruit size, shape, texture, and colors has been observed in *Capsicum* fruits (Fig. 1). The researchers succeeded in the identification of several QTLs associated with fruit shape, size, and color in *Capsicum* (Ben-Chaim et al. 2001, 2003; Rao et al. 2003; Zygier et al. 2005; Barchi et al. 2009; Borovsky and Paran 2011; Tsaballa et al. 2011; Dwivedi et al. 2013) using different mapping populations. Tsaballa et al. (2011) confirmed the conservation of *OVATE*-like genes (for fruit shape) in Solanaceae by cloning and characterizing the *CaOvate* genes in pepper. Downregulation of *CaOvate* resulted in oval-shaped fruit. In *Capsicum*,

Fig. 1 *Capsicum* fruits belonging to different species (*Capsicum annuum*, *C. chinense*, and *C. frutescence*) showing diverse morphology and color

CaGLK2 GOLDEN2-like transcription factor gene was found to be important for chloroplast development and was colocalized with a QTL *pc10* (Brand et al. 2014). For carotenoids, several genomic region harboring QTLs/genes *cl*, *c2*, *psy*, *Ccs*, *Pds*, *CrtZ-2*, and *y* have also been reported in *Capsicum* (Hurtado-Hernandez and Smith 1985; Popovsky and Paran 2000; Huh et al. 2001; Ha et al. 2007; Kim et al. 2010a; Rodriguez-Uribe et al. 2012). Recently, Liu et al. (2017) identified several miRNAs associated with fruit development and fruit quality.

The pungency trait, the unique property of *Capsicum*, which distinguishes it from other Solanaceae plants, is considered to be one among the most economically valuable quality traits. Naga chili has been identified as a potential source of capsaicinoids (Meghvansi et al. 2010). Ben-Chaim et al. (2006) identified six QTLs for three capsaicinoids (capsaicin, dihydrocapsaicin, and nordihydrocapsaicin) and found that genes playing a key role in the valine catabolism, *BCAT* and *3A2*, were associated with the observed QTL. In a separate study, Yarnes et al. (2013) reported 12 QTLs for capsaicinoids. Curry et al. (1999) reported that *Pal*, *Ca4h*, and *Comt* genes are involved in the capsaicinoid biosynthesis pathway. Later, other genes like *Amt*, *Cas*, and *Pun1* are also found to be involved in same pathway (Kim et al. 2001; Blum et al. 2002; Stewart et al. 2005). Aluru et al. (2003) identified and mapped three placental-specific genes *Acl*, *Fat*, and *Kas* which positively regulates the pungency trait. Further, Stewart et al. (2007) analyzed the *At3* gene in a nonpungent *C. chinense* NMCA 30036 chili pepper and observed a 4-bp deletion in the first exon of *At3* gene, and this allele was named *Pun1*. Involvement of the abovementioned genes in capsaicin biosynthesis pathway for development of pungent flavor in pepper fruit was further confirmed (Kim et al. 2014b; Qin et al. 2014). Reddy et al. (2014b) performed an association mapping study and identified SNPs in *Pun1*, *KAS*, *HCT*, and *CCR* genes and revalidated that *PUN1*, *CCR*, and *KAS* act as important candidates in capsaicinoid production. Furthermore, it was observed that *Pun1*gene in the capsaicin biosynthesis pathway directly correlated with the accumulation capsaicinoids. They also identified six SNPs in the upstream promoter region of *Pun1* and proposed that the capsaicinoid accumulation correlates with the degree of expression of *Pun1* gene.

4 Future Prospects

The progress made in the conventional breeding and identification of QTL/genes governing economically important traits were successful to some extent toward translational research in *Capsicum* species. For biotic and abiotic stresses, several genes/QTLs have been identified; however, majority of them are having only minor effect since these traits are complex in nature and controlled by many minor genes/ QTLs. Further studies are required to dissect genetic architecture of stress biology for precision breeding. For this purpose, genome-wide association studies (GWAS) using a large number of natural accessions of *Capsicum* supplemented with historic recombinations with contrasting phenotypes would help to identify minor gene/ QTLs through higher-resolution mapping. To achieve this, high-throughput genotyping using next-generation sequencing techniques and phenotyping could be used. Most of the studies have used less number of genotypes for screening metabolite contents, fruit morphology, and biotic and abiotic tolerance/resistance traits. Therefore, breeders should also focus on the systematic phenotyping of a large number of germplasm for every economically important trait including biotic and abiotic traits in *Capsicum*, which will generate a large data. Once generated, this data could be shared among breeders. It is well-known fact that different biotic and abiotic stresses are sometimes co-activated and plants show simultaneous response to multiple stresses. To better understand the genetic mechanism of correlated stresses, multiparental population like NAM (nested association mapping) and MAGIC (multiparental advanced generation intercross) population should be developed and used. The Asian Vegetable Research and Development Center (AVRDC, Taiwan) has the largest collection of *Capsicum* species with recorded number of accessions totaling 8170, followed by the United States Department of Agriculture (USDA) with 6067 accessions, both from wild and cultivated *Capsicum* species representing germplasm from around the world. The National Bureau of Plant Genetic Resources (NBPGR), New Delhi, India, has collections of 2774 accessions. The global initiative to collect and share and systematic evaluation of phenotypes of *Capsicum* genetic material for biotic and abiotic stress resistance/tolerance, biochemical composition, and other traits would greatly enhance the understanding of genetic mechanism regulating those traits besides improving *Capsicum* varieties for sustainable production in addition to fulfilling the worldwide demand and increase the income of the farmers. The epigenetic mechanism has also been reported to control biotic and abiotic stresses which should also be explored by *Capsicum* researchers.

Acknowledgment This work was supported by the research grant from the University Grants Commission, India, to Dr. Nirala Ramchiary, from the UGC Resource Networking Project of the School of Life Sciences, Jawaharlal Nehru University, New Delhi, India.

References

Abuqamar S, Luo H, Laluk K, Mickelbart MV, Mengiste T (2009) Crosstalk between biotic and abiotic stress responses in tomato is mediated by the AIM1 transcription factor. Plant J 58:347–360

Ahn YK, Tripathi S, Kim JH, Cho YI, Lee HE, Kim DS, Woo JG, Cho MC (2013) Transcriptome analysis of *Capsicum annuum* varieties Mandarin and Blackcluster: assembly, annotation and molecular marker discovery. Gene 533:494–499

Ahn YK, Tripathi S, Kim JH, Cho YI, Lee HE, Kim DS, Woo JG, Yoon MK (2014) Microsatellite marker information from high-throughput next-generation sequence data of *Capsicum annuum* varieties Mandarin and Blackcluster. Sci Hortic 170:23–130

Al-Harbi AR, Saleh AM, Al-Omran AM, Wahb-Allah MA (2014) Response of bell-pepper (*Capsicum annuum* L.) to salt stress and deficit irrigation strategy under greenhouse conditions. Acta Hortic 1034:443–445

Al-Hattab ZN, Al-Ajeel SA, El-Kaaby EA (2015) Effect of salinity stress on *Capsicum annuum* callus growth, regeneration and callus content of capsaicin, phenylalanine, proline and ascorbic acid. Life Sci 9:304–310

Alimi NA, Bink MCAM, Dieleman JA et al (2013) Genetic and QTL analyses of yield and a set of physiological traits in pepper. Euphytica 190:181–201

Aluru MR, Mazourek M, Landry LG, Curry J, Jahn M, O'Connell MA (2003) Differential expression of fatty acid synthase genes, *Acl*, *Fat* and *Kas*, in *Capsicum* fruit. J Exp Bot 54:1655–1664

An SH, Sohn KH, Choi HW, Hwang IS, Lee SC, Hwang BK (2008) Pepper pectin methylesterase inhibitor protein *CaPMEI1* is required for antifungal activity, basal disease resistance and abiotic stress tolerance. Planta 228:61–78

Armita D, Arumyngtyas EL, Mastuti R (2017) Tolerance level of three genotypes of cayenne pepper (*Capsicum frutescens L.*) toward drought stress of vegetative phase based on morphological and physiological responses. Int J Chem Tech Res 10:183–192

Arnedo-Andres MS, Gil-Ortega R, Luis-Arteaga M, Hormaza JI (2002) Development of RAPD and SCAR markers linked to the *Pvr4* locus for resistance to PVY in pepper (*Capsicum annuum* L.) Theor Appl Genet 105:1067–1074

Babu BS, Pandravada SR, Prasad Rao RDVJ, Anitha K, Chakrabarty SK, Varaprasad KS (2011) Global source of pepper genetic resources against arthropods, nematodes and pathogens. Crop Prot 30:389–400

Banerjee A, Dutta R, Roy S, Ngachan SV (2014) First report of Chilliveinal mottle virus in Naga chilli (*Capsicum chinense*) in Meghalaya, India. Virus Dis 25:142–143

Barchi L, Bonnet J, Boudet C, Signoret P, Nagy I, Lanteri S, Palloix A, Lefebvre V (2007) A high-resolution, intraspecific linkage map of pepper (*Capsicum annuum* L.) and selection of reduced recombinant inbred line subsets for fast mapping. Genome 50:51–60

Barchi L, Lefebvre V, Sage-Palloix AM, Lanteri S, Palloix A (2009) QTL analysis of plant development and fruit traits in pepper and performance of selective phenotyping. Theor Appl Genet 118:1157–1171

Bartoszewski G, Waszczak C, Gawronski P et al (2012) Mapping of the *ms8* male sterility gene in sweet pepper (*Capsicum annuum* L.) on the chromosome P4 using PCR-based markers useful for breeding programmes. Euphytica 186:453–461

Ben-Chaim A, Grube RC, Lapidot M, Jahn M, Paran I (2001) Identification of quantitative trait loci associated with resistance to cucumber mosaic virus in *Capsicum annuum*. Theorn Appl Genet 102:1213–1220

Ben-Chaim A, Borovsky Y, Rao GU, Tanyolac B, Paran I (2003) *fs3*.1: a major fruit shape QTL conserved in *Capsicum*. Genome 46:1–9

Ben-Chaim A, Brodsky Y, Falise M, Mazourek M, Kang BC, Paran I, Jahn M (2006) QTL analysis for capsaicinoid content in *Capsicum*. Theor Appl Genet 113:1481–1490

Berger A, Henderson M, Nadoolman W, Duffy V, Cooper D, Saberski L, Bartoshuk L (1996) Oral capsaicin provides temporary relief for oral mucositis pain secondary to chemotherapy/radiation therapy. J Pain Symptom Manag 10:243–248. Erratum in: J Pain Symptom Manage. 11:331

Blum E, Liu K, Mazourek M, Yoo EY, Jahn M, Paran I (2002) Molecular mapping of the C locus for presence of pungency in *Capsicum*. Genome 45:702–705

Boiteux LS, Nagata T, Dutra WP, Fonseca MEN (1993) Sources of resistance to tomato spotted wilt virus (TSWV) in cultivated and wild species of *Capsicum*. Euphytica 67:89–94

Borovsky Y, Paran I (2011) Characterization of *fs10.1*, a major QTL controlling fruit elongation in *Capsicum*. Theor Appl Genet 123:657–665

Bosland PW (1996) Capsicums: innovative uses of an ancient crop. In: Janick J (ed) Progress in new crops. Ashs Press, Arlington, pp 479–487

Bosland PW (2008) Think global, breed local: specificity and complexity of *Phytophthoracapsici*. In: 19th Int. Pepper Conf. Atlantic City, NJ

Brand A, Borovsky Y, Hill T, Rahman KAA, Bellalou A, Deynze AV (2014) CaGLK2 regulates natural variation of chlorophyll content and fruit color in pepper fruit. Theor Appl Genet 127:2139–2148

Caranta C, Palloix A, Lefebvre V, Daube ze AM (1997) QTLs for a component of partial resistance to cucumber mosaic virus in pepper: restriction of virus installation in host cells. Theor Appl Genet 94:431–438

Caranta C, Thabuis A, Palloix A (1999) Development of a CAPS marker for the *Pvr4* locus: a tool for pyramiding potyvirus resistance genes in pepper. Genome 42:1111–1116

Caranta C, Pflieger S, Lefebvre V, Daubeze AM, Thabuis A, Palloix A (2002) QTLs involved in the restriction of cucumber mosaic virus (CMV) long-distance movement in pepper. Theor Appl Genet 104:586–591

Chatzidimitriadou K, Nianiou-Obeidat I, Madesis P, Perl-Treves R, Tsaftaris A (2009) Expression of SOD transgene in pepper confer stress tolerance and improve shoot regeneration. Electron J Biotechnol 12. https://doi.org/10.2225/vol12-issue4-fulltext-10

Chen RG, Li HX, Zhang LY, Zhang JH, Xiao JH, Ye ZB (2007) *CaMi*, a root-knot nematode resistance gene from hot pepper (*Capsicum annuum* L.) confers nematode resistance in tomato. Plant Cell Rep 26:895–905

Chen C, Chen G, Hao Z, Cao B, Chen Q, Liu S, Lei J (2011) *CaMF2*, an anther-specific lipid transfer protein (LTP) gene, affects pollen development in *Capsicum annuum* L. Plant Sci 181:439–448

Chen CM, Hao XF, Chen GJ, Cao BH, Chen QH, Liu SQ, Lei JJ (2012) Characterization of a new male sterility-related gene *Camf1* in *Capsicum annum* L. Mol Biol Rep 39:737–744

Chen C, Chen G, Cao B, Lei J (2015) Transcriptional profiling analysis of genic male sterile–fertile *Capsicum annuum* reveal candidate genes for pollen development and maturation by RNA-Seq technology. Plant Cell Tissue Organ Cult 122:465–476

Cheng J, Qin C, Tang X, Zhou H, Hu Y, Zhao Z, Cui J, Li B, Wu Z, Yu, Hu K (2016) Development of a SNP array and its application to genetic mapping and diversity assessment in pepper (*Capsicum* spp.) Sci Rep 6. https://doi.org/10.1038/srep33293

Cheong YH, Chang HS, Gupta R, Wang X, Zhu T, Luan S (2002) Transcriptional profiling reveals novel interactions between wounding, pathogen, abiotic stress, and hormonal responses in *Arabidopsis*. Plant Physiol 129:661–677

Cho SK, Kim JE, Park JA, Eom TJ, Kim WT (2006) Constitutive expression of abiotic stress-inducible hot pepper *CaXTH3*, which encodes a xyloglucanendotransglucosylase/hydrolase homolog, improves drought and salt tolerance in transgenic *Arabidopsis* plants. FEBS Lett 580:3136–3144

Choi HW, Hwang BK (2012) The pepper extracellular peroxidase *CaPO2*is required for salt, drought and oxidative stress tolerance as well as resistance to fungal pathogens. Planta 235:1369–1382

Choi HW, Kim YJ, Lee SC, Hong JK, Hwang BK (2007) Hydrogen peroxide generation by the pepper extracellular peroxidase *CaPO2* activates local and systemic cell death and defense response to bacterial pathogens. Plant Physiol 145:890–904

Csillery G, Szarka E, Sardi E, Mityko J, Kapitany J, Nagy B, Szarka J (2004) The unity of plant defense: genetics, breeding and physiology. In Proceedings 12th Eucarpia meeting on genetics and breeding of *Capsicum* and Egg-plant, Noordwijkerhout, the Netherlands, pp 147–153

Curry J, Aluru M, Mendoza M, Nevarez J, Melendrez M, O'Connell MA (1999) Transcripts for possible capsaicinoid biosynthetic genes are differentially accumulated in pungent and non-pungent *Capsicum* spp. Plant Sci 148:47–57

Devran Z, Kahveci E, Ozkaynak E, Studholome DJ, Tor M (2015) Development of molecular markers tightly linked to *Pvr4* gene in pepper using next-generation sequencing. Mol Breed 35:101

Djian-Caporalino C, Pijarowski L, Januel A, Lefebvre V, Daubeze A, Palloix A, Dalmasso A, Abad P (1999) Spectrum of resistance to root-knot nematodes and inheritance of heat stable resistance in pepper (*Capsicum annuum* L.) Theor Appl Genet 99:496–502

Djian-Caporalino C, Pijarowski L, Fazari A et al (2001) High-resolution genetic mapping of the pepper (*Capsicum annuum* L.) resistance loci Me_3 and Me_4 conferring heat-stable resistance to root-knot nematodes (*Meloidogyne*spp.) Theor Appl Genet 103:592–600

Djian-Caporalino C, Fazari A, Arguel MJ et al (2007) Root-knot nematode (*Meloidogyne* spp.) Me resistance genes in pepper (*Capsicum annuum* L.) are clustered on the P9 chromosome. Theor Appl Genet 114:473–486

Driedonks N, Rieu I, Vriezen WH (2016) Breeding for plant heat tolerance at vegetative and reproductive stages. Plant Reprod 29:67–79

Duan MM (2014) Construction of intraspecific genetic linkage map and qtl analysis of phytological traits and phytophthoracapsici resistance in pepper (*Capsicum* L.): Chinese Academy of Agricultural Sciences (in Chinese with English Abstract). 46

Dwivedi N, Kumar R, Paliwal R, Kumar U, Kumar S, Singh M, Singh RK (2013) QTL mapping for important horticultural traits in pepper (*Capsicum annuum* L.) J Plant Biochem Biotechnol. https://doi.org/10.1007/s13562-013-0247-1

Fazari A, PalloixA WL, Hua YM, Sage-Palloix AM, Zhang BX, Djian-Caporalino C (2012) The root-knot nematode resistance *N*-gene co-localizes in the *Me*-genes cluster on the pepper (*Capsicum annuum* L.) P9 chromosome. Plant Breed 131:665–673

FAOSTAT (2015) http://faostat3.fao.org/download/Q/QC/E

Fraire-Velzquez S, Rodriguez-Guerra R, Sanchez-Calderon L (2011) Abiotic and biotic stress response crosstalk in plants. In: Shanker A, Venkateswarlu B (eds) Abiotic stress response in plants – physiological, biochemical and genetic perspectives. Intech, New York, pp 1–24

Fujita M, Fujita Y, Noutoshi Y, Takahashi F, Narusaka Y, Yamaguchi-Shinozaki K, Shinozaki K (2006) Crosstalk between abiotic and biotic stress responses: a current view from the points of convergence in the stress signaling networks. Curr Opin Plant Biol 9:436–442

Gao F, Chang F, Shen J, Shi F, Xie L, Zhan J (2014) Complete genome analysis of a novel recombinant isolate of potato virus Y from China. Arch Virol 159:3439–3442

Garcia SNJ, Vazquez Cruz MA, Gonzalez RGG, Pacheco IT, Angelica AA, Perez F (2016) Influence of salicylic acid application on oxidative and molecular responses and functional properties of *Capsicum annuum L.* cultvated in greenhouse conditions 1st electronic conference on metabolite chaired by Dr. Peter Meikle

Garruna-Hernandez R, Orellana R, Larque-Saavedra A, Canto A (2014) Understanding the physiological responses of a tropical crop (*Capsicumchinense Jacq.*) at high temperature. PLoS One 9:e111402

Gongora-Castillo E, Fajardo-Jaime R, Fernández-Cortes A et al (2012) The *Capsicum* transcriptome DB: a "hot" tool for genomic research. Bioinformation 8:043–047

Green SK, Kim JS (1991) Characteristics and control of viruses infecting peppers, Technical Bulletin No. 18. Asian Vegetable Research and Development Centre, Taipei

Guo W, Chen R, Du XZ, Yin Y, Gong Z, Wang G (2014) Reduced tolerance to abiotic stress in transgenic *Arabidopsis* overexpressing a *Capsicum annuum* multiprotein bridging factor. BMC Plant Biol 14:138

Gupta B, Huang B (2014) Mechanism of salinity tolerance in plants: physiological, biochemical, and molecular characterization. Int J Geno. https://doi.org/10.1155/2014/701596

Ha SH, Kim JB, Jong-Sug Park JS, Lee SW, Cho KJ (2007) A comparison of the carotenoid accumulation in *Capsicum* varieties that show different ripening colours: deletion of the capsanthin-capsorubin synthase gene is not a prerequisite for the formation of a yellow pepper. J Exp Bot 58:3135–3144

Hayat Q, Hayat S, Alyemeni MN, Ahmad A (2012) Salicylic acid mediated change in growth, photosynthesis, nitrogen metabolism and antioxidant defense system in *Cicer orietinum* L. Plant Soil Environ 58:417–423

Hernan VA, Rosa U, Luisa OL, Dominique R, Orlene G, Yereni M, Oscar M (2013) A new virus-induced gene silencing vector based on Euphorbia mosaic virus-Yucatan peninsula for NPR1 silencing in *Nicotiana benthamiana* and *Capsicum annuum* var. Anaheim Biotechnol Lett 35:811

Herting SH, Fock HP (2002) Oxygen exchange in relation to carbon assimilation in water-stressed leaves during photosynthesis. Ann Bot 89:851–859

Hill TA, Ashrafi H, Reyes-Chin-Wo S et al (2013) Characterization of *Capsicum annuum* genetic diversity and population structure based on parallel polymorphism discovery with a 30K unigene Pepper GeneChip. PLoS One 8:e56200

Hong JP, Kim WT (2005) Isolation and functional characterization of the *Ca-DREBLP1* gene encoding a dehydration-responsive element binding-factor-like protein 1 in hot pepper (*Capsicum annuum* L. cv. Pukang). Planta 220:875–888

Hong Truong HT, Kim JH, Cho MC, Chae SY, Lee HE (2013) Identification and development of molecular markers linked to Phytophthora root rot resistance in pepper (*Capsicum annuum* L.) Eur J Plant Pathol 135:289–297

Hu WH, Xiao YA, Zeng JJ, Hu XH (2010) Photosynthesis, respiration and antioxidant enzymes in pepper leaves under drought and heat stresses. Biol Plant 54:761–765

Huez-Lopez MA, Ulery AL, Samani Z, Picchioni G, Flynn RP (2011) Response of chille pepper (*Capsicum annuum L.*) to salt stress and organic and inorganic nitrogen sources: I. growth and yield. Trop Subtrop Agroecosyst 14:757–763

Huh J, Kang B, Nahm S, Kim S, Ha K, Lee M, Kim B (2001) A candidate gene approach identified phytoene synthase as the locus for mature fruit colour in red pepper (*Capsicum* spp.) Theor Appl Genet 102:524–530

Huh SU, Kim KJ, Paek KH (2012a) *Capsicum annuum* basic transcription factor 3 (*CaBtf3*) regulates transcription of pathogenesis-related genes during hypersensitive response upon Tobacco mosaic virus infection. Biochem Biophys Res Commun 417:910–917

Huh SU, Choi LM, Lee GJ, Kim YJ, Paek KH (2012b) *Capsicum annuum* WRKY transcription factor d (*CaWRKYd*) regulates hypersensitive response and defense response upon Tobacco mosaic virus infection. Plant Sci 197:50–58

Hurtado-Hernandez H, Smith P (1985) Inheritance of mature fruit colour in *Capsicum annuum* L. J Hered 76:211–213

Hwang IS, Hwang BK (2010) The pepper 9-Lipoxygenase gene *CaLOX1* functions in defense and cell death responses to microbial pathogens. Plant Physiol 152:948–967

Hwang EW, Kim KA, Park SC, Jeong MJ, Byun MO, Kwon HB (2005) Expression profiles of hot pepper (*Capsicumannuum*) genes under cold stress conditions. J Biosci 30:657–667

Hwang JN, Li J, Liu WY, An SJ, Cho H, Her NH, Yeam I, Kim D, Kang B (2009) Double mutations in *eIF4E* and *eIFiso4E* confer recessive resistance to *Chilliveinal mottle virus* in pepper. Mol Cells 27:329–336

Isbat M, Zeba N, Kim SR, Hong CB (2009) A BAX inhibitor-1 gene in *Capsicum annuum* is induced under various abiotic stresses and endows multi-tolerance in transgenic tobacco. J Plant Physiol 166(2009):1685–1693

Jeong K, Choi D, Lee J (2017) Fine mapping of the genic male-sterile *ms1* gene in *Capsicum annuum* L. Theor Appl Genet. https://doi.org/10.1007/s00122-017-2995-0

Jing H, Li C, Ma F et al (2016) Genome-wide identification, expression diversication of dehydrin gene family and characterization of *CaDHN3* in pepper (*Capsicum annum L.*) PLoS One 11(8): e0161073

Jones JB, Stall RE (1998) Diversity among xanthomonads pathogenic on pepper and tomato. Annu Rev Phytopathol 36:41–58

Jones JB, Minsavage GV, Roberts PD, Johnson RR, Kousik CS, Subramanian S, Stall RE (2002) A non-hypersensitive resistance in pepper to the bacterial spot pathogen is associated with two recessive genes. Phytopathology 92:273–277

Jones JB, Lacy GH, Bouzar H, Stall RE, Schaad NW (2004) Reclassification of the Xanthomonads associated with bacterial spot disease of tomato and pepper. Syst Appl Microbiol 27:755–762

Kang BC, Nahm SH, Huh JH, Yoo HS, Yu JW, Lee MH, Kim BD (2001) An interspecific (*Capsicum annuum* x C. *Chinense*) F2 linkage map in pepper using RFLP and AFLP markers. Theor Appl Genet 102:531–539

Kang BC, Yeam I, Frantz JD, Murphy JF, Jahn MM (2005) The*pvr1* locus in *Capsicum* encodes a translation initiation factor eIF4E that interacts with tobacco etch virus VPg. Plant J42:392–405

Kang WH, Hoang NH, Yang HB et al (2010) Molecular mapping and characterization of a single dominant gene controlling CMV resistance in peppers (*Capsicum annuum* L.) Theor Appl Genet 120:1587–1596

Kaur N, Singh DJ, Singh KD (2011) Physiological and biochemical traits analysis of *Capsicum annuum L.* germplasm for resistance to *Colletotrichum capsici*. J Cell Plant Sci 2:12–21

Khan AL, Kang S, Dhakal KH, Hussain J, Adnan M, Kim J, Lee I (2013) Flavonoids and amino acid regulation in *Capsicum annuum* L. by endophytic fungi under different heat stress regimes. Sci Hortic 155:1–7

Kim DS, Hwang BK (2012) The pepper MLO gene, CaMLO2, is involved in the susceptibility cell-death response and bacterial and oomycete proliferation. Plant J 72:843–855

Kim M, Kim S, Kim S, Kim BD (2001) Isolation of cDNA clones differentially accumulated in the placenta of pungent pepper by suppression subtractive hybridization. Mol Cells 11:213–219

Kim SH, Hong JK, Lee SC, Sohn KH, Jung HW, Hwang BK (2004) *CAZFP1*, Cys2/His-type zinc-finger transcription factor gene functions as a pathogen-induced early-sene in *Capsicum annuum*. Plant Mol Biol 55:883–904

Kim HJ, Nahm SH, Lee HR et al (2008) BAC-derived markers converted from RFLP linked to *Phytophthora capsici* resistance in pepper (*Capsicum annuum* L.) Theor Appl Genet 118:15–27

Kim OR, Cho MC, Kim BD, Huh JH (2010a) A splicing mutation in the gene encoding phytoene synthase causes orange coloration in Habanero pepper fruits. Mol Cells 30:569–574

Kim S, Kim KT, Kim DH et al (2010b) Identification of quantitative trait loci associated with anthracnose resistance in chili pepper (*Capsicum* spp.) Korean J Hortic Sci Technol 28:1014–1024

Kim MK, Seo JK, Kwak HR, Kim JS, Kim KH, Cha BJ, Choi HS (2014a) Molecular genetic analysis of cucumber mosaic virus populations infecting pepper suggests unique patterns of evolution in Korea. Phytopathology 104:993–1000

Kim S, Park M, Yeom SI et al (2014b) Genome sequence of the hot pepper provides insights into the evolution of pungency in *Capsicum* species. Nat Genet 46(3):270–278. https://doi.org/10.1038/ng.2877

Kim DS, Choi HW, Hwang BK (2014c) Pepper mildew resistance locus O interacts with pepper calmodulin and suppresses *Xanthomonas* AvrBsT-triggered cell death and defense responses. Planta 240:827–839

Kumar S, Kumar R, Kumar S, Singh M, Rai AB, Rai M (2011) Incidences of leaf curl disease on *Capsicum* germplasm under field conditions. Ind J Agric Sci 81:187–189

Lee S, Choi D (2013) Comparative transcriptome analysis of pepper (*Capsicum annuum*) revealed common regulons in multiple stress conditions and hormone treatments. Plant Cell Rep 32:1351–1359

Lee SC, Kim SH, An SH, Yi SY, Hwang BK (2006) Identification and functional expression of the pepper pathogen induced gene, *CAPIP2*, involved in disease resistance and drought and salt stress tolerance. Plant Mol Biol 62:151–164

Lee HR, Bae IH, Park SW, Kim HJ, Min WK, Han JH, Kim KT, Kim BD (2008a) Construction of an integrated pepper map using RFLP, SSR, CAPS, AFLP, WRKY, rRAMP, and BAC end sequences. Mol Cells 27:21–37

Lee J, Yoon JB, Park HG (2008b) Linkage analysis between the partial restoration (*pr*) and the restorer-of-fertility (*Rf*) loci in pepper cytoplasmic male sterility. Theor Appl Genet 117:383–389

Lee J, Han JH, An CG, Lee WP, Yoon JB (2010a) A CAPS marker linked to a genic male-sterile gene in the colored sweet pepper, 'Paprika' (*Capsicum annuum* L.) Breed Sci 60:93–98

Lee J, Yoon JB, Han JH, Lee WP, Do JW, Ryu H, Kim SH, Park HG (2010b) A codominant SCAR marker linked to the genic male sterility gene (*ms1*) in chili pepper (*Capsicum annuum*). Plant Breed 129:35–38

Lee J, Yoon JB, Han JH, Lee WP, Kim SH, Park HG (2010c) Three AFLP markers tightly linked to the genic male sterility *ms3* gene in chili pepper (*Capsicum annuum* L.) and conversion to a CAPS marker. Euphytica 173:55–61

Lee SC, Hwang BK, Choi DS, Hwang IS (2010d) The pepper oxidoreductase *CaOXR1* interacts with the transcription factor *CaRAV1* and is required for salt and osmotic stress tolerance. Plant Mol Biol 73:409–424

Lee HR, An HJ, You YJ, Lee J, Kim HJ, Kang BC, Harn CH (2013) Development of a novel codominant molecular marker for chiliveinal mottle virus resistance in *Capsicum annuum* L. Euphytica 193:197–205

Lefebvre V, Palloix A, Caranta C, Pochard E (1995) Construction of an intra-specific integrated linkage map of pepper using molecular markers and doubled-haploid progenies. Genome 38:112–121

Lefebvre V, Pflieger S, Thabuis A, Caranta C, Blattes A, Chauvet JC, Daubeze AM, Palloix A (2002) Towards the saturation of the pepper linkage map by alignment of three intraspecific maps including known-function genes. Genome 45:839–854

Lefebvre V, Daubeze AM, Voort RJ, Peleman J, Bardin M, Palloix A (2003) QTLs for resistance to powdery mildew in pepper under natural and artificial infections. Theor Appl Genet 107:661–666

Lei Y, Yin C, Li C (2006) Differences in some morphological, physiological, and biochemical responses to drought stress in two contrasting populations of Populus przewalskii. Physiol Plant 127(2):182–191

Li T, Xu X, Li Y, Wang H, Li Z, Li Z (2015a) Comparative transcriptome analysis reveals differential transcription in heat-susceptible and heat-tolerant pepper (*Capsicumannum L.*) cultivars under heat stress. J Plant Biol 58:411–424

Li W, Cheng J, Wu Z, Qin C, Tan S, Tang X, Cui J, Zhang L, Hu K (2015b) An InDel-based linkage map of hot pepper (*Capsicum annuum*). Mol Breed 35:32

Li J, Yang P, Kang J et al (2016) Transcriptome analysis of pepper (*Capsicumannuum*) revealed a role of 24-epibrassinolide in response to chilling. Front Plant Sci 7:1281

Lim JH, Park CJ, Huh SU, Choi LM, Lee GJ, Kim YJ, Paek KH (2011) *Capsicum annuum* WRKYb transcription factor that binds to the *CaPR*-10 promoter functions as a positive regulator in innate immunity upon TMV infection. Biochem Biophys Res Commun 411:613–619

Lim CW, Han SW, Hwang IS, Kim DS, Hwang BK, Lee SC (2015) The pepper lipoxygenase *CaLOX1* plays a role in osmotic, drought and high salinity stress response. Plant Cell Physiol 56:930–942

Liu WY, Kang JH, Jeong HS, Choi HJ, Yang HB, Kim KT, Choi D, Choi GJ, Jahn M, Kang BC (2014) Combined use of bulked segregant analysis and microarrays reveals SNP markers pinpointing a major QTL for resistance to *Phytophthora capsici* in pepper. Theor Appl Genet 127:2503–2513

Liu Z, Zhang Y, Ou L, Kang L, Liu Y, Lv J, Wei G, Yang B, Yang S, Chen W, Dai X, Li X, Zhou S, Zhang Z, Ma Y, Zou X (2017) Identification and characterization of novel microRNAs for fruit development and quality in hot pepper (*Capsicum annuum* L.) Gene 15:66–72

Livingstone KD, Lackney VK, Blauth J, Wijk VR, Jahn MK (1999) Genome mapping in *Capsicum* and the evolution of genome structure in the *Solanaceae*. Genetics 152:1183–1202

Lovato FA, Inoue-Nagata AK, Nagata T, de Avila AC, Pereira LA, Resende RO (2008) The N protein of Tomato spotted wilt virus (TSWV) is associated with the induction of programmed cell death (PCD) in *Capsicum chinense* plants, a hypersensitive host to TSWV infection. Virus Res 137:245–252

Lu FH, Cho MC, Park YJ (2012) Transcriptome profiling and molecular marker discovery in red pepper, *Capsicum annuum* L. TF68. Mol Biol Rep 39:3327–3335

Ma Y, Huang W, Ji JJ, Gong ZH, Yin CC, Ahmed SS, Zhao ZL (2013) Maintaining and restoration cytoplasmic male sterility systems in pepper (*Capsicum annuum* L.). Genet Mol Res 4:12

Maga JA (1975) *Capsicum*. Crit Rev Food Sci Nutr 6:177–199

Mahasuk P, Struss D, Mongkolporn O (2016) QTLs for resistance to anthracnose identified in two *Capsicum* sources. Mol Breed 36:1–10

Margaria P, Ciuffo M, Pacifico D, Turina M (2007) Evidence that the nonstructural protein of Tomato spotted wilt virus is the avirulence determinant in the interaction with resistant pepper carrying the TSW gene. Mol Plant-Microbe Interact 20:547–558

Matichenkov VV, Calvert DV (2002) Silicon as a beneficial element for sugarcane. J Am Soc Sugarcane Technol 22:21–30

Matsunaga H, Saito T, Hirai M, Nunome T, Yoshida T (2003) DNA markers linked to Pepper mild mottle virus (PMMoV) resistant locus (*L4*) in *Capsicum*. J Jpn Soc Hort Sci 72:218–220

Maurya VK, Srinivasan R, Nalini E, Ramesh N, Gothandam KM (2014) Analysis of stress responsive genes in *Capsicum* for salinity responses. ARRB. https://doi.org/10.9734/ARRB/2015/14107

Meghvansi MK, Siddiqui S, Khan MH, Gupta VK, Vairale MG, Gogoi HK, Singh L (2010) Naga chilli: a potential source of capsaicinoids with broad-spectrum ethnopharmacological applications. J Ethnopharmacol 132:1–14

Mimura Y, Minamiyama Y, Sano H (2010) Mapping for axillary shooting, flowering date, primary axis length, and number of leaves in pepper (*Capsicum annuum*). J Jap Soc Hort Sci 79:56–63

Mimura Y, Inoue T, Minamiyama Y, Kubo N (2012) An SSR-based genetic map of pepper (*Capsicum annuum* L.) serves as an anchor for the alignment of major pepper maps. Breed Sci 62:93–98

Mondal CK, Acharyya O, Hazra P (2013) Biochemical basis of plant defense for leaf curl virus of chilli (*Capsicum annuum*). In Proceeding XV EUCARPIA meeting on genetics and breeding of *Capsicum* and Eggplant, 2–4 Sept, Turin, Italy, pp 315–322

Moodley V, Ibaba JD, Naidoo R, Gubba A (2014) Full-genome analyses of a Potato virus Y (PVY) isolate infecting pepper (*Capsicum annuum* L.) in the Republic of South Africa. Virus Genes 49:466–476

Moon SJ, Han SY, Kim DY, Yoon IS, Shin D, Byun MO, Kwon HB, Kim BG (2015) Ectopic expression of a hot pepper bZIP-like transcription factor in potato enhances drought tolerance without decreasing tuber yield. Plant Mol Biol 89:421

Mou S, Liu Z, Gao F, Yang S, Su M, Shen L, Wu Y, He S (2017) *CaHDZ27*, a homeodomain-leucine zipper i protein, positively regulates the resistance to *Ralstonia solanacearum* infection in pepper. Mol Plant-Microbe Interact. https://doi.org/10.1094/MPMI-06-17-0130-R

Moulin MM, Rodrigues R, Bento CS, Gonçalves LSA, Santos JO, Sudre CP, Viana AP (2015) Genetic dissection of agronomic traits in *Capsicum baccatum* var. pendulum. Genet Mol Res 14:2122–2132

Moury B, Pflieger S, Blattes A, Lefebvre V, Palloix A (2000) A CAPS marker to assist selection of tomato spotted wilt virus (TSWV) in pepper. Genome 43:943–951

Munns R, Tester M (2008) Mechanism of salinity tolerance. Annu Rev Plant Niol 59:651–681

Murphy JF, Blauth JR, Livingstone KD, Lackney VK, Jahn MK (1998) Genetic mapping of the *pvr1* locus in *Capsicum* spp. and evidence that distinct potyvirus resistance loci control responses that differ at the whole plant and cellular levels. Mol Plant-Microbe Interact 11:943–951

Naegele RP, Ashrafi H, Hill TA, Reyes Chin-Wo S, Van Deynze AE, Hausbeck MK (2014) QTL mapping of fruit rot resistance to the plant pathogen *Phytophthora capsici* in a recombinant inbred line *Capsicum annuum* population. Phytopathology 104:479–483

Neto AD, Prisco JT, Eńeas-Filho J, Lacerda CF, Silva CF, PHA C, Gomes-Filho E (2004) Effects of salt stress on plant growth, stomatal response and solute accumulation of different maize genotypes. Brz J Plant Physiol 16:31–34

Nicolai M, Pisani C, Bouchet JP, Vuylsteke M, Palloix A (2012) Discovery of a large set of SNP and SSR genetic markers by high-throughput sequencing of pepper (*Capsicum annuum*). Genet Mol Res 11:2295–2300

Ogundiwin EA, Berke M, Massoudi TF, Black LL, Huestis G, Choi D, Lee S, Prince JP (2005) Construction of 2 intraspecific linkage maps and identification of resistance QTLs for *Phytophthoracapsici*root-rot and foliar-blight diseases of pepper (*Capsicum annuum* L.) Genome 48:698–711

Oh SK, Baek KH, Seong ES et al (2010) *CaMsrB2*, pepper methionine sulfoxide reductase *B2*, is a novel defense regulator against oxidative stress and pathogen attack. Plant Physiol 154:245–261

Okunlola GO, Olatunji OA, Akinwale RO, Tariq A, Adelusi AA (2017) Physiological response of the three most cultivated pepper species (*Capsicumspp.*) in Africa to drought stress imposed at three stages of growth and development. Sci Hortic 224:198–205

Pandravada SR, Varaprasad KS, Reddy KJ, Rao ES (2010) Screening and identification of sources of resistance against root-knot nematode (*Meloidogyne javanica*) in chilli (*Capsicum annuum*) germplasm. Ind J Agri Sci 80:92–94

Park SW, Jung JK, Choi EA, Kwon JK, Kang JH, Jahn M, Kang BC (2014) An EST-based linkage map reveals chromosomal translocation in *Capsicum*. Mol Breed. https://doi.org/10.1007/s11032-014-0089-0

Pernezny K, Roberts PD, Murphy JF, Goldberg NP (2003) Compendium of pepper diseases. The American Phytopathological Society, Minnesota

Perry L, Dickau R, Zarrillo S, Holst I, Pearsal DM, Piperno DR, Berman MJ, Cooke RG, Rademaker K, Ranere AJ, Raymond JS, Sandweiss DH, Scaramelli F, Tarble K, Zeidler JA (2007) Starch fossils and the domestication and dispersal of chili peppers (*Capsicum spp.* L.) in the Americas. Science 315:986–988

Phimchan P, Techawongstien S (2012) Impact of drought stress on the accumulation of capsaicinoids in *Capsicum* cultivars with different initial capsaicinoid levels. HortSci 47:1204–1209

Pierre M, Noel L, Lahaye T, Ballvora A, Veuskens J, Ganal M, Bonas U (2000) High-resolution genetic mapping of the pepper resis-tance locus Bs3governing recognition of the *Xanthomonas campestris* pv *vesicatora* AvrBs3 protein. Theor Appl Genet 101:255–263

Pochard E, Dumas de Valuix R, Florent A (1983) Linkage between partial resistance to CMV and susceptibility to TMV in the line Perennial: analysis on androgenetic homozygous lines. *Capsicum* Eggplant News 2:34–35

Popovsky S, Paran I (2000) Molecular analysis of the Y locus in pepper: its relation to capsanthin-capsorubin synthase and to fruit color. Theor Appl Genet 101:86–89

Portis E, Nagy I, Sasva Z, Stagelri A, Barchi L, Lanteri S (2007) The design of *Capsicum* spp. SSR assays via analysis of *In silico* DNA sequence, and their potential utility for genetic mapping. Plant Sci 172:640–648

Potato Genome Sequencing Consortium (2011) Genome sequence and analysis of the tuber crop potato. Nature 475:189–195

Prince JP, Pochard E, Tanksley SD (1993) Construction of a molecular linkage map of pepper and a comparison of synteny with tomato. Genome 36:404–417

Qados A, Amira MS (2011) Effect of salt stress on plant growth and metabolism of bean plant *Vicia faba* (L.) J Sau Soci Agric Sci 10:7–15

Qin C, Yu C, Shen Y, Fang X, Chen L, Min J et al (2014) Whole-genome sequencing of cultivated and wild peppers provides insights into *Capsicum* domestication and specialization. Proc Natl Aca Sci USA 111:5135–5140

Quirin EA, Ogundiwin EA, Prince JP et al (2005) Development of sequence characterized amplified region (SCAR) primers for the detection of *Phyto.5.2*, a major QTL for resistance to *Phytophthoracapsici*Lon. in pepper. Theor Appl Genet 110:605–612

Ramchiary N, Mechuselie K, Brahma V, Kumaria S, Tandon P (2013) Application of genetics and genomics towards *Capsicum* translational research. Plant Biotechnol Rep 8:101–123

Rao GU, Chaim AB, Borovsky E, Paran I (2003) Mapping of yield related QTLs in pepper in an interspecific cross of *Capsicum annuum* and *C. frutescens*. Theor Appl Genet 106:1457–1466

Reddy KM, Reddy MK (2010) Breeding for virus resistance. In: Kumar R, Rai AB, Rai M, Singh HP (eds) Advances in chilli research. Studium Press Pvt. Ltd., New Delhi, pp 119–132

Reddy MK, Srivastava A, Kumar S, Kumar R, Chawda N, Ebert AW, Vishwakarma M (2014a) Chilli (*Capsicum annuum* L.) breeding in India: an overview. J Breed Genet 46:160–173

Reddy UK, Almeida A, Abburi VL et al (2014b) Identification of gene-specific polymorphisms and association with capsaicin pathway metabolites in *Capsicum annuum* L. collections. PLoS One 9:e86393

Reeves G, Monroy-Barbosa A, Bosland PW (2013) A novel Capsicum gene inhibits host-specific disease resistance to *Phytophthoracapsici*. Phytopathology 103:472–478

Rodriguez-Uribe L, Guzman I, Rajapakse W, Richins RD, O'Connell MA (2012) Carotenoid accumulation in orange-pigmented *Capsicum annuum* fruit, regulated at multiple levels. J Exp Bot 63:517–526

Romer P, Jordan T, Lahaye T (2010) Identification and application of a DNA-based marker that is diagnostic for the pepper (*Capsicum annuum*) bacterial spot resistance gene *Bs3*. Plant Breed 129:737–740

Rubio M, Caranta C, Palloix A (2008) Functional markers for selection of potyvirusresistance alleles at the pvr2-eIF4E locus in pepper using tetra-primer ARMS-PCR. Genome 51:767–771

Rubio M, Nicolaï M, Caranta C, Palloix A (2009) Allele mining in the pepper gene pool provided new complementation effects between *pvr2-eIF4E* and *pvr6-eIF(iso)4E* alleles for resistance to pepperveinal mottle virus. J Gen Virol 90:2808–2814

Ruffel S, Dussault MH, Palloix A, Moury B, Bendahmane A, Robaglia C, Caranta C (2002) A natural recessive resistance gene against potato virus Y in pepper corresponds to the eukaryotic initiation factor 4E (eIF4E). Plant J 32:1067–1075

Ruffel S, Gallois JL, Moury B, Robaglia C, Palloix A, Caranta C (2006) Simultaneous mutations in translation initiation factors *eIF4E* and *eIF*(iso)*4E* are required to prevent pepperveinal mottle virus infection of pepper. J Gen Virol 87:2089–2098

Sanchita SA (2016) Computational gene expression profiling under salt stress reveals patterns of co-expression. Genom Data. https://doi.org/10.1016/j.gdata.2016.01.009

Semiz GD, Suarez DL, Unlukara A, Yurtseven E (2014) Interactive effects of salinity and n on pepper *(Capsicum annuum l.*) yield, water use efficiency and root zone and drainage salinity. J Plant Nutr 37:595–610

Shivakumara TN, Sreevathsa R, Dash PK, Sheshshayee MS, Papolu PK, Rao U, Tuteja N, Kumar U (2017) Overexpression of *Pea DNA Helicase 45 (PDH45)* imparts tolerance to multiple abiotic stresses in chili (*Capsicum annum L.*) Sci Rep 7:2760

Silvar C, Garcia-Gonzalez (2017) Screening old peppers (*Capsicum* spp.) for disease resistance and pungency-related traits. Scientia Hort 220:20–26

Simonne AH, Simonne EH, Eitenmiller RR, Mills HA, Green NR. Ascorbic acid and provitamin (1997) A contents in unusually colored bell peppers (*Capsicum annuum* L.). J Food Compos Anal 10:299–311

Sohn KH, Lee SC, Jung HW, Hong JK, Hwang BK (2006) Expression and functional roles of the pepper pathogen-induced transcription factor RAV1 in bacterial disease resistance, and drought and salt stress tolerance. Plant Mol Biol 61:897

Stewart C, Kang BC, Liu K et al (2005) The Pun1 gene for pungency in pepper encodes a putative acyltransferase. Plant J 42:675–688

Stewart C, Mazourek M, Stellari GM, O'Connell M, Jahn M (2007) Genetic control of pungency in *Capsicum chinense* via the *Pun1* locus. J Exp Bot 58:979–991

Sugita T, Semi Y, Sawada H, Utoyama Y et al (2013) Development of simple sequence repeat markersand construction of a high-density linkage map of *Capsicum annuum*. Mol Breed 31:909–920

Suwor P, Sanitchona J, Thummabenjapone P, Kumar S, Techawongstien S (2017) Inheritance analysis of anthracnose resistance and marker-assisted selection in introgression populations of chili (*Capsicum annuum* L.) Scientia Hort 220:20–26

Swindell WR, Huuebner M, Weber AP (2007) Transcriptional profiling of *Arabidopsis* heat shock protein and transcription factor reveals extensive overlap between heat and non heat stress response pathway. BMC Genomics 8:125

Tai T, Dahlbeck D, Stall RE, Peleman J, Staskawicz BJ (1999) High-resolution genetic and physical mapping of the region containing the Bs2 resistance gene of pepper. Theor Appl Genet 99:1201–1206

Tan S, Cheng J-W, Zhang L, Qin C, Nong D-G, Li W-P et al (2015) Construction of an Interspecific Genetic Map Based on InDel and SSR for Mapping the QTLs Affecting the Initiation of Flower Primordia in Pepper (*Capsicum* spp.) PLoS One 10(3):e0119389

Tanksley SD, Bernatzky R, Lapitan NL, Prince JP (1988) Conservation of gene repertoire but not gene order in pepper and tomato. Proc Natl Acad Sci USA 85:6419–6423

Tao L, Zeba N, Ashrafuzzaman M, Hong CB (2011) Heavy metal stress-inducible early light-inducible gene *CaELIP* from hot pepper (*Capsicum annuum*) shows broad expression patterns under various abiotic stresses and circadian rhythmicity. Environ Exp Bot 72:297–303

Thabuis A, Palloix A, Servin B, Daubèze AM, Signoret P, Hospital F, Lefebvre V (2004) Marker-assisted introgression of 4 *Phytophthora capsici* resistance QTL alleles into a bell pepper line: validation of additive and epistatic effects. Mol Breed 14:9–20. https://doi.org/10.1023/B:MOLB.0000037991.38278.82

The Tomato Genome Consortium (2012) The tomato genome sequence provides insights into fleshy fruit evolution. Nature 485:635–641

Thies JA, Fery RL (2000) Characterization of resistance conferred by the N gene to *Meloidogyne arenaria* races 1 and 2, *M. hapla*, and *M. javanica* in two sets of isogenic lines of *Capsicum annuum*. J Am Soc Hort Sci 125:71–75

Thies JA, Fery RL (2002) Heat stability of resistance to Southern root-knot nematode in bell pepper genotypes homozygous and heterozygous for the N gene. J Am Soc Hort Sci 127:371–375

Tomita R, Murai J, Miura Y et al (2008) Fine mapping and DNA fiber FISH analysis locates the tobamovirus resistance gene *L3* of *Capsicum chinense* in a 400-kb region of R-like genes cluster embedded in highly repetitive sequences. Theor Appl Genet 117:1107–1118

Tsaballa A, Pasentsis K, Darzentas N, Tsaftaris A (2011) Multiple evidence for the role of an *Ovate*-like gene in determining fruit shape in pepper. BMC Plant Biol 11:46

Vallejos CE, Jones V, Stall RE et al (2010) Characterization of two recessive genes controlling resistance to all races of bacterial spot in peppers. Theor Appl Genet 121:37–46

Voorips RE, Finkers R, Lia S (2004) QTL mapping of anthracnose (*Colletotrichum* spp.) resistance in a cross between. *Capsicum annuum* and *C. chinense*. Theor Appl Genet 109:1275–1282

Vurukonda SSKP, Vardharajula S, Shrivastava M, Sk ZA (2016) Enhancement of drought stress tolerance in crops by plant growth promoting rhizobacteria. Microbiol Res 184:13–24

Wahid A, Farooq M, Hussain I, Rasheed R, Galani S (2012) Responses and management of heat stress in plants. In: Ahmad P, Prasad MNV (eds) Environmental adaptations and stress tolerance of plants in the era of climate change. Springer, New York, pp 135–157

Wang LH, Zhang BX, Lefebvre V, Huang SW, Daubeze AM, Palloix A (2004) QTL analysis of fertility restoration in cytoplasmic male sterile pepper. Theor Appl Genet 109:1058–1063

Wang JZ, Cui LJ, Wang Y, Li JL (2009) Growth, lipid peroxidation and photosynthesis in two tall fescue cultivars differing in heat tolerance. Biol Plant 53:237–242

Wang JE, Liu KK, Li DW, Zhang YL, Zhao Q, Gong ZH, He YM (2013) A novel peroxidase *CanPOD* gene of pepper is involved in defense responses to *Phytophtora capsici* infection as well as abiotic stress tolerance. Int J Mol Sci 14(2):3158–3177

Wang P, Liu X, Guo J, Liu C, Fu N, Shen H (2015) Identification and expression analysis of candidate genes associated with defense responses to phytophthoracapsici in pepper line "PI 201234". Int J Mol Sci 16:11417–11438

Wang H, Niu H, Zhai Y, Lu M (2017) Characterization of *BiP* genes from pepper (*Capsicum annuum* L.) and the role of *CaBip1* in response to endoplasmic reticulum and multiple abiotic stresses. Front Plant Sci 8:1122

Warren G, McKown R, Marin AL, Teutonico R (1996) Isolation of mutations affecting the development of freezing tolerance in *Arabidopsis thaliana* (L.) Heynh. Plant Physiol 111:1011–1019

Widana Gamage SMK, McGrath DJ, Persley DM, Dietzgen RG (2016) Transcriptome Analysis of *Capsicum* chlorosis virus-induced hypersensitive resistance response in bell capsicum. PLoS One 11(7):e0159085

Wollenweber, Porter JR, Schellber J (2003) Lack of interaction between extreme high temperature events at vegetative and reproductive growth stages in wheat. J Agron Crop Sci 189:142–150

Wu L, Zhang Z, Zhang H, Wang XC, Huang R (2008) Transcriptional modulation of ethylene response factor protein JERF3 in the oxidative stress response enhances tolerance of tobacco seedlings to salt, drought, and freezing. Plant Physiol 148:1953–1963

Wu F, Eannetta NT, Xu Y, Durrett R, Mazourek M, Jahn MM, Tanksley SD (2009) A COSII genetic map of the pepper genome provides a detailed picture of synteny with tomato and new insights into recent chromosome evolution in the genus *Capsicum*. Theor Appl Genet 118:1279–1293

Xu D, Duan X, Wang B, Hong B, Ho T, Wu R (1996) Expression of a late embryogenesis abundant protein gene, HVA1, from barley confers tolerance to water deficit and salt stress in transgenic rice. Plant Physiol 110:249–257

Xu X, Chao J, Cheng X et al (2016) Mapping of a novel race specific resistance gene to Phytophthora root rot of pepper (*Capsicum annuum*) using bulked segregant analysis combined with specific length amplified fragment sequencing strategy. PLoS One 11(3):e0151401

Yadav SK (2010) Cold stress tolerance mechanisms in plants. A review. Agron Sustain Dev 30:515–527

Yao M, Li N, Wang F, Ye Z (2013) Genetic analysis and identification of QTLs for resistance to cucumber mosaic virus in chili pepper (*Capsicum annuum* L.) Euphytica 193:135–145

Yarnes SC, Ashrafi H, Reyes-Chin-Wo S, Hill TA, Stoffel KM, Van Deynze A (2013) Identification of QTLs for capsaicinoids, fruit quality, and plant architecture-related traits in an interspecific *Capsicum* RIL population. Genome 56(1):61–74

Yi SY, Kim JH, Joung YH, Lee S, Kim WT, Yu SH, Choi D (2004) The Pepper Transcription Factor *CaPF1* Confers Pathogen and Freezing Tolerance in *Arabidopsis*. Plant Physiol 136:2862–2874

Yi G, Lee J, Lee S, Choi D, Kim B-D (2006) Exploitation of pepper EST–SSRs and an SSR-based linkage map. Theor Appl Genet 114:113–130

Ying SC, Li MS, Hai ZZ, Alain P, Hao WL, Xi ZB (2015) Resistances to anthracnose (*Colletotrichum acutatum*) of *Capsicum* mature green and ripe fruit are controlled by a major dominant cluster of QTLs on chromosome P5. Sci Hortic 181:81–88

Zhang HX, Hodson JN, Williams JP, Blumwald E (2001) Engineering salt-tolerant *Brassica* plants: characterization of yield and seed oil quality in transgenic plants with increased vacuolar sodium accumulation. Proc Natl Acad Sci U S A 98:12832–12836

Zhang YL, Jia QL, Li DW, Wang JE, Yin YX, Gong ZH (2013) Characteristic of the pepper CaRGA2 gene in defense responses against *Phytophthora capsici* Leonian. Int J Mol Sci 14:8985–9004

Zheng Z, Nonomura T, Appiano M et al (2013) Loss of function in *Mlo* orthologs reduces susceptibility of pepper and tomato to powdery mildew disease caused by *Leveillula taurica*. PLoS ONE 8(7):e70723

Zhu JK, (2007) Plant salt stress. Wiley

Zong HX (2013) Construction of a molecular linkage map and QTL analysis on fruit-related traits in pepper. Jiangxi Agricultural University (in Chinese with English Abstract)

Zygier S, Ben Chaim A, Efrati A, Kaluzky G, Borovsky Y, Paran I (2005) QTLs mapping for fruit size and shape in chromosomes 2 and 4 in pepper and a comparison of the pepper QTL map with that of tomato. Theor Appl Genet 111:437–445

MicroRNA (miRNA) and Small Interfering RNA (siRNA): Biogenesis and Functions in Plants

Parul Chowdhury

Abstract Small RNA was first identified in 1981 in the genetic screening of *Caenorhabditis elegans*. Functions of these RNA are to repress gene expression by base pairing with complementary sequences within gene. Therefore, regulation by these small RNAs is called as RNA silencing, gene silencing or RNA interference (RNAi). Till date various kinds of small RNA have been discovered and categorized on the basis of their origin, biogenesis and functions. RNase III type of ribonuclease enzymes, i.e. dicers, is involved in small RNA processing, along with many other enzymes. Small RNAs are classified broadly into two classes, microRNAs (miRNAs) and small interfering RNAs (siRNAs) according to their origin. These small RNAs are further classified on the basis of their mechanism of gene silencing, cleavage of complementary mRNA, translational repression, transcriptional repression and DNA elimination through histone modification. These small-sized RNAs have bigger and vital roles to play in plants, which pertain to gene regulation during biotic stress and abiotic stress and development. Small RNA also plays a role in the plant defence against viruses and transposable elements.

Keywords Small RNA · MicroRNA · siRNA · Dicer · Development · Biotic and abiotic stress

1 Introduction

The information necessary for an organism's growth and development is carried by DNA (deoxyribonucleic). This information is stored in genes, which are specific sequences of nucleotides, the building blocks of DNA. This information is communicated to the living organism with the help of mRNA (messenger RNA) and further through protein, which is called the 'central dogma'. Various classes of RNA, viz. tRNA, rRNA and mRNA, are thought to play functional roles in plants. With the

P. Chowdhury (✉)
Dr B Lal Institute of Biotechnology, Jaipur, Rajasthan, India

S. Vats (ed.), *Biotic and Abiotic Stress Tolerance in Plants*,
https://doi.org/10.1007/978-981-10-9029-5_12

discovery of an increasing number of large and small nonprotein-coding RNAs having specific regulatory roles, the view of gene expression has changed.

Small RNA was first identified in genetic screening of *C. elegans* in 1981 (Chalfie and Sulston 1981). In 1993, a group of investigators (Rosalind C. Lee, Rhonda L. Feinbaum and Victor Ambros) from Harvard University was studying the *C. elegans* larval development. They discovered that *lin* 4 gene that is involved in the regulation of larval development codes for transcripts of 22 and 61 nucleotide in length instead of coding of protein. These transcripts bind to 3′untranslated region (3′ UTR) of the *lin 14* mRNA and regulate the expression of genes by RNA-RNA interactions (Lee et al. 1993). This RNA-RNA interaction was opposite to the central dogma. After years of discovery of other small RNAs in *Drosophila melanogaster* (Kennerdell and Carthew 1998), *Trypanosoma* (Ngo et al. 1998) and vertebrates (Aravin et al. 2003; Yang et al. 2001; Ruby et al. 2006; Landgraf et al. 2007), these groups of RNA were started to be called as small regulatory RNA. These regulatory 19–28 nucleotide, non-coding RNAs comprise of a family of small RNA. The basic function of these small RNA is to down regulate the expression of gene called as gene silencing. It is also called as post-transcriptional gene silencing (PTGS) in plants and 'quelling' in fungi. RNAi (RNA interference) can also refer to the technology in which small RNA is used as an experimental tool to shut off gene expression. These small regulatory RNA plays an important role in biotic stress, abiotic stress and developmental regulation. They also play a role in plant defence against viruses.

2 Types of Small RNA

Small RNA-mediated regulation is often referred to as RNA silencing, gene silencing or RNAi, because small RNAs repress gene expression. Although there are some basic differences in the origin of both microRNA (miRNAs) and small interfering RNAs (siRNAs), eventually they are related in the biogenesis mechanism as both are excised from the precursor by dicer and Argonaute (Ago) proteins helping in the silencing effect. Dicer, Ago and 21–23 nucleotide duplex RNA are key components of silencing complex (Carthew and Sontheimer 2009). Various types of small RNAs have been discovered and categorized based on their mode of biogenesis and origin (Fig. 1).

3 Small Interfering RNAs (siRNAs)

Gene silencing mostly causes inhibition of transcription (transcriptional gene silencing (TGS)) or RNA degradation (post-transcriptional gene silencing (PTGS)). RNAi is a mechanism that specifically silences genes through exogenous double-stranded RNA (dsRNA). SiRNAs are derived from long dsRNAs. On the other hand, dsRNAs

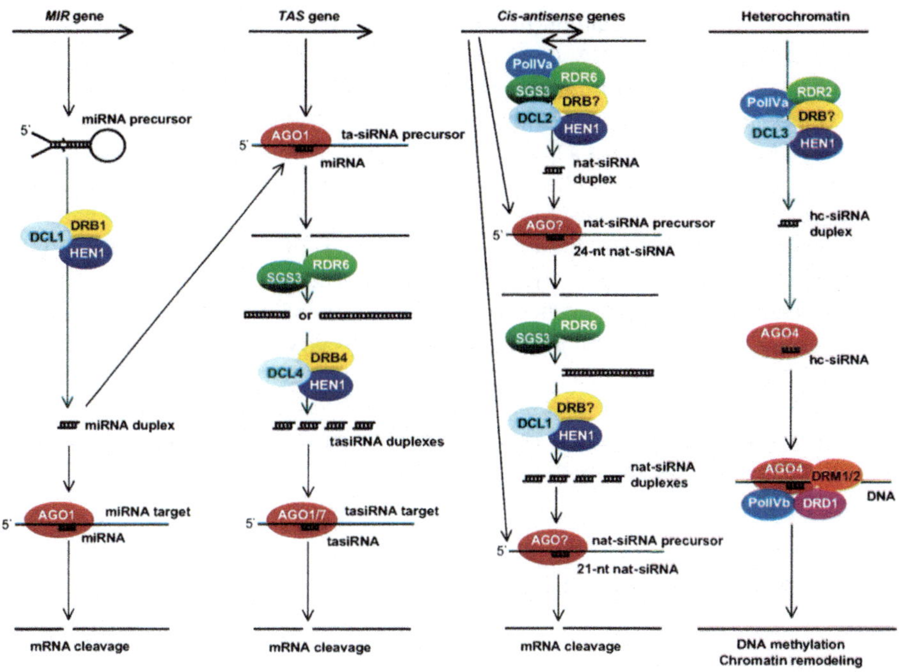

Fig. 1 Pathway of microRNA, transacting-siRNA, natural antisense-siRNA, and heterochromatic-siRNA in plants. siRNAs are derived from long dsRNAs. dsRNAs are produced by viral infection or inverted repeats or RNA-dependent RNA polymerases (RDRs) which are cellular-encoded. Apart from exogenous source, some endogenous source like transposons, centromere and other repetitive elements are other source of siRNA. miRNA is required for the processing of ta-siRNA, which is indicated by an arrow between these two pathways. In nat-siRNA pathway, multiple arrows indicate that the same product of the gene is used in various steps. The same gene family is indicated in the same colour. (Adapted from Vaucheret 2006)

are produced by viral infection or inverted repeats or cellular-encoded RNA-dependent RNA polymerases (RDRs) and are processed to siRNAs by cellular DICER-like enzymes (DCLs) and Argonaute proteins (Hannon 2002; MacRae et al. 2006). Apart from an exogenous source, some endogenous source like transposons, centromere and other repetitive elements are also source of siRNA (Lippman and Martienssen 2004). They are involved in both post-transcriptional forms of RNAi and transcriptional silencing through chromatin modification (Finnegan and Matzke 2003).

Exogenously derived siRNA can act in the following ways:

Source of double-stranded RNA in plants is by viral infection, which is further processed by viral or cellular RDRs and DCLs. According to Mallory and Vaucheret (2006), these siRNAs which are derived from exogenous viral infection can result in:

Viral recovery: In this the mature siRNAs (negative strand) guide the cleavage of the complementary viral RNA (positive strand). This results in decrease in viral accumulation, a kind of immune response which is called as viral recovery.

Viral synergism: It is the interaction of two independent viruses in the same host. One of the virus produces proteins which can suppress siRNA pathways at various levels and allow the successful infection. Whereas, infection of the second virus, at the same time, cannot produce proteins for siRNA pathway suppression and results in successful infection of both the viruses. This co-infection phenomenon is referred to as viral synergism.

Co-suppression: It is a phenomenon where overexpressed transgene of an organism causes suppression of homologous transcripts by inducing mRNA degradation (De Paoli et al. 2009). Transgene can also produce structure-like viral RNA, producing dsRNA and siRNA. And this mature siRNA can degrade the complementary endogenous mRNA. This phenomenon of transgene-mediated degradation is called as co-suppression.

3.1 Types of Endogenous siRNA

There are three types of endogenous siRNA found in plants.

3.1.1 Repeat-Associated siRNAs (Ra-siRNAs)

Ra-siRNAs arise from loci with repeat sequences and are involved in DNA methylation of loci from which they are derived. It helps in the establishment or maintenance of transcriptionally silent chromatin (Lippman et al. 2004; Xie et al. 2004). DNA methylation in plants takes place through a process called RNA-directed DNA methylation (RdDM). The process of RdDM starts with the 24-nucleotide siRNAs (ra-siRNAs) synthesized from RdDM target loci. These ra-siRNAs methylate the cytosine content of the homologous sequence with the help of DNA methyltransferase, DOMAINS REARRANGED METHYLTRANSFERASE 2 (DRM2) (Xie et al. 2012).

3.1.2 Transacting siRNAs (ta-siRNAs)

Ta-siRNAs are a class of secondary siRNAs derived from non-coding transacting SiRNA (TAS) transcripts. Ta-siRNAs are initially targeted for cleavage by miRNA in a phased manner. Phased manner indicates that RNAs are generated from cleavage of dsRNA at a specific site by dicer. A unique first cleavage site is required for creating the small RNA in phased manner (Johnson et al. 2009). Cleavage product of miRNA is converted into dsRNA by RDR 6 and suppressor of gene silencing 3 (SGS3); subsequently it is cleaved by DCL4 into 21-nt ta-siRNAs. So it produces a phase of siRNA starting from a specific cleavage site, and these in phase siRNA are termed as ta-siRNA (Adenot et al. 2006). Ta-siRNA can act both in *cis* and *trans*, guiding the cleavage of mRNA. The primary proteins that participate in ta-siRNA biogenesis include RDR6, SGS3, DCL4, AGO1, AGO7 and DOUBLE-STRANDED RNA BINDING FACTOR 4 (Peragine et al. 2004; Vaucheret et al. 2004; Allen and Howell 2010).

For cleavage of TAS transcript, AGO1, DCL1, HEN1 and HYL1 are required. Ta-siRNA was first identified in *Arabidopsis*, and now it is found in many plants like wheat, maize and *Brassica* (Allen et al. 2005; Nagasaki et al. 2007; Nogueira et al. 2009).

3.1.3 Natural Antisense siRNAs (nat-siRNAs)

Endogenous transcripts that can form double-stranded RNA structures are called as natural antisense siRNAs (nat-siRNAs). Partially overlapping genes on opposite strands of DNA from the same locus (cis-antisense genes) can anneal and form dsRNAs and give rise to nat-siRNAs. Nat-siRNAs are divided into two categories: (1) *cis-nat-siRNAs* are transcribed from opposite DNA strands from the same genomic loci, whereas (2) *trans-nat-siRNAs* are transcribed from different genomic loci (Zhang et al. 2012; Held et al. 2008). Nat-siRNAs are found commonly in eukaryotes including humans and mouse (Faghihi and Wahlestedt 2009; Tosic et al. 1990). Sixty-nine plant species were identified having nat-siRNAs, and a plant natural antisense transcripts database (PlantNATsDB) has been developed (Zhang et al. 2011). Many nat-siRNAs are specific to a tissue or developmental stage or to an environmental condition. For example, nat-siRNASRO5 is induced by salt stress (Borsani et al. 2005). Nat-siRNAATGB2 is induced by infection of bacterial pathogen (Katiyar-Agarwal et al. 2006).

3.2 Biogenesis of siRNA

RNAi is generated from long, linear and perfectly base-paired dsRNAs, introduced directly into the cytoplasm or taken up from the environment. The precursors for siRNAs are usually long and double-stranded RNA (Mello and Conte 2004).

In *Arabidopsis thaliana*, four DCLs are involved in different types of small RNA biosynthesis (siRNA and miRNA) (Chapman and Carrington 2007). For ra-siRNA, 24nt long siRNAs are produce by DCL3. Their function is to silence gene by DNA methylation via RdDM pathway. This secondary siRNA-generation machinery via RdDM is further supported by RNA polymerases IV and V, AGO4, RDR 2, chromatin remodelling proteins and DNA and histone methylases (Verdel et al. 2009) (Fig. 2).

DCL4 helps in generation of long transacting 21-nucleotide ta-siRNAs from TAS RNA precursors, along with RDR 6, SGS3 and DRB4 (dsRNA BINDING PROTEIN 4). Low-abundant 22nt long siRNAs are generated from DCL2 from different

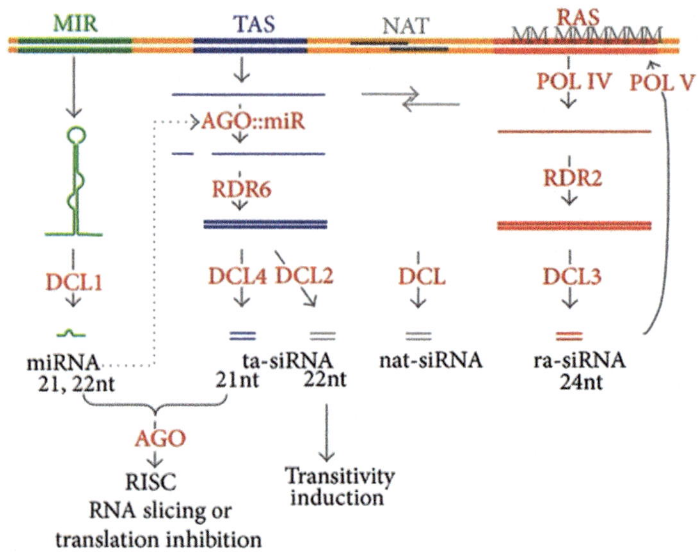

Fig. 2 Different DICER-LIKE enzymes (DCLs) involved in different types of small RNA biosynthesis (siRNA and miRNA). For ra-siRNA, 24nt long siRNAs for gene silencing by DNA methylation via RNA-dependent DNA methylation (RdDM) pathway are produced from DCL3. Long transacting (ta-siRNAs) 21-nucleotide from TAS RNA precursors are generated from DCL4. Low-abundant 22nt long siRNAs from different precursors are generated from DCL2, and it functions only when there is mutation in DCL4 or DCL3. 21–22nt miRNAs are generated from DCL1, 24nt-long miRNAs are generated from DCL3, 21–22nt miRNAs are generated from DCL1, and 23–24nt-long miRNAs are generated from DCL3 from pri-miRNAs. (Adapted from Vazquez and Hohn 2013)

precursors, and they function only when *DCL4* or *DCL3* is mutated or suppressed (Zhang et al. 2006; Blevins et al. 2006).

Pri-miRNAs are processed to mature 21–22nt miRNAs by DCL1 and 24nt-long miRNAs by DCL3 (Papp et al. 2003). siRNAs from long hairpins can be generated from DCL2, DCL3 and DCL4 (Zhang et al. 2010). Silencing by RNA-induced silencing complex (RISC) occurs by two types of activities. Target RNA is recognized and either cleaved (sliced), and its translation is inhibited, like miRNA silencing.

The two types of siRNAs, ta-siRNAs and ra-siRNAs, differ in their precursors and in the process of synthesis. Ta-siRNAs are processed from nuclear RNA precursors, TAS transcripts, while ra-siRNAs are generated from transposable and repetitive elements. Requirement of miRNA-dependent generation of ssRNA precursor is essential for ta-siRNA generation (Blevins et al. 2006), while for ra-siRNA generation, a DNA-dependent RNA polymerase RNA pol IV transcribes ssRNA precursor from the heterochromatic locus, which is then followed by RDR2-catalyzed synthesis of dsRNAs. These dsRNAs are processed by DCL3, and resulting siRNA is assembled in duplexes in AGO4-clade AGOs. The rest of siRNA biogenesis pathway is as same as miRNA, although different types of protein and enzymes are involved (Fig. 1).

Nat-siRNAs do not rely on RDRs to synthesize dsRNAs; they are formed from separately transcribed complementary mRNAs, which hybridize with each other to form mRNA duplexes. If nat-siRNAs are transcribed from opposite strands of the same locus, then these are called as cis-nat-siRNAs. Biogenesis of different nat-siRNA requires individual types of RDRs, DCLs and other enzymes and proteins for their accumulation (Borsani et al. 2005).

4 MicroRNA (miRNA)

MiRNAs are endogenously generated non-coding small RNAs (Bartel 2004), which originate from ssRNA usually consisting of 20–22 nucleotides in animals and 20–24 nucleotides in plants. All miRNA precursors have stem-loop structure. This stem-loop is like foldback hairpin structure, which can be well predicted (Fig. 3) and has a low free energy (Bonnet et al. 2004; Reinhart et al. 2002). They are processed and generated from primary transcripts known as *pri-miRNA* to short stem-loop structures called *pre-miRNA* and finally to functional miRNA. Mature miRNA molecules are partially complementary to messenger RNA (mRNA) molecules, and their main function is to downregulate the gene expression.

4.1 Identification and Conservation of Plant miRNA

Since the discovery of the first miRNA, evolution of miRNA genes and their functions are enhanced by the small RNA sequencing data. Usually miRNAs are identified by some common characteristics like the following: all miRNAs are small non-coding RNA usually 22–24 nucleotides in length (Bartel 2004), mature miRNA arises from predicted stem-loop structure which is having low free energy as compared to tRNA and rRNA shown by using statistical methods (Zuker 2003), and last but not the least, miRNAs are evolutionary conserved. In animals hairpin loop structure is conserved, and in plants only mature miRNA sequence is conserved (Altuvia et al. 2005; Axtell and Bartel 2005). All these criteria should be kept in mind before identification and confirmation of miRNA along with mechanism of biogenesis of microRNA. There are various methods for the identification of miRNA, viz. small RNA cloning (Sunkar et al. 2005) and computational approach, which includes filter-based approaches of identification of all potential hairpins (Wang et al. 2004). Web-based programmes used for identification of miRNA are MIR check, miRFinder and Find-miRNA (Huang et al. 2007; Adai et al. 2005).

Nowadays, sequencing technology has achieved great progresses. Based on the high degree, depth and coverage of sequencing, the high-throughput sequencing (HTS), which is also called next-generation sequencing (NGS) technology, helps in

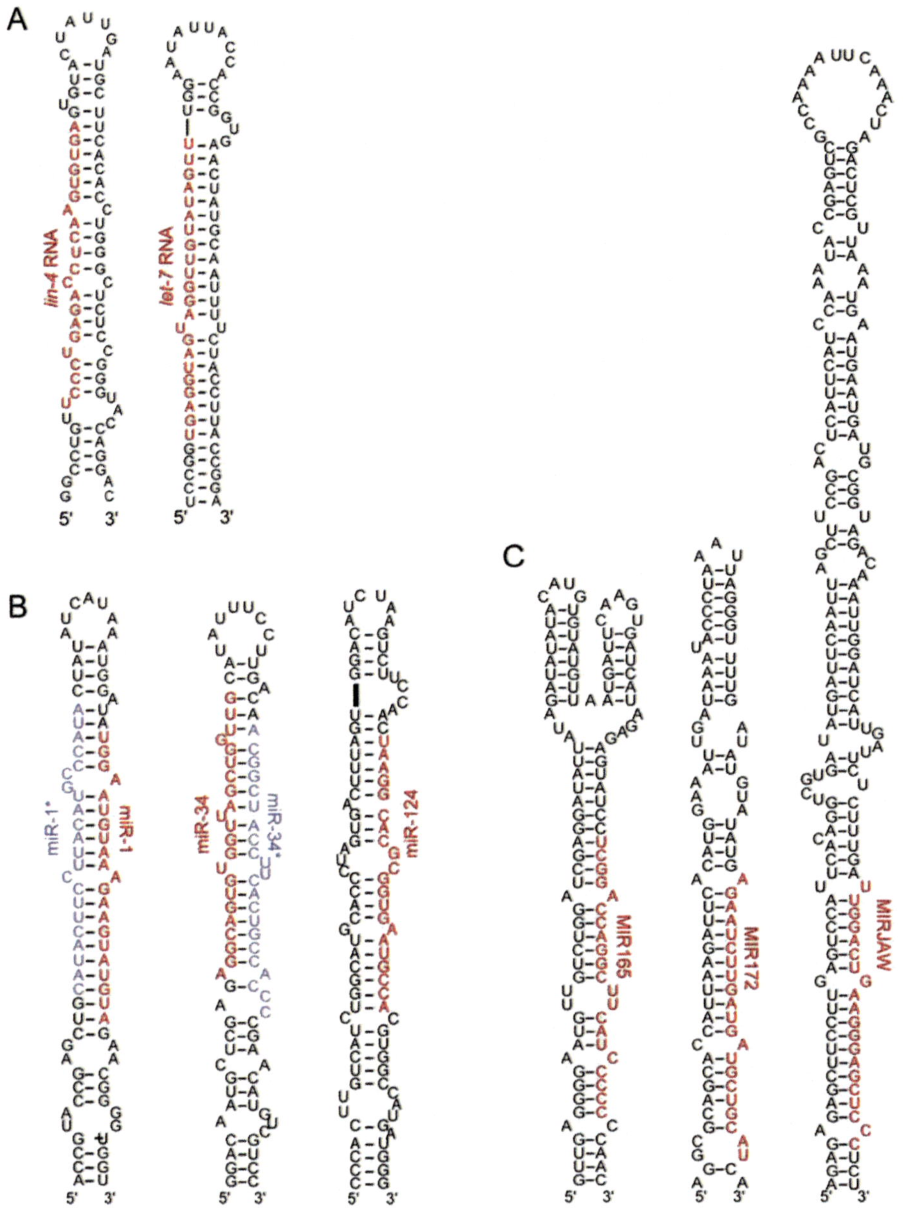

Fig. 3 Hairpin loop structure of some microRNA: examples. Stem-loop hairpin structure of all miRNA can be well predicted, which has low free energy. Sequence highlighted in red shows the mature miRNA sequence. (Adapted from Bartel 2004)

genome-wide or transcriptome analysis (Fox et al. 2009). HTS method includes RNA-seq (RNA sequencing), RNA-PET-seq (paired-end tag sequencing of RNAs), sRNA-seq (small RNA sequencing), dsRNA-seq (double-stranded RNA sequencing), ssRNA-seq (single-stranded RNA sequencing) and degradome-seq (degradome sequencing) (Ma et al. 2015).

The identification and analyses of several miRNA are revealed that miRNAs are evolutionary conserved, and they are conserved between monocots and dicots (Sunkar and Zhu 2004; Wang et al. 2004; Adai et al. 2005). In *Arabidopsis*, Allen et al. (2004) have shown that miRNAs originate from their target by the duplication of gene. Apart from this, there are many regions in genome from where microRNAs originate.

4.2 Biogenesis of miRNA

Plant miRNAs are endogenously expressed RNAs usually processed from one arm of foldback precursors, which are evolutionary conserved, and originate from regions of the genome which doesn't include previously annotated genes (Reinhart et al. 2002). Most of the miRNAs arise from their own transcriptional unit, i.e. they have their own promoter. After being transcribed from their own promoter, these primary transcripts are called pri-miRNA (Lee et al. 2002). Some miRNAs, especially in animals, reside in the intronic regions and are transcribed from the promoter of the host gene (Baskerville and Bartel 2005). Plant pri-miRNAs can be more than 1 kb in length, and they have typical TATA box motifs upstream to the miRNA gene. RNA polymerase II is probably responsible for transcribing most plant miRNAs, because of the fact that pri-miRNAs can undergo splicing, polyadenylation and capping (Lee et al. 2002). An essential step in miRNA maturation is mature miRNA excision from the pri-miRNA by RNaseIII-type endonucleases, such as Drosha (a nuclear-localized RNaseIII enzyme) and Dicer (a cytoplasmic-localized RNaseIII enzyme), which are present in animals. Drosha produces the first cut and liberate the stem-loop of miRNA, which is exported to the cytoplasm. The second cut is produced by Dicer, which liberates the mRNA duplex (the miRNA) with its near reverse complement, leaving behind the loop structure (Lee et al. 2003).

Instead of Drosha and Dicer, DCL is present in plants. There are four *Arabidopsis* Dicer-like genes, DCL1, DCL2, DCL3 and DCL4 (Schauer et al. 2002). DCL belongs to the family of RNase III endonucleases. Short double-stranded RNAs (dsRNAs) are generated from pri-miRNA stem-loops consisting of miRNA (*guide RNA*) and miRNA* (*passenger RNA*), (miRNA* refers to the antisense sequence to the mature miRNA). miRNA* strands have 2-nucleotide 3′ overhangs by a family of four DCL (Werner 2010).

DCL1 and DCL4 generate small RNA with 21 nucleotides, DCL2 generates 22 nucleotides, and 24 nucleotides are by DCL3 (Cuperus et al. 2011). DCL1 possesses functions of both Drosha and Dicer (Kurihara and Watanabe 2004) and that is supported by the fact that the two sets of cuts that liberate the miRNA/miRNA* are duplex in plants and occur in the nucleus, which is the predominant location of DCL1 (Yoshikawa et al. 2005). miRNA stem-loop precursor is cut at the specific position by DCL1 that results in accumulation of the appropriate mature miRNA (Reinhart et al. 2002). Mutation of DCL1 greatly reduced the accumulation of DCL1 and exhibited the deleterious developmental effect, thus proving the role of DCL1 in miRNA accumulation (Nodine and Bartel 2010). The dsRNA molecules with 2-nt 3′ overhangs are the in vitro products of Dicer cleavage, but only one of the two cleavage products accumulates as the miRNA matures. miRNA/miRNA* duplex is unwound by a helicase, and only the mature microRNA is transferred to RISC (Hammond 2005).

Along with DCL1, dsRNA binding domain (DRB) protein HYPONASTIC LEAVES1 (HYL1/DRB1, HUA ENHANCER1 (HEN1)) (Lu and Fedoroff 2000), G-patch domain protein TOUGH (TGH) (Ren et al. 2012) and zinc finger protein SERRATE (SE) (Grigg et al. 2005) are also important for miRNA maturation. All these RNA-binding proteins bind to various regions of pri-dsRNA along with DCL1 and form a small microprocessor complex (Kurihara et al. 2006) and ultimately help in miRNA maturation.

Other proteins required for miRNA processing in nucleus are:

- C-TERMINAL DOMAIN PHOSPHATASE-LIKE1 (CPL1) is required for exact miRNA processing by maintaining phosphorylated state of HYL1 (Manavella et al. 2012).
- DAWDLE (DDL), phosphothreonine binding forkhead-associated domain associate with DCL1 for miRNA processing (Yu et al. 2008).
- SICKLE (SIC) proline-rich protein required for miRNA accumulation (Zhan et al. 2012).
- MODIFIER OF SNC1 (MOS2) is required for HYL1 and pri-miRNA association (Copeland et al. 2013).
- CAP-BINDING PROTEINs (CBP20 and CBP80) are required for splicing of introns (Laubinger et al. 2008).
- STABILIZED1 (STA1) binds with CBP and regulates the miRNA splicing (Chaabane et al. 2013).

HEN1 can methylate 3′ nucleotides of miRNA/miRNA* duplexes with the help of methyltransferase domain. HEN1 deposits a methyl group on the 2′ OH of the 3′ terminal nucleotide (Yang et al. 2006). As HEN1 is located both in nucleus and cytoplasm, the site of methylation is not clear. If there is no methylation, then it results in the degradation of miRNA (Li et al. 2005).

Pre-miRNAs are thought to be exported to cytoplasm by HASTY (HST), a member of the exportin-β family of nucleocytoplasmic transporters (Bollman et al. 2003), but concentration of miRNA in the nucleus and cytoplasm is unchanged in *hst* mutants. Export of miRNA from the nucleus to cytoplasm is unclear, but Hasty may be required for the stability of miRNA as suggested by cytoplasmic exportin5 function in animals

(Zeng et al. 2005). When miRNA-RNA is unwound, there are two ssRNAs called as the passenger strand and the guide strand. The passenger strand is degraded, and the guide strand is incorporated into the RISC. HYL1 and CPL1 help in the strand selection (Eamens et al. 2009). Passenger strand is degraded with the help of AGO, which forms complex with HEAT SHOCK PROTEIN90 (HSP90) and SQUINT (SQN) (Iki et al. 2010).

The miRNA* is degraded, when miRNA/miRNA* duplex enters into the *RISC*, as the strand which is less stable thermodynamically, i.e. less stable 5′ possessing 5′ uridine enters the RISC and silence the genes (Khvorova et al. 2003; Schwarz et al. 2003). The RISC complex contains the Argonaute protein family (Hammond et al. 2001). Argonaute and its homologs are approximately 100 kDa proteins. Argonaute are also called as PPD proteins because they all share the PAZ and PIWI domains (Cerutti et al. 2000).

Now it is known that the miRNA* sequences of some miRNAs are also functional under certain circumstances. The miRNA and miRNA* are now more commonly named as *miRNA-5p* and *miRNA-3p*, respectively, according to their positions on the hairpin-shaped precursor (Liu et al. 2014).

4.3 *Effect of miRNA on Targets*

There are two ways by which various microRNAs repress the target.

4.3.1 Cleavage of Target

mRNA cleavage is the most common mechanism of miRNA-guided regulation in plants (Palatnik et al. 2003). Plant microRNA slices its target mRNA in the middle of complementary sequence which has high degree of sequence complementary with target mRNA. Target mRNA is cleaved by AGO1 proteins between positions 10 and 11 of the alignment, i.e. exactly in the middle of the miRNA. This was demonstrated by the cleavage end detection. 3′ cleavage products have the 5′ end which starts at the middle of complementary region (Mi et al. 2008). The AGO proteins of PIWI domain form an RNaseH-like fold with an endonuclease activity capable of cleaving RNA targets that are complementary to the loaded guide strand (Liu et al. 2004).

3′ cleavage products are further degraded by 5–3′ exonuclease, exoribonuclease 4 (XRN4), as it was demonstrated by *xrn* 4 mutants that 3′ cleavage product accumulates in these mutants (Souret et al. 2004). Other 3′ cleavage products are degraded independent of exonuclease XRN4 as they do not accumulate in *xrn* 4 mutants. 5′ cleavage product acquires poly U tail, and this cleavage product starts shortening from 5′ end. According to Shen and Goodman (2004), oligo U tail causes 5–3′ exonucleolytic degradation of the 5′ cleavage products.

4.3.2 Translational Inhibition

Translation repression is another mode of regulation by miRNA, where the mRNA level is unaffected but protein level is reduced. One of the examples is overexpression of miRNA172 which leads to reduced level of AP2 protein rather than AP2 mRNA (Chen 2004). Apart from mir172, it was also shown that miR156/157 and miR854 lead to reduced protein but not mRNA levels of their target genes (Arteaga-Vázquez et al. 2006; Gandikota et al. 2007). Actual mechanism of translation inhibition in plants is still unknown and requires more investigation to clear the mechanisms. It was found that ALTERED MERISTEM PROGRAM1 (AMP1) gene is involved in translation repression, AMP1 plays an important role in exclusion of miRNA target mRNAs from membrane-bound polysomes (Helliwell et al. 2001). There are multiple mechanisms of translation repression in animal like ribosome stalling, ribosome dropoff, etc. (Fabian et al. 2010).

4.4 MiRNAs Database

Research on microRNA and their function is increasing day by day. With the genome sequencing projects, lots of miRNAs are being identified in plants' genome. So there is a need of a public database where all the known miRNA can be grouped and submitted. As the new information is generated, these databases are updated from time to time.

Some of the miRNA databases are:

(a) miRBase (http://www.mirbase.org/)
(b) deepBase (deepbase.sysu.edu.cn/)
(c) MicroRNA.org (www.microrna.org/)
(d) PMTED (pmted.agrinome.org/)

Most of these databases are based upon the characteristic of miRNA and their target that are evolutionary conserved. There should be other approaches to identify and maintain database of the miRNAs which are not conserved. This can be achieved by predicting the novel miRNA from the small RNA library constructed, with the help of stem-loop structure and their free energy. Promoter study of the miRNA gene should be included so that further studies on the miRNA gene can be done Moreover, there is a need of functional characterization of all these miRNAs identified so that their exact function can be known.

5 Function of Small RNA in Plants

Plant development, metabolism and their response to different stresses are dependent on proper gene expression and proper regulation of gene expression. Countless number of events is involved in proper function and metabolism of plants. Mostly

Table 1 Difference between miRNA and siRNA

	MicroRNA (miRNA)	Small interfering RNA (siRNA)
1.	They are regulators of endogenous genes	Defend the genome in response to foreign DNA such as viruses and transposons
2.	miRNAs are derived from long, single-stranded RNAs (ssRNAs)	siRNAs are derive from long, perfect dsRNAs
3.	Originate from distinct genomic location, of their own gene	Originate from virus, transposone, dsRNA and heterochromatin region
4.	One miRNA locus produces only one miRNA duplex	One siRNA locus produces many siRNA
5.	miRNAs can form imperfectly double-stranded RNAs (dsRNAs) because they have the ability to fold back	siRNAs are originated from transcription by RNA-dependent RNA polymerases (RDRs) of sense-antisense gene pairs, or they are originated from inverted repeat sequences
6.	Processed by RNase III proteins of the Drosha/Dicer family (animals) and DCL1 (plants)	Different types of siRNA are generated from DCL2, DCL3 and DCL4
7.	miRNAs always act in trans by regulating mRNAs that exhibit strong complementarity to the 5′ end of the miRNA sequence	siRNA can act both in trans and cis form; it can regulate expression of gene from which it originate as well as other elements that exhibit complementarity to their sequence
8.	miRNAs usually repress gene expression at post-transcriptional level	siRNA can act both at transcriptional and post-transcriptional level of gene repression
9.	miRNA doesn't play a role in cell immunity	siRNA gene silencing leads to cell immunity
10.	miRNAs are conversed across the species	siRNAs are rarely conserved

gene regulation takes place at the level of transcription, but there are many RNA-binding proteins (RBP) that regulate stability and localization of mRNA in cell. These proteins bind to untranslated regions (UTRs) and regulate the gene expression at post-transcriptional level. Small RNAs (miRNA and siRNA) also regulate the gene expression at post-transcriptional level. As already discussed, these miRNAs usually silence the gene expression by binding to complementary sequences leading to degrading of mRNA or translational repression (Mallory and Vaucheret 2006; Jones-Rhoades et al. 2006). Difference between miRNA and siRNA is listed in Table 1.

Regulation of gene expression by transcriptional, post-transcriptional, translational and post-translation mechanisms determines appropriate course of plant development, metabolism and stress responses. The most common form of RNA, mRNA, is used for the protein production, but there are many genes whose final products are RNA,i.e. they do not code for the protein. These are called as non-coding RNA. Apart from miRNA and siRNA, other non-coding RNAs range from transfer and ribosomal RNAs to regulatory RNAs (Finnegan and Matzke 2003; Wirth and Crespi 2009). Regulatory RNA includes miRNA and siRNA.

5.1 Role of MicroRNA in Plant Development

Several experiments demonstrate that microRNAs are involved in various developmental process (Rhoades et al. 2002; Llave et al. 2002; Aukerman and Sakai 2003; Vaucheret et al. 2004; Sunkar and Zhu 2004; Mallory et al. 2004; Floyd and Bowman 2004), along with response to abiotic (Achard et al. 2004; Chiou 2007; Sunkar et al. 2006; Zhou et al. 2007) and biotic stress (Navarro et al. 2006; Rosa et al. 2009).

Functions of some miRNAs that are involved in developmental processes are described below.

5.1.1 miR156/157

miR156/157 targets Squamosa-Promoter Binding Protein-Like (SPL) plant-specific transcription factors. SPL is a family of transcription factors, which includes SPL2, SPL3, SPL4 and SPL10 members. MiR156-SPL regulatory modules are known to play a central role in the regulation of diverse developmental processes. MiR156-SPL are involved in flower development and phase change (Wang et al. 2009) and also regulate trichome distribution (Yu et al. 2010). MiR156-overexpressing plants have delayed flowering time, increased leaf initiation and decreased apical dominance, resulting in dramatically bushier plant (Schwab et al. 2005; Wang et al. 2009; Shikata et al. 2012). Flowering time in response to temperature in *Arabidopsis* is mediated by miR156-SPL, by downregulation of *FLOWERING LOCUS T (FT)* and *FRUITFULL* expression. This indicates that *miR156-SPL3* module and *FT* are involved in regulatory mechanism of temperature-dependent flowering time.

5.1.2 miR159

miR159 is known to regulate MYB class of transcription factors that in turn regulates LEAFY which eventually is involved in flower development (Gocal et al. 2001). Overexpression of miR159 results in male sterility, as anther development is affected (Achard et al. 2004). Gibberellic acid (GA) and abscisic acid (ABA) regulate miR159 expression and control the floral organ development (Sunkar and Zhu 2004). In a recent study, miR159 was shown to promote programmed cell death in *Arabidopsis* (Alonso-Peral et al. (2010)). In another study, tomato mir159 (Sl-miR159) targets unigene (SGN-U567133), which is not related to MYB and represents novel target of mir159. Resistant form of unigene for mir159, when overexpressed in tomato, results in overaccumulation of SGN-U567133 and shows defects in leaf and flower development (Buxdorf et al. 2010).

5.1.3 miR319/JAW

miR319 targets TCP family of transcription factors, overexpression of which leads to aberrant seedling; fused cotyledon, with no apical meristem; and abnormal leaf development (Palatnik et al. 2003). Overexpression of miR319/JAW results in jaw-D phenotype characterized by uneven leaf curvature and shape (Nag et al. 2009). Expression of miR319 is also induced by cold and drought stress (Sunkar and Zhu 2004).

5.1.4 miR160

Expression of auxin-inducible genes, like as GH3 and auxin/indole-3-acetic acid (Aux/IAA), is regulated by auxin response factor (ARF) by binding to auxin response promoter elements (AREs). mirl60 targets ARF transcription factors (Hagen and Guilfoyle 2002). Overexpression of miR160 leads to gravitropic roots with disorganized root caps and increased lateral rooting (Wang et al. 2005). Along with this, overexpression of miR160 leads to severe developmental defects like premature inflorescence development, sterility due to abnormal stamen and root growth defects (Mallory et al. 2005; Liu et al. 2011).

5.1.5 miR164

miR164 targets NAC domain transcription factors, such as NAM, CUC1 and CUC2 (Mallory et al. 2004). miR164 plays an important role in controlling shoot and root development which is shown by overexpression of miR164. Overexpression of miR164 shows morphological abnormalities like fusion of vegetative and floral organs, unbalanced floral organ numbers and reduced lateral root formation (Laufs et al. 2004; Guo et al. 2005; Berger et al. 2009).

5.1.6 miR171

miR171 targets GRAS-domain transcription factor, GRAS proteins are an important family of plant-specific proteins named after the first three members, GIBBERELLIC-ACID INSENSITIVE (GAI), REPRESSOR of GAI (RGA) and SCARECROW (SCR), which include SCL (scarecrow-like). miR171 plays a role in the floral development as it shows high expression in inflorescence and flower tissues, and low expression can be detected in leaf or stem tissues (Llave et al. 2002).

5.1.7 miR172

miR172 targets AP2-like transcription factors. Overexpression of miR172 leads to delayed phase change from vegetative to reproductive stages (Lauter et al. 2005). Overexpression leads to translation inhibition of AP2- and AP2-like genes and results in early flowering and phenotypical changes similar to the loss-of-function *ap2* mutants (Navarro et al. 2006; Zhu and Helliwell 2010).

5.1.8 miR167

Plant hormone auxin plays a critical role in regulating plant growth and development. It influences ARF, a plant-specific family of DNA-binding proteins, by binding to auxin response promoter elements (AREs) (Hagen and Guilfoyle 2002). miR167 targets members of the ARF family of transcription factors, ARF6 and ARF8, which regulate gynoecium and stamen development in immature flowers, and overexpression of miR167 induces various developmental changes that mainly regulate flower development (Wu et al. 2006; Ru et al. 2006). It has been shown that mutations in ARF8 affects fruit initiation post fertilization, resulting in the formation of seedless, parthenocarpic fruit (Goetz et al. 2007)

5.2 Role of Small RNA in Abiotic and Biotic Stress Tolerance

Plants are continuously exposed to different biotic and abiotic stress conditions. Therefore, they develop different mechanisms such as altered physiology, metabolism and gene expression to withstand different stress effects. Plants respond to abiotic stress, both at transcriptional and post-transcriptional level of gene expression (Sunkar and Zhu 2004). miRNAs are differentially regulated under various stresses and play important role in abiotic stress response. Although miRNAs are conserved, they can be species, tissue, any physiological stage and stress specific (Jagadeeswaran et al. 2009; Sunkar et al. 2006). In abiotic stress, miRNA has been identified and characterized to decipher their role in abiotic stress by binding to the transcription factor. 49 conserved miRNAs and 22 novel salt stress-responsive miRNAs were identified in radish (*Raphanus sativus*) (Sun et al. 2015), 29 new miRNAs belonging to 24 novel families and 15 miRNAs belonging to 6 conserved families in other plant species have been identified in *Glycine max* by Solexa sequencing in response to drought stress (Kulcheski et al. 2011). miRNA in response to cold stress has been identified in rice (Lv et al. 2010) and in tea (*Camellia sinensis*) (Zhang et al. 2014). miRNAs are also identified in heavy metal stress. Heavy metals like copper (Cu), iron (Fe), zinc (Zn), cobalt (Co) and manganese (Mn) are beneficial to plants at low concentration, but at higher stress, these metal ions are toxic to plants. Heavy metals from contaminated water includes cadmium (Cd), aluminium

(Al) and mercury (Hg), which are highly toxic to plants. Differentially expressed microRNAs were identified in *Medicago truncatula* (Zhou et al. 2012), *Phaseolus vulgaris* (Valdés-López et al. 2010), etc. Understanding the role of miRNA during abiotic stress will help in identifying their role in abiotic stress tolerance and generating more resistant plants. Some of the microRNAs are overexpressed and their roles have been determined; few examples include the following.

5.2.1 miR399

Nutrient requirement is one of the basic necessities of the cell for sustaining itself. Plants require nutrients for normal growth. These must be in a form that can be used by the plants and in concentrations that allow optimum plant growth. Furthermore, the concentrations of the various soluble soil nutrients are very important for the plants. mir399 target ubiquitin-conjugating E2 enzyme in *Arabidopsis*. mir399 expression is regulated by phosphate level. The lesser the phosphate, the more is the expression of mir399 and the level of target UBC24 mRNA is reduced in roots. Ubiquitin-conjugating E2 enzyme (UBC) helps in degradation of proteins, so when the phosphate level is high, UCB in the roots is high, and it downregulates the specific Pi uptake, therefore preventing excess Pi accumulation in plants. As there is starvation of Pi, it induces the mir399 expression and downregulates UCB. In this way phosphate homeostasis is maintained in plant (Hackenberg et al. 2013).

5.2.2 miR398

Metabolic balance of the plant is disturbed by the various abiotic and biotic stresses. Reactive oxygen species (ROS) are produced as a normal product of plant cellular metabolism. Different stresses lead to excessive production of ROS (reactive oxygen species) leading to cell damage and ultimately cell death. ROS include free radicals such as superoxide anion (O_2•−), hydroxyl radical (•OH), hydrogen peroxide (H_2O_2) and singlet oxygen ($^1O_2$). Superoxide dismutases (SODs) convert the highly toxic superoxide radicals (2 O^-) into less toxic hydrogen peroxide (H_2O_2). Superoxide radicals are detoxified by cytosolic CSD1 and plastidic CSD2 (Cu/Zn-superoxide dismutase genes). miRNA 398 target CSD gene. Therefore, the expression of mir398 is downregulated by oxidative stress, which is an important post-translation event. Two mir398 T-DNA mutants of *Arabidopsis* were studied for the target recognition of mir398, so along with CSD (*CSD1* and *CSD2*) mRNAs, *CCS1* mRNAs encoding the chaperone that delivers copper to the Cu/Zn SODs of *Arabidopsis* were also identified. Transgenic plants carrying mutated CCS1 DNA, resistant to cleavage, accumulate CCS1 proteins in plants, so along with direct silencing of CSD transcripts by mir398, it indirectly affects CSD by cleavage of the mRNA encoding the copper chaperone necessary for their activation, when copper is limiting (Bouché 2010).

5.2.3 miR159

Plant hormone regulates plant development and generates various responses to changing environmental conditions. These hormones play an important role in the abiotic stress response because with the help of these hormones, plants are able to modify their physiology, which is very important for their survival. miR159 plays an important role in abscisic acid (ABA) signalling pathway, by targeting MYB group of transcription factor (Achard et al. 2004). ABA-induced accumulation of mir159 was observed in the germinating *Arabidopsis* seedling in an ABI3 (abscisic acid-insensitive)-dependent fashion, and miRNA159 mediates cleavage of *MYB101* and *MYB33* transcripts in vitro and in vivo. These MYB transcription factors are positive regulators of ABA response. Therefore, when mir159 is overexpressed, plants become less sensitive to ABA, and when transgenic forms of MYB33 and MYB101 are made, which are resistant form, they become more sensitive (Reyes and Chua 2007).

5.2.4 mir393

It has been reported that abiotic stress causes increased TIR1 (transport inhibitor response 1) mRNA degradation or translational repression (Jin 2008). miR393 targets TIR1 (transport inhibitor response1), an F-box auxin receptor gene. This helps in proteolysis of ubiquitin ligase by auxin-dependent manner (Vierstra 2003). MicroRNA393 is also upregulated in bacterial infection, fungal infection and other abiotic stress (Navarro et al. 2006; Weiberg et al. 2013).

5.2.5 mir319

mi319 has been characterized in flax, tomato, bentgrass and sugarcane and is reported to play a role in cold, drought, salt and heavy metal toxicity. Rice *mir319* gene when overexpressed in bentgrass showed morphological changes and salt and drought tolerance, which were demonstrated by the downregulated expression of the target gene (TCP) in transgenic plants (Zhou et al. 2013; Zhou and Luo 2014). Similarly, mir319 in tomato and sugarcane was upregulated in cold stress, and when this miRNA was overexpressed, downregulation of target genes was observed (Valiollahi et al. 2014; Thiebaut et al. 2012). miRNA319 has also been reported to be upregulated in aluminium stress in flax, indicating their crucial role in heavy metal toxicity as well (Dmitriev et al. 2017).

5.3 siRNA in Abiotic and Biotic Stress

In *Arabidopsis*, nat-siRNA, plays an important role in oxidative stress. Gene pair of SRO5 and P5CDH (pyrroline-5-carboxylate dehydrogenase) produces the dsRNA, which is in turn induced by salt stress. This dsRNA is further processed by normal siRNA pathway, and the mature 24-nt nat-siRNA guides the cleavage of the P5CDH transcript, suppressing proline degradation and, thus, allowing proline accumulation (Borsani et al. 2005).

Furini et al. (1997) characterized dehydration-related, ABA-inducible gene CDT-1 from *Craterostigma plantagineum.* CDT-1 mRNA does not code for polypeptide, and its genomic structure reveals the presence of short interspersed element retrotransposon. Therefore, it was predicted that it functions as non-coding RNA rather than coding RNA, i.e. protein. It was experimentally shown that CTD-1 can code for siRNA. The expression level of the siRNA was altered by inducing desiccation tolerance in callus of *C. plantagineum.* Thus, it was proven that small RNAs are needed for the adaptation of the stressful environmental condition. Small RNAs also play a role in plant defence by initiating the small RNA production and defence against the pathogen attack. Therefore, these small RNAs control almost all important functioning of a plant cell in development and abiotic and biotic stress, and further characterization and study will help in dissecting their role in details.

6 Conclusion

Small RNAs, i.e. miRNA and siRNA, are endogenously expressed non-coding RNA ranging from 21 to 24 nucleotides in length, and they are transcribed by specific region by RNA polymerase II. Further these are processed by Dicer enzymes, which produce dsRNAs. Plant small RNAs are discovered by either direct sequencing of small RNA library or through comparative genome analysis. All the small RNAs discovered are deposited in public databases generated for small RNAs. Small RNAs function by cleavage of the target gene or by other mechanisms and repress the target gene. These small RNAs have an important role in plant development and biotic and abiotic stress by regulating the expression of the gene which they target. Further, these small RNAs are being characterized, and there is a need for characterization of more novel and unique miRNA to explore new dimensions in regulating various process of the plant cell.

References

Achard P, Herr A, Baulcombe DC, Harberd NP (2004) Modulation of floral development by a gibberellin-regulated microRNA. Development 131(14):3357–3365

Adai A, Johnson C, Mlotshwa S, Archer-Evans S, Manocha V, Vance V, Sundaresan V (2005) Computational prediction of miRNAs in *Arabidopsis thaliana*. Genome Res 15(1):78–91

Adenot X, Elmayan T, Lauressergues D, Boutet S, Bouché N, Gasciolli V, Vaucheret H (2006) DRB4-dependent TAS3 trans-acting siRNAs control leaf morphology through AGO7. Curr Biol 16(9):927–932

Allen E, Howell MD (2010) miRNAs in the biogenesis of trans-acting siRNAs in higher plants. Semin Cell Dev Biol 21:798–804

Allen E, Xie Z, Gustafson AM, Sung GH, Spatafora JW, Carrington JC (2004) Evolution of microRNA genes by inverted duplication of target gene sequences in *Arabidopsis thaliana*. Nat Genet 36(12):1282–1290

Allen E, Xie Z, Gustafson AM, Carrington JC (2005) microRNA-directed phasing during trans-acting siRNA biogenesis in plants. Cell 121(2):207–221

Alonso-Peral MM, Li J, Li Y et al (2010) The microRNA159-regulated GAMYB-like genes inhibit growth and promote programmed cell death in *Arabidopsis*. Plant Physiol 154(2):757–771

Altuvia Y, Landgraf P, Lithwick G et al (2005) Clustering and conservation patterns of human microRNAs. Nucleic Acids Res 33(8):2697–2706

Aravin AA, Lagos-Quintana M, Yalcin A et al (2003) The small RNA profile during *Drosophila melanogaster* development. Dev Cell 5(2):337–350

Arteaga-Vázquez M, Caballero-Pérez J, Vielle-Calzada JP (2006) A family of microRNAs present in plants and animals. Plant Cell 18(12):3355–3369

Aukerman MJ, Sakai H (2003) Regulation of flowering time and floral organ identity by a microRNA and its APETALA2-like target genes. Plant Cell 15(11):2730–2741

Axtell MJ, Bartel DP (2005) Antiquity of microRNAs and their targets in land plants. Plant Cell 17 (6):1658–1673

Bartel DP (2004) MicroRNAs: genomics, biogenesis, mechanism, and function. Cell 116 (2):281–297

Baskerville S, Bartel DP (2005) Microarray profiling of microRNAs reveals frequent coexpression with neighboring miRNAs and host genes. RNA 11(3):241–247

Berger Y, Harpaz-Saad S, Brand A et al (2009) The NAC-domain transcription factor GOBLET specifies leaflet boundaries in compound tomato leaves. Development 136(5):823–832

Blevins T, Rajeswaran R, Shivaprasad PV et al (2006) Four plant dicers mediate viral small RNA biogenesis and DNA virus induced silencing. Nucleic Acids Res 34(21):6233–6246

Bollman KM, Aukerman MJ, Park MY, Hunter C, Berardini TZ, Poethig RS (2003) HASTY, the *Arabidopsis* ortholog of exportin 5/MSN5, regulates phase change and morphogenesis. Development 130(8):1493–1504

Bonnet E, Wuyts J, Rouzé P, Van de Peer Y (2004) Evidence that microRNA precursors, unlike other non-coding RNAs, have lower folding free energies than random sequences. Bioinformatics 20(17):2911–2917

Borsani O, Zhu J, Verslues PE, Sunkar R, Zhu JK (2005) Endogenous siRNAs derived from a pair of natural cis-antisense transcripts regulate salt tolerance in *Arabidopsis*. Cell 123 (7):1279–1291

Bouché N (2010) New insights into miR398 functions in *Arabidopsis*. Plant Signal Behav 5 (6):684–686

Buxdorf K, Hendelman A, Stav R, Lapidot M, Ori N, Arazi T (2010) Identification and characterization of a novel miR159 target not related to MYB in tomato. Planta 232(5):1009–1022

Carthew RW, Sontheimer EJ (2009) Origins and mechanisms of miRNAs and siRNAs. Cell 136 (4):642–655

Cerutti L, Mian N, Bateman A (2000) Domains in gene silencing and cell differentiation proteins: the novel PAZ domain and redefinition of the Piwi domain. Trends Biochem Sci 25 (10):481–482

Chaabane SB, Liu R, Chinnusamy V et al (2013) STA1, an *Arabidopsis* pre-mRNA processing factor 6 homolog, is a new player involved in miRNA biogenesis. Nucleic Acids Res 41 (3):1984–1997

Chalfie M, Sulston J (1981) Developmental genetics of the mechanosensory neurons of *Caenorhabditis elegans*. Dev Biol 82(2):358–370

Chapman EJ, Carrington JC (2007) Specialization and evolution of endogenous small RNA pathways. Nat Rev Genet 8(11):884–896

Chen X (2004) A microRNA as a translational repressor of APETALA2 in *Arabidopsis* flower development. Science 303(5666):2022–2025

Chiou TJ (2007) The role of microRNAs in sensing nutrient stress. Plant Cell Environ 30 (3):323–332

Copeland C, Xu S, Qi Y, Li X (2013) MOS2 has redundant function with its homolog MOS2H and is required for proper splicing of SNC1. Plant Signal Behav 8:e25372

Cuperus JT, Fahlgren N, Carrington JC (2011) Evolution and functional diversification of MIRNA genes. Plant Cell 23(2):431–442

De Paoli E, Dorantes-Acosta A, Zhai J et al (2009) Distinct extremely abundant siRNAs associated with cosuppression in petunia. RNA 15(11):1965–1970

Dmitriev AA, Kudryavtseva AV, Bolsheva NL et al (2017) miR319, miR390, and miR393 are involved in aluminum response in flax (*Linum usitatissimum* L.) Biomed Res Int 2017:4975146. https://doi.org/10.1155/2017/4975146

Eamens AL, Smith NA, Curtin SJ, Wang MB, Waterhouse PM (2009) The *Arabidopsis thaliana* double-stranded RNA binding protein DRB1 directs guide strand selection from microRNA duplexes. RNA 15(12):2219–2235

Fabian MR, Sonenberg N, Filipowicz W (2010) Regulation of mRNA translation and stability by microRNAs. Annu Rev Biochem 79:351–379

Faghihi MA, Wahlestedt C (2009) Regulatory roles of natural antisense transcripts. Nat Rev Mol Cell Biol 10(9):637–643

Finnegan EJ, Matzke MA (2003) The small RNA world. J Cell Sci 116(23):4689–4693

Floyd SK, Bowman JL (2004) Gene regulation: ancient microRNA target sequences in plants. Nature 428(6982):485–486

Fox S, Sergei F, Mockler TC (2009) Applications of ultra-high-throughput sequencing. In: Belostotky DA (ed) Plant systems biology. Humana Press, New York, pp 79–108

Furini A, Koncz C, Salamini F, Bartels D (1997) High level transcription of a member of a repeated gene family confers dehydration tolerance to callus tissue of *Craterostigma plantagineum*. EMBO J 16(12):3599–3608

Gandikota M, Birkenbihl RP, Höhmann S, Cardon GH, Saedler H, Huijser P (2007) The miRNA156/157 recognition element in the 3′ UTR of the *Arabidopsis* SBP box gene SPL3 prevents early flowering by translational inhibition in seedlings. Plant J 49(4):683–693

Gocal GF, Sheldon CC, Gubler F et al (2001) GAMYB-like genes, flowering, and gibberellin signaling in *Arabidopsis*. Plant Physiol 127(4):1682–1693

Goetz M, Hooper LC, Johnson SD, Rodrigues JCM, Vivian-Smith A, Koltunow AM (2007) Expression of aberrant forms of AUXIN RESPONSE FACTOR8 stimulates parthenocarpy in *Arabidopsis* and tomato. Plant Physiol 145(2):351–366

Grigg SP, Canales C, Hay A, Tsiantis M (2005) SERRATE coordinates shoot meristem function and leaf axial patterning in *Arabidopsis*. Nature 437(7061):1022–1026

Guo HS, Xie Q, Fei JF, Chua NH (2005) MicroRNA directs mRNA cleavage of the transcription factor NAC1 to downregulate auxin signals for *Arabidopsis* lateral root development. Plant Cell 17(5):1376–1386

Hackenberg M, Shi BJ, Gustafson P, Langridge P (2013) Characterization of phosphorus-regulated miR399 and miR827 and their isomirs in barley under phosphorus-sufficient and phosphorus-deficient conditions. BMC Plant Biol 13(1):214

Hagen G, Guilfoyle T (2002) Auxin-responsive gene expression: genes, promoters and regulatory factors. Plant Mol Biol 49(3–4):373–385

Hammond SM (2005) Dicing and slicing. FEBS Lett 579(26):5822–5829

Hammond SM, Boettcher S, Caudy AA, Kobayashi R, Hannon GJ (2001) Argonaute2, a link between genetic and biochemical analyses of RNAi. Science 293(5532):1146–1150

Hannon GJ (2002) RNA interference. Nature 418(6894):244–251

Held MA, Penning B, Brandt AS, Kessans SA, Yong W, Scofield SR, Carpita NC (2008) Small-interfering RNAs from natural antisense transcripts derived from a cellulose synthase gene modulate cell wall biosynthesis in barley. Proc Natl Acad Sci U S A 105(51):20534–20539

Helliwell CA, Chin-Atkins AN, Wilson IW, Chapple R, Dennis ES, Chaudhury A (2001) The *Arabidopsis* AMP1 gene encodes a putative glutamate carboxypeptidase. Plant Cell 13 (9):2115–2125

Huang TH, Fan B, Rothschild MF, Hu ZL, Li K, Zhao SH (2007) MiRFinder: an improved approach and software implementation for genome-wide fast microRNA precursor scans. BMC Bioinform 8(1):1

Iki T, Yoshikawa M, Nishikiori M et al (2010) In vitro assembly of plant RNA-induced silencing complexes facilitated by molecular chaperone HSP90. Mol Cell 39(2):282–291

Jagadeeswaran G, Saini A, Sunkar R (2009) Biotic and abiotic stress down-regulate miR398 expression in *Arabidopsis*. Planta 229(4):1009–1014

Jin H (2008) Endogenous small RNAs and antibacterial immunity in plants. FEBS Lett 582 (18):2679–2684

Johnson C, Kasprzewska A, Tennessen K et al (2009) Clusters and superclusters of phased small RNAs in the developing inflorescence of rice. Genome Res 19(8):1429–1440

Jones-Rhoades MW, Bartel DP, Bartel B (2006) MicroRNAs and their regulatory roles in plants. Annu Rev Plant Biol 57:19–53

Katiyar-Agarwal S, Morgan R, Dahlbeck D et al (2006) A pathogen-inducible endogenous siRNA in plant immunity. Proc Natl Acad Sci U S A 103(47):18002–18007

Kennerdell JR, Carthew RW (1998) Use of dsRNA-mediated genetic interference to demonstrate that frizzled and frizzled 2 act in the wingless pathway. Cell 95(7):1017–1026

Khvorova A, Reynolds A, Jayasena SD (2003) Functional siRNAs and miRNAs exhibit strand bias. Cell 115(2):209–216

Kulcheski FR, de Oliveira LF, Molina LG et al (2011) Identification of novel soybean microRNAs involved in abiotic and biotic stresses. BMC Genomics 12(1):307

Kurihara Y, Watanabe Y (2004) *Arabidopsis* micro-RNA biogenesis through dicer-like 1 protein functions. Proc Natl Acad Sci U S A 101(34):12753–12758

Kurihara Y, Takashi Y, Watanabe Y (2006) The interaction between DCL1 and HYL1 is important for efficient and precise processing of pri-miRNA in plant microRNA biogenesis. RNA 12 (2):206–212

Landgraf P, Rusu M, Sheridan R et al (2007) A mammalian microRNA expression atlas based on small RNA library sequencing. Cell 129(7):1401–1414

Laubinger S, Sachsenberg T, Zeller G, Busch W, Lohmann JU, Rätsch G, Weigel D (2008) Dual roles of the nuclear cap-binding complex and SERRATE in pre-mRNA splicing and microRNA processing in *Arabidopsis thaliana*. Proc Natl Acad Sci U S A 105(25):8795–8800

Laufs P, Peaucelle A, Morin H, Traas J (2004) MicroRNA regulation of the CUC genes is required for boundary size control in *Arabidopsis* meristems. Development 131(17):4311–4322

Lauter N, Kampani A, Carlson S, Goebel M, Moose SP (2005) microRNA172 down-regulates glossy15 to promote vegetative phase change in maize. Proc Natl Acad Sci U S A 102 (26):9412–9417

Lee RC, Feinbaum RL, Ambros V (1993) The *C. elegans* heterochronic gene lin-4 encodes small RNAs with antisense complementarity to lin-14. Cell 75(5):843–854

Lee Y, Jeon K, Lee JT, Kim S, Kim VN (2002) MicroRNA maturation: stepwise processing and subcellular localization. EMBO J 21(17):4663–4670

Lee Y, Ahn C, Han J et al (2003) The nuclear RNase III Drosha initiates microRNA processing. Nature 425(6956):415–419

Li J, Yang Z, Yu B, Liu J, Chen X (2005) Methylation protects miRNAs and siRNAs from a 3′-end uridylation activity in *Arabidopsis*. Curr Biol 15(16):1501–1507

Lippman Z, Martienssen R (2004) The role of RNA interference in heterochromatic silencing. Nature 431(7006):364–370

Lippman Z, Gendrel AV, Black M et al (2004) Role of transposable elements in heterochromatin and epigenetic control. Nature 430(6998):471–476

Liu J, Carmell MA, Rivas FV et al (2004) Argonaute2 is the catalytic engine of mammalian RNAi. Science 305(5689):1437–1441

Liu Z, Jia L, Wang H, He Y (2011) HYL1 regulates the balance between adaxial and abaxial identity for leaf flattening via miRNA-mediated pathways. J Exp Bot 62:4367–4381

Liu YX, Wang M, Wang XJ (2014) Endogenous small RNA clusters in plants. Genomics Proteom Bioinform 12(2):64–71

Llave C, Xie Z, Kasschau KD, Carrington JC (2002) Cleavage of scarecrow-like mRNA targets directed by a class of *Arabidopsis* miRNA. Science 297(5589):2053–2056

Lu C, Fedoroff N (2000) A mutation in the *Arabidopsis* HYL1 gene encoding a dsRNA binding protein affects responses to abscisic acid, auxin, and cytokinin. Plant Cell 12(12):2351–2365

Lv DK, Bai X, Li Y et al (2010) Profiling of cold-stress-responsive miRNAs in rice by microarrays. Gene 459(1):39–47

Ma X, Tang Z, Qin J, Meng Y (2015) The use of high-throughput sequencing methods for plant microRNA research. RNA Biol 12(7):709–719

MacRae IJ, Zhou K, Li F et al (2006) Structural basis for double-stranded RNA processing by dicer. Science 311(5758):195–198

Mallory AC, Vaucheret H (2006) Functions of microRNAs and related small RNAs in plants. Nat Genet 38:S31–S36

Mallory AC, Reinhart BJ, Jones-Rhoades MW, Tang G, Zamore PD, Barton MK, Bartel DP (2004) MicroRNA control of PHABULOSA in leaf development: importance of pairing to the microRNA 5′ region. EMBO J 23(16):3356–3364

Mallory AC, Bartel DP, Bartel B (2005) MicroRNA-directed regulation of *Arabidopsis* AUXIN RESPONSE FACTOR17 is essential for proper development and modulates expression of early auxin response genes. Plant Cell 17(5):1360–1375

Manavella PA, Hagmann J, Ott F, Laubinger S, Franz M, Macek B, Weigel D (2012) Fast-forward genetics identifies plant CPL phosphatases as regulators of miRNA processing factor HYL1. Cell 151(4):859–870

Mello CC, Conte D (2004) Revealing the world of RNA interference. Nature 431(7006):338–342

Mi S, Cai T, Hu Y et al (2008) Sorting of small RNAs into *Arabidopsis* argonaute complexes is directed by the 5′ terminal nucleotide. Cell 133(1):116–127

Nag A, King S, Jack T (2009) miR319a targeting of TCP4 is critical for petal growth and development in *Arabidopsis*. Proc Natl Acad Sci U S A 106(52):22534–22539

Nagasaki H, Itoh JI, Hayashi K et al. Y(2007) The small interfering RNA production pathway is required for shoot meristem initiation in rice. Proc Natl Acad Sci U S A 104(37):14867–14871

Navarro L, Dunoyer P, Jay F et al. JD(2006) A plant miRNA contributes to antibacterial resistance by repressing auxin signaling. Science 312(5772):436–439

Ngo H, Tschudi C, Gull K, Ullu E (1998) Double-stranded RNA induces mRNA degradation in Trypanosoma brucei. Proc Natl Acad Sci U S A 95(25):14687–14692

Nodine MD, Bartel DP (2010) MicroRNAs prevent precocious gene expression and enable pattern formation during plant embryogenesis. Genes Dev 24(23):2678–2692

Nogueira FT, Chitwood DH, Madi S, Ohtsu K, Schnable PS, Scanlon MJ, Timmermans MC (2009) Regulation of small RNA accumulation in the maize shoot apex. PLoS Genet 5(1):e1000320

Palatnik JF, Allen E, Wu X, Schommer C, Schwab R, Carrington JC, Weigel D (2003) Control of leaf morphogenesis by microRNAs. Nature 425(6955):257–263

Papp I, Mette MF, Aufsatz W et al (2003) Evidence for nuclear processing of plant micro RNA and short interfering RNA precursors. Plant Physiol 132(3):1382–1390

Peragine A, Yoshikawa M, Wu G, Albrecht HL, Poethig RS (2004) SGS3 and SGS2/SDE1/RDR6 are required for juvenile development and the production of trans-acting siRNAs in *Arabidopsis*. Genes Dev 18(19):2368–2379

Reinhart BJ, Weinstein EG, Rhoades MW, Bartel B, Bartel DP (2002) MicroRNAs in plants. Genes Dev 16(13):1616–1626

Ren G, Xie M, Dou Y, Zhang S, Zhang C, Yu B (2012) Regulation of miRNA abundance by RNA binding protein TOUGH in *Arabidopsis*. Proc Natl Acad Sci U S A 109(31):12817–12821

Reyes JL, Chua NH (2007) ABA induction of miR159 controls transcript levels of two MYB factors during *Arabidopsis* seed germination. Plant J 49(4):592–606

Rhoades MW, Reinhart BJ, Lim LP, Burge CB, Bartel B, Bartel DP (2002) Prediction of plant microRNA targets. Cell 110(4):513–520

Rosa M, Prado C, Podazza G, Interdonato R, González JA, Hilal M, Prado FE (2009) Soluble sugars: metabolism, sensing and abiotic stress: a complex network in the life of plants. Plant Signal Behav 4(5):388–393

Ru P, Xu L, Ma H, Huang H (2006) Plant fertility defects induced by the enhanced expression of microRNA167. Cell Res 16(5):457–465

Ruby JG, Jan C, Player C et al (2006) Large-scale sequencing reveals 21U-RNAs and additional microRNAs and endogenous siRNAs in *C. elegans*. Cell 127(6):1193–1207

Schauer SE, Jacobsen SE, Meinke DW, Ray A (2002) DICER-LIKE1: blind men and elephants in *Arabidopsis* development. Trends Plant Sci 7(11):487–491

Schwab R, Palatnik JF, Riester M, Schommer C, Schmid M, Weigel D (2005) Specific effects of microRNAs on the plant transcriptome. Dev Cell 8(4):517–527

Schwarz DS, Hutvágner G, Du T, Xu Z, Aronin N, Zamore PD (2003) Asymmetry in the assembly of the RNAi enzyme complex. Cell 115(2):199–208

Shen B, Goodman HM (2004) Uridine addition after microRNA-directed cleavage. Science 306 (5698):997–997

Shikata M, Yamaguchi H, Sasaki K, Ohtsubo N (2012) Overexpression of *Arabidopsis* miR157b induces bushy architecture and delayed phase transition in *Torenia fournieri*. Planta 236 (4):1027–1035

Souret FF, Kastenmayer JP, Green PJ (2004) AtXRN4 degrades mRNA in *Arabidopsis* and its substrates include selected miRNA targets. Mol Cell 15(2):173–183

Sun X, Xu L, Wang Y et al (2015) Identification of novel and salt-responsive miRNAs to explore miRNA-mediated regulatory network of salt stress response in radish (*Raphanus sativus* L.) BMC Genomics 16(1):197

Sunkar R, Zhu JK (2004) Novel and stress-regulated microRNAs and other small RNAs from *Arabidopsis*. Plant Cell 16(8):2001–2019

Sunkar R, Girke T, Jain PK, Zhu JK (2005) Cloning and characterization of microRNAs from rice. Plant Cell 17(5):1397–1411

Sunkar R, Kapoor A, Zhu JK (2006) Posttranscriptional induction of two Cu/Zn superoxide dismutase genes in *Arabidopsis* is mediated by downregulation of miR398 and important for oxidative stress tolerance. Plant Cell 18(8):2051–2065

Thiebaut F, Rojas CA, Almeida KL et al (2012) Regulation of miR319 during cold stress in sugarcane. Plant Cell Environ 35(3):502–512

Tosic M, Roach A, de Rivaz JC, Dolivo M, Matthieu JM (1990) Post-transcriptional events are responsible for low expression of myelin basic protein in myelin deficient mice: role of natural antisense RNA. EMBO J 9(2):401

Valdés-López O, Yang SS, Aparicio-Fabre R, Graham PH, Reyes JL, Vance CP, Hernández G (2010) MicroRNA expression profile in common bean (*Phaseolus vulgaris*) under nutrient deficiency stresses and manganese toxicity. New Phytol 187(3):805–818

Valiollahi E, Farsi M, Kakhki AM (2014) Sly-miR166 and Sly-miR319 are components of the cold stress response in *Solanum lycopersicum*. Plant Biotechnol Rep 8(4):349–356

Vaucheret H (2006) Post-transcriptional small RNA pathways in plants: mechanisms and regulations. Genes Dev 20(7):759–771

Vaucheret H, Vazquez F, Crété P, Bartel DP (2004) The action of ARGONAUTE1 in the miRNA pathway and its regulation by the miRNA pathway are crucial for plant development. Genes Dev 18(10):1187–1197

Vazquez F, Hohn T (2013) Biogenesis and biological activity of secondary siRNAs in plants. Scientifica 2013

Verdel A, Vavasseur A, Le Gorrec M, Touat-Todeschini L (2009) Common themes in siRNA-mediated epigenetic silencing pathways. Int J Dev Biol 53(2):245

Vierstra RD (2003) The ubiquitin/26S proteasome pathway, the complex last chapter in the life of many plant proteins. Trends Plant Sci 8(3):135–142

Wang XJ, Reyes JL, Chua NH, Gaasterland T (2004) Prediction and identification of *Arabidopsis thaliana* microRNAs and their mRNA targets. Genome Biol 5:R65

Wang JW, Wang LJ, Mao YB, Cai WJ, Xue HW, Chen XY (2005) Control of root cap formation by microRNA-targeted auxin response factors in *Arabidopsis*. Plant Cell 17(8):2204–2216

Wang JW, Czech B, Weigel D (2009) miR156-regulated SPL transcription factors define an endogenous flowering pathway in *Arabidopsis thaliana*. Cell 138(4):738–749

Weiberg A, Wang M, Lin FM et al (2013) Fungal small RNAs suppress plant immunity by hijacking host RNA interference pathways. Science 342(6154):118–123

Werner S (2010) MicroRNA processing in *Arabidopsis thaliana*. Doctoral dissertation, Universität Tübingen

Wirth S, Crespi M (2009) Non-protein coding RNAs, a diverse class of gene regulators, and their action in plants. RNA Biol 6(2):161–164

Wu MF, Tian Q, Reed JW (2006) *Arabidopsis* microRNA167 controls patterns of ARF6 and ARF8 expression, and regulates both female and male reproduction. Development 133(21):4211–4218

Xie Z, Johansen LK, Gustafson AM et al (2004) Genetic and functional diversification of small RNA pathways in plants. PLoS Biol 2(5):e104

Xie M, Ren G, Zhang C, Yu B (2012) The DNA-and RNA-binding protein FACTOR of DNA METHYLATION 1 requires XH domain-mediated complex formation for its function in RNA-directed DNA methylation. Plant J 72(3):491–500

Yang Z, Zhu Q, Luo K, Zhou Q (2001) The 7SK small nuclear RNA inhibits the CDK9/cyclin T1 kinase to control transcription. Nature 414(6861):317–322

Yang Z, Ebright YW, Yu B, Chen X (2006) HEN1 recognizes 21–24 nt small RNA duplexes and deposits a methyl group onto the 2′ OH of the 3′ terminal nucleotide. Nucleic Acids Res 34 (2):667–675

Yoshikawa M, Peragine A, Park MY, Poethig RS (2005) A pathway for the biogenesis of trans-acting siRNAs in *Arabidopsis*. Genes Dev 19(18):2164–2175

Yu B, Bi L, Zheng B et al (2008) The FHA domain proteins DAWDLE in *Arabidopsis* and SNIP1 in humans act in small RNA biogenesis. Proc Natl Acad Sci U S A 105(29):10073–10078

Yu N, Cai WJ, Wang S, Shan CM, Wang LJ, Chen XY (2010) Temporal control of trichome distribution by microRNA156-targeted SPL genes in *Arabidopsis thaliana*. Plant Cell 22 (7):2322–2335

Zeng Y, Yi R, Cullen BR (2005) Recognition and cleavage of primary microRNA precursors by the nuclear processing enzyme Drosha. EMBO J 24(1):138–148

Zhan X, Wang B, Li H, Liu R, Kalia RK, Zhu JK, Chinnusamy V (2012) *Arabidopsis* proline-rich protein important for development and abiotic stress tolerance is involved in microRNA biogenesis. Proc Natl Acad Sci U S A 109(44):18198–18203

Zhang X, Yazaki J, Sundaresan A et al (2006) Genome-wide high-resolution mapping and functional analysis of DNA methylation in Arabidopsis. Cell 126(6):1189–1201

Zhang W, Gao S, Zhou X et al (2010) Multiple distinct small RNAs originate from the same microRNA precursors. Genome Biol 11(8):R81

Zhang H, Jin J, Tang L, Zhao Y, Gu X, Gao G, Luo J (2011) PlantTFDB 2.0: update and improvement of the comprehensive plant transcription factor database. Nucleic Acids Res 39: D1114–D1117

Zhang X, Xia J, Lii YE et al (2012) Genome-wide analysis of plant nat-siRNAs reveals insights into their distribution, biogenesis and function. Genome Biol 13(3):R20

Zhang Y, Zhu X, Chen et al (2014) Identification and characterization of cold-responsive microRNAs in tea plant (*Camellia sinensis*) and their targets using high-throughput sequencing and degradome analysis. BMC Plant Biol 14(1):271

Zhou M, Luo H (2014) Role of microRNA319 in creeping bentgrass salinity and drought stress response. Plant Signal Behav 9(4):1375–1391

Zhou X, Wang G, Zhang W (2007) UV-B responsive microRNA genes in *Arabidopsis thaliana*. Mol Syst Biol 3(1):103

Zhou ZS, Zeng HQ, Liu ZP, Yang ZM (2012) Genome-wide identification of Medicago truncatula microRNAs and their targets reveals their differential regulation by heavy metal. Plant Cell Environ 35(1):86–99

Zhou M, Li D, Li Z, Hu Q, Yang C, Zhu L, Luo H (2013) Constitutive expression of a miR319 gene alters plant development and enhances salt and drought tolerance in transgenic creeping bentgrass. Plant Physiol 161(3):1375–1391

Zhu QH, Helliwell CA (2010) Regulation of flowering time and floral patterning by miR172. J Exp Bot 62:487–495. https://doi.org/10.1093/jxb/erq295

Zuker M (2003) Mfold web server for nucleic acid folding and hybridization prediction. Nucleic Acids Res 31(13):3406–3415

Bryomonitoring of Environmental Pollution

Afroz Alam

Abstract Biological monitoring has become an important tool for evaluating the negative blow of human activities on the atmosphere. Due to ever-increasing population alongside other environmental problems, introduction of heavy metals in our surroundings is a huge drawback to the sustainable environment. Heavy metal pollution of the biosphere has augmented piercingly since 1900. These metals, though being deposited constantly in minute amounts, may build up in the surroundings over extended periods of time and will most likely create potential ecological and human wellbeing hazards in upcoming future. Thus, it appears very imperative to develop and perk up an enduring reflexive monitoring method to evaluate the nature and intensity of heavy metal and gaseous pollutions. In this review, the potential of bryophytes has been discussed in light of notable researches in this direction worldwide.

Keywords Atmosphere · Air quality · Bryophyta · Heavy metals · Indicator

1 Introduction

For the last few decades, the estimation of ecological contaminants and the hazard they create to the entire natural system has been an imperative confront in ecological sciences. Biomonitoring and bioindication have been verified to be exceptional and economical methods to monitor these blows of secondary factors (Markert et al. 2003; Krommer et al. 2007; Singh et al. 2017).

The intention of the Man and Biosphere Program (MAB), initiated by the UNESCO (2016), is the protection and the ever-lasting extension of the preferred areas worldwide. These areas are confined protected zones, coalescing core protected areas with other zones where cautious development is promoted by local inhabitants and endeavours. So far, 482 areas have been established as 'biosphere

A. Alam (✉)
Department of Bioscience and Biotechnology, Banasthali Vidyapith, Banasthali, Rajasthan, India

S. Vats (ed.), *Biotic and Abiotic Stress Tolerance in Plants*,
https://doi.org/10.1007/978-981-10-9029-5_13

reserves'. However, among these protected areas, air pollution always plays a main role due to the tourist influx. Though substantial optimistic inclinations could be pragmatic within the last few decades pertaining to the discharge of various atmospheric pollutants, pollutants like sulphur dioxide (SO_2), heavy metals, carbon dioxide (CO_2), suspended particulate material (SPM), PAHs and ozone have shown ever-increasing trend. These particulars may have harmful effects on the total biota of the concerning biosphere reserve (Umweltbundesamt 2004).

2 Biomonitoring

Biomonitoring can be explained as the employ of biological unit to get an indication on specific distinctiveness of the biosphere. The life form used for this purpose is frequently termed as 'bioindicator'. Another term 'biological monitor' is also used invariably for the same purpose; however, there is a substantial distinction between these terms. Bioindicator typically refers to all organisms that make accessible information on the ambiance or the stipulation of environmental changes, and these biomonitors give quantitative data-based information on the distinctiveness of the environment (Markert et al. 2003). Accordingly, with an apposite choice of organisms, scrutinization of various pollutions is achievable still in remote localities since samples can be obtained, and evaluation of contaminants can also be done in laboratory, away from the site of study.

Biomonitoring is a spontaneous technique and calculates the built-in exposure over a time phase. The key advantages of this method include (a) cost-effectiveness and less time taking, (b) a lot easier compared to any other traditional method, (c) better precision of quantification and (d) more suitable for seasonal monitoring of various pollutants.

2.1 Fundamental Characteristics of the Biomonitors

Supposedly, biomonitoring species for micro- or trace-element contamination are chosen on the criterion of specificity (Ruhling 1994). Conversely, virtually, an apposite biomonitor should possess a few defined necessities such as the life form has to be prevalent in the chosen study area; it has to be accessible for seasonal sampling round the year; if not, then a few uncomplicated unique tactics have to be devised to nurture it for a complete year; and the organism must be tolerable to pollutants during the considerable stages. In addition to these vital criterions, auxiliary requirements include that the element uptake must be free from the neighbouring circumstances, the biological distinction of the life form should be restricted, the buildup levels of concentration should be experimental, lack of noticeable quantities of element uptake from origins excluding the ambiance, physiological machinery for uptake of various elements must be identified to make feasible the explanation of the obtained results, the species must standardize the metal concentrations over a time period in consequence of combined exposure, the species must have relatively low localized concentrations of these noxious elements and the procedure of sampling and method for trial groundwork for measurement must be straightforward and fast. Appropriate biomonitors, that assemble these necessities, put up uninterrupted monitoring and even the retrospective examination of contamination attainable at comparatively inexpensive.

In the present scenario, biological monitoring is a progressively more noteworthy tool for evaluating the hammer of man-made activities on the atmosphere. Due to escalating populace alongside other ecological tribulations, a rapid introduction of heavy metals and noxious gases in the surroundings is a great ecological concern. The pollution level of the biosphere has augmented piercingly since 1900 (Nriagu 1979). These pollutants yet in traces, if settled down continually in diminutive quantities over extended time periods, mount up in the atmosphere and will possibly pretence an escalating foremost ecological and individual fitness hazard in the future (Walkenhorst et al. 1993). Consequently, it appears crucial to build up and develop a long-standing reflexive examination method to evaluate the nature and intensity of heavy metal contamination of any area.

3 Bryophytes As Biomonitor (Bryomonitor)

With escalating disruption, the composition and microclimate of tropical forests, particularly light, temperature and dampness conditions, alter drastically. Hence, species budding entirely in the tree canopies in undisturbed tropical rain forest are able to inhabit the tree trunks in concerned and open forests. Species growing completely in moist conditions in close proximity to the trunk base vanish quickly in disrupted forests. The bryoflora responds to alterations in forest composition, free of their cause; therefore, the biomonitoring will not differentiate between human being and innate upheaval.

Bryophytes, owing to their incomparable morphological and physiological organization, are essential tools within the field of monitoring atmospheric contamination: they are deficient in a proper root organization and cuticle; therefore the uptake of water, nutrients and noxious substances mainly by means of atmospheric deposition (Bates 1992) occurs directly through their gametophytic plant bodies. Similar to lichens, bryophytes' response to alteration in the atmosphere is rapid and simpler than those of the widespread higher plants with the well-developed stellar organization. Furthermore, bryophytes exhibit amazing sensitivity to specific toxic substances, for instance, sulphurous or nitrogenous amalgamation, thus responding with changes in vigour, density and reproduction system. On the contrary, they are strongly differing to several polluted substances and even gathering them, for instance, heavy metals and a wide-ranging collection of insistent organic contaminants. Owing to their unmatched uptake machinery of these hazardous substances that permit a relationship amid input and absorption, and the chance of determining the accurate period of deposition, they verify as excellent buildup indicators (Zechmeister et al. 2003a, 2007).

The use of bryophytes for biological monitoring was initiated by Ruhling and Tyler (1968, 1970). Consequently, the use of indigenous terricolous moss species for biomonitoring is now a very much familiar technique in studies of pollution in ambiance (Fernández and Carballeira 2001) and is efficient as a handy tool in instituting and characterization of various sources that are responsible for metal deposition. The competence of the bryophytes as bioindicators is mainly related to their capacity to soak up and to fix the pollutants, in addition to their interdependence concerning the ground mineral input (Brown and Brûmelis 1996). These amphibian plants are untied to biomonitoring since they are invasive and uncomplicated to use and they devoid of cuticle, vasculature and root system, hence reflecting straight aerial deposition of heavy metals and gases. Their incomparable cation substitution ability and high surface-to-volume ratio support the buildup of the hefty concentrations of heavy metals across the cell wall for considerably long time (Brown 1982; Tyler 1990; Sawidis et al. 1993; Thöni et al. 1996; Markert et al. 1997, 1999; Fernández et al. 2000; Gerdol et al. 2002). Bryophytes flourish well in a muggy climate. Usually, ectohydric moss species have been commonly employed as biomonitors for air pollutant's monitoring because they are capable of acquiring nutrients from dry and wet deposition, they cannot take minerals from soil or substratum, their faintly formed cuticle promotes metal uptake, and large surface-to-weight ratio also perks up the adsorption (Onianwa 2001; Zeichmeister et al. 2003a).

The impending of the moss species to be utilized as bioindicators principally relies on their ability to take up and to attach with metallic pollutants in addition to their interdependence pertaining to the soil mineral input (Brown and Brûmelis 1996). Even though the biomonitoring procedure is commonly recognized, there are dreadfully only few records of utilization of this practice for examining trace elements that are known. Earlier, the task related to biomonitoring was based on lichens, the pioneers of ecological succession, and amazingly, currently, many of the native moss species have not been assessed for their biomonitoring capacity (Alam 2014).

3.1 Entrapment Mechanism of Heavy Metals

Various pollutants are set down on bryophytes basically in three forms: aqueous solution, gaseous and suspended particles. The buildup of these pollutants in bryophytes happens by several diverse mechanisms which comprise entrapment on the surface of the cells, as layers of particles, and inclusion into the outer wall of cells in the course of ion exchange procedures, metabolically governed route into the cells (Brown and Bates 1990).

The entrapment of the element is managed by the particle size of the element and the surface structure of the bryophytes. The ion exchange process is a rapid physiological-chemical method that is governed by the quantity and kind of free cation swapping sites, the current age of the participating cells, their response to aridness, budding condition, ambient temperature, rainfall, pH, pollutant composition and leaching (Tyler 1990). During the process of ion exchange, anions and cations become closely bound to the functional organic groups that are present in the cell wall mainly by the chelation (Rao 1982).

The chemistry of deposition has a huge upshot over the buildup of various pollutants, as the uptake competence of the bryophytes for each element varies significantly (Berg et al. 1995). The efficiency of uptake for most frequently found heavy metals follows typical assortment as Pb > Co, Cr > Cu> Cd>Mo> Ni> V > Zn > As (Zeichmeister et al. 2003b). A towering amount of the pollutant stack builds up in bryophytes through the common process of wet deposition. The extent, period and concentration of the rainfall affect buildup and leaching (Berg et al. 1995). The role of desiccated deposition augments on shifting from damp to parched climates (Couto et al. 2004). There are substantial demarcations in the leaching out process of elements, which depend on whether these elements are closely attached to the cell wall or just hoarded on the exterior of the cells (Čeburnis and Valiulis 1999). The efficacy of the metal uptake is also pretentious by antagonism for free cation replacement sites, such as the existence of the oceanic salts, and acidic accumulation has been reported to have a definite effect on the absorption of metals by cells of the bryophytes (Gjengedal and Steinnes 1990). The flora composition and earthly dirt have also been found to produce local variations in uptake competence (Čeburnis et al. 1999). Overall, the preeminent relationship amid the concentration of pollutants in mosses and in sopping deposition has been reported for those elements only (e.g. Pb, Cd, Co, Cu) that encompass a soaring uptake effectiveness from soggy deposition (Ross 1990).

3.2 *Factors Affecting the Concentrations of Trace Metals in Bryophytes*

Epiphytic or epiphyllous bryophytes are possibly considered for widespread utilization as biomonitors. The actuality is primarily related to the absence of root system. Therefore, they attain their elemental supply merely from the airborne supplies and not from the rhizoidal system (Martin and Coughtrey 1982).

Besides the pollutants that instigate from the man-made sources, their accumulation in bryophytes is pretentious by several native causes linked with them, such as morphological nature and physiological possessions of the mosses, and the site wherever the bryophytes are burgeoning and their instantaneous surroundings.

There are innate variations in chemistry among particular species with diverse growths and conditions, as well as among separate components of the specific bryophyte. There are certain innate variations in biochemistry among individual bryophytic species and also among the populations of the identical species, amid individuals with dissimilar increase and ambient conditions, and among the separate parts of the specific species of bryophytes (Thöni et al. 1996). Minute quantities of nutrients may perhaps pass into the plants from their substrate (Økland et al. 1999), and these nutrients also have the ability to be translocated from receiving part of the bryophytes to the remaining parts of the thallus (Brŭmelis and Brown 1997). These nutrients instigate from topmost soil and also from rock bases that increase the concentrations of elements like Fe, Cr, Al and Ti particularly in the regions which have sparse flora, a dry climate or bare mineral soil (Mäkinen 1994). Beside these, few other factors, for instance, stand through fall, leaching from vegetation cover positioned above the bryophytes (Steinnes 1993, 1995), the nutrient level of the region, ice melt water (Ford et al. 1995), foliage zone (De Caritat et al. 2001) and elevation, have an upshot owing to alterations in rainfall, dirt or biomass creation (Zechmeister 1995; Gerdol et al. 2002) and also have certain effects on the metal concentration. Additionally, the sampling and evaluating procedures that are used also have a significant persuasion on the reasoned outcomes in studies related to biomonitoring (Markert and Weckert 1989). The age of the bryophyte also has definite effects on the procedure of biomonitoring. According to a finding (Singh et al. 2017), usually older parts of the plant have elevated concentration of metal; it has led to the hypothesis that the bryophytes offer a chronological and interactive data regarding the metal supply in the atmosphere.

3.3 Biomonitoring Through Mosses: The Global Scenario

The employ of moss taxa as potential biomonitor for evaluation of atmospheric contamination has happened to be a widespread practice since appropriate techniques for sampling and evaluating them have been first ever developed in Sweden (Tyler 1970). Since then, mosses have repeatedly been utilized to determine the intensity of metal deposition in the areas adjacent to industrial set up, for instance, coal-fired power plants (Mankovska 1994; Palmieri et al. 1997), geothermal power plants (Bargagli et al. 1997), chlor-alkali plants (Calasans and Malm 1997; Lodenius 1998; Loppi and Bonini 2000; Fernändez et al. 2000) and municipal solid waste disposal (Carpi et al. 1994). Carballeira and Fernández (2002) studied the moss *Scleropodium purum* as model plant for Hg monitoring near the power plant. Other studies on the use of bryophyte for environmental monitoring have been provided in Table 1 and the representative bryophytes in Plate 1.

In Europe, mosses have especially been employed in the assessment of regional heavy metal contamination. For instance, in Scandinavia, the crucial all-embracing survey was performed during the 1960s; subsequently, in the next few years, extensive surveys at the national level were done in Sweden, Norway and Denmark (Steinnes 1977). As a result, during the 1980s, these types of survey were performed all over the Nordic nations, and by the end of the 1990s, most of the European

Table 1 List of important bryophytes used as bryomonitor for the assessment of gaseous and heavy metal pollution

Name of the taxa	Metals	References
Cololejeunea minutissima (Plate 1; Fig. c), *Frullania dilatata* (Plate 1; Fig. b), *Lejeunea cavifolia*, *Porella platyphylla*, *Metzgeria furcata*, *Radula complanata*, *Targionia hypophylla (Plate 1; Fig. e)*	As, Ni, Cr, Cu, Pb	Giordana et al. (2004)
Scapania undulata	Ni, Cr, Co, V, Ba, Sr, Fe, Zn, Mn, Pb, Cd, Cu	Samecka-Cymerman and Kempers (1995)
Plagiochila porelloides and *Scapania undulata*	Co, Cu, Cr	Gana and Yurukova (2006)
Anthoceros fusiformis and *A. punctatus*	Pb, Zn, Cu, Mn, Ni	Hutten et al. (2005)
Fabronia ciliaris and Leskea angustata	Cr, Zn, Pb and Cd	Macedo-Miranda et al. (2016)
Pseudoscleropodium purum	Zn, Pb and Cd	Fernández et al. (2007)
Pleurozium schreberi and *Polytrichum formosum*	Ce, Fe, Ga, G	Wappelhorst et al. (2000)
Pleurozium schreberi	Mn, Ni, Cu, Zn, Cd, Pb	Kolon et al. (2010)
Hylocomium splendens		
Hypnum cupressiforme	Na, Al, Cl, K, Ca, Sc, Ti, V, Cr, Mn, Fe, Co, Ni, Cu, Zn, As, Se, Br, Rd, Sr, Mo, Cd, Sb, I, Cs, Ba, La, Ce, Nd, Sm, Tb, Dy, Tm, Yb, Hf, Ta, W, Au, Pb, Th, U	Marinova et al. (2010)
Sphagnum russowii	Al, B, Ca, Cd, Co, Cu, Cr, Fe, Mg, Mn, Ni, S, Zn	Makholm and Mladenoff (2005)
Pleurozium spp., *Polytrichum* spp. and *Rhytidiadelphus* spp.	Zn, Ni, Cd, Cr, Fe, Pb, Mn	Zawadzki et al. (2016)
Bazzania trilobata, *Conocephalum conicum*, *Mnium punctatum* and *Polytrichum commune*	Cu, Cr, Pb, S, Ni, Zn	Zeichmeister et al. (2007)
Sphagnum girgensohnii	Na, Al, Cl, K, Ca, Sc, Ti, V, Cr, Mn, Fe, Co, Ni, Cu, Zn, As, Se, Br, Rd, Sr, Mo, Cd, Sb, I, Cs, Ba, La, Ce, Nd, Sm, Tb, Dy, Tm, Yb, Hf, Ta, W, Au, Pb, Th, U	Culicov et al. (2005)
Amblystegium riparium, *Atrichum undulatum* (Plate Zeichmeister; Fig. g), *Brachythecium velutinum*, *B. plumosum*, *Bryum pseudotriquetrum*, *Fontinalis antipyretica*, *Rhizomnium*	K, Ca, Sc, S, Ti, V, Cr, Mn, Fe, Co, Ni, Cu, Zn, Mo, Cd	Gana and Yurukova (2006)

(continued)

Table 1 (continued)

Name of the taxa	Metals	References
punctatum, *Rhynchostegium riparioides*, *Sanionia uncinata* and *Warnstorfia exannulata*		
Leucobryum glaucum and *Thuidium tamariscinum*	Ni, Cr, Pb, Zn, Cu	Saxena et al. (2000)
Rhodobryum, *Racomitrium*, *Pleurozium*, *Pseudoscleropodium*, *Aerobryopsis longissima* (Plate 1; Fig. f) and *Rhytidiadelphus*	Cr, Cu, Pb, Cr, Zn	Gecheva and Yurukova (2014), Singh et al. (2017), and Chakrabortty et al. (2004 and 2006)
Hypnum cupressiforme		Saxena et al. (2008)
Lunularia cruciata	S, Cu, Cr, Cd, Pb, Zn	Alam and Sharma (2012)
Marchantia paleacea	S, P, Fe, Co, Cr, Cu, Ca	Alam and Srivastava (2009)
Anthoceros fusiformis, *Cyathodium cavernarum* (Plate 1; Fig. d) and *A. punctatus*	Cd, Co, Cr, Cu, Mn, Ni, Pb, Zn	Hutten et al. (2005)
Bryum argentemum (Plate 1; Fig. i)., *Pterobryopsis flexiceps*, *Pinnatella alopecuroides*, *Leucobryum glaucum* and *Thuidium tamariscinum*, *Pleurozium*, *Pseudoscleropodium* and *Rhytidiadelphus*	Cu, Cr, Pb, Ni, Zn	Saxena et al. (2008) and Govindapyari et al. (2010)
Hypnum cupressiforme, *Barbula* spp.	SO_2	Saxena et al. (2008)
Brachythecium plumosum (Plate 1; Fig. h), *Erythrodontium julaceum*, *Trachyphyllum inflexum*, *Herpetineuron toccoae*, *Fabronia matsumurae*, *Octoblepharum albidum*, *Brothera leana*, *Fabronia matsumurae*, *Entodon viridulus*, *Campylodontium flavescens*	Cd, Co, Cr, Cu, Mn, Ni, Pb, Zn	Shakya et al. (2004)
Conocephalum conicum and *Marchantia polymorpha*	Gaseous pollution	Govindapyari et al. (2010)
Plagiochasma rupestre	S, Cu, Cr, Cd, Zn, Pb and Cr	Alam (2013)

nations are well equipped with this new handy tool for biomonitoring (Buse et al. 2003). Several moss species, viz. *Fabronia ciliaris*, *Leskea angustata*, *Sphagnum girgensohnii* and *Pseudoscleropodium purum*, were employed to get information on the local buildup of heavy metals, changes in the accumulation patterns and the wide spread of such emissions and neighbouring sources of emission.

Plate 1 Few important taxa frequently used as bryomonitor: (**a**) *Anthoceros fusiformis* sp., (**b**) *Frullania dialata* sp., (**c**) *Cololejeunea minutissima* sp., (**d**) *Cyathodium cavernarum* sp., (**e**) *Targionia hypophylla* sp., (**f**) *Aerobryopsis longissima* sp., (**g**) *Atrichum undulatum* sp., (**h**) *Brachymnium plumosum* sp., (**i**). *Bryum argenteum* sp. (Courtesy: J-P. Frahm)

A substantial quantity of other local studies related to heavy metal and other noxious element concentrations has been performed using moss species which are prevalent in the studied area, mainly in North American region (Pott and Turpin 1996). Besides nationwide surveys, several provincial surveys have also been done on other factors distressing concentration, including mosses as accumulators of these depositions in relation to other biomonitors, in addition to find all-inclusive deposition values (Wolterbeek et al. 1996; Zeichmeister 1998; Bargagli 1998; Reimann et al. 1999; Berg et al. 2001; Sucharová and Suchara 2004). An interesting 'moss bag technique' was used to examine heavy metal, sulphur and nitrogen using mesh bags containing *Sphagnum russowii* Warnst., in Wisconsin, USA (Makholm and Mladenoff 2005). Parallel studies were also performed in Bulgaria, Russia and

Romania using *Sphagnum girgensohnii* Russow; interestingly, a total of 36 elements were investigated (Puckett 1988; Culicov et al. 2002, 2005; Poikolainen 2004). The aquatic mosses like *Amblystegium riparium* (Hedw.) Schimp., *Atrichum undulatum* (Hedw.) P. Beauv., *Brachythecium velutinum* (Hedw.) Schimp., *B. plumosum* (Hedw.) Schimp., *Bryum pseudotriquetrum* (Hedw.) P. Gaertn., B. Mey. and Scherb., *Fontinalis antipyretica* Hedw., *Rhizomnium punctatum* (Hedw.) T. J. Kop., *Rhynchostegium riparioides* (Hedw.) Cardot, *Sanionia uncinata* (Hedw.) Loeske and *Warnstorfia exannulata* (Schimp.) Loeske were sampled and estimated for heavy metal accumulation in Bulgaria as bryomonitors from 23 locations (Gana and Yurukova 2006).

3.3.1 Liverworts As Biomonitors

Like mosses, many liverworts have also been used for biomonitoring which include *Cololejeunea minutissima* (Sm.) Schiffner, *Frullania dilatata* (L.) Dumort., *Lejeunea cavifolia* (Ehrh.) Lindb., *Porella platyphylla* (L.) Pfeiff., *Metzgeria furcata* (L.) Corda, *Radula complanata* (L.) Dumort. etc. (Giordana et al. 2004). Samecka-Cymerman and Kempers (1995) investigated *Scapania undulata* as a potent bioindicator plant to assess the accumulation of mercury. Two more liverworts, viz. *Plagiochila porelloides* (Torr. ex Nees) Lindenb. and *Scapania undulata* (L.) Dumort., were systematically assessed for heavy metals, Co, Cd, Cu, Pb, Ni and Zn; for toxic element, As; and for macronutrients, Al, Ca, Fe, K, Mg, Mn, N, Na, and P along with S (Gana and Yurukova 2006).

3.3.2 Use of Hornworts As Biomonitors

Unlike mosses and liverworts, hornworts are rarely studied for their pollution-monitoring potential; hence, limited literature is available regarding hornworts. Till date only *Anthoceros fusiformis* and *A. punctatus* have been used in biomonitoring of pollution in Olympic National Park, Washington (Hutten et al. 2005).

3.4 The Indian Scenario

3.4.1 Mosses As Biomonitors

India, one of the hot spots of biodiversity, hosts a great range of geographic distinction that supports a wide variety of flora. Preliminary studies were conducted basically to examine trace element contamination via the use of few mosses as biomonitors (Pant and Tewari 1998; Chakrabortty et al. 2004, 2006; Gupta 1995). Substantial work has been done in various parts of the country by using mosses like

Bryum spp., *Pterobryopsis flexiceps*, *Pinnatella alopecuroides* etc. These taxa have been proved as effective biomonitors for this region. Few terrestrial forms like *Leucobryum glaucum* and *Thuidium tamariscinum* were also assessed as an accumulator of soil pollution. Species of *Pleurozium*, *Pseudoscleropodium* and *Rhytidiadelphus* were used as biomonitors to study the atmospheric deposition of metals around the lead and copper-zinc smelters (Culicov et al. 2002). Ectohydric mosses, *Thuidium* spp., were also used as accumulators of atmospheric heavy metals, and Cu, Cr, Pb and Ni concentrations were quantified in the tissue of these mosses (Jonathan and Lehman 2002). Biomonitoring of metal discharge at filling stations and their effect on moss *Sphagnum cuspidatum* was studied by Saxena et al. (2000). Later on, Saxena (2006) further studied the seasonal pattern of metal bioaccumulation and their toxicity on *Sphagnum squarrosum*. Recently, Saxena et al. (2008) also studied the biomonitoring of metal deposition by using the moss transplant method through *Hypnum cupressiforme* and *Racomitrium crispulum* (Rasmussen 1978; Saxena et al. 2010).

3.4.2 Liverworts As Biomonitors

Alam (2013) studied the uptake of heavy metal by *Plagiochasma* sp. and showed its significance as bioaccumulator. Alam and Srivastava (2009) studied the potential of *Marchantia paleacea* as bioindicator species for heavy metal pollution. *Riccia aravalliensis* was also assessed for its bryomonitoring potential (Alam 2014). Govindapyari et al. (2010) reported the occurrence of Bi in the thallose liverworts like *Conocephalum conicum* and *Marchantia polymorpha*. Alam and Sharma (2012) also assessed seasonal variation in accumulation of heavy metals in *Lunularia cruciata* and found interesting results. Similar to earlier research on few Indian bryophytes (Nath et al. 2010; Sahu et al. 2014)

No report is available till date regarding use of hornworts as biomonitor from India.

4 Conclusion

In comparison to other sampling methods, sampling with bryophytes is relatively easier due to the lack of any call for intricate and exclusive procedural equipment, and the all-inclusive and time-integrative performance of the bryophytes as biomonitor offers easy biomonitoring of atmospheric trace elements with their prolong practice in the prospect, particularly in larger-scaled surveys. Above all, bryomonitoring is becoming an imperative system for the precise identification of the source.

The ascertaining of consistent procedure for sampling, sample groundwork and elemental analysis with the intention of obtaining analogous results is one of the most imperative confronts in studies related to biomonitoring. With the advancement

Plate 2 The technique of bryomonitoring to assess air pollution. (Adapted with permission from © Aničić Urošević M. http://www.envpl.ipb.ac.rs/bio2.htm)

in bryology now, it is feasible to collect bryophytes in selected areas ranging from unpolluted localities to extremely polluted areas. By collecting bryophytes in pre-monsoon, during monsoon and post-monsoon seasons, it is feasible to distinguish the specific trace element pollution locale. Furthermore, the knowledge of the suitable bryophyte species as a biomonitor for a solitary trace element or an assemblage of trace elements increases the chances of precise prediction (Plate 2).

It is only possible with bryophytes to acquire such a comprehensive depiction of differences in time and space at a rational expenditure (Giordano et al. 2005).

Additionally the exacting spatial composition of the contamination sources intermingles with the spatial arrangement of the samples, ensuing in data sets with disruptions that are incredibly dissimilar from the typically understood usual distribution. The gateway in investigation of this category of data is required to test out for the occurrence of spatial arrangement on scales larger than the sampling lattice, to evade mapping clatter. Then the obtained map should not contain information about contamination sources with a spatial scale less significant than the spatial scale of the sampling grid (Aboal et al. 2006; UNEP 2016; Lukáš et al. 2017). The data can be collected by cluster analytical method that permits for an analysis of the discharge structures that stay unchanged over time or by percentile statistics that exemplify equally spatial and temporal inclinations of elemental buildup (Pesch and Schroeder 2006).

Finally, it is clear that bryophyte can delicately respond to the atmospheric heavy metal contamination and easily claimed the status of bryo-indicators. With the intention to encourage more studies on bryophyte's ecological role, the linked research advances are requisite (Wang et al. 2015). A comprehensive dialogue on the practicability of diverse diagnostic procedures to measure the trace element concentration is necessary in the future. Several multielement procedures, like ICP-MS, EDXRF and INAA, can be employed to resolve various aspects of trace and noxious elements. Advancement in multidisciplinary plans that would bring together all the data concerning atmospheric pollution caused by these elements can make it achievable to illustrate ecosystem-level models describing the spreading and possessions of air pollutants through bryophytes.

References

Aboal JR, Real C, Fernandez JA, Carballeira A (2006) Mapping the results of extensive surveys: the case of atmospheric biomonitoring and terrestrial mosses. Sci Total Environ 356:256–274

Alam A (2013) Bio-monitoring of metal deposit ion in Ranthambore National Park (Rajasthan), India using *Plagiochasma rupestre* (G. Frost) Stephani. Arch Bryol 186:1–10

Alam A (2014) Bio-monitoring of metal deposition in Ranthambore National Park (Rajasthan), India using *Riccia aravalliensis* Pande et Udar. Elixir Bio Tech 69:22838–22842

Alam A, Sharma V (2012) Seasonal variation in accumulation of heavy metals in *Lunularia cruciata* (Linn.) Dum. at Nilgiri hills, Western Ghats. Int J Biol Sci Eng 3(2):91–99

Alam A, Srivastava SC (2009) *Marchantia paleacea* Bert.- as an indicator of heavy metal pollution. Indian J For 32(3):465–470

Aničić UM (n.d.) http://www.envpl.ipb.ac.rs/bio2.htm. Accessed on 25th Oct 2017

Bargagli R (1998) Trace element in terrestrial plants. A ecophysiological approach to biomonitoring and Biorecovery. Springer, Berlin, p 324

Bargagli R, Cateni D, Nelli L, Olmastroni S, Zagarese B (1997) Environmental impact of trace element emissions from geothermal power plants. Arch Environ Contam Toxicol 33:172–181

Bates JW (1992) Influence of chemical and site factors on Quercus and Fraxinus epiphytes at Loch Sunart, western Scotland: a multivariate analysis. J Ecol 80:163–179

Berg T, Røyset O, Steinnes E (1995) Moss hylocomium splendens used as biomonitor of atmospheric trace element deposition: estimation of uptake efficiencies. Atmos Environ 29:353–360

Berg T, Hjebrekke AG, Larseen R (2001) Heavy metals and POPs within the EMEP region 1999. EMEP/CCC Report 9/2001. Norwegian Institute for Air Research

Brown DH (1982) Mineral nutrition. In: Smith AJE (ed) Bryophyte ecology. Chapman and Hall, London, pp 383–444

Brown DH, Bates JW (1990) Bryophyte and nutrient cycling. Bot J Linn Soc 104:129–147

Brown DH, Brûmelis G (1996) A biomonitoring method using the cellular distribution of metals in mosses. Sci Total Environ 187:153–161

Brûmelis G, Brown DH (1997) Movement of metals to new growing tissues on the Moss *Hylocomium Splendens* (Hedw). BSG. Ann Bot 79:679–686

Buse A, Norris D, Harmens H (2003) Heavy metal in European Mosses: 2000/2001 Survey. UNECE ICP Vegetation. Centre for Ecology and Hydrology, Bangor, UK p 45

Calasans C, Malm O (1997) Elemental mercury contamination survey in a chlor-alkali plant by the use of transplanted Spanish moss, *Tillandsia usneoides* (L.) Sci Total Environ 208:165–177

Carballeira A, Fernández JA (2002) Bioconcentration of metals in the moss *Scleropodium purum* in the area surrounding a power plant A geotopographical predictive model for mercury. Chemosphere 47:1041–1048

Carpi A, Weinstein LH, Ditz DW (1994) Bioaccumulation of mercury by *Sphagnum* moss near a municipal solid waste incinerator. Air Waste 44:669–672

Čeburnis D, Valiulis D (1999) Investigation of absolute metal uptake efficiency from precipitation in Moss. Sci Total Environ 226:247–253

Čeburnis D, Steinnes E, Kveitkus K (1999) Estimation of metal uptake efficiencies from assessment-a review. Environ Pollut 114:471–492

Chakrabortty S, Jha SK, Paratkar GT, Puranik VD (2004) Distribution of trace elements in Moss biomonitors near Mumbai. Evansia 21(4):180–188

Chakrabortty S, Jha SK, Puranik VD, Paratkar GT (2006) Use of Mosses and Lichens as biomonitors in the study of air pollution near Mumbai. Evansia 23:1–8

Couto JA, Fernandez J, Aboal JR, Carballeira A (2004) Active biomonitorng of element uptake with terrestrial mosses: a comparison of bulk and dry deposition. Sci Total Environ 324:211–222

Culicov OA, Frontasyeva MV, Steinnes E, Okina OS, Santa Z, Todoran R (2002) Atmospheric deposition of heavy metals around the lead and copper-zinc smelters in Baia Mare, Romania, studied by the moss biomonitoring technique, neutron activation analysis and flame atomic absorption spectrometry. J Radioanal Nucl Chem 254(1):109–115

Culicov OA, Mocanu R, Frontasyeva MV, Yurukova L, Steinnes E (2005) Active Moss biomonitoring applied to an industrial site in Romania: relative accumulation of 36 elements in Moss-bags. Environ Monit Assess 108:22

De Caritat P, Reimann C, Bogatyrev I, Chekuskin V, Finne TE, Halleraker JH, Kashulina G, Niskavaara H, Pavlov V (2001) Regional distribution of Al, B, Ba, Ca, K, La, Mg, Mn, Na, P, Rb, Si, Sr, Th, U and Y in terrestrial Moss within a 188,000 km^2 area of the central barents region: influence of geology, Seaspray, and Uuman activity. Appl Geochem 16:137–159

Fernández JA, Carballeira A (2001) A comparison of indigenous mosses and topsoils for use in monitoring atmospheric heavy metal deposition in Galicia (Northwest Spain). Environ Pollut 114(3):431–441

Fernández JA, Aboal JR, Carballeira A (2000) Use of native and transplanted mosses as complementary techniques for biomonitoring mercury around an industrial facility. Sci Total Environ 256(2–3):51–61

Fernández JÁ, Aboal JR, Real C, Carballeira A (2007) A new moss biomonitoring method for detecting sources of small scale pollution. Atmos Environ 41(10):2098–2110

Ford J, Landers D, Kugler D, Lasorsa B, Crecelius E, Martinson J (1995) Inorganic contaminants in Arctic Alaskan ecosystem: long range atmospheric transport or local point sources. Sci Total Environ 160:323–335

Gana MG, Yurukova LD (2006) Biomonitoring in running river water with aquatic bryophytes. Scientific Articles. Ecology, Part 2. 209–216

Gecheva G, Yurukova L (2014) Water pollutant monitoring with aquatic bryophytes: a review. Environ Chem Lett 12(1):49–61

Gerdol R, Bragazza L, Marchesini R (2002) Element concentrations in the forest moss *Hylocomium splendens*: variations associated with altitude, net primary production and soil chemistry. Environ Pollut 116:129–135

Giordana S, Sorbo S, Adamo P, Basile A, Spagnuola V, Castaldo Cobianchi R (2004) Biodiversity and trace element content of epiphytic bryophytes in urban and extraurban sites of southern Italy. Plant Ecol 170:1–14

Giordano S, Adamo P, Sorbo S, Vingiani S (2005) Atmospheric trace metal pollution in the Naples urban area based on results from Moss and Lichen bags. Environ Pollut 136(3):431–442

Gjengedal E, Steinnes E (1990) Uptake of metal ions in Moss from artificial precipitation. Environ Monit Assess 14:77–87

Govindapyari H, Leleeka M, Nivedita M, Uniyal PL (2010) Bryophytes: indicators and monitoring agents of pollution. NeBIO 1(1):35–41

Gupta A (1995) Heavy metal accumulation by three species of mosses in Shillong, North-Eastern India. Water Air Soil Pollut 82(3–4):751–756. https://doi.org/10.1007/BF00479424

Hutten M, Woodward A, Hutten K (2005) Inventory of the mosses, liverworts, hornworts, and lichens of Olympic National Park, Washington: species list, Scientific Investigations Report. 1–5240. U.S. Geological Survey, Reston, pp 1–86

Jonathan SS, Lehman ME (2002) Bioindication of atmospheric heavy metal deposition in the Southeastern US using the moss *Thuidium delicatulum*. Atmos Environ 36:1611–1618

Kolon K, Samecka-Cymerman A, Kempers AJ, Alexander JK, Lucyna M (2010) *Pleurozium schreberi* of the Tatra mountains (Poland) used as a bioindicational system for observing long range atmospheric transport of chemical elements. J Atmos Chem 66:157–166. https://doi.org/10.1007/s10874-011-9198-x

Krommer V, Zechmeister HG, Roder I, Scharf S, Hanus-Illnar A (2007) Monitoring atmospheric pollutants in the biosphere reserve Wienerwald by a combined approach of biomonitoring methods and technical measurements. Chemosphere 67:1956–1966

Lodenius M (1998) Dry and wet deposition near a chlor-alkali plant. Sci Total Environ 213:53–56

Loppi S, Bonini I (2000) Lichens and mosses as biomonitors of trace elements in areas with thermal springs and fumarole activity (Mt. Amiata, Central Italy). Chemosphere 41:1333–1336

Lukáš Č, Oto K, Vítězslav P (2017) Modeling the distribution of rare and interesting moss species of the family Orthotrichaceae (Bryophyta) in Tajikistan and Kyrgyzstan. Acta Soc Bot Pol 86 (2):35–43

Macedo-Miranda G, Avila-Pérez P, Gil-Vargas P, Zarazúa G, Sánchez-Meza JC, Zepeda-Gómez C, Tejeda S (2016) Accumulation of heavy metals in mosses: a biomonitoring study. Springerplus 5(1):715

Makholm MM, Mladenoff DJ (2005) Efficacy of a biomonitoring (moss bag) technique for determining element deposition on a mid-range (375) km scale. Environ Monit Assess 104 (1–3):1–18

Mäkinen A (1994) Biomonitoring of atmospheric deposition in the Kola Peninsula (Russia) and Finnish Lapland, based on the chemical analysis of mosses. Minist Environ Rapp 4:1–83

Mankovska B (1994) Airborne sulphur and heavy metal pollution in the environment of a thermal plant. Ekologia (Bratislava) 13(2):207–217

Marinova S, Yurukova L, Frotasyeva MV et al (2010) Air pollution studies in Bulgaria using the moss biomonitoring technique. Ecol Chem Eng S 17:37–52

Markert B, Weckert V (1989) Time and site integrated long-term biomonitoring of chemicals by means of mosses. Toxicol Environ Chem 40:177–189

Markert B, Oehlmann J, Roth M (1997) General aspects of heavy metal monitoring by plants and animals. In: Subramanian G, Iyengar V (eds) Environmental biomonitoring – exposure

assessment and specimen banking, ACS Symposium Series 654. American Chemical Society, Washington, DC
Markert B, Wappelhorst O, Weckert V, Herpin U, Siewers U, Friese K, Breulmann G (1999) The use of bioindicators for monitoring the heavy-metal status of the environment. J Radioanal Nucl Chem 240(2):425–429
Markert BA, Breure AM, Zechmeister HG (2003) In: Markert BA, Breure AM, Zechmeister HG (eds) Definitions, strategies, and principles for Bioindication/biomonitoring of the environment. Elsevier, Oxford, pp 3–39
Martin MH, Coughtrey PJ (1982) Biological monitoring of heavy metal pollution. Land, and Air Appl Sci Publishers, London, pp 136–142
Nath V, Sinha S, Asthana AK, Sahu V (2010) A study on metal accumulation in two selected bryophytes. Env Sci Ind J 5(1):42–45
Nriagu JO (1979) Global inventory of natural and anthropogenic emissions of trace metals to the atmosphere. Nature 279(5712):409–411
Økland T, Økland RH, Steinnes E (1999) Element concentrations in the Boreal Forest Moss hylocomium splendens: variations related to gradients in vegetation and local environmental factors. Plant Soil 209:71–83
Onianwa PC (2001) Monitoring atmospheric metal pollution: a review of the use of mosses as indicators. Environ Monit Assess 71:13–50
Palmieri F, Neri R, Benco C, Serracca L (1997) Lichens and moss as bioindicators and bioaccumulators in air pollution monitoring. J Environ Path Toxicol Oncol 16(2–3):175–190
Pant G, Tewari SD (1998) Bryophytes as biogeoindicators: Bryophytic Associations of Mineral enriched substrates in Kumaon Himalaya. In: Chopra RN (ed) Topics in Bryology. Allied Publishers Ltd, New Delhi, p 202
Pesch R, Schroeder W (2006) Mosses as Bioindicators for metal accumulation: statistical aggregation of measurement data to exposure indices. Ecol Indic 6(1):137–152
Poikolainen J (2004) Mosses, epiphytic lichens and tree bark as biomonitors for air pollutants—specially for heavy metals in regional surveys. Dissertation, University of Oulu, Oulu
Pott U, Turpin D (1996) Changes in atmospheric trace element deposition in Fraser Valley, B.C., Canada from 1960–1993 measured by Moss monitoring with *Isothecium Stoloniferum*. Can J Bot 74:1345–1353
Puckett KJ (1988) Bryophytes and lichens as monitor of metal deposition. Bibl Lichenol 30:231–267
Rao DN (1982) Responses of bryophytes to air pollution. In: Smith AJE (ed) Bryophyte ecology. Springer, Dordrecht, pp 445–471
Rasmussen L (1978) Element content of epiphytic *Hypnum cupressiforme* related to element content of the bark of different species of phorophytes. Lindbergia 4:209–218
Reimann C, Halleraker JH, Kashulina G, Bogatyrev I (1999) Comparison of plant and precipitation chemistry in catchments with different levels of pollution in Kola Peninsula, Russia. Sci Total Environ 243(244):169–191
Ross HB (1990) On the use of the mosses *Hylocomium Splendens* and *Pleurozium schreberi* for estimating atmospheric trace metal deposition. Water Air Soil Poll 50:63–76
Ruhling A (ed) (1994) Atmospheric heavy metal deposition in Europe – estimations based on moss analysis. Nordic Council of Ministers, Copenhagen, p 9
Ruhling A, Tyler G (1968) An ecological approach to the lead problem. Bot Notiser 122:248–342
Ruhling A, Tyler G (1970) Sorption and retention of heavy metals in the woodland Moss *Hylocomium splendens*. Oikos 21:92–97
Sahu V, Nath V, Asthana AK, Yunus M (2014) *Marchantia paleacea* Bertol. as quantitative biomonitor of atmospheric heavy metals deposition. J Recent Adv Appl Sci 29:22–27
Samecka-Cymerman A, Kempers AJ (1995) Preliminary investigations into the bioaccumulation of mercury by the liverwort *Scapania undulata* (L.) Dum. Ecotoxicol Env Safety 31:57–61
Sawidis T, Zachariadis G, Stratis J, Eukakis L (1993) Mosses as biological indicators for monitoring of heavy metal pollution. Environ Bull 2:26–229

Saxena A (2006) Seasonal pattern of metal bioaccumulation and their toxicity on *Sphagnum squarrosum*. J Environ Biol 27:71–75

Saxena DK, Saxena A, Srivastava HS (2000) Biomonitoring of metal precipitation at petrol pumps and their effect on moss *Sphagnum cuspidatum* Hoffm. J Environ Stud Policy 3:95–102

Saxena DK, Srivastava K, Singh S (2008) Biomonitoring of metal deposition by using moss transplant method through *Hypnum cupressiforme* (Hedw.) in Mussoorie. J Env Biol 29 (5):683–688

Saxena DK, Tuba Z, Arfeen MS (2010) Seasonal passive metal monitoring during year 2003 to 2006 in Nainital of Kumaon hills (INDIA) by moss *Racomitrium crispulum*. Acta Bot Hung 52 (1–2):273–297

Shakya K, Chettri MK, Swidis T (2004) Appraisal of some mosses for biomonitoring airborne heavy metals in Kathmandu valley. An Intl J Ecol 11(1). https://doi.org/10.3126/eco.v11i1.143

Singh S, Srivastava K, Gahtori D, Saxena DK (2017) Bryomonitoring of atmospheric elements in *Rhodobryum giganteum* (Schwaegr.) Par., growing in Uttarakhand region of Indian Himalayas. Aerosol Air Qual Res 17:810–820

Steinnes E (1977) Atmospheric deposition of trace elements in Norway studied by means of Moss analysis, Kjeller Report, KR 154. Institute for Atomenegri, Kjeller

Steinnes E (1993) Some aspects of biomonitoring of air pollutants using mosses as illustrated by the 1976 Norwegian survey. In: Markert B (ed) Plants as biomonitors, indicators for heavy metals of the terrestrial environment. VCH Publishers, Weinheim, pp 381–339

Steinnes E (1995) A critical evaluation of the use of naturally growing moss to monitor the deposition of atmospheric metals. Sci Total Environ 160(161):243–249

Sucharová J, Suchara I (2004) Distribution of 36 element deposition rates in a historic mining and smelting area as determined through fine-scale biomonitoring techniques. Part I: relative and absolute current atmospheric deposition levels detected by Moss analysis. Water Air Soil Poll 153:205–228

Thöni L, Schnyder N, Kreig F (1996) Comparisons of metal concentrations in three species of mosses and metal freights in bulk precipitations. Fresenius J Anal Chem 354:703–708

Tyler G (1970) Moss analysis-a method for surveying heavy metal deposition. In: Englaund HM, Berry WT (eds) Proceedings of the Second International Clean Air Congress. Academic, New York, pp 129–132

Tyler G (1990) Bryophyte and heavy metals: a literature review. Bot J Linn Soc 104:231–253

Umweltbundesamt (2004) Recycling von phosphor verbessurn Presse-Information Nr. 103/2004. UBA-Berlin

UNESCO (2016) United Nations environment programme, UNEP/POPS/POPRC. 12/11(2016) Report of the persistent organic pollutants review committee on the work of its twelfth meeting, 19–23 Sept 2016, Rome

Walkenhorst A, Hagemeyer J, Breckle WS (1993) Passive monitoring of air borne pollutants, peculiarly trace metals, with tree bark. In: Markert B (ed) Plants as biomonitors. Indicators for heavy metals in the terrestrial environment. VCH, Weinheim, pp 523–540

Wang S, Zhang Z, Wang Z (2015) Bryophyte communities as biomonitors of environmental factors in the Goujiang karst bauxite, southwestern China. Sci Total Environ 15(538):270–278. https://doi.org/10.1016/j.scitotenv.2015.08.049

Wappelhorst O, Kuhn I, Oehlmann J, Markert B (2000) Deposition and disease: a moss monitoring project as an approach to ascertaining potential connections. Sci Total Environ 249 (1–3):243–256

Wolterbeek HT, Bode P, Verburg TG (1996) Assessing the quality of biomonitoring via signal-to-noise ratio analysis. Sci Total Environ 180:107–116

Zawadzki K, Samecka-Cymerman A, Kolon K, Wojtuń B, Mróz L, Kempers AJ (2016) Metals in Pleurozium schreberi and Polytrichum commune from areas with various levels of pollution. Environ Sci Pollut Res Int 23:11100–11108

Zechmeister HG (1995) Correlation between altitude and heavy metal deposition in Alps. Environ Pollut 89:73–80

Zechmeister HG, Dirnböck T, Hülber K, Mirtl M (2007) Assessing airborne pollution effects on bryophytes lessons learned through long-term integrated monitoring in Austria. Environ Pollut 147:696–705

Zeichmeister HG (1998) Annual growth of four pleurocarpous Moss species and their applicability for biomonitoring heavy metals. Environ Monit Assess 52:441–451

Zeichmeister HG, Grodzinska K, Szarek-Lukaszewska G (2003a) In: Markert BA, Breure AM, Zeichmeister HG (eds) Bryophytes. Elsevier, Oxford, pp 329–375

Zeichmeister HG, Hohenwallner D, Riss A, Hanus-Illnar A (2003b) Variation in heavy metal concentrations in the Moss species Abietinella abietina (Hedw.) Fleisch according to sampling time, within site variability and increase in biomass. Sci Total Environ 301:55–65

Zechmeister HG, Dirnböck T, Hülber K, Mirtl M (2007) Assessing airborne pollution effects on bryophytes – lesson learned through long-term integrated monitoring in Austria. Environ Pollut 147:696–705

Bioinformatics Resources for the Stress Biology of Plants

Sonu Kumar and Asheesh Shanker

Abstract Bioinformatics play an invaluable role in many areas of biological research including stress biology. In the present global scenario, almost every organism faces stress as a response to stressors (biotic or abiotic). Any stress has serious impact on the overall growth and development of organisms. Moreover, productivity of plants is also affected by stress. Due to these reasons, stress biology has been the focus of research for many scientists, and the massive data generated by them require appropriate management and analysis tools. The availability of bioinformatics tools including software, databases, and web resources has brought a major change in the stress-related research. These resources help in the analysis and better interpretation of the data generated through experiments. This chapter deals with various general and specialized bioinformatics resources useful for the stress biology community working on plants.

Keywords Bioinformatics · Plant · Stress · Abiotic · Biotic · Databases · Software

1 Introduction

Stress is a biological or physiological response of an organism to a stress factor or stressor. In nature, plants are continuously confronted with a wide variety of stress factors during their life cycle, which can be broadly categorized into biotic or abiotic stress. Biotic stress occurs in plants due to damage done by plant pathogens like bacteria, fungi, insects, parasites, viruses, and other living organisms (Leonberger et al. 2016), whereas abiotic stress is caused by environmental stimuli such as temperature, radiation, chemical toxicity, drought, salinity, etc.

Both biotic and abiotic stresses have serious impact on the growth, development, and productivity of plants. Plants are able to recognize unfavorable climatic variations and accordingly respond to it. Being sessile in nature, plants are unable

S. Kumar · A. Shanker (✉)
Bioinformatics Programme, Center for Biological Sciences, Central University of South Bihar, Patna, India

S. Vats (ed.), *Biotic and Abiotic Stress Tolerance in Plants*,
https://doi.org/10.1007/978-981-10-9029-5_14

to escape these stresses; however, they have developed mechanisms to survive unfavorable conditions. Response of plants against these stresses is much complex and causes changes from genomic to physiological levels (Atkinson and Urwin 2012), which occur at cellular, biochemical, and molecular level.

In spite of the extensive research on the response of plants to abiotic and biotic stresses, knowledge gap about the molecular mechanisms that control different functions of plant genes and proteins associated with response to stress still persist (Hirayama and Shinozaki 2010). In the last few decades, our understanding of plant response to stress has increased significantly mainly because of the availability of genomic sequences of plants along with high-throughput bioinformatics resources (Mochida and Shinozaki 2010; Tatusova et al. 2007).

Recent advances in bioinformatics resources including software, databases, and web servers have brought a major change in plant stress research. In last few decades, massive data emerging from stress associated researches in plants require appropriate management and analysis. These resources help researchers in better interpretation of data generated through several experiments and prove useful for stress biology community. In this chapter we describe various general and specific bioinformatics resources which play invaluable role in plant stress biology.

2 General Bioinformatics Resources

Viewing the immense significance of bioinformatics in the field of plant science, different bioinformatics resources like tools, servers, and biological databases were developed. Some general resources are described here.

2.1 Biological Databases

Biological databases are the collection of biological information from different scientific research laboratories worldwide (Attwood et al. 2011). These databases have been developed considering type of biological data like genomics, proteomics, metabolomics, gene expression, phylogenetic, etc. (Altman 2004). Establishment of the National Center for Biotechnology Information (NCBI; http://www.ncbi.nlm.nih.gov/) in 1988 marked as an opening of bioinformatics infrastructure for public in the field of biological research (Smith 2013). The International Nucleotide Sequence Database Collaboration (INSDC; http://www.insdc.org) mentioned three major databases, viz., GenBank (http://www.ncbi.nlm.nih.gov/genbank/), European Molecular Biology Laboratory's European Bioinformatics Institute (EMBL-EBI; http://www.ebi.ac.uk/embl/), and DNA Data Bank of Japan (DDBJ; http://www.ddbj.nig.ac.jp/), as primary databases for nucleotide sequence. These resources share data on daily basis (Cochrane et al. 2015). Apart from this, other biological databases

Table 1 List of commonly used biological databases

S. no.	Database	Description	Uniform Resource Locator (URL)	Reference
1.	GenBank	It contains collection of all publicly available DNA sequences	http://www.ncbi.nlm.nih.gov/genbank/	Benson et al. (2000)
2.	EMBL-EBI	A nucleotide sequence data library	http://www.ebi.ac.uk/embl/	Amid et al. (2011)
3.	DDBJ	It provides freely available nucleotide sequence data to support research activities in biological research	http://www.ddbj.nig.ac.jp/	Kaminuma et al. (2010)
4.	UniProtKB/Swiss-Prot	It provides high-quality and freely accessible protein sequences along with their functional information	http://www.uniprot.org/uniprot/	Boutet et al. (2007)
5.	PIR	Protein Information Resource contains functionally annotated protein sequences data	http://pir.georgetown.edu/	Barker et al. (1998)
6.	Entrez protein database	It is a collection of protein sequences from several sources	https://www.ncbi.nlm.nih.gov/protein/	Wheeler et al. (2007)
7.	PDB	Protein Data Bank provides three-dimensional structures of proteins, nucleic acids, and complex assemblies determined experimentally	http://www.rcsb.org/pdb/home/home.do/	Berman et al. (2002)
8.	MMDB	The Molecular Modeling Database contains experimentally determined three-dimensional biomolecular structures	http://www.ncbi.nlm.nih.gov/Structure/MMDB/mmdb.shtml/	Chen et al. (2003)
9.	KEGG	The Kyoto Encyclopedia of Genes and Genomes provides information about biological pathways, genomes, chemical substances, diseases, and drugs	https://www.genome.jp/kegg/	Kanehisa and Goto (2000)
10.	Ensembl Plants	Ensembl Plants is a genome-centric portal for plant species of scientific interest	http://plants.ensembl.org/index.html/	Bolser et al. (2016)

have also been developed to store biological information. A list of common biological databases widely used in biological studies including plant stress is shown in Table 1.

The following examples help to understand how some of the aforesaid databases can be of use to relate the wet lab work available bioinformatic resources. In the recent past, cold acclimation of the model legume *Lotus japonicus* by transcriptome profiling under cold stress was analyzed. The Illumina reads generated were deposited in the NCBI BioProject. To analyze transcripts, annotation information was obtained from various databases including GenBank and KEGG. Additionally, sequence similarity search was used to assign functional annotations (Calzadilla et al. 2016). Similarly, KEGG database was utilized to perform pathway enrichment analysis of differentially expressed genes under varied nitrate stress in leaves of Tibetan hulless barley (Wei et al. 2016) and salinity tolerance in a spaceflight-induced wheat mutant (Xiong et al. 2017).

2.2 *Bioinformatics Tools and Techniques*

Bioinformatics play a significant role in the development of tools and techniques, which helps in understanding the structural and functional relationship of the biological macromolecules. Till now different tools and techniques were developed in the field of bioinformatics to resolve many biological problems including plants stress.

2.2.1 Sequence Alignment

Sequence alignment is one of the most important techniques used to compare two or more biological sequences to identify sequence similarity. It helps to infer evolutionary relationships between the sequences that assist in function and structure prediction (Mount 2004; Sharma et al. 2016). Sequence alignment is generally categorized into global and local alignment. Different computational algorithms like dynamic programming have been developed for sequence alignment (Needleman and Wunsch 1970; Smith and Waterman 1981). Global alignment algorithms are used to optimize overall alignment of two or more sequences from beginning to end. Due to this, global alignment of sequences may include more gaps resulted in low similarity. Local alignment algorithms align relatively conserved portions between the sequences. Apart from this, efficient heuristic algorithms including FASTA (Pearson 1990) and BLAST (Altschul et al. 1990) designed to search large-scale databases were also developed. Alignment between two sequences to find out the best-matching regions is predicted by pairwise sequence alignment methods. Multiple sequence alignment (MSA) methods incorporate more than two sequences and used to detect conservation among sequences imagined to be evolutionarily related (Mount 2004). Earlier, different sequence alignment tools were applied in various researches to understand the response of plants in stress condition, like involvement of omega-3 fatty acid desaturases enzyme in plants (Palm and Jojoba) stress responses was observed with the help of sequence alignment (Sham and Aly 2012). The sequence alignment tools commonly used in various biological researches including plant stress are shown in Table 2.

Table 2 Various sequence alignment tools

S. no.	Resource	Description	URL	Reference
1.	BLAST	Used to find out nucleotide or protein sequence similarity	http://ncbi.nlm.nih.gov/BLAST/	Altschul et al. (1990)
2.	FASTA	A pairwise sequence alignment tool used to align DNA and protein sequences	https://www.ebi.ac.uk/Tools/sss/fasta/	Pearson (1990)
3.	ClustalW	A multiple sequence alignment program, used to align more than two DNA or protein sequences	http://ebi.ac.uk/Tools/msa/clustalw2/	Thompson et al. (1994)
4.	T-COFFEE	It is a collection of tools to perform MSA	http://tcoffee.crg.cat/	Notredame et al. (2000)
5.	MUSCLE	Multiple Sequence Comparison by Log-Expectation generates MSA	http://ebi.ac.uk/Tools/msa/muscle/	Edgar (2004)
6.	MAFFT	A multiple sequence alignment program based on fast Fourier transform algorithm	http://mafft.cbrc.jp/alignment/software/	Katoh et al. (2005)
7.	DIALIGN 2	Uses segment-to-segment approach of MSA	http://bibiserv.techfak.uni-bielefeld.de/dialign/	Morgenstern (1999)
8.	EMBOSS Needle	A tool for optimal global alignment of two sequences based on Needleman-Wunsch algorithm	https://www.ebi.ac.uk/Tools/psa/emboss_needle/	Rice et al. (2000)
9.	EMBOSS Water	A tool to calculate local alignment of two sequences based on Smith-Waterman algorithm	https://www.ebi.ac.uk/Tools/psa/emboss_water/	Rice et al. (2000)

2.2.2 Protein Structure Prediction

Proteins play a key role in governing several molecular processes in an organism. The function of proteins depends on their unique three-dimensional structure. A minor change or alteration in protein structure may alter their function and cause harmful effects. Biotic and abiotic stresses are one of the reasons that alter the structure and function of protein (Rodziewicz et al. 2014; Zhu 2016). For example, in acclimation process of plants to abiotic stress, reactive oxygen species (ROS) decode its signals by the cell which is thought to occur via different redox reactions. In these reactions ROS like hydrogen peroxide (H_2O_2) will oxidize sulfur-containing residues of proteins and cause alteration of structure and function (Choudhury et al. 2017). Protein structure determination has been successfully applied in different biotic and abiotic stress response study in plants (Wang et al. 2017; Moraes-Filho et al. 2017). To predict three-dimensional structure of proteins, different methods have been developed like homology or comparative or template-based modeling, fold recognition, and ab initio protein structure prediction.

Homology modeling refers to build a three-dimensional (3D) structure of protein using a known experimental structure of a homologous protein. The qualities of both sequence alignment and template structure have direct impact on the accuracy of predicted model. Protein threading, also called as fold recognition method, predicts whether protein of interest possess the similar fold, but lacks significant sequence similarity, as proteins of known structures (Schwede et al. 2003). De novo (ab initio) protein structure prediction is an energy-based method; it can be used to build protein structure when no suitable template structure is identified.

Recently, the protein structure of plant salinity stress-responsive phosphoserine phosphatase (PSP) of *Brassica juncea* was predicted using template-based homology modeling (Purty et al. 2017). In an another study, abiotic stress tolerance of Oriental hybrid lily cultivar Sorbonne gene (*LhSorP5CS*) was observed with the help of homology modeling-based 3D structure prediction (Wang et al. 2017). These studies explain the significance of protein structure prediction technique in plant stress studies. The commonly used protein structure prediction tools and servers are listed in Table 3.

2.2.3 Molecular Docking and Molecular Dynamics Simulation

Molecular docking is a well-established method to determine the binding between two molecules along with their preferred positioning. It helps to predict the affinity of the small molecule (ligand) to their protein targets. Several computational tools are available that used to study molecular interaction in molecular biology including plant stress. Additionally, molecular dynamics simulation technique also applied in stress associated research in plants to study molecular behavior and to refine the predicted structure. Table 4 shows list of available tools for molecular docking and molecular dynamics simulation. Earlier, binding of dehydration-responsive element binding (DREB) proteins with the dehydration-responsive element/C repeat (DRE/CRT) of stress-inducible gene promoters were observed using homology modeling and molecular docking approach (Nawaz et al. 2014).

3 Bioinformatics Resources for Plant Stress

Several databases have been developed that specifically store data related to plant stress including gene sequences, functional and experimental validation of stress proteins. These resources directly provide access to data or information involved in different stresses in plants. Here we explore various important resources associated with plant stress that helps researchers to retrieve required information.

Table 3 Web servers and tools for protein structure prediction

S. no.	Resource	Description	URL	Reference
1.	SWISS-MODEL	An automated homology modeling server to predict 3D structures of protein	http://swissmodel.expasy.org/	Guex and Peitsch (1997)
2.	MODELLER	It is a homology modeling program used to predict 3D structures of protein	http://www.salilab.org/modeller/	Webb and Sali (2014)
3.	ESyPred3D	It is an automated homology modeling program and uses MODELLER to build the 3D structure	http://www.unamur.be/sciences/biologie/urbm/bioinfo/esypred/	Lambert et al. (2002)
4.	CPHmodels	A web server to predict 3D structure of protein through single template homology modeling	http://www.cbs.dtu.dk/services/CPHmodels/	Nielsen et al. (2010)
5.	LOMETS	Local Meta-Threading-Server is an online web server to build 3D models of protein. It generates high-scoring target-to-template alignments using threading approach	http://zhanglab.ccmb.med.umich.edu/LOMETS/	Wu and Zhang (2007)
6.	RaptorX	It is a web server to predict protein structure along with function. It also predicts secondary structure, template-based tertiary structure, and probabilistic alignment sampling of protein	http://raptorx.uchicago.edu/	Källberg et al. (2014)
7.	I-TASSER	Iterative Threading ASSEmblyRefinement uses hierarchical approach to predict protein structure and function	http://zhanglab.ccmb.med.umich.edu/I-TASSER/	Yang et al. (2015)
8.	ROBETTA	It is a web server which provides automated tools based on either comparative modeling or de novo structure prediction methods for structure prediction and analysis of proteins	http://robetta.bakerlab.org/	Kim et al. (2004)
9.	Bhageerath-H	It is a hybrid web server of homology and ab initio methods for protein tertiary structure prediction	http://www.scfbio-iitd.res.in/bhageerath/bhageerath_h.jsp/	Jayaram et al. (2014)
10.	CABS-fold	It is a web server which includes tools to predict protein structure	http://biocomp.chem.uw.edu.pl/CABSfold/	Blaszczyk et al. (2013)
11.	PEP-FOLD	It is an online service which uses amino acid sequences to predict peptide structures through de novo approach	http://bioserv.rpbs.univ-paris-diderot.fr/services/PEP-FOLD/	Shen et al. (2014)

Table 4 Molecular docking and molecular dynamics simulation tools

S. no.	Resource	Description	URL	Reference
1.	AutoDock	It is a suite of automated docking tools used to predict interactions between small molecules and receptor of known 3D structure of protein	http://autodock.scripps.edu/	Morris et al. (2009)
2.	AutoDock Vina	It is faster than AutoDock and accomplishes significant perfections in average prediction of the binding mode	http://vina.scripps.edu/	Trott and Olson (2010)
3.	SANJEEVINI	It is known as complete drug designing software and helps in lead molecule discovery	http://www.scfbio-iitd.res.in/sanjeevini/sanjeevini.jsp/	Jayaram et al. (2012)
4.	Glide	It is an exhaustive search-based program for ligand-receptor docking	http://www.schrodinger.com/glide/	Friesner et al. (2004)
5.	GEMDOCK	Generic Evolutionary Method for molecular DOCKing software computes conformation and orientation of ligand to the active site of receptor	http://gemdock.life.nctu.edu.tw/dock/	Yang and Chen (2004)
6.	AMBER	Assisted Model Building and Energy Refinement is a collection of programs for molecular dynamics simulations of nucleic acids and proteins	http://ambermd.org/	Pearlman et al. (1995)
7.	CHARMM	Chemistry at Harvard Macromolecular Mechanics is a commonly used molecular simulation program	https://www.charmm.org/	Brooks et al. (1983)
8.	GROMACS	GROningen MAchine for Chemical Simulations is an open-source and extremely high-performance package for molecular dynamics simulation	http://www.gromacs.org/	Hess et al. (2008)

3.1 Plant Environmental Stress Transcript Database

Plant Environmental Stress Transcript Database (http://intranet.icrisat.org/gt1/tog/homepage.htm; Fig. 1) contains stress transcripts from crop. It provides annotated tentative orthologous sequence information of 16 plant species which includes six cereal crops (wheat, rice, maize, barley, rye, and *Sorghum*) and ten dicots (*Arabidopsis thaliana*, *Medicago*, *Glycine max*, chickpea, potato, tomato, *Phaseolus*, *Pennisetum*, groundnut, and cowpea) across abiotic stress conditions of four types. It also contains expressed sequence tags from stress cDNA libraries. This database allows searching for different queries like annotated transcripts that are expressed across stress conditions, microsatellites containing transcripts,

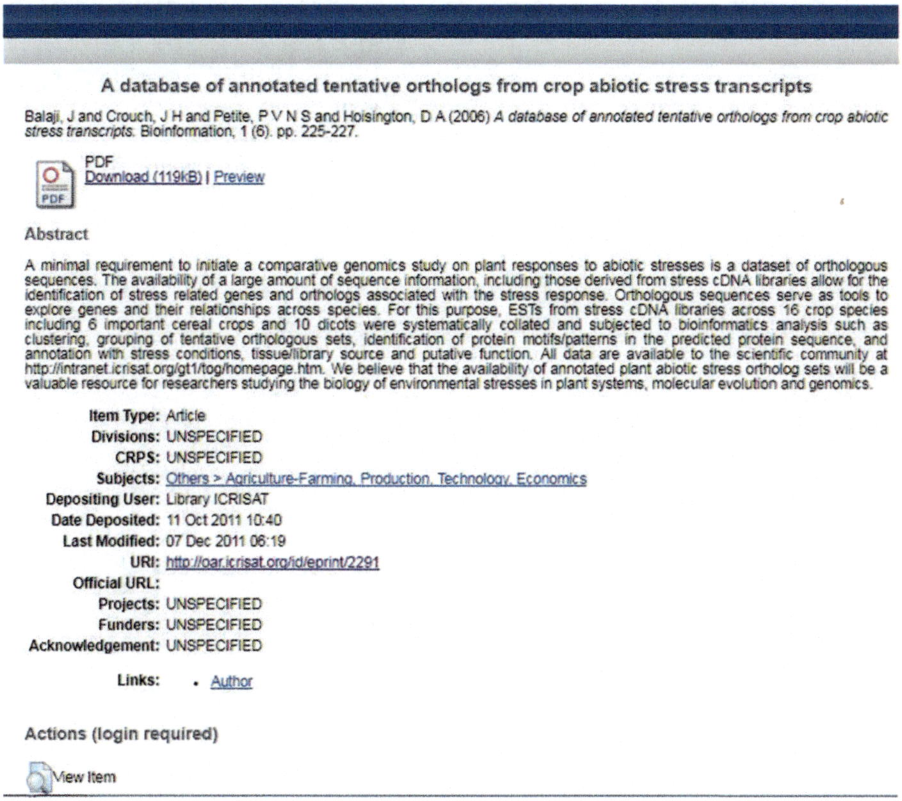

A database of annotated tentative orthologs from crop abiotic stress transcripts

Balaji, J and Crouch, J H and Petite, P V N S and Hoisington, D A (2006) *A database of annotated tentative orthologs from crop abiotic stress transcripts.* Bioinformation, 1 (6). pp. 225-227.

PDF
Download (119kB) | Preview

Abstract

A minimal requirement to initiate a comparative genomics study on plant responses to abiotic stresses is a dataset of orthologous sequences. The availability of a large amount of sequence information, including those derived from stress cDNA libraries allow for the identification of stress related genes and orthologs associated with the stress response. Orthologous sequences serve as tools to explore genes and their relationships across species. For this purpose, ESTs from stress cDNA libraries across 16 crop species including 6 important cereal crops and 10 dicots were systematically collated and subjected to bioinformatics analysis such as clustering, grouping of tentative orthologous sets, identification of protein motifs/patterns in the predicted protein sequence, and annotation with stress conditions, tissue/library source and putative function. All data are available to the scientific community at http://intranet.icrisat.org/gt1/tog/homepage.htm. We believe that the availability of annotated plant abiotic stress ortholog sets will be a valuable resource for researchers studying the biology of environmental stresses in plant systems, molecular evolution and genomics.

Item Type: Article
Divisions: UNSPECIFIED
CRPS: UNSPECIFIED
Subjects: Others > Agriculture-Farming, Production, Technology, Economics
Depositing User: Library ICRISAT
Date Deposited: 11 Oct 2011 10:40
Last Modified: 07 Dec 2011 06:19
URI: http://oar.icrisat.org/id/eprint/2291
Official URL:
Projects: UNSPECIFIED
Funders: UNSPECIFIED
Acknowledgement: UNSPECIFIED

Links: • Author

Actions (login required)

View Item

Fig. 1 Home page of plant environmental stress transcript database

hypothetical genes which are conserved, and sequence alignment of ortholog sets based on cluster size, stress conditions, or annotation (Balaji et al. 2006).

3.2 *STIFDB (Stress-Responsive Transcription Factor Database)*

STIFDB (http://caps.ncbs.res.in/stifdb/; Fig. 2) database contains abiotic stress-responsive genes along with their predicted abiotic transcription factor binding sites in *Arabidopsis thaliana*. STIFDB is a valuable database for scientists to understand the different abiotic stress in plant system (Shameer et al. 2009).

Another version of STIFDB called STIFDB V2.0 (http://caps.ncbs.res.in/stifdb2/) is also available with additional information related to biotic and abiotic stress-responsive genes in *Oryza sativa* L. and *A. thaliana*. STIFDB2 contains information of 5984 stress-

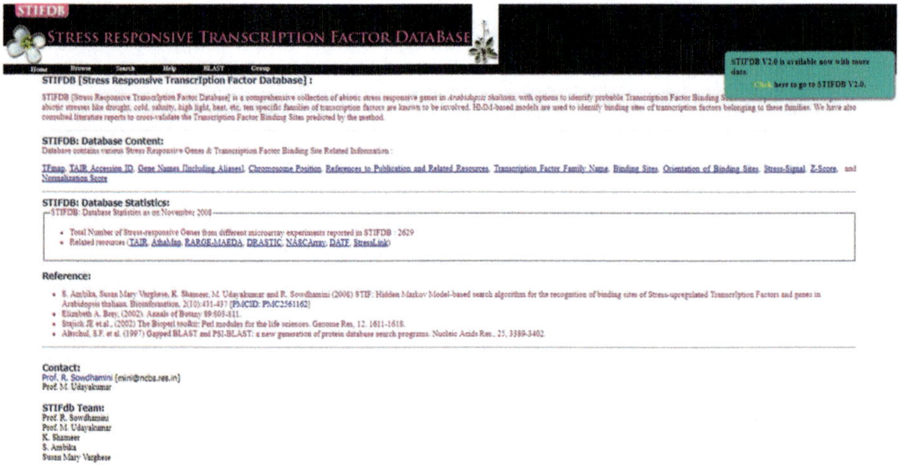

Fig. 2 Home page of stress-responsive transcription factor database

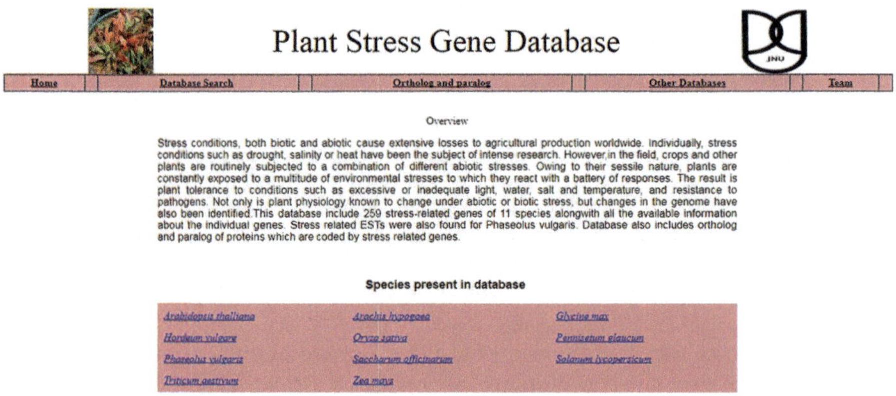

Fig. 3 Home page of plant stress gene database

responsive genes from *A. thaliana*, *O. sativa* subsp. *japonica*, and *O. sativa* subsp. *indica*. Moreover, it provides data related to 31 transcription factors and 15 stress signals (Naika et al. 2013).

3.3 *Plant Stress Gene Database*

Plant Stress Gene Database (http://ccbb.jnu.ac.in/stressgenes/frontpage.html; Fig. 3) contains information about 259 genes involved in stress conditions from *Arachis hypogaea*, *Arabidopsis thaliana*, *Hordeum vulgare*, *Glycine max*, *Oryza sativa*,

Phaseolus vulgaris, *Pennisetum*, *Solanum lycopersicum*, *Saccharum officinarum*, *Zea mays*, and *Triticum aestivum*. The database provides information of orthologs and paralogs of stress-related genes along with other data (Prabha et al. 2011).

3.4 *QlicRice: A Web Interface for Abiotic Stress-Responsive QTL and Loci Interaction Channel in Rice*

The QlicRice (http://nabg.iasri.res.in:8080/qlic-rice/; Fig. 4) database and search engine was developed to incorporate the most of publicly available rice quantitative trait loci (QTLs) responsive to abiotic stresses along with their corresponding gene (TIGR/MSU) loci. The database contains abiotic stress-related QTLs (974) with overlapping TIGR/MSU loci (460). The gene ontology (GO) and KEGG orthology (KO) terms were used to functionally characterized QTLs. Additionally, it provides mined genomic data in rice, specifically in *O. sativa* ssp. *japonica* cv. Nipponbare (Smita et al. 2011).

Fig. 4 Home page of QlicRice database

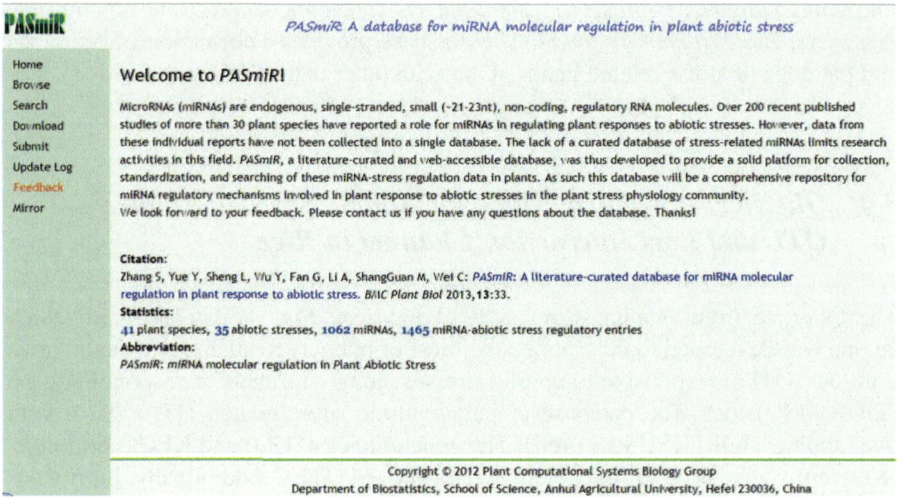

Fig. 5 Home page of PASmiR database

3.5 *PASmiR: A Database for miRNA Molecular Regulation in Plant Abiotic Stress*

PASmiR (http://pcsb.ahau.edu.cn:8080/PASmiR/; Fig. 5) is a web-accessible database and curated using literature. It contains description of microRNA (miRNA) molecular regulation in various abiotic stresses of plant. The database contains information from around 200 published research studies in 33 plant species and represents 1038 regulatory relationships between 682 miRNAs and 35 abiotic stresses. The database also facilitates keyword search to retrieve miRNA-stress regulatory entries by using miRNA identifier, abiotic stress, and plant species (Zhang et al. 2013).

3.6 *The* Arabidopsis *Information Resource (TAIR)*

TAIR (http://arabidopsis.org; Fig. 6) is a database which provides molecular biology and genetic data of *A. thaliana*. TAIR provides centralized access of data over 30,000 *Arabidopsis* genes. TAIR also provides tools for analysis and data visualization (Berardini et al. 2015).

Fig. 6 Home page of TAIR database

3.7 *Rice Stress-Responsive Transcription Factor Database (Rice SRTFDB)*

Rice SRTFDB (http://www.nipgr.res.in/RiceSRTFDB.html; Fig. 7) contains detailed information of rice transcription factor expression patterns in salinity and water-deficit stress conditions at different phases of development. Stress-responsive expression data is stored in the database representing a curated set of 99 Affymetrix

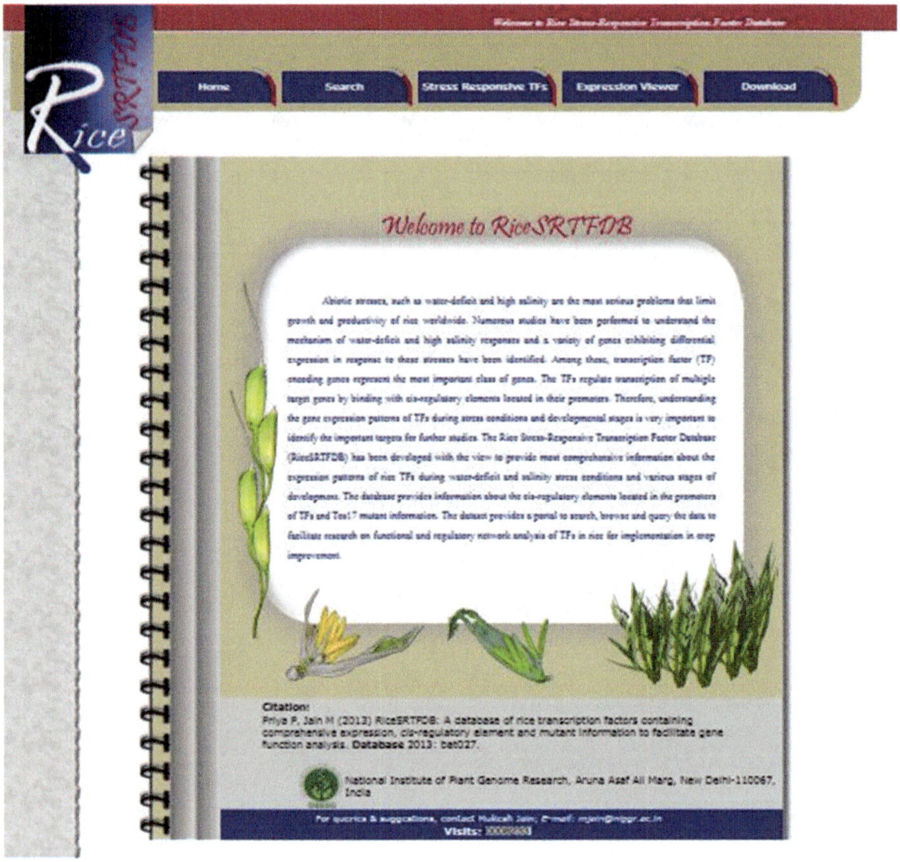

Fig. 7 Home page of Rice SRTFDB database

GeneChip Rice Genome arrays from 18 different salinity and drought stress treatments. The expression viewer provided in database gives stress and developmental stages specific expression of selected transcription factor with graphical representations. Salt- and drought-induced differential expression data of all the transcription factors or specific transcription factor family in rice can be retrieved. The data about the cis-regulatory elements in the promoters of transcription factors and mutant information of Tos17 have also been provided (Priya and Jain 2013).

3.8 Drought Stress Gene Database (DroughtDB)

DroughtDB (http://pgsb.helmholtz-muenchen.de/droughtdb/; Fig. 8) provides information of genes related to drought stress mechanism like drought avoidance and drought tolerance. Orthologous genes identified in nine crop and model plant species

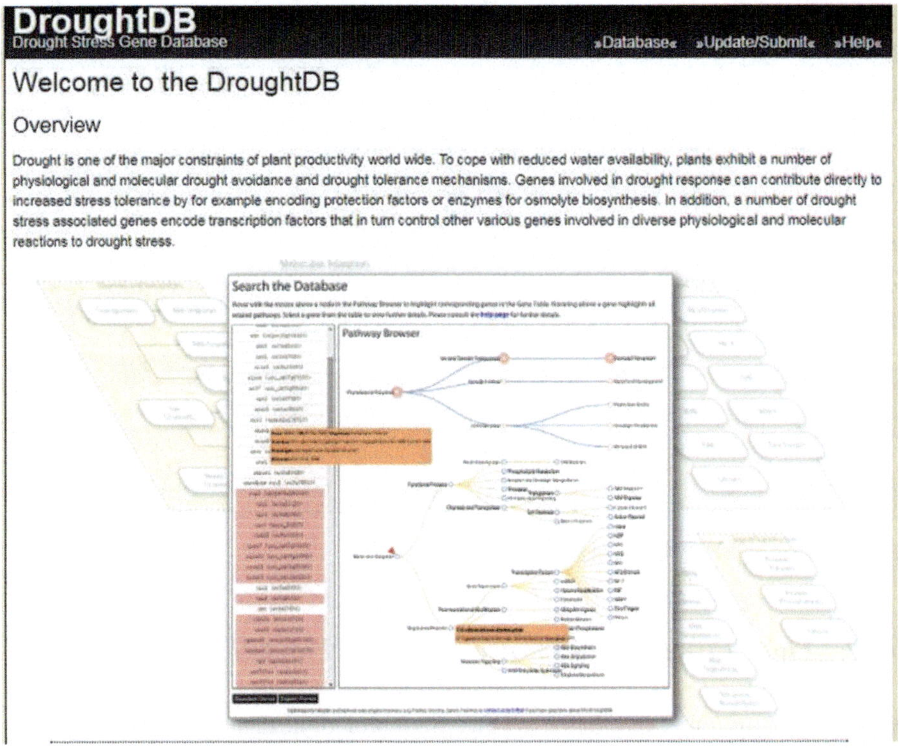

Fig. 8 Home page of DroughtDB database

including barley and maize are also provided. DroughtDB contains crucial information of genes like their identifier, organism with reference to publication, function, phenotype, PubMed reference number, orthologous groups, and sequence information. The database also provides interactive pathway browser which facilitates simplified interactive interface to the pathways for molecular and physiological adaptation (Alter et al. 2015).

3.9 *PSPDB: Plant Stress Protein Database*

PSPDB (http://bioclues.org/pspdb/; Fig. 9) is curated manually and contains annotations of abiotic and biotic stress-related proteins. It includes 2064 proteins involved in 30 different types of stress conditions from 134 plant species. In PSPDB each peptide is given an accession number to uniquely identify it and further cross-linked to other databases. The inclusion of protein in PSPDB is exclusively based on their experimental and functional information associated with abiotic and biotic stress conditions (Kumar et al. 2014).

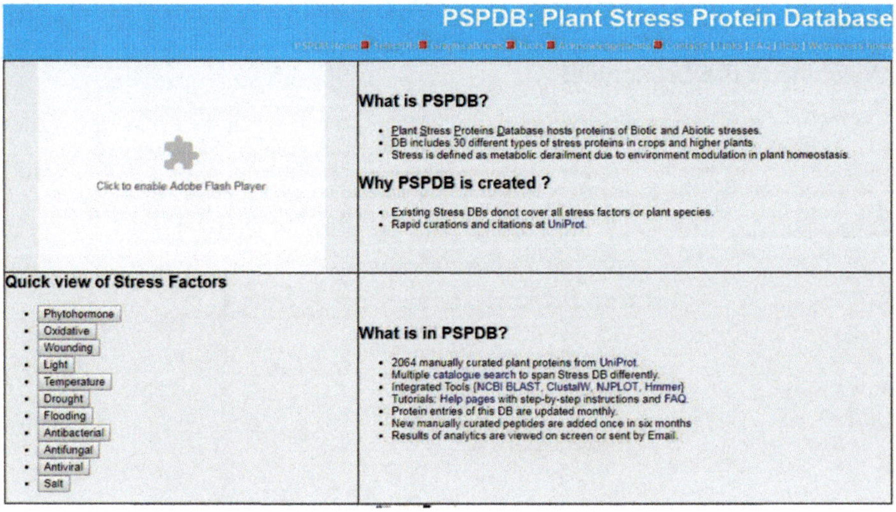

Fig. 9 Home page of PSPDB database

3.10 Plant Proteome Response to Stress (PlantPReS)

PlantPReS (http://www.proteome.ir/; Fig. 10) a proteomic database contains greater than 35086 entries of stress-responsive proteins from 577 manually curated articles. These entries consist of greater than 10600 unique stress-responsive proteins. It provides information related to plant type, accession number, protein name, stress types, tissue, and developmental stage of all stress-responsive proteins. PlantPReS also provides customized version of BLAST tool to search sequences with common ancestory. Moreover, a filtration mode is provided in PlantPReS to perform multiple analyses. The text or graphical format of results can be displayed (Mousavi et al. 2016).

4 Conclusion

As a concluding remark, we suggest to create a common platform which can help researchers to access the data related to plant stress. Although individual resources have their own utility, however, a user can easily get confused which one to use and in which condition. An integrated platform incorporating the different bioinformatics resources described here will help in disseminating knowledge and information to persons interested in abiotic and biotic stress-related studies of plants.

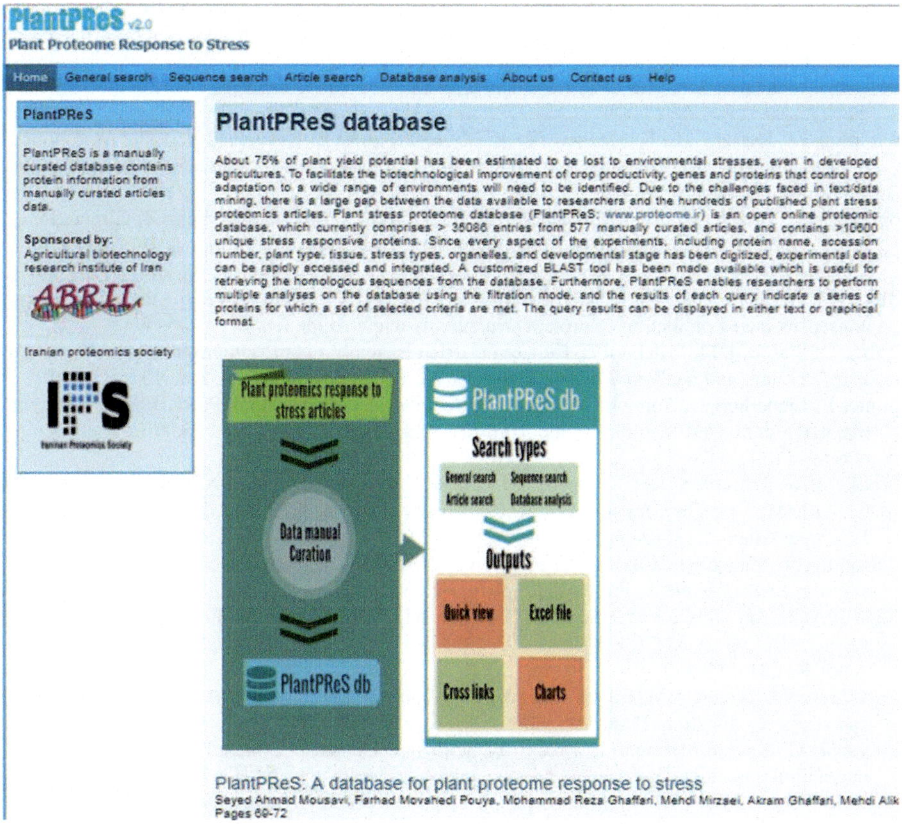

Fig. 10 Home page of PlantPReS database

References

Alter S, Bader KC, Spannagl M, Wang Y, Bauer E, Schön CC, Mayer KF (2015) DroughtDB: an expert-curated compilation of plant drought stress genes and their homologs in nine species. Database 2015:bav046

Altman RB (2004) Building successful biological databases. Brief Bioinform 5:4–5

Altschul SF, Gish W, Miller W, Myers EW, Lipman DJ (1990) Basic local alignment search tool. J Mol Biol 215:403–410

Amid C, Birney E, Bower L, Cerdeño-Tárraga A, Cheng Y, Cleland I, Faruque N, Gibson R, Goodgame N, Hunter C, Jang M (2011) Major submissions tool developments at the European nucleotide archive. Nucleic Acids Res 40:D43–D47

Atkinson NJ, Urwin PE (2012) The interaction of plant biotic and abiotic stresses: from genes to the field. J Exp Bot 63:3523–3543

Attwood TK, Gisel A, Eriksson NE, Bongcam-Rudloff E (2011) Concepts, historical milestones and the central place of bioinformatics in modern biology: a European perspective. In Bioinfo Tren Meth InTech

Balaji J, Crouch JH, Petite PV, Hoisington DA (2006) A database of annotated tentative orthologs from crop abiotic stress transcripts. Bioinformation 1:225–227

Barker WC, Garavelli JS, Haft DH, Hunt LT, Marzec CR, Orcutt BC, Srinivasarao GY, Yeh LS, Ledley RS, Mewes HW, Pfeiffer F (1998) The PIR-international protein sequence database. Nucleic Acids Res 26:27–32

Benson DA, Karsch-Mizrachi I, Lipman DJ, Ostell J, Rapp BA, Wheeler DL (2000) GenBank. Nucleic Acids Res 28:15–18

Berardini TZ, Reiser L, Li D, Mezheritsky Y, Muller R, Strait E, Huala E (2015) The *Arabidopsis* information resource: making and mining the "gold standard" annotated reference plant genome. Genesis 53:474–485

Berman HM, Battistuz T, Bhat TN, Bluhm WF, Bourne PE, Burkhardt K, Feng Z, Gilliland GL, Iype L, Jain S, Fagan P (2002) The protein data bank. Acta Crystallogr D Biol Crystallogr 58:899–907

Blaszczyk M, Jamroz M, Kmiecik S, Kolinski A (2013) CABS-fold: server for the de novo and consensus-based prediction of protein structure. Nucleic Acids Res 41:W406–W411

Bolser D, Staines DM, Pritchard E, Kersey P (2016) Ensembl plants: integrating tools for visualizing, mining, and analyzing plant genomics data. Plant Bioinfo: Meth Proto 1374:115–40

Boutet E, Lieberherr D, Tognolli M, Schneider M, Bairoch A (2007) UniProtKB/Swiss-Prot: the manually annotated section of the UniProt Knowledge Base. Plant Bioinfo: Meth Proto 406:89–112

Brooks BR, Bruccoleri RE, Olafson BD, States DJ, Swaminathan SA, Karplus M (1983) CHARMM: a program for macromolecular energy, minimization, and dynamics calculations. J Comput Chem 4:187–217

Calzadilla PI, Maiale SJ, Ruiz OA, Escaray FJ (2016) Transcriptome response mediated by cold stress in *Lotus japonicus*. Front Plant Sci 7:374

Chen J, Anderson JB, DeWeese-Scott C, Fedorova ND, Geer LY, He S, Hurwitz DI, Jackson JD, Jacobs AR, Lanczycki CJ, Liebert CA (2003) MMDB: Entrez's 3D-structure database. Nucleic Acids Res 31:474–477

Choudhury FK, Rivero RM, Eduardo B, Mittler R (2017) Reactive oxygen species, abiotic stress and stress combination. Plant J 90:856–867

Cochrane G, Karsch-Mizrachi I, Takagi T, Sequence Database Collaboration IN (2015) The international nucleotide sequence database collaboration. Nucleic Acids Res 44:D48–D50

Edgar RC (2004) MUSCLE: multiple sequence alignment with high accuracy and high throughput. Nucleic Acids Res 32:1792–1797

Friesner RA, Banks JL, Murphy RB, Halgren TA, Klicic JJ, Mainz DT, Repasky MP, Knoll EH, Shelley M, Perry JK, Shaw DE (2004) Glide: a new approach for rapid, accurate docking and scoring. 1. Method and assessment of docking accuracy. J Med Chem 47:1739–1749

Guex N, Peitsch MC (1997) SWISS-MODEL and the Swiss-Pdb viewer: an environment for comparative protein modeling. Electrophoresis 18:2714–2723

Hess B, Kutzner C, Van Der Spoel D, Lindahl E (2008) GROMACS 4: algorithms for highly efficient, load-balanced, and scalable molecular simulation. J Chem Theory Comput 4:435–447

Hirayama T, Shinozaki K (2010) Research on plant abiotic stress responses in the post-genome era: past, present and future. Plant J 61:1041–1052

Jayaram B, Singh T, Mukherjee G, Mathur A, Shekhar S, Shekhar V (2012) Sanjeevini: a freely accessible web-server for target directed lead molecule discovery. BMC Bioinforma 13:S7

Jayaram B, Dhingra P, Mishra A, Kaushik R, Mukherjee G, Singh A, Shekhar S (2014) Bhageerath-H: a homology/ab initio hybrid server for predicting tertiary structures of monomeric soluble proteins. BMC Bioinforma 15:S7

Källberg M, Margaryan G, Wang S, Ma J, Xu J (2014) RaptorX server: a resource for template-based protein structure modeling. Prot Str Prediction 1137:17–27

Kaminuma E, Kosuge T, Kodama Y, Aono H, Mashima J, Gojobori T, Sugawara H, Ogasawara O, Takagi T, Okubo K, Nakamura Y (2010) DDBJ progress report. Nucleic Acids Res 39:D22–D27

Kanehisa M, Goto S (2000) KEGG: kyoto encyclopedia of genes and genomes. Nucleic Acids Res 28:27–30

Katoh K, Kuma KI, Toh H, Miyata T (2005) MAFFT version 5: improvement in accuracy of multiple sequence alignment. Nucleic Acids Res 33:511–518

Kim DE, Chivian D, Baker D (2004) Protein structure prediction and analysis using the Robetta server. Nucleic Acids Res 32:W526–W531

Kumar SA, Kumari PH, Sundararajan VS, Suravajhala P, Kanagasabai R, Kishor PK (2014) PSPDB: plant stress protein database. Plant Mol Biol Report 32:940–942

Lambert C, Leonard N, De Bolle X, Depiereux E (2002) ESyPred3D: prediction of proteins 3D structures. Bioinformatics 18:1250–1256

Leonberger K, Jackson K, Smith R, Ward Gauthier N (2016) Plant diseases [2016]. Agric Nat Res Pub. 182

Mochida K, Shinozaki K (2010) Genomics and bioinformatics resources for crop improvement. Plant Cell Physiol 51:497–523

Moraes Filho RM, Menezes AF, Martins LS (2017) In silico modeling and characterization of phytoparasitic nematodes translationally-controlled tumor proteins. Genet Mol Res:16

Morgenstern B (1999) DIALIGN 2: improvement of the segment-to-segment approach to multiple sequence alignment. Bioinformatics (Oxford, England) 15:211–218

Morris GM, Huey R, Lindstrom W, Sanner MF, Belew RK, Goodsell DS, Olson AJ (2009) AutoDock4 and AutoDockTools4: automated docking with selective receptor flexibility. J Comput Chem 30:2785–2791

Mount DM (2004/2003) Bioinformatics: sequence and genome analysis (2). CSHL Press, New York: 1–8

Mousavi SA, Pouya FM, Ghaffari MR, Mirzaei M, Ghaffari A, Alikhani M, Ghareyazie M, Komatsu S, Haynes PA, Salekdeh GH (2016) PlantPReS: a database for plant proteome response to stress. J Proteome 143:69–72

Naika M, Shameer K, Mathew OK, Gowda R, Sowdhamini R (2013) STIFDB2: an updated version of plant stress-responsive transcription factor database with additional stress signals, stress-responsive transcription factor binding sites and stress-responsive genes in *Arabidopsis* and rice. Plant Cell Physiol 54:e8

Nawaz M, Iqbal N, Idrees S, Ullah I (2014) DREB1A from Oryza sativa var. IR6: homology modelling and molecular docking. Turk J Bot 38:1095–1102

Needleman SB, Wunsch CD (1970) A general method applicable to the search for similarities in the amino acid sequence of two proteins. J Mol Biol 48:443–453

Nielsen M, Lundegaard C, Lund O, Petersen TN (2010) CPHmodels-3.0-remote homology modeling using structure-guided sequence profiles. Nucleic Acids Res 38:W576–W581

Notredame C, Higgins DG, Heringa J (2000) T-coffee: a novel method for fast and accurate multiple sequence alignment. J Mol Biol 302:205–217

Pearlman DA, Case DA, Caldwell JW, Ross WS, Cheatham TE, DeBolt S, Ferguson D, Seibel G, Kollman P (1995) AMBER, a package of computer programs for applying molecular mechanics, normal mode analysis, molecular dynamics and free energy calculations to simulate the structural and energetic properties of molecules. Comput Phys Commun 91:1–41

Pearson WR (1990) Rapid and sensitive sequence comparison with FASTP and FASTA. Methods Enzymol 183:63–98

Prabha R, Ghosh I, Singh DP (2011) Plant stress gene database: a collection of plant genes responding to stress condition. ARPN J Sci Techno 1:28–31

Priya P, Jain M (2013) RiceSRTFDB: a database of rice transcription factors containing comprehensive expression, cis-regulatory element and mutant information to facilitate gene function analysis. Database 2013:bat027

Purty RS, Sachar M, Chatterjee S (2017) Structural and expression analysis of salinity stress responsive phosphoserine phosphatase from *Brassica juncea* (L.) J Proteomics Bioinform 10:119–127

Rice P, Longden I, Bleasby A (2000) EMBOSS: the European molecular biology open software suite. Trends Genet 16:276–277

Rodziewicz P, Swarcewicz B, Chmielewska K, Wojakowska A, Stobiecki M (2014) Influence of abiotic stresses on plant proteome and metabolome changes. Acta Physiol Plant 36:1–9
Schwede T, Kopp J, Guex N, Peitsch MC (2003) SWISS-MODEL: an automated protein homology-modeling server. Nucleic Acids Res 31:3381–3385
Sham A, Aly MA (2012) Bioinformatics based comparative analysis of omega-3 fatty acids in desert plants and their role in stress resistance and tolerance. Int J Plant Sci 2:80–89
Shameer K, Ambika S, Varghese SM, Karaba N, Udayakumar M, Sowdhamini R (2009) STIFDB-*Arabidopsis* stress responsive transcription factor dataBase. Int J Plant Genomics 2009:583429
Sharma V, Munjal A, Shanker A (2016) A text book of bioinformatics, 2nd edn. Rastogi Publications, Meerut, p 350
Shen Y, Maupetit J, Derreumaux P, Tufféry P (2014) Improved PEP-FOLD approach for peptide and miniprotein structure prediction. J Chem Theory Comput 10:4745–4758
Smita S, Lenka SK, Katiyar A, Jaiswal P, Preece J, Bansal KC (2011) QlicRice: a web interface for abiotic stress responsive QTL and loci interaction channels in rice. Database 2011:bar037
Smith K (2013) A brief history of NCBI's formation and growth. The NCBI Handbook
Smith TF, Waterman MS (1981) Identification of common molecular subsequences. J Mol Biol 147:195–197
Tatusova T, Smith-White B, Ostell J (2007) A collection of plant-specific genomic data and resources at NCBI. Plant Bioinfo: Meth Proto 406:61–87
Thompson JD, Higgins DG, Gibson TJ (1994) CLUSTAL W: improving the sensitivity of progressive multiple sequence alignment through sequence weighting, position-specific gap penalties and weight matrix choice. Nucleic Acids Res 22:4673–4680
Trott O, Olson AJ (2010) AutoDock Vina: improving the speed and accuracy of docking with a new scoring function, efficient optimization, and multithreading. J Comput Chem 31:455–461
Wang L, Guo Z, Zhang Y, Wang Y, Yang G, Yang L, Wang R, Xie Z (2017) Characterization of LhSorP5CS, a gene catalyzing proline synthesis in Oriental hybrid lily Sorbonne: molecular modelling and expression analysis. Bot Stud 58:10
Webb B, Sali A (2014) Protein structure modeling with MODELLER. Protein Struct Prediction 2014:1–15
Wei Z, Zeng X, Qin C, Wang Y, Bai L, Xu Q, Yuan H, Tang Y, Nyima T (2016) Comparative transcriptome analysis revealed genes commonly responsive to varied nitrate stress in leaves of Tibetan hulless barley. Front Plant Sci 7:1067
Wheeler DL, Barrett T, Benson DA, Bryant SH, Canese K, Chetvernin V, Church DM, DiCuccio M, Edgar R, Federhen S, Feolo M (2007) Database resources of the national center for biotechnology information. Nucleic Acids Res 36:D13–D21
Wu S, Zhang Y (2007) LOMETS: a local meta-threading-server for protein structure prediction. Nucleic Acids Res 35:3375–3382
Xiong H, Guo H, Xie Y, Zhao L, Gu J, Zhao S, Li J, Liu L (2017) RNAseq analysis reveals pathways and candidate genes associated with salinity tolerance in a spaceflight-induced wheat mutant. Sci Rep 7:2731
Yang JM, Chen CC (2004) GEMDOCK: a generic evolutionary method for molecular docking. Proteins: Struct Funct Bioinfo 55:288–304
Yang J, Yan R, Roy A, Xu D, Poisson J, Zhang Y (2015) The I-TASSER suite: protein structure and function prediction. Nat Methods 12:7–8
Zhang S, Yue Y, Sheng L, Wu Y, Fan G, Li A, Hu X, ShangGuan M, Wei C (2013) PASmiR: a literature curated database for miRNA molecular regulation in plant response to abiotic stress. BMC Plant Biol 13:33
Zhu JK (2016) Abiotic stress signaling and responses in plants. Cell 167:313–324